CALCULUS
Theory and Applications
Volume I

CALCULUS
Theory and Applications

Volume I

Kenneth Kuttler

Brigham Young University, USA

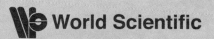 **World Scientific**

NEW JERSEY · LONDON · SINGAPORE · BEIJING · SHANGHAI · HONG KONG · TAIPEI · CHENNAI

Published by

World Scientific Publishing Co. Pte. Ltd.

5 Toh Tuck Link, Singapore 596224

USA office: 27 Warren Street, Suite 401-402, Hackensack, NJ 07601

UK office: 57 Shelton Street, Covent Garden, London WC2H 9HE

British Library Cataloguing-in-Publication Data
A catalogue record for this book is available from the British Library.

CALCULUS — Volume 1
Theory and Applications

ISBN-13 978-981-4324-26-7
ISBN-10 981-4324-26-4
ISBN-13 978-981-4329-69-9 (pbk)
ISBN-10 981-4329-69-X (pbk)

Printed in Singapore by B & Jo Enterprise Pte Ltd

Preface

Calculus consists of the study of limits of various sorts and the systematic exploitation of the completeness axiom. It was developed over a period of several hundred years in order to solve problems from the physical sciences. It is the language by which precision and quantitative predictions for many complicated problems are obtained. It also has significant applications in pure mathematics. For example, it is used to define and find lengths of curves, areas and volumes of three dimensional shapes. It is essential in order to solve many maximization problems and it is prerequisite material in order to understand models based on differential equations. These and other applications are discussed to some extent in this book.

It is assumed the reader has a good understanding of algebra on the level of college algebra or what used to be called algebra II along with some exposure to geometry and trigonometry although the book does contain a review of these things.

If the optional sections and nonstandard sections are not included, this book is fairly short. However, there is a lot of nonstandard material, including the big theorems of advanced calculus.

I have tried to give complete proofs of all theorems somewhere in the book because I believe that calculus is part of mathematics and that in mathematics the validity of some assertion is typically established by giving a proof. This is certainly true in algebra so it seems to me it should also be true in calculus. Mathematics is not about accepting on faith unproved assertions presumably understood by someone else.

I expect the reader to be able to use a calculator whenever it would be helpful to do so. In addition, a minimal introduction to the use of computer algebra systems is presented. Having said this, calculus is not about using calculators or any other form of technology. Weierstrass could do calculus quite well without the benefit of modern gadgets. I believe that when the syntax and arcane notation associated with technology are presented too prominently, these things become the topic of study rather than the concepts of calculus and this is a real shame. This is a book on calculus.

Pictures are often helpful in seeing what is going on, and there are many pictures in this book for this reason. However, calculus is not art and ultimately rests on logic and definitions, like the rest of mathematics. Algebra plays a central role in gaining the sort of understanding which generalizes to higher dimensions where pictures are not available.

A star next to an exercise indicates it is either technically difficult or perhaps a little bit challenging to figure out. Do not be frightened by these exercises. They just might take a little longer to work but some are very worth while and ought to be attempted.

Supplementary material for this text including routine exercise sets is available at http://www.math.byu.edu/~klkuttle/CalculusMaterials. I have made these exercise sets using Scientific Workplace Exam Builder. The source file is available at this site and you can modify it to include more types of exercises if you desire.

The topics discussed in this book are arranged in a typical order for most calculus courses I have seen, with the notable exception of the early introduction of sequences and their limits. The concepts of limit of a function of a real variable and limit of a sequence were made rigorous at around the same time in the nineteenth century and also, a sequence is a type of function, so I think this order makes good sense. Another advantage is that continuity and various theorems about continuous functions can be understood by some people, more easily in terms of convergent sequences than in terms of the traditional epsilon delta definition.

Instead of the order listed in the table of contents, one could begin with the chapter on Page 323 and do all the topics about vectors in this and the remaining chapters of the book before beginning the usual topics of one variable calculus in the chapter which begins on Page 39. Just leave out the material on the parabolic mirror and an occasional exercise which depend on the derivative. This approach may be better because students often encounter these topics in their physics classes at the time they begin calculus.

I am grateful to World Scientific for publishing this and the second volume of this calculus book. I am also grateful to Kate Phillips for the many drawings which I could not have done.

Contents

Chapter 1

A Short Review Of Precalculus

1.1 Set Notation

A set is just a collection of things called elements. Often these are also referred to as points in calculus. For example $\{1, 2, 3, 8\}$ would be a set consisting of the elements 1,2,3, and 8. To indicate that 3 is an element of $\{1, 2, 3, 8\}$, it is customary to write $3 \in \{1, 2, 3, 8\}$. $9 \notin \{1, 2, 3, 8\}$ means 9 is not an element of $\{1, 2, 3, 8\}$. Sometimes a rule specifies a set. For example you could specify a set as all integers larger than 2. This would be written as $S = \{x \in \mathbb{Z} : x > 2\}$. This notation says: the set of all integers x, such that $x > 2$.

If A and B are sets with the property that every element of A is an element of B, then A is a subset of B. For example, $\{1, 2, 3, 8\}$ is a subset of $\{1, 2, 3, 4, 5, 8\}$, in symbols, $\{1, 2, 3, 8\} \subseteq \{1, 2, 3, 4, 5, 8\}$. The same statement about the two sets may also be written as $\{1, 2, 3, 4, 5, 8\} \supseteq \{1, 2, 3, 8\}$.

The union of two sets is the set consisting of everything which is contained in at least one of the sets A or B. As an example of the union of two sets, $\{1, 2, 3, 8\} \cup \{3, 4, 7, 8\} = \{1, 2, 3, 4, 7, 8\}$ because these numbers are those which are in at least one of the two sets. In general

$$A \cup B \equiv \{x : x \in A \text{ or } x \in B\}.$$

Be sure you understand that something which is in both A and B is in the union. It is not an exclusive or.

The intersection of two sets A and B consists of everything which is in both of the sets. Thus $\{1, 2, 3, 8\} \cap \{3, 4, 7, 8\} = \{3, 8\}$ because 3 and 8 are those elements the two sets have in common. In general,

$$A \cap B \equiv \{x : x \in A \text{ and } x \in B\}.$$

The symbol $[a, b]$ denotes the set of real numbers x, such that $a \le x \le b$ and $[a, b)$ denotes the set of real numbers such that $a \le x < b$. (a, b) consists of the set of real numbers x such that $a < x < b$ and $(a, b]$ indicates the set of numbers x such that $a < x \le b$. $[a, \infty)$ means the set of all numbers x such that $x \ge a$ and $(-\infty, a]$ means the set of all real numbers which are less than or equal to a. These

1

sorts of sets of real numbers are called intervals. The two points a and b are called endpoints of the interval. Other intervals such as $(-\infty, b)$ are defined by analogy to what was just explained. In general, the curved parenthesis indicates the end point it sits next to is not included while the square parenthesis indicates this end point is included. The reason that there will always be a curved parenthesis next to ∞ or $-\infty$ is that these are not real numbers. Therefore, they cannot be included in any set of real numbers. The symbol is called "infinity" or minus "infinity".

A special set which needs to be given a name is the empty set also called the null set, denoted by $\emptyset$. Thus $\emptyset$ is defined as the set which has no elements in it. Mathematicians like to say the empty set is a subset of every set. The reason they say this is that if it were not so, there would have to exist a set A, such that $\emptyset$ has something in it which is not in A. However, $\emptyset$ has nothing in it and so the least intellectual discomfort is achieved by saying $\emptyset \subseteq A$.

If A and B are two sets, $A \setminus B$ denotes the set of things which are in A but not in B. Thus

$$A \setminus B \equiv \{x \in A : x \notin B\}.$$

Set notation is used whenever convenient.

1.2   Completeness Of The Real Numbers

I assume the reader is familiar with the usual algebraic properties of the real numbers. However, they have another property known as completeness.

Definition 1.1. A nonempty set $S \subseteq \mathbb{R}$ is bounded above (below) if there exists $x \in \mathbb{R}$ such that $x \geq (\leq) s$ for all $s \in S$. If S is a nonempty set in $\mathbb{R}$ which is bounded above, then a number l which has the property that l is an upper bound and that every other upper bound is no smaller than l is called a least upper bound, $l.u.b.\,(S)$ or often $\sup(S)$. If S is a nonempty set bounded below, define the greatest lower bound, $g.l.b.\,(S)$ or $\inf(S)$ similarly. Thus g is the $g.l.b.\,(S)$ means g is a lower bound for S and it is the largest of all lower bounds. If S is a nonempty subset of $\mathbb{R}$ which is not bounded above, this information is expressed by saying $\sup(S) = +\infty$ and if S is not bounded below, $\inf(S) = -\infty$.

Every existence theorem in calculus depends on some form of the completeness axiom.

Axiom 1.1. *(completeness) Every nonempty set of real numbers which is bounded above has a least upper bound and every nonempty set of real numbers which is bounded below has a greatest lower bound.*

It is this axiom which distinguishes Calculus from Algebra.

A fundamental result about sup and inf is the following.

Proposition 1.1. Let S be a nonempty set and suppose $\sup(S)$ exists. Then for every $\delta > 0$,

$$S \cap (\sup(S) - \delta, \sup(S)] \neq \emptyset.$$

If $\inf(S)$ exists, then for every $\delta > 0$,

$$S \cap [\inf(S), \inf(S) + \delta) \neq \emptyset.$$

Proof: Consider the first claim. If the indicated set equals $\emptyset$, then $\sup(S) - \delta$ is an upper bound for S which is smaller than $\sup(S)$, contrary to the definition of $\sup(S)$ as the least upper bound. In the second claim, if the indicated set equals $\emptyset$, then $\inf(S) + \delta$ would be a lower bound which is larger than $\inf(S)$ contrary to the definition of $\inf(S)$. ∎

1.3 A Few Algebraic Conventions And Techniques

Summation notation is a convenient way to specify a sum.

Definition 1.2. For $i = m, \cdots, n$ let a_i be specified. Then

$$\sum_{i=m}^{n} a_i \equiv a_m + a_{m+1} + \cdots + a_n$$

I will use this whenever convenient.

Example 1.1. Find $\sum_{i=1}^{3} 2i - 1$.

From the definition, it equals $(2 - 1) + (2 \times 2 - 1) + (2 \times 3 - 1) = 9$.

An important technique is the technique of proof by **induction**. I will illustrate with a simple example which is useful for its own sake.

Example 1.2. For $n = 1, 2, 3, \cdots$ and $\alpha > 0$ it is always the case that

$$(1 + \alpha)^n \geq 1 + n\alpha + \frac{n(n-1)}{2}\alpha^2$$

Here is why. The statement is true if $n = 1$. Now suppose I can show that whenever the statement is true for some value of n it follows that it must be true for the next value of n. Then it must be the case that it is true for each value of n. Consider why this is. Since it is true for $n = 1$, and whenever it is true for some n, it is true for $n + 1$, it follows that it must be true for 2. Since it is true for 2, it must be true for 3 by the same reasoning, and so forth. Thus it suffices to show that **if** it is true for n **then** it is true for $n + 1$. I haven't done this yet but I am about to do it.

Suppose then that the inequality is true for n. I need to verify that with this assumption, it holds for $n + 1$. That is, the same formula needs to hold with n replaced everywhere with $n + 1$.

Using the assumption that it is true for n,

$$(1 + \alpha)^{n+1} = (1 + \alpha)(1 + \alpha)^n$$

$$\geq (1 + \alpha)\left(1 + n\alpha + \frac{n(n-1)}{2}\alpha^2\right)$$

$$= 1 + n\alpha + \frac{n(n-1)}{2}\alpha^2 + \alpha + n\alpha^2 + \frac{n(n-1)}{2}\alpha^3$$

$$\geq 1 + (n+1)\alpha + \frac{n(n-1)}{2}\alpha^2 + n\alpha^2$$

where I simply threw out the last term in going to the last line. This equals

$$1 + (n+1)\alpha + \frac{n(n-1) + 2n}{2}\alpha^2$$

$$= 1 + (n+1)\alpha + \frac{(n+1)n}{2}\alpha^2$$

Thus the inequality holds with n replaced with $n + 1$. This proves the desired inequality.

There are many other examples where induction is useful.

Also of use is the concept of **absolute value** of a number. This is defined as follows.

$$|x| \equiv \text{ the distance from } x \text{ to } 0 \text{ on the number line.}$$

An equivalent way of defining it is to say $|x| = x$ if $x \geq 0$ and $|x| = -x$ if $x < 0$.

1.4   The Circular Arc Subtended By An Angle

How can angles be measured? This will be done by considering arcs on a circle. To see how this will be done, let θ denote an angle and place the vertex of this angle at the center of the circle. Next, extend its two sides till they intersect the circle. Note the angle could be opening in any of infinitely many different directions. Thus this procedure could yield any of infinitely many different circular arcs. Each of these arcs is said to **subtend** the angle. In fact each of these arcs has the same length. When this has been shown, it will be easy to measure angles. Angles will be measured in terms of lengths of arcs subtended by the angle. Of course it is also necessary to define what is meant by the length of a circular arc in order to do any of this. First I will describe an intuitive way of thinking about this and then give a rigorous definition and proof. If the intuitive way of thinking about this satisfies you, no harm will be done by skipping the more technical discussion which follows.

Take an angle and place its **vertex** (the point) at the center of a circle of radius r. Then, extending the sides of the angle if necessary till they intersect the circle, this determines an arc on the circle. If r were changed to R, this really amounts to a change of units of length. Think for example, of keeping the numbers the same but changing centimeters to meters in order to produce an enlarged version of the same picture. Thus the picture looks exactly the same, only larger. It is reasonable to suppose, based on this reasoning, that the way to measure the angle is to take the length of the arc subtended in whatever units being used, and divide this length by the radius measured in the same units, thus obtaining a number which is independent of the units of length used, just as the angle itself is independent of units of length. After all, it is the same angle regardless of how far its sides are extended. This is in fact how to define the radian measure of an angle, and the definition is well defined. Thus in particular, the ratio between the circumference (length) of a circle and its radius is a constant which is independent of the radius of the circle[1]. Since the time of Euler in the 1700's, this constant has been denoted by 2π. In summary, if θ is the radian measure of an angle, the length of the arc subtended by the angle on a circle of radius r is $r\theta$.

This is a little sloppy right now because no precise definition of the length of an arc of a circle has been given. For now, imagine taking a string, placing one end of it on one end of the circular arc and then wrapping the string till you reach the other end of the arc. Stretching this string out and measuring it would then give you the length of the arc. Such intuitive discussions involving string may or may not be enough to convey understanding. If you need to see more discussion, read on. Otherwise, skip to the next section.

To give a precise description of what is meant by the length of an arc, consider the following picture.

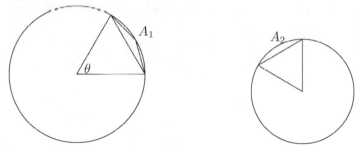

In this picture, there are two circles, a big one having radius R and a little one having radius r. The angle θ is situated in two different ways subtending the arcs

---

[1]In 2 Chronicles 4:2 the "molten sea" found in Solomon's temple is described. It sat on 12 oxen, was round, 5 cubits high, 10 across and 30 around. Thus its radius was 5 and the Bible, taken literally, gives the value of π as 3. This is not too far off but is not correct. Other incorrect values of π can be found in the Indiana pi bill of 1897. Later, methods will be given which allow one to calculate π more precisely. A better value is 3.1415926535 and presently this number is known to thousands of decimal places. It was proved by Lindeman in 1882 that π is transcendental which is the worst sort of irrational number.

A_1 and A_2 as shown.

Letting A be an arc of a circle like those shown in the above picture, a subset of A $\{p_0, \cdots, p_n\}$ is a partition of A if p_0 is one endpoint p_n is the other end point, and the points are encountered in the indicated order as one moves in the counter clockwise direction along the arc. To illustrate, see the following picture.

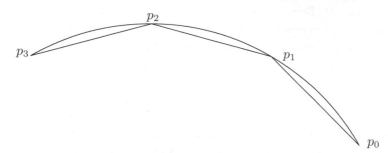

Also denote by $\mathcal{P}(A)$ the set of all such partitions. For $P = \{p_0, \cdots, p_n\}$, denote by $|p_i - p_{i-1}|$ the distance between p_i and p_{i-1}. Then for $P \in \mathcal{P}(A)$, define

$$|P| \equiv \sum_{i=1}^{n} |p_i - p_{i-1}|$$

Thus $|P|$ consists of the sum of the lengths of the little lines joining successive points of P and appears to be an approximation to the length of the circular arc A. By geometry, the length of any of the straight line segments joining successive points in a partition is smaller than the sum of the two sides of a right triangle having the given straight line segment as its hypotenuse. For example, see the following picture.

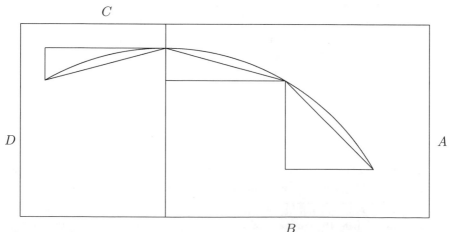

The sum of the lengths of the straight line segments in the part of the picture found in the right rectangle above is less than $A + B$ and the sum of the lengths of the straight line segments in the part of the picture found in the left rectangle above is less than $C + D$ and this would be so for any partition. Therefore, for any $P \in \mathcal{P}(A)$, $|P| \leq M$ where M is the perimeter of a rectangle containing the arc, A. To be a little sloppy, simply pick M to be the perimeter of a rectangle containing the whole circle of which A is a part. The only purpose for doing this is to obtain the existence of an upper bound to the lengths of these polygonal curves. Therefore, $\{|P| : P \in \mathcal{P}(A)\}$ is a set of numbers which is bounded above by M, and by completeness of $\mathbb{R}$, it is possible to define the length of $A, l(A)$, by $l(A) \equiv \sup\{|P| : P \in \mathcal{P}(A)\}$.

A fundamental observation following from the theorem in geometry which says the sum of the lengths of two sides of a triangle is always at least as large as the length of the third side, is that if $P, Q \in \mathcal{P}(A)$ and $P \subseteq Q$, then $|P| \leq |Q|$. To see this, add in one point at a time to P. This effect of adding in one point is illustrated in the following picture.

Also, letting $\{p_0, \cdots, p_n\}$ be a partition of A, specify angles, θ_i as follows. The angle θ_i is formed by the two lines, one from the center of the circle to p_i and the other line from the center of the circle to p_{i-1}. Furthermore, a specification of these angles yields the partition of A in the following way. Place the vertex of θ_1 on the center of the circle, letting one side lie on the line from the center of the circle to p_0 and the other side extended resulting in a point further along the arc in the counter clockwise direction. When the angles, $\theta_1, \cdots, \theta_{i-1}$ have produced points $p_0, \cdots, p_{i-1}$ on the arc, place the vertex of θ_i on the center of the circle and let one side of θ_i coincide with the side of the angle θ_{i-1} which is most counter clockwise, the other side of θ_i when extended, resulting in a point further along the arc, A in the counterclockwise direction as shown below.

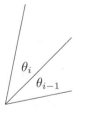

Now let $\varepsilon > 0$ be given and pick $P_1 \in \mathcal{P}(A_1)$ such that $|P_1| + \varepsilon > l(A_1)$. Then determining the angles as just described, use these angles to produce a corresponding partition of A_2, P_2. If $|P_2| + \varepsilon > l(A_2)$, then stop. Otherwise, pick $Q \in \mathcal{P}(A_2)$

such that $|Q| + \varepsilon > l(A_2)$ and let $P_2' = P_2 \cup Q$. Then use the angles determined by P_2' to obtain $P_1' \in \mathcal{P}(A_1)$. Then $|P_1'| + \varepsilon > l(A_1), |P_2'| + \varepsilon > l(A_2)$, and both P_1' and P_2' determine the same sequence of angles. Using plane geometry, the theorem which says that two isosceles triangles are similar if they have the same central angle, $\frac{|P_1'|}{|P_2'|} = \frac{R}{r}$ and so

$$l(A_2) < |P_2'| + \varepsilon = \frac{r}{R}|P_1'| + \varepsilon \le \frac{r}{R}l(A_1) + \varepsilon.$$

Since ε is arbitrary, this shows $Rl(A_2) \le rl(A_1)$. But now reverse the argument and write

$$l(A_1) < |P_1'| + \varepsilon = \frac{R}{r}|P_2'| + \varepsilon \le \frac{R}{r}l(A_2) + \varepsilon$$

which implies, since ε is arbitrary, that $Rl(A_2) \ge rl(A_1)$ and this has proved the following theorem.

Theorem 1.1. *Let θ be an angle which subtends two arcs, A_R on a circle of radius R and A_r on a circle of radius r each circle having the same center. Then denoting by $l(A)$ the length of a circular arc as described above, $Rl(A_r) = rl(A_R)$.*

Before proceeding further, note the proof of the above theorem involved showing $l(A_1) < \frac{R}{r}l(A_2) + \varepsilon$ where $\varepsilon > 0$ was arbitrary and from this, the conclusion that $l(A_1) \le \frac{R}{r}l(A_2)$. This is a very typical way of showing one number is no larger than another. To show $a \le b$ first show that for every $\varepsilon > 0$ it follows that $a < b + \varepsilon$. This implies $a - b < \varepsilon$ for all positive ε and so it must be the case that $a - b \le 0$, since otherwise, you could take $\varepsilon = \frac{a-b}{2}$ and conclude $0 < a - b < \frac{a-b}{2}$, a contradiction.

With this preparation, here is the definition of the measure of an angle.

Definition 1.3. Let θ be an angle. The measure of θ is defined to be the length of the circular arc subtended by θ on a circle of radius r divided by r. This is also called the radian measure of the angle.

You should note again that the measure of θ is independent of dimension. This is because the units of length cancel when the division takes place.

Proposition 1.2. *The above definition is well defined and also, if A is an arc subtended by the angle θ on a circle of radius r then the length of A, denoted by $l(A)$ is given by $l(A) = r\theta$.*

Proof: That the definition is well defined follows from Theorem 1.1. The formula also follows from Theorem 1.1 and letting $R = 1$. ∎

Recall from trigonometry the following definition of the cosine and sine.

Definition 1.4. Consider the unit circle graphed in terms of the x and y axes. Position an angle θ such that its vertex is at the origin, and one side is on the positive x axis as shown in the following picture. Then the cosine of the angle is the x coordinate of the point obtained from the intersection of the other side with the circle and the sine of the angle is the y coordinate of this point.

Now is a good time to present a useful inequality which may or may not be self evident. Here is a picture which illustrates the conclusion of this corollary.

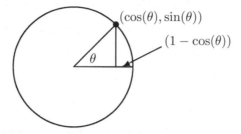

The following corollary states that the length of the subtended arc shown in the picture is longer than the vertical side of the triangle and smaller than the sum of the vertical side with the segment having length $1 - \cos\theta$. To me, this seems abundantly clear but in case it is hard to believe, the following corollary gives a proof.

Corollary 1.1. *Let $0 <$ radian measure of $\theta < \pi/4$. Then letting A be the arc on the unit circle resulting from situating the angle with one side on the positive x axis and the other side pointing up from the positive x axis,*

$$(1 - \cos\theta) + \sin\theta \geq l(A) \geq \sin\theta \tag{1.1}$$

Proof: Situate the angle θ such that one side is on the positive x axis and extend the other side till it intersects the unit circle at the point $(\cos\theta, \sin\theta)$. Then denoting the resulting arc on the circle by A, it follows that for all $P \in P(A)$ the inequality $(1 - \cos\theta) + \sin\theta \geq |P| \geq \sin\theta$ holds. It follows that $(1 - \cos\theta) + \sin\theta$ is an upper bound for all the $|P|$ where $P \in P(A)$ and so $(1 - \cos\theta) + \sin\theta$ is at least as large as the sup or least upper bound of the $|P|$. This proves the top half of the inequality. The bottom half follows because $l(A) \geq L$ where L is the length of the line segment joining $(\cos\theta, \sin\theta)$ and $(1, 0)$ due to the definition of $l(A)$. However, $L \geq \sin\theta$ because L is the length of the hypotenuse of a right triangle having $\sin\theta$ as one of the sides. ∎

1.5 The Trigonometric Functions

Now the Trigonometric functions will be defined as functions of an arbitrary real variable. Up till now they have been defined as functions of pointy things called angles. The following theorem will make possible the definition.

Theorem 1.2. *Let $b \in \mathbb{R}$. Then there exists a unique integer p and real number r such that $0 \leq r < 2\pi$ and $b = p2\pi + r$.*

Proof: Consider the equally spaced points $k(2\pi)$ where k is an integer. Now the disjoint half open intervals $[k(2\pi), (k+1)(2\pi))$ cover the whole real line and for different values of k these intervals are disjoint. Let p be the value of k such that $b \in [p(2\pi), (p+1)(2\pi))$. Then for some $r \in [0, 2\pi), b = p2\pi + r$. ∎

The following definition is for $\sin b$ and $\cos b$ for $b \in \mathbb{R}$.

Definition 1.5. Let $b \in \mathbb{R}$. Then $\sin b \equiv \sin r$ and $\cos b \equiv \cos r$ where $b = 2\pi p + r$ for p an integer, and $r \in [0, 2\pi)$.

Several observations are now obvious.

Observation 1.1. Let $k \in \mathbb{Z}$, then the following formulas hold.

$$\sin b = -\sin(-b), \cos b = \cos(-b), \tag{1.2}$$

$$\sin(b + 2k\pi) = \sin b, \cos(b + 2k\pi) = \cos b \tag{1.3}$$

$$\cos^2 b + \sin^2 b = 1 \tag{1.4}$$

The other trigonometric functions are defined in the usual way provided they make sense.

From the observation that the x and y axes intersect at right angles the four arcs on the unit circle subtended by these axes are all of equal length. Therefore, the measure of a right angle must be $2\pi/4 = \pi/2$. The measure of the angle which is determined by the arc from $(1, 0)$ to $(-1, 0)$ is seen to equal π by the same reasoning. From the definition of the trig functions, $\cos(\pi/2) = 0$ and $\sin(\pi/2) = 1$. You can easily find other values for cos and sin at all the other multiples of $\pi/2$.

The next topic is the important formulas for the trig. functions of sums and differences of numbers. For $b \in \mathbb{R}$, denote by r_b the element of $[0, 2\pi)$ having the property that $b = 2\pi p + r_b$ for p an integer.

Lemma 1.1. *Let $x, y \in \mathbb{R}$. Then $r_{x+y} = r_x + r_y + 2k\pi$ for some $k \in \mathbb{Z}$.*

Proof: By definition,

$$x + y = 2\pi p + r_{x+y}, \ x = 2\pi p_1 + r_x, \ y = 2\pi p_2 + r_y.$$

From this the result follows because

$$0 = ((x + y) - x) - y = 2\pi \overbrace{((p - p_1) - p_2)}^{\equiv -k} + r_{x+y} - (r_x + r_y). \quad ∎$$

Using the above lemma, let $z \in \mathbb{R}$ and let $p(z)$ denote the point on the unit circle determined by the length r_z whose coordinates are $\cos z$ and $\sin z$. Thus, starting at $(1, 0)$ and moving counter clockwise a distance of r_z on the unit circle yields $p(z)$. Note also that $p(z) = p(r_z)$.

Lemma 1.2. *Let $x, y \in \mathbb{R}$. Then the length of the arc between $p(x + y)$ and $p(x)$ is equal to the length of the arc between $p(y)$ and $(1, 0)$.*

Proof: The length of the arc between $p(x + y)$ and $p(x)$ is $|r_{x+y} - r_x|$. There are two cases to consider here.

First assume $r_{x+y} \geq r_x$. Then $|r_{x+y} - r_x| = r_{x+y} - r_x = r_y + 2k\pi$ for some integer k. Since both r_{x+y} and r_x are in $[0, 2\pi)$, their difference is also in $[0, 2\pi)$ and so $k = 0$. Therefore, the arc joining $p(x)$ and $p(x + y)$ is of the same length as the arc joining $p(y)$ and $(1, 0)$. In the other case, $r_{x+y} < r_x$ and in this case $|r_{x+y} - r_x| = r_x - r_{x+y} = -r_y - 2k\pi$. Since r_x and r_{x+y} are both in $[0, 2\pi)$ their difference is also in $[0, 2\pi)$ and so in this case $k = -1$. Therefore, in this case, $|r_{x+y} - r_x| = 2\pi - r_y$. Now since the circumference of the unit circle is 2π, the length of the arc joining $p(2\pi - r_y)$ to $(1, 0)$ is the same as the length of the arc joining $p(r_y) = p(y)$ to $(1, 0)$. ■

The following theorem is the fundamental identity from which all the major trig. identities involving sums and differences of angles are derived.

Theorem 1.3. *Let $x, y \in \mathbb{R}$. Then*

$$\cos(x + y)\cos y + \sin(x + y)\sin y = \cos x. \tag{1.5}$$

Proof: Recall that for a real number z, there is a unique point $p(z)$ on the unit circle and the coordinates of this point are $\cos z$ and $\sin z$. Now from the above lemma, the length of the arc between $p(x + y)$ and $p(x)$ has the same length as the arc between $p(y)$ and $p(0)$. As an illustration see the following picture.

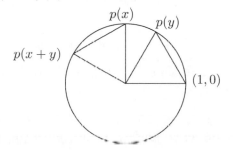

It follows from the definition of the radian measure of an angle that the two angles determined by these arcs are equal and so the distance between the points $p(x + y)$ and $p(x)$ must be the same as the distance from $p(y)$ to $p(0)$. Writing this in terms of the definition of the trig functions and the distance formula,

$$(\cos(x + y) - \cos x)^2 + (\sin(x + y) - \sin x)^2 = (\cos y - 1)^2 + \sin^2 x.$$

$$\cos^2(x + y) + \cos^2 x - 2\cos(x + y)\cos x + \sin^2(x + y) + \sin^2 x - 2\sin(x + y)\sin x$$

$$= \cos^2 y - 2\cos y + 1 + \sin^2 y$$

From Observation 1.1, this implies (1.5). ■

Letting $y = \pi/2$, this shows that

$$\sin(x + \pi/2) = \cos x. \tag{1.6}$$

Now let $u = x + y$ and $v = y$. Then (1.5) implies

$$\cos u \cos v + \sin u \sin v = \cos(u - v) \tag{1.7}$$

Also, from this and (1.2),

$$\cos\left(u+v\right) = \cos\left(u - \left(-v\right)\right) = \cos u \cos\left(-v\right) + \sin u \sin\left(-v\right)$$
$$= \cos u \cos v - \sin u \sin v \qquad (1.8)$$

Thus, letting $v = \pi/2$,

$$\cos\left(u + \frac{\pi}{2}\right) = -\sin u \qquad (1.9)$$

It follows

$$\sin\left(x+y\right) = -\cos\left(x + \frac{\pi}{2} + y\right) = -\left[\cos\left(x + \frac{\pi}{2}\right)\cos y - \sin\left(x + \frac{\pi}{2}\right)\sin y\right]$$
$$= \sin x \cos y + \sin y \cos x \qquad (1.10)$$

Then using Observation 1.1 again, this implies

$$\sin\left(x - y\right) = \sin x \cos y - \cos x \sin y \qquad (1.11)$$

In addition to this, Observation 1.1 implies

$$\cos 2x = \cos^2 x - \sin^2 x \qquad (1.12)$$
$$= 2\cos^2 x - 1 \qquad (1.13)$$
$$= 1 - 2\sin^2 x \qquad (1.14)$$

Therefore, making use of the above identities and Observation 1.1,

$$\cos\left(3x\right) = \cos 2x \cos x - \sin 2x \sin x = \left(2\cos^2 x - 1\right)\cos x - 2\cos x \sin^2 x$$
$$= 4\cos^3 x - 3\cos x \qquad (1.15)$$

With these fundamental identities, it is easy to obtain the cosine and sine of many special angles, called **reference angles**. First, cosider $\cos\left(\frac{\pi}{4}\right)$.

$$0 = \cos\left(\frac{\pi}{2}\right) = \cos\left(\frac{\pi}{4} + \frac{\pi}{4}\right) = 2\cos^2\left(\frac{\pi}{4}\right) - 1$$

and so $\cos\left(\frac{\pi}{4}\right) = \sqrt{2}/2$. (Why is not it equal to $-\sqrt{2}/2$? **Hint:** Draw a picture.) Thus $\sin\left(\frac{\pi}{4}\right) = \sqrt{2}/2$ also. (Why?) Here is another one. From (1.15),

$$0 = \cos\left(\frac{\pi}{2}\right) = \cos 3\left(\frac{\pi}{6}\right) = 4\cos^3\left(\frac{\pi}{6}\right) - 3\cos\left(\frac{\pi}{6}\right).$$

Therefore, $\cos\left(\frac{\pi}{6}\right) = \frac{\sqrt{3}}{2}$ and consequently, $\sin\left(\frac{\pi}{6}\right) = \frac{1}{2}$. Here is a short table for the reference angles. In the table, θ refers to the radian measure of the angle.

θ	0	$\frac{\pi}{6}$	$\frac{\pi}{4}$	$\frac{\pi}{3}$	$\frac{\pi}{2}$
$\cos\theta$	1	$\frac{\sqrt{3}}{2}$	$\frac{\sqrt{2}}{2}$	$\frac{1}{2}$	0
$\sin\theta$	0	$\frac{1}{2}$	$\frac{\sqrt{2}}{2}$	$\frac{\sqrt{3}}{2}$	1

Remember that in calculus, angles are considered as real numbers.

1.6   Exercises

(1) The factorial symbol is defined as $n! \equiv n(n-1)\cdots 1$ whenever n is a positive integer. $0! \equiv 1$. Now the binomial coefficients are of the form
$$\frac{n!}{k!(n-k)!} \equiv \binom{n}{k}, 0 \le k \le n$$
Show that
$$\binom{n}{k} + \binom{n}{k-1} = \binom{n+1}{k}$$

(2) The binomial theorem says that
$$(a+b)^n = \sum_{k=0}^{n} \binom{n}{k} a^{n-k} b^k$$
Using the identity of Problem 1, prove the binomial theorem by induction.

(3) Find $\cos\theta$ and $\sin\theta$ for $\theta \in \left\{ \frac{2\pi}{3}, \frac{3\pi}{4}, \frac{5\pi}{6}, \pi, \frac{7\pi}{6}, \frac{5\pi}{4}, \frac{4\pi}{3}, \frac{3\pi}{2}, \frac{5\pi}{3}, \frac{7\pi}{4}, \frac{11\pi}{6}, 2\pi \right\}$.

(4) Prove $\cos^2\theta = \frac{1+\cos 2\theta}{2}$ and $\sin^2\theta = \frac{1-\cos 2\theta}{2}$.

(5) $\pi/12 = \pi/3 - \pi/4$. Therefore, from Problem 4, $\cos(\pi/12) = \sqrt{\frac{1+(\sqrt{3}/2)}{2}}$. On the other hand,
$$\cos(\pi/12) = \cos(\pi/3 - \pi/4) = \cos\pi/3\cos\pi/4 + \sin\pi/3\sin\pi/4$$
and so $\cos(\pi/12) = \sqrt{2}/4 + \sqrt{6}/4$. Is there a problem here? Please explain.

(6) Prove $1 + \tan^2\theta = \sec^2\theta$ and $1 + \cot^2\theta = \csc^2\theta$.

(7) Prove that $\sin x \cos y = \frac{1}{2}(\sin(x+y) + \sin(x-y))$.

(8) Prove that $\sin x \sin y = \frac{1}{2}(\cos(x-y) - \cos(x+y))$.

(9) Prove that $\cos x \cos y = \frac{1}{2}(\cos(x+y) + \cos(x-y))$.

(10) Using Problem 7, find an identity for $\sin x - \sin y$.

(11) Suppose $\sin x = a$ where $0 < a < 1$. Find all possible values for

   (a) $\tan x$

   (b) $\cot x$

   (c) $\sec x$

   (d) $\csc x$

   (e) $\cos x$

(12) Solve the equations and give all solutions.

   (a) $\sin(3x) = \frac{1}{2}$

   (b) $\cos(5x) = \frac{\sqrt{3}}{2}$

   (c) $\tan(x) = \sqrt{3}$

   (d) $\sec(x) = 2$

   (e) $\sin(x+7) = \frac{\sqrt{2}}{2}$

   (f) $\cos^2(x) = \frac{1}{2}$

   (g) $\sin^4(x) = 4$

(13) Sketch a graph of $y = \sin x$.

(14) Sketch a graph of $y = \cos x$.

(15) Sketch a graph of $y = \sin 2x$.

(16) Sketch a graph of $y = \tan x$.

(17) Find a formula for $\sin x \cos y$ in terms of sines and cosines of $x + y$ and $x - y$.

(18) Using Problem 4 graph $y = \cos^2 x$.

(19) If $f(x) = A \cos \alpha x + B \sin \alpha x$, show there exists ϕ such that

$$f(x) = \sqrt{A^2 + B^2} \sin(\alpha x + \phi).$$

Show that there also exists ψ such that $f(x) = \sqrt{A^2 + b^2} \cos(\alpha x + \psi)$. This is a very important result, enough that some of these quantities are given names. $\sqrt{A^2 + B^2}$ is called the amplitude and ϕ or ψ are called phase shifts.

(20) Using Problem 19 graph $y = \sin x + \sqrt{3} \cos x$.

(21) Give all solutions to $\sin x + \sqrt{3} \cos x = \sqrt{3}$. **Hint:** Use Problem 20.

(22) ABC is a triangle where the capitol letters denote vertices of the triangle and the angle at the vertex. Let a be the length of the side opposite A and b is the length of the side opposite B and c is the length of the side opposite the vertex C. The law of sines says $\sin(A)/a = \sin(B)/b = \sin(C)/c$. Prove the **law of sines** from the definition of the trigonometric functions.

(23) In the picture, $a = 5, b = 3$, and $\theta = \frac{2}{3}\pi$. Find c.

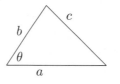

(24) In the picture, $\theta = \frac{1}{4}\pi$, $\alpha = \frac{2}{3}\pi$ and $c = 3$. Find a.

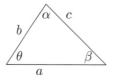

(25) An isosceles triangle is one which has two equal sides. For example the following picture is of an isosceles triangle

the two equal sides having length a. Show the "base angles" θ and α are equal. **Hint:** You might want to use the law of sines.

(26) Find a formula for $\tan(\theta + \beta)$ in terms of $\tan \theta$ and $\tan \beta$

(27) Find a formula for $\tan(2\theta)$ in terms of $\tan \theta$.

(28) Find a formula for $\tan\left(\frac{\theta}{2}\right)$ in terms of $\tan\theta$.

(29) Explain why, through any point in the plane, there exists a line parallel to a given line in the plane.

(30) Explain why the sum of the radian measures of the angles in any triangle equals π. **Hint:** Consider the following picture and use the result of Problem 29.

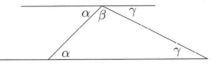

(31) The following picture is of an "inscribed angle", denoted by θ in a circle of radius a. Drawing a line from the center as shown in the picture, it follows since this forms an isosceles triangle, the two base angles are equal. These are denoted as θ in the picture.

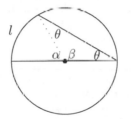

Now the radian measure of α is l/a. Using the result of Problem 30, show that the radian measure of θ equals $l/2a$.

(32) The inscribed angle in Problem 31 has the special property that one side is a diameter of the circle. A general inscribed angle is just like the one shown in this problem but without the requirement that either of the sides of the angle are a diameter. Show that for a completely arbitrary inscribed angle a similar result holds to the one in Problem 31.

1.7 Parabolas, Ellipses, And Hyperbolas

1.7.1 The Parabola

A parabola is a collection of points P in the plane such that the distance from P to a fixed line, called the **directrix**, is the same as the distance from P to a given point P_0, called the **focus**, as shown in the following picture.

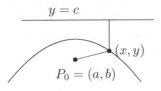

From this geometric description, it is possible to obtain a formula for a parabola. For simplicity, assume the top (vertex) of the parabola is $(0,0)$ and the line is $y = c$. It follows the focus must be $(0, -c)$. Then if (x, y) is a point on the parabola, the geometric description requires

$$\left(x^2 + (y+c)^2\right)^{1/2} = |y - c|.$$

Squaring both sides yields

$$x^2 + y^2 + 2yc + c^2 = y^2 - 2yc + c^2$$

and so

$$x^2 = -4yc \qquad (1.16)$$

If $c > 0$ then the largest y can be is 0. This follows from solving the above equation for y. Thus $y = -x^2/4c$ which is in this case is always less than or equal to 0. Therefore, the largest y can be and still satisfy the equation for the parabola is 0 when $x = 0$. Thus the vertex is $(0, 0)$. Similarly, if $c < 0$, the smallest y can be is 0 and the vertex is still $(0, 0)$.

This yields the following equation for a parabola.

Equation Of A Parabola: The equation for a parabola which has directrix $y = c$, focus $(0, -c)$ and vertex $(0, 0)$ is $x^2 = -4cy$.

This simple description can be used to consider more general situations in which, for example, the vertex is not at $(0, 0)$.

Example 1.3. Find the focus and directrix of the parabola $2x^2 - 3x + 1 = 5y$.

First complete the square on the left. Thus $2\left(x^2 - \frac{3}{2}x + \frac{9}{16}\right) - \frac{9}{8} + 1 = 5y$. this yields $2\left(x - \frac{3}{4}\right)^2 = 5y + \frac{1}{8} = 5\left(y + \frac{1}{40}\right)$. Dividing both sides by 2, $\left(x - \frac{3}{4}\right)^2 = \frac{5}{2}\left(y + \frac{1}{40}\right)$. Now if this were just $x^2 = -4cy$, you would know the directrix is $y = c$ and the focus is $(0, -c)$. It does not look like this, however. Therefore, change the variables, letting $u = x - \frac{3}{4}$ and $v = y + \frac{1}{40}$. Then the equation in terms of u and v is of the form $u^2 = -4\left(\frac{-5}{8}\right)v$ and so the focus for this parabola in the uv plane is $\left(0, \frac{5}{8}\right)$ and its directrix is $v = -\frac{5}{8}$. Since $x = u + \frac{3}{4}$ and $y = v - \frac{1}{40}$, if follows the focus of the parabola in the xy plane is $\left(\frac{3}{4}, \frac{5}{8} - \frac{1}{40}\right) = \left(\frac{3}{4}, \frac{3}{5}\right)$. Similarly, the directrix is $y = \frac{-5}{8} - \frac{1}{40} = -\frac{13}{20}$.

Sometimes the y variable is squared and the x variable is not squared. This corresponds to having a directrix of the form $x = c$ and to the parabola opening either to the right or the left. The same reasoning given above yields the following equation of a sideways parabola.

Equation Of A Parabola: The equation for a parabola which has directrix $x = c$, focus $(-c, 0)$ and vertex $(0, 0)$ is $y^2 = -4cx$.

Example 1.4. Find the focus and directrix of the parabola $x = 3y^2 + 2y + 1$.

Complete the square on the right to get

$$x = 3\left(y^2 + \frac{2}{3}y + \frac{1}{9}\right) - \frac{1}{3} + 1 = 3\left(y + \frac{1}{3}\right)^2 + \frac{2}{3}$$

and so $-4\left(\frac{-1}{12}\right)\left(x - \frac{2}{3}\right) = \left(y + \frac{1}{3}\right)^2$. Changing variables, letting $u = x - \frac{2}{3}$ and $v = y + \frac{1}{3}$, this gives $-4\left(\frac{-1}{12}\right)u = v^2$. Therefore, in the uv plane, the directrix is $u = \frac{-1}{12}$ and the focus is $\left(\frac{1}{12}, 0\right)$ while the vertex is $(0,0)$. In terms of the original variables the directrix is $x = \frac{2}{3} - \frac{1}{12} = \frac{7}{12}$, the focus is $\left(\frac{1}{12} + \frac{2}{3}, 0 - \frac{1}{3}\right) = \left(\frac{3}{4}, -\frac{1}{3}\right)$ while the vertex is $\left(\frac{2}{3}, \frac{-1}{3}\right)$.

This illustrates the following procedure.

Procedure 1.4. To find the focus, directrix and vertex of a parabola of the form $x = ay^2 + by + c$, you complete the square and otherwise massage things to get it in the form

$$-4c\left(x - p\right) = \left(y - q\right)^2$$

and then the vertex is at (p, q) and the focus is at $(p - c, q)$ while the directrix is $x = p + c$. To find the focus directrix and vertex of a parabola of the form $y = ax^2 + bx + c$, you complete the square and otherwise massage things to obtain an equation of the form

$$\left(x - p\right)^2 = -4c\left(y - q\right)$$

Then the vertex is (p, q) the focus is $(p, q - c)$ and the directrix is $y = q + c$.

1.7.2   The Ellipse

With an ellipse, there are two points P_1 and P_2 which are fixed and the ellipse consists of the set of points P such that $d\left(P, P_1\right) + d\left(P, P_2\right) = d$, where d is a fixed positive number. These two points are called the **foci** of the ellipse. Each is called a focus point by itself. Here is a picture in case the two points are of the form $(\alpha - h, \beta)$ and $(\alpha + h, \beta)$.

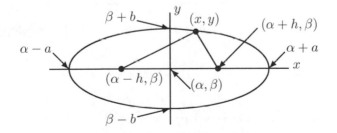

Note in the picture $2a$ is the length of this ellipse and $2b$ is its height. The major axis has length $2a$ and the minor axis has length $2b$. You can think of a string stretched tight and the pencil being at the point on the graph of the ellipse.

Thus you can see from the picture that $d = 2a$. Just imagine the point is at $(a + \alpha, \beta)$, the right vertex of the ellipse.

For simplicity assume the foci are at $(-c, 0)$ and $(c, 0)$ where $c \geq 0$. Then from the description, and letting $2a$ be the length of the ellipse between the two farthest vertices, the equation of the ellipse is given by $\sqrt{(x + c)^2 + y^2} + \sqrt{(x - c)^2 + y^2} = 2a$. Therefore, $\sqrt{(x + c)^2 + y^2} = 2a - \sqrt{(x - c)^2 + y^2}$, and so squaring both sides,

$$x^2 + 2xc + c^2 + y^2 = 4a^2 + x^2 - 2xc + c^2 + y^2 - 4a\sqrt{(x - c)^2 + y^2}$$

Subtracting like terms from both sides and then dividing by 4, $xc = a^2 - a\sqrt{(x - c)^2 + y^2}$. It follows $xc - a^2 = -a\sqrt{(x - c)^2 + y^2}$. Now square both sides of this to obtain

$$x^2 c^2 - 2xca^2 + a^4 = a^2 \left(x^2 - 2cx + c^2 + y^2 \right).$$

Next simplify this some more to obtain

$$a^2 \left(a^2 - c^2 \right) = \left(a^2 - c^2 \right) x^2 + a^2 y^2.$$

Therefore,

$$1 = \frac{x^2}{a^2} + \frac{y^2}{b^2}$$

where $b^2 = a^2 - c^2$. This verifies the following equation of an ellipse.

Equation Of An Ellipse: Suppose the ellipse is centered at $(0, 0)$ has major axis parallel to the x axis and minor axis parallel to the y axis and that the major axis is of length $2a$ while the minor axis is of length $2b$. Then the equation of the ellipse is

$$1 = \frac{x^2}{a^2} + \frac{y^2}{b^2}$$

and the focus points are at $\left(-\sqrt{a^2 - b^2}, 0 \right)$ and $\left(\sqrt{a^2 - b^2}, 0 \right)$. Switching the variables, if the ellipse is centered at $(0, 0)$ and the major axis of length $2a$ is parallel to the y axis while the minor axis of length $2b$ is parallel to the x axis, then the equation of the ellipse is

$$1 = \frac{x^2}{b^2} + \frac{y^2}{a^2}$$

and the focus points are at $\left(0, -\sqrt{a^2 - b^2} \right)$ and $\left(0, \sqrt{a^2 - b^2} \right)$.

As in the case of the parabola, you can use this to find the foci of ellipses which are not centered at $(0, 0)$.

Example 1.5. Find the focus points for the ellipse $\frac{(x-1)^2}{9} + \frac{(y-2)^2}{4} = 1$.

Let $u = x - 1$ and $v = y - 2$. Then the focus points of the identical ellipse in the uv plane would be $\left(\sqrt{5}, 0 \right)$ and $\left(-\sqrt{5}, 0 \right)$. Now $x = u + 1, y = v + 2$ so the focus points of the original ellipse in the xy plane are $\left(\sqrt{5} + 1, 2 \right)$ and $\left(-\sqrt{5} + 1, 2 \right)$.

Example 1.6. Show the following is the equation of an ellipse and find its major and minor axes along with the focus points. $2x^2 + 4x + y^2 + 2y + 2 = 0$.

As usual you have to complete the square in this. Thus $2\left(x^2 + 2x + 1\right) + y^2 + 2y + 1 = 1$. Therefore,

$$\frac{(x+1)^2}{\left(1/\sqrt{2}\right)^2} + (y+1)^2 = 1$$

The major axis has length 2 and is parallel to the y axis and the minor axis has length $\sqrt{2}$ and is parallel to the x axis. Let $x + 1 = u$ and $y + 1 = v$. Then in terms of u, v this equation is of the form

$$\frac{u^2}{\left(1/\sqrt{2}\right)^2} + v^2 = 1.$$

The focus points in the uv plane are $\left(0, \pm\sqrt{1 - (1/2)}\right) = \left(0, \pm\frac{1}{\sqrt{2}}\right)$. Therefore, the focus points of the ellipse in the xy plane are $\left(-1, -1 \pm \frac{1}{\sqrt{2}}\right)$.

This illustrates the following procedure.

Procedure 1.5. For $a, b > 0$ and an ellipse in the form $Ax^2 + By^2 + Cx + Dy + E = 0$ you first complete the square and then massage to obtain something of the form

$$\frac{(x-p)^2}{a^2} + \frac{(y-q)^2}{b^2} = 1$$

where a, b are positive numbers. If $a > b$ then the major axis has length $2a$ and is parallel to the x axis. In this case the focus points are at $\left(p - \sqrt{a^2 - b^2}, q\right)$ and $\left(p + \sqrt{a^2 - b^2}, q\right)$ and the minor axis has length $2b$ and is parallel to the y axis. If $a < b$ then the major axis is parallel to the y axis and has length $2b$. In this case the focus points are $\left(p, q + \sqrt{b^2 - a^2}\right)$ and $\left(p, q - \sqrt{b^2 - a^2}\right)$ and the minor axis is parallel to the x axis and has length $2a$. If $a = b$, the ellipse is a circle and it has center at (p, q) with radius equal to $a = b$.

1.7.3 The Hyperbola

With a hyperbola, there are two points P_1 and P_2 which are fixed and the hyperbola consists of the set of points P such that $d\left(P, P_1\right) - d\left(P, P_2\right) = d$, where d is a fixed positive number. These two points are called the foci of the hyperbola. Each is called a focus point by itself. The following picture is descriptive of the above situation.

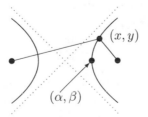

Now one can obtain an equation which will describe a hyperbola. For simplicity, consider the case where the focus points are $(-c,0)$ and $(c,0)$ and let d denote the difference between the two distances. Then from the above description of the hyperbola, $\sqrt{(x+c)^2 + y^2} - \sqrt{(x-c)^2 + y^2} = d$. Then

$$x^2 + 2cx + y^2 = d^2 + 2d\sqrt{(x-c)^2 + y^2} + x^2 - 2xc + y^2.$$

Subtracting like terms, $4cx - d^2 = 2d\sqrt{(x-c)^2 + y^2}$. Therefore, squaring both sides again,

$$16c^2 x^2 + d^4 - 8xcd^2 = 4d^2\left(x^2 - 2cx + c^2 + y^2\right)$$

Another simplification gives $d^2\left(d^2 - 4c^2\right) = 4\left(d^2 - 4c^2\right)x^2 + 4d^2 y^2$. Therefore,

$$1 = \frac{x^2}{\left(d^2/4\right)} - \frac{y^2}{\left(4c^2 - d^2\right)/4}$$

Therefore, the vertices of this hyperbola, one of which is labeled (α,β) in the above picture are in this case at $\left(\frac{d}{2},0\right)$ and $\left(-\frac{d}{2},0\right)$ so d is also the distance between the two vertices which is less than the distance between the two focus points $2c$. Therefore, $4c^2 - d^2 > 0$. Let $a^2 = d^2/4$ and let $b^2 = \left(4c^2 - d^2\right)/4$, so the equation of the hyperbola is of the form

$$1 = \frac{x^2}{a^2} - \frac{y^2}{b^2}.$$

Observe also the relation between a, b and c. $a^2 + b^2 = c^2$.

Another interesting thing about hyperbolas is the asymptotes. Solve the equation above for x.

$$x = \pm\left(a^2 + \frac{a^2 y^2}{b^2}\right)^{1/2}$$

Now for large $|y|$, this is essentially equal to $\pm\frac{ay}{b}$. The asymptotes of the hyperbola are $x = \frac{a}{b}y$ and $x = -\frac{a}{b}y$. These are the dotted lines in the above picture.

Equation For A Hyperbola: Suppose you have the relation,

$$1 = \frac{x^2}{a^2} - \frac{y^2}{b^2}.$$

Then this is a hyperbola having vertices at $(a,0)$ and $(-a,0)$ and foci at $\left(-\sqrt{a^2 + b^2},0\right)$ and $\left(\sqrt{a^2 + b^2},0\right)$. Its asymptotes are given by

$$x = \pm\frac{a}{b}y.$$

Switching the variables, suppose now you have the relation

$$1 = \frac{y^2}{b^2} - \frac{x^2}{a^2}.$$

Then it is a hyperbola having vertices at $(0,b)$ and $(0,-b)$ and foci at $\left(0, -\sqrt{a^2 + b^2}\right)$ and $\left(0, \sqrt{a^2 + b^2}\right)$. Its asymptotes are

$$y = \pm\frac{bx}{a}.$$

Example 1.7. Find the focus points and vertices for the hyperbola $\frac{(x-1)^2}{4} - \frac{(y-2)^2}{9} = 1$.

Change the variables. Let $u = x - 1$ and $v = y - 2$. Then the focus points of the identical hyperbola in the uv plane would be $\left(\sqrt{13}, 0\right)$ and $\left(-\sqrt{13}, 0\right)$. Now $x = u + 1, y = v + 2$ so the focus points of the original hyperbola in the xy plane are $\left(\sqrt{13} + 1, 2\right)$ and $\left(-\sqrt{13} + 1, 2\right)$. The vertices are obtained by taking $y = 2$, $(3, 2)$ and $(-1, 2)$.

Example 1.8. Show that the relation, $2y^2 + 4y - 4x^2 + 8x = 3$ is a hyperbola and find its foci.

As usual, you complete the square. Thus
$$2\left(y^2 + 2y + 1\right) - 4\left(x^2 - 2x + 1\right) = 3 + 2 - 4 = 1.$$
Therefore,
$$\frac{(y+1)^2}{(1/\sqrt{2})^2} - \frac{(x-1)^2}{(1/2)^2} = 1.$$

Therefore, the vertices are $\left(1, -1 + \frac{1}{\sqrt{2}}\right)$ and $\left(1, -1 - \frac{1}{\sqrt{2}}\right)$. The foci are at $\left(1, -1 - \frac{\sqrt{3}}{2}\right)$ and $\left(1, -1 + \frac{\sqrt{3}}{2}\right)$.

This illustrates the following procedure for a hyperbola.

Procedure 1.6. Suppose $Ax^2 + By^2 + Cx + Dy + E = 0$ and A and B have different signs. Complete the square and massage to obtain an expression of one of the following forms.

$$\text{(a)} \qquad \frac{(x-p)^2}{a^2} - \frac{(y-q)^2}{b^2} = 1$$

$$\text{(b)} \qquad \frac{(y-q)^2}{b^2} - \frac{(x-p)^2}{a^2} = 1$$

where a, b are positive numbers. In the case of (a) the focus points are at $\left(p + \sqrt{a^2 + b^2}, q\right)$ and $\left(p - \sqrt{a^2 + b^2}, q\right)$, the vertices are at $(p + a, q), (p - a, q)$, and the asymptotes are of the form

$$\frac{(x - p)}{a} \pm \frac{(y - q)}{b} = 1$$

In the case of (b) the focus points are at $\left(p, q + \sqrt{a^2 + b^2}\right)$ and $\left(p, q - \sqrt{a^2 + b^2}\right)$, the vertices are at $(p, q + b), (p, q - b)$, and the asymptotes are of the form

$$\frac{(y - q)}{b} \pm \frac{(x - p)}{a} = 1.$$

1.8 Exercises

(1) Consider $y = 2x^2 + 3x + 7$. Find the focus and the directrix of this parabola.

(2) Sketch a graph of the ellipse whose equation is $\frac{(x-1)^2}{4} + \frac{(y-2)^2}{9} = 1$.

(3) Sketch a graph of the ellipse whose equation is $\frac{(x-1)^2}{9} + \frac{(y-2)^2}{4} = 1$.

(4) Sketch a graph of the hyperbola, $\frac{x^2}{4} - \frac{y^2}{9} = 1$.

(5) Sketch a graph of the hyperbola, $\frac{y^2}{4} - \frac{x^2}{9} = 1$.

(6) Find the focus points for the hyperbola $\frac{(x-1)^2}{4} - \frac{(y-2)^2}{9} = 1$.

(7) Consider the hyperbola, $\frac{y^2}{4} - \frac{x^2}{9} = 1$. Show that $y = \pm\sqrt{b^2 + \frac{b^2 x^2}{a^2}}$. The straight lines $y = \frac{bx}{a}$ and $y = -\frac{bx}{a}$ are called the asymptotes of the hyperbola. Show that for large x, $\sqrt{b^2 + \frac{b^2 x^2}{a^2}} - \frac{bx}{a}$ is very small.

(8) What is the diameter of the ellipse $\frac{(x-1)^2}{9} + \frac{(y-2)^2}{4} = 1$? The diameter of a set, S is defined as $\sup\{|(x_1, y_1) - (x_2, y_2)| : (x_1, y_1), (x_2, y_2) \in S\}$.

(9) Consider $9x^2 - 36x + 32 - 4y^2 - 8y = 36$. Identify this as either an ellipse, a hyperbola or a parabola. Then find its focus point(s) and its directrix if it is a parabola. **Hint:** First complete the square.

(10) Consider $4x^2 - 8x + 68 + 16y^2 - 64y = 64$. Identify this as either an ellipse, a hyperbola or a parabola. Then find its focus point(s) and its directrix if it is a parabola. **Hint:** First complete the square.

(11) Consider $-8x + 8 + 16y^2 - 64y = 64$. Identify this as either an ellipse, a hyperbola or a parabola. Then find its focus point(s) and its directrix if it is a parabola. **Hint:** First complete the square.

(12) Consider $5x + 3y^2 + 2y = 7$. Identify this as either an ellipse, a hyperbola or a parabola. Then find its focus point(s) and its directrix if it is a parabola.

(13) Consider the two points $P_1 = (0, 0)$ and $P_2 = (1, 1)$. Find the equation of the ellipse defined by $d(P, P_1) + d(P, P_2) = 4$ which has these as focus points.

(14) Find the equation of the parabola which has focus $(0, 0)$ and directrix $x + y = 1$. This is pretty hard. To do it you need to figure out how to find the distance between a point and the given line.

(15) As explained earlier, $(\cos t, \sin t)$ for $t \in \mathbb{R}$ is a point on the circle of radius 1. Find a formula for the coordinates of a point on the ellipse $\frac{(x-2)^2}{4} + \frac{(y+1)^2}{8} = 1$. **Hint:** This says $\left(\frac{x-2}{2}, \frac{y+1}{\sqrt{8}}\right)$ is a point on the unit circle.

1.9   The Complex Numbers

This is a brief treatment of the complex numbers. Complex numbers will not be needed very much in Calculus but you will need it when you take differential equations and various other subjects, so it is a good idea to consider these numbers. However, if you are in a hurry to get to calculus, you can skip this section.

Just as a real number should be considered as a point on the line, a complex number is considered a point in the plane which can be identified in the usual way using the Cartesian coordinates of the point. Thus (a, b) identifies a point whose x coordinate is a and whose y coordinate is b. In dealing with complex numbers, such a point is written as $a + ib$. For example, in the following picture, I have graphed the point $3 + 2i$. It corresponds to the point in the plane whose coordinates are $(3, 2)$.

Multiplication and addition are defined in the most obvious way subject to the convention that $i^2 = -1$. Thus,

$$(a + ib) + (c + id) = (a + c) + i(b + d)$$

and

$$(a + ib)(c + id) = ac + iad + ibc + i^2 bd$$
$$= (ac - bd) + i(bc + ad).$$

Every nonzero complex number $a + ib$, with $a^2 + b^2 \neq 0$, has a unique multiplicative inverse.

$$\frac{1}{a + ib} = \frac{a - ib}{a^2 + b^2} = \frac{a}{a^2 + b^2} - i\frac{b}{a^2 + b^2}.$$

You should prove the following theorem.

Theorem 1.7. *The complex numbers with multiplication and addition defined as above form a field satisfying all the field axioms,*

Axiom 1.2. $x + y = y + x$, *(commutative law for addition)*.

Axiom 1.3. $x + 0 = x$, *(additive identity)*.

Axiom 1.4. *For each* $x \in \mathbb{R}$, *there exists* $-x \in \mathbb{R}$ *such that* $x + (-x) = 0$, *(existence of additive inverse)*.

Axiom 1.5. $(x + y) + z = x + (y + z)$, *(associative law for addition)*.

Axiom 1.6. $xy = yx$, *(commutative law for multiplication)*.

Axiom 1.7. $(xy)z = x(yz)$, *(associative law for multiplication)*.

Axiom 1.8. $1x = x$, *(multiplicative identity)*.

Axiom 1.9. *For each* $x \neq 0$, *there exists* x^{-1} *such that* $xx^{-1} = 1$, *(existence of multiplicative inverse)*.

Axiom 1.10. $x(y + z) = xy + xz$, *(distributive law)*.

The field of complex numbers is denoted as $\mathbb{C}$. An important construction regarding complex numbers is the complex conjugate denoted by a horizontal line above the number. It is defined as follows.

$$\overline{a + ib} \equiv a - ib.$$

What it does is reflect a given complex number across the x axis. The following formula is easy to obtain.

$$\left(\overline{a + ib}\right)(a + ib) = a^2 + b^2.$$

Definition 1.6. Define the absolute value of a complex number as follows.

$$|a + ib| \equiv \sqrt{a^2 + b^2}.$$

Thus, denoting by z the complex number $z = a + ib$,

$$|z| = (z\bar{z})^{1/2}.$$

With this definition, it is important to note the following. Be sure to verify this. It is not too hard but you need to do it.

Remark 1.1. Let $z = a + ib$ and $w = c + id$. Then $|z - w| = \sqrt{(a - c)^2 + (b - d)^2}$. Thus the distance between the point in the plane determined by the ordered pair (a, b) and the ordered pair (c, d) equals $|z - w|$ where z and w are as just described.

For example, consider the distance between $(2, 5)$ and $(1, 8)$. From the distance formula this distance equals $\sqrt{(2 - 1)^2 + (5 - 8)^2} = \sqrt{10}$. On the other hand, letting $z = 2 + i5$ and $w = 1 + i8$, $z - w = 1 - i3$ and so $(z - w)(\overline{z - w}) = (1 - i3)(1 + i3) = 10$ so $|z - w| = \sqrt{10}$, the same thing obtained with the distance formula.

Proposition 1.3. Let z, w be real or complex numbers. Then the triangle inequality holds.

$$|z + w| \leq |z| + |w|, \quad ||z| - |w|| \leq |z - w|.$$

Proof: From the definition,

$$|z + w|^2 = (z + w)(\bar{z} + \bar{w})$$

$$= z\bar{z} + w\bar{w} + z\bar{w} + \bar{z}w = |z|^2 + 2\mathrm{Re}\,(z\bar{w}) + |w|^2$$
$$\leq |z|^2 + 2|z||w| + |w|^2 = (|z| + |w|)^2$$

and taking the square root, this gives the first inequality above. To get the second,

$$z = z - w + w, \quad w = w - z + z,$$

and so by the first form of the inequality,

$$|z| \leq |z - w| + |w|, \quad |w| \leq |z - w| + |z|.$$

It follows that both $|z| - |w|$ and $|w| - |z|$ are no larger than $|z - w|$ which proves the second version because $||z| - |w||$ is one of $|z| - |w|$ or $|w| - |z|$. ∎

Complex numbers, are often written in the so called polar form which is described next. Suppose $x + iy$ is a complex number. Then

$$x + iy = \sqrt{x^2 + y^2} \left(\frac{x}{\sqrt{x^2 + y^2}} + i \frac{y}{\sqrt{x^2 + y^2}} \right).$$

Now note that

$$\left(\frac{x}{\sqrt{x^2 + y^2}} \right)^2 + \left(\frac{y}{\sqrt{x^2 + y^2}} \right)^2 = 1$$

and so

$$\left(\frac{x}{\sqrt{x^2 + y^2}}, \frac{y}{\sqrt{x^2 + y^2}} \right)$$

is a point on the unit circle. Therefore, there exists a unique angle $\theta \in [0, 2\pi)$ such that

$$\cos \theta = \frac{x}{\sqrt{x^2 + y^2}}, \quad \sin \theta = \frac{y}{\sqrt{x^2 + y^2}}.$$

The polar form of the complex number is then

$$r(\cos \theta + i \sin \theta)$$

where θ is this angle just described, and $r = \sqrt{x^2 + y^2}$.

A fundamental identity is the formula of De Moivre which follows.

Theorem 1.8. *Let $r > 0$ be given. Then if n is a positive integer,*

$$[r(\cos t + i \sin t)]^n = r^n (\cos nt + i \sin nt).$$

Proof: It is clear the formula holds if $n = 1$. Suppose it is true for n.

$$[r(\cos t + i \sin t)]^{n+1} = [r(\cos t + i \sin t)]^n [r(\cos t + i \sin t)]$$

which by induction equals

$$= r^{n+1} (\cos nt + i \sin nt)(\cos t + i \sin t)$$

$$= r^{n+1} ((\cos nt \cos t - \sin nt \sin t) + i (\sin nt \cos t + \cos nt \sin t))$$

$$= r^{n+1} (\cos (n+1) t + i \sin (n+1) t)$$

by the formulas for the cosine and sine of the sum of two angles. ∎

Corollary 1.2. *Let z be a nonzero complex number. Then there are always exactly k k^{th} roots of z in $\mathbb{C}$.*

Proof: Let $z = x + iy$ and let $z = |z|(\cos t + i \sin t)$ be the polar form of the complex number. By De Moivre's theorem, a complex number

$$r(\cos \alpha + i \sin \alpha),$$

is a kth root of z if and only if

$$r^k (\cos k\alpha + i \sin k\alpha) = |z|(\cos t + i \sin t).$$

This requires $r^k = |z|$ and so $r = |z|^{1/k}$ and also both $\cos(k\alpha) = \cos t$ and $\sin(k\alpha) = \sin t$. This can only happen if

$$k\alpha = t + 2l\pi$$

for l an integer. Thus

$$\alpha = \frac{t + 2l\pi}{k}, \ l \in \mathbb{Z}$$

and so the kth roots of z are of the form

$$|z|^{1/k} \left(\cos \left(\frac{t + 2l\pi}{k} \right) + i \sin \left(\frac{t + 2l\pi}{k} \right) \right), \ l \in \mathbb{Z}.$$

Since the cosine and sine are periodic of period 2π, there are exactly k distinct numbers which result from this formula. ∎

Example 1.9. Find the three cube roots of i.

First note that $i = 1 \left(\cos \left(\frac{\pi}{2} \right) + i \sin \left(\frac{\pi}{2} \right) \right)$. Using the formula in the proof of the above corollary, the cube roots of i are

$$1 \left(\cos \left(\frac{(\pi/2) + 2l\pi}{3} \right) + i \sin \left(\frac{(\pi/2) + 2l\pi}{3} \right) \right)$$

where $l = 0, 1, 2$. Therefore, the roots are

$$\cos \left(\frac{\pi}{6} \right) + i \sin \left(\frac{\pi}{6} \right), \ \cos \left(\frac{5}{6}\pi \right) + i \sin \left(\frac{5}{6}\pi \right),$$

and

$$\cos \left(\frac{3}{2}\pi \right) + i \sin \left(\frac{3}{2}\pi \right).$$

Thus the cube roots of i are $\frac{\sqrt{3}}{2} + i \left(\frac{1}{2} \right), \frac{-\sqrt{3}}{2} + i \left(\frac{1}{2} \right)$, and $-i$.

The ability to find k^{th} roots can also be used to factor some polynomials.

Example 1.10. Factor the polynomial $x^3 - 27$.

First find the cube roots of 27. By the above procedure using De Moivre's theorem, these cube roots are $3, 3 \left(\frac{-1}{2} + i\frac{\sqrt{3}}{2} \right)$, and $3 \left(\frac{-1}{2} - i\frac{\sqrt{3}}{2} \right)$. Therefore, $x^3 - 27 =$

$$(x - 3) \left(x - 3 \left(\frac{-1}{2} + i\frac{\sqrt{3}}{2} \right) \right) \left(x - 3 \left(\frac{-1}{2} - i\frac{\sqrt{3}}{2} \right) \right).$$

Note also $\left(x - 3 \left(\frac{-1}{2} + i\frac{\sqrt{3}}{2} \right) \right) \left(x - 3 \left(\frac{-1}{2} - i\frac{\sqrt{3}}{2} \right) \right) = x^2 + 3x + 9$ and so

$$x^3 - 27 = (x - 3)(x^2 + 3x + 9)$$

where the quadratic polynomial, $x^2 + 3x + 9$ cannot be factored without using complex numbers.

1.10 Exercises

(1) Let $z = 5 + i9$. Find z^{-1}.

(2) Let $z = 2 + i7$ and let $w = 3 - i8$. Find zw, $z + w$, z^2, and w/z.

(3) Give the complete solution to $x^4 + 16 = 0$.

(4) Graph the complex cube roots of 8 in the complex plane. Do the same for the four fourth roots of 16.

(5) If z is a complex number, show there exists w a complex number with $|w| = 1$ and $wz = |z|$.

(6) De Moivre's theorem says $[r(\cos t + i\sin t)]^n = r^n(\cos nt + i\sin nt)$ for n a positive integer. Does this formula continue to hold for all integers n, even negative integers? Explain.

(7) You already know formulas for $\cos(x + y)$ and $\sin(x + y)$ and these were used to prove De Moivre's theorem. Now using De Moivre's theorem, derive a formula for $\sin(5x)$ and one for $\cos(5x)$.

(8) If z and w are two complex numbers and the polar form of z involves the angle θ while the polar form of w involves the angle ϕ, show that in the polar form for zw the angle involved is $\theta + \phi$. Also, show that in the polar form of a complex number z, $r = |z|$. Use this to give another proof of De Moivre's theorem.

(9) Factor $x^3 + 8$ as a product of linear factors.

(10) Write $x^3 + 27$ in the form $(x + 3)(x^2 + ax + b)$ where $x^2 + ax + b$ cannot be factored any more using only real numbers.

(11) Completely factor $x^4 + 16$ as a product of linear factors.

(12) Factor $x^4 + 16$ as the product of two quadratic polynomials each of which cannot be factored further without using complex numbers.

(13) If z, w are complex numbers prove $\overline{zw} = \bar{z}\bar{w}$ and then show by induction that $\overline{z_1 \cdots z_m} = \bar{z_1} \cdots \bar{z_m}$. Also verify that $\overline{\sum_{k=1}^m z_k} = \sum_{k=1}^m \bar{z_k}$. In words this says the conjugate of a product equals the product of the conjugates and the conjugate of a sum equals the sum of the conjugates.

(14) Suppose $p(x) = a_n x^n + a_{n-1} x^{n-1} + \cdots + a_1 x + a_0$ where all the a_k are real numbers. Suppose also that $p(z) = 0$ for some $z \in \mathbb{C}$. Show it follows that $p(\bar{z}) = 0$ also.

(15) I claim that $1 = -1$. Here is why.

$$-1 = i^2 = \sqrt{-1}\sqrt{-1} = \sqrt{(-1)^2} = \sqrt{1} = 1.$$

This is clearly a remarkable result but is there something wrong with it? If so, what is wrong?

(16) De Moivre's theorem is really a grand thing. I plan to use it now for rational exponents, not just integers.

$$1 = 1^{(1/4)} = (\cos 2\pi + i\sin 2\pi)^{1/4} = \cos(\pi/2) + i\sin(\pi/2) = i.$$

Therefore, squaring both sides it follows $1 = -1$ as in the previous problem. What does this tell you about De Moivre's theorem? Is there a profound difference between raising numbers to integer powers and raising numbers to noninteger powers?

(17) Review Problem 6 at this point. Now here is another question: If n is an integer, is it always true that $(\cos\theta - i\sin\theta)^n = \cos(n\theta) - i\sin(n\theta)$? Explain.

(18) Suppose you have any polynomial in $\cos\theta$ and $\sin\theta$. By this I mean an expression of the form $\sum_{\alpha=0}^{m}\sum_{\beta=0}^{n} a_{\alpha\beta}\cos^\alpha\theta\sin^\beta\theta$ where $a_{\alpha\beta}\in\mathbb{C}$. Can this always be written in the form $\sum_{\gamma=-(n+m)}^{m+n} b_\gamma\cos\gamma\theta + \sum_{\tau=-(n+m)}^{n+m} c_\tau\sin\tau\theta$? Explain.

1.11   Solving Systems Of Equations

It is often necessary to solve simultaneous systems for equations. When the equations are nonlinear, there is no way to do it in general, but when they are linear equations, there are easy methods available to find the solution.

Definition 1.7. A system of linear equations is a set of p equations for the n variables $x_1, \cdots, x_n$ which is of the form

$$\sum_{k=1}^{n} a_{mk}x_k = d_m, m = 1, 2, \cdots, p$$

Written less compactly it is a set of equations of the following form

$$a_{11}x_1 + a_{12}x_2 + \cdots + a_{1n}x_n = d_1$$
$$a_{21}x_1 + a_{22}x_2 + \cdots + a_{2n}x_n = d_2$$
$$\vdots$$
$$a_{p1}x_1 + a_{p2}x_2 + \cdots + a_{pn}x_n = d_p$$

The problem is to find the values of $x_1, x_2 \cdots, x_n$ which satisfy all p equations. This is called the **solution set** of the system of equations. In other words, $(a_1, \cdots, a_n)$ is in the solution set of the system of equations if, when you plug a_1 in place of x_1, a_2 in place of x_2 etc., each equation in the system is satisfied.

Consider the following example.

Example 1.11. Find x and y such that

$$x + y = 7 \text{ and } 2x - y = 8. \tag{1.17}$$

The set of ordered pairs, (x, y) which solve both equations is called the **solution set**.

You can verify that $(x, y) = (5, 2)$ is a solution to the above system. The interesting question is this: If you were not given this information to verify, how could you determine the solution? You can do this by using the following basic operations on the equations, none of which change the set of solutions of the system of equations.

Definition 1.8. Elementary operations are those operations consisting of the following.

(1) Interchange the order in which the equations are listed.
(2) Multiply any equation by a **nonzero** number.
(3) Replace any equation with itself added to a multiple of another equation.

Example 1.12. To illustrate the third of these operations on this particular system, consider the following.

$$x + y = 7$$
$$2x - y = 8$$

The system has the same solution set as the system

$$x + y = 7$$
$$-3y = -6$$

To obtain the second system, take the second equation of the first system and add -2 times the first equation to obtain

$$-3y = -6.$$

Now, this clearly shows that $y = 2$ and so it follows from the other equation that $x + 2 = 7$ and so $x = 5$.

Of course a linear system may involve many equations and many variables. The solution set is still the collection of solutions to the equations. In every case, the above operations of Definition 1.8 do not change the set of solutions to the system of linear equations.

Theorem 1.9. *Suppose you have two equations, involving the variables* $\{x_1, \cdots, x_n\}$

$$E_1 = f_1, E_2 = f_2 \tag{1.18}$$

where E_1 and E_2 are expressions involving the variables and f_1 and f_2 are constants. (In the above example there are only two variables x and y and $E_1 = x + y$ while $E_2 = 2x - y$.) Then the system $E_1 = f_1, E_2 = f_2$ has the same solution set as

$$E_1 = f_1, \ E_2 + aE_1 = f_2 + af_1. \tag{1.19}$$

Also the system $E_1 = f_1, E_2 = f_2$ has the same solutions as the system $E_2 = f_2, E_1 = f_1$. The system $E_1 = f_1, E_2 = f_2$ has the same solution as the system $E_1 = f_1, aE_2 = af_2$ provided $a \neq 0$.

Proof: If $\{x_1, \cdots, x_n\}$ solves $E_1 = f_1, E_2 = f_2$ then it solves the first equation in $E_1 = f_1, \ E_2 + aE_1 = f_2 + af_1$. Also, it satisfies $aE_1 = af_1$ and so, since it also solves $E_2 = f_2$ it must solve $E_2 + aE_1 = f_2 + af_1$. Therefore, if $\{x_1, \cdots, x_n\}$ solves $E_1 = f_1, E_2 = f_2$ it must also solve $E_2 + aE_1 = f_2 + af_1$. On the other hand, if it solves the system $E_1 = f_1$ and $E_2 + aE_1 = f_2 + af_1$, then $aE_1 = af_1$ and so you can subtract these equal quantities from both sides of $E_2 + aE_1 = f_2 + af_1$ to obtain $E_2 = f_2$ showing that it satisfies $E_1 = f_1, E_2 = f_2$.

The second assertion of the theorem which says that the system $E_1 = f_1, E_2 = f_2$ has the same solution as the system $E_2 = f_2, E_1 = f_1$ is seen to be true because it involves nothing more than listing the two equations in a different order. They are the same equations.

The third assertion of the theorem which says $E_1 = f_1, E_2 = f_2$ has the same solution as the system $E_1 = f_1, aE_2 = af_2$ provided $a \neq 0$ is verified as follows: If $(x_1, \cdots, x_n)$ is a solution of $E_1 = f_1, E_2 = f_2$, then it is a solution to $E_1 = f_1, aE_2 = af_2$ because the second system only involves multiplying the equation, $E_2 = f_2$ by a. If $(x_1, \cdots, x_n)$ is a solution of $E_1 = f_1, aE_2 = af_2$, then upon multiplying $aE_2 = af_2$ by the number, $1/a$, you find that $E_2 = f_2$. ∎

Stated simply, the above theorem shows that the elementary operations do not change the solution set of a system of equations.

Here is an example in which there are three equations and three variables. You want to find values for x, y, z such that each of the given equations are satisfied when these values are plugged in to the equations.

Example 1.13. Find the solutions to the system

$$x + 3y + 6z = 25$$
$$2x + 7y + 14z = 58 \tag{1.20}$$
$$2y + 5z = 19$$

To solve this system replace the second equation by (-2) times the first equation added to the second. This yields the system

$$x + 3y + 6z = 25$$
$$y + 2z = 8 \tag{1.21}$$
$$2y + 5z = 19$$

Now take (-2) times the second and add to the third. More precisely, replace the third equation with (-2) times the second added to the third. This yields the system

$$x + 3y + 6z = 25$$
$$y + 2z = 8 \tag{1.22}$$
$$z = 3$$

At this point, you can tell what the solution is. This system has the same solution as the original system and in the above, $z = 3$. Then using this in the second equation, it follows $y + 6 = 8$ and so $y = 2$. Now using this in the top equation yields $x + 6 + 18 = 25$ and so $x = 1$. This process is called **back substitution**.

Alternatively, in (1.22) you could have continued as follows. Add (-2) times the bottom equation to the middle and then add (-6) times the bottom to the top. This yields

$$x + 3y = 7$$
$$y = 2$$
$$z = 3$$

Now add (-3) times the second to the top. This yields

$$x = 1$$
$$y = 2 \ ,$$
$$z = 3$$

a system which has the same solution set as the original system. This avoided back substitution and led to the same solution set.

1.11.1 Gauss Elimination

A less cumbersome way to represent a linear system is to write it as an **augmented matrix.** For example the linear system (1.20) can be written as

$$\begin{pmatrix} 1 & 3 & 6 & | & 25 \\ 2 & 7 & 14 & | & 58 \\ 0 & 2 & 5 & | & 19 \end{pmatrix} .$$

It has exactly the same information as the original system but here it is understood there is an x column $\begin{pmatrix} 1 \\ 2 \\ 0 \end{pmatrix}$, a y column $\begin{pmatrix} 3 \\ 7 \\ 2 \end{pmatrix}$ and a z column $\begin{pmatrix} 6 \\ 14 \\ 5 \end{pmatrix}$. The rows correspond to the equations in the system. Thus the top row in the augmented matrix corresponds to the equation,

$$x + 3y + 6z - 25.$$

Now when you replace an equation with a multiple of another equation added to itself, you are just taking a row of this augmented matrix and replacing it with a multiple of another row added to it. Thus the first step in solving (1.20) would be to take (-2) times the first row of the augmented matrix above and add it to the second row,

$$\begin{pmatrix} 1 & 3 & 6 & | & 25 \\ 0 & 1 & 2 & | & 8 \\ 0 & 2 & 5 & | & 19 \end{pmatrix} .$$

Note how this corresponds to (1.21). Next take (-2) times the second row and add to the third,

$$\begin{pmatrix} 1 & 3 & 6 & | & 25 \\ 0 & 1 & 2 & | & 8 \\ 0 & 0 & 1 & | & 3 \end{pmatrix}$$

This augmented matrix corresponds to the system

$$x + 3y + 6z = 25$$
$$y + 2z = 8$$
$$z = 3$$

which is the same as (1.22). By back substitution you obtain the solution $x = 1, y = 6$, and $z = 3$.

In general a linear system is of the form

$$a_{11}x_1 + \cdots + a_{1n}x_n = b_1$$
$$\vdots$$
$$a_{m1}x_1 + \cdots + a_{mn}x_n = b_m$$

$$(1.23)$$

where the x_i are variables and the a_{ij} and b_i are constants. This system can be represented by the augmented matrix,

$$\begin{pmatrix} a_{11} & \cdots & a_{1n} & | & b_1 \\ \vdots & & \vdots & | & \vdots \\ a_{m1} & \cdots & a_{mn} & | & b_m \end{pmatrix}.$$

$$(1.24)$$

Changes to the system of equations in (1.23) as a result of an elementary operations translate into changes of the augmented matrix resulting from a row operation. Note that Theorem 1.9 implies that the row operations deliver an augmented matrix for a system of equations which has the same solution set as the original system.

Definition 1.9. The **row operations** consist of the following

(1) Switch two rows.
(2) Multiply a row by a nonzero number.
(3) Replace a given row by a multiple of another row added to the given row.

The idea is to do row operations, until you obtain an augmented matrix for a system of equations whose solution is easy to see.

Example 1.14. Give the solution to the system of equations $3x - y - z = 2, -x + y = 1, x - z = 1$.

Write the augmented matrix.

$$\begin{pmatrix} 3 & -1 & -1 & | & 2 \\ -1 & 1 & 0 & | & 1 \\ 1 & 0 & -1 & | & 1 \end{pmatrix}$$

Add the second row to the bottom row. This yields

$$\begin{pmatrix} 3 & -1 & -1 & | & 2 \\ -1 & 1 & 0 & | & 1 \\ 0 & 1 & -1 & | & 2 \end{pmatrix}$$

Next take three times the second row added to the top row.

$$\begin{pmatrix} 0 & 2 & -1 & | & 5 \\ -1 & 1 & 0 & | & 1 \\ 0 & 1 & -1 & | & 2 \end{pmatrix}$$

Take -1 times the bottom row and add it to the second row and then take -1 times the bottom row and add to the top. This yields

$$\begin{pmatrix} 0 & 1 & 0 & | & 3 \\ -1 & 0 & 1 & | & -1 \\ 0 & 1 & -1 & | & 2 \end{pmatrix}$$

The system of equations for which this is the augmented matrix is now easy to solve. The top equation says $y = 3$. Then placing 3 in the bottom equation, $3 - z = 2$ and so $z = 1$. Now use this information in the middle row to obtain $-x + 1 = -1$ and so $x = 2$.

Example 1.15. Give the complete solution to the system of equations $5x + 10y - 7z = -2$, $2x + 4y - 3z = -1$, and $3x + 6y + 5z = 9$.

The augmented matrix for this system is

$$\begin{pmatrix} 2 & 4 & -3 & | & -1 \\ 5 & 10 & -7 & | & -2 \\ 3 & 6 & 5 & | & 9 \end{pmatrix}$$

Multiply the second row by 2, the first row by 5, and then take (-1) times the first row and add to the second. Then multiply the first row by $1/5$. This yields

$$\begin{pmatrix} 2 & 4 & -3 & | & -1 \\ 0 & 0 & 1 & | & 1 \\ 3 & 6 & 5 & | & 9 \end{pmatrix}$$

Now, combining some row operations, take (-3) times the first row and add this to 2 times the last row and replace the last row with this. This yields.

$$\begin{pmatrix} 2 & 4 & -3 & | & -1 \\ 0 & 0 & 1 & | & 1 \\ 0 & 0 & 1 & | & 21 \end{pmatrix}.$$

One more row operation, taking (-1) times the second row and adding to the bottom yields.

$$\begin{pmatrix} 2 & 4 & -3 & | & -1 \\ 0 & 0 & 1 & | & 1 \\ 0 & 0 & 0 & | & 20 \end{pmatrix}.$$

This is impossible because the last row indicates the need for a solution to the equation

$$0x + 0y + 0z = 20$$

and there is no such thing because $0 \neq 20$. This shows there is no solution to the three given equations. When this happens, the system is called **inconsistent**. In this case it is very easy to describe the solution set. The system has no solution.

Here is another example based on the use of row operations.

Example 1.16. Give the complete solution to the system of equations $3x - y - 5z = 9$, $y - 10z = 0$, and $-2x + y = -6$.

The augmented matrix of this system is

$$\begin{pmatrix} 3 & -1 & -5 & | & 9 \\ 0 & 1 & -10 & | & 0 \\ -2 & 1 & 0 & | & -6 \end{pmatrix}$$

Replace the last row with 2 times the top row added to 3 times the bottom row, thus doing a succession of two row operations. This gives

$$\begin{pmatrix} 3 & -1 & -5 & | & 9 \\ 0 & 1 & -10 & | & 0 \\ 0 & 1 & -10 & | & 0 \end{pmatrix}.$$

Next take -1 times the middle row and add to the bottom.

$$\begin{pmatrix} 3 & -1 & -5 & | & 9 \\ 0 & 1 & -10 & | & 0 \\ 0 & 0 & 0 & | & 0 \end{pmatrix}$$

Take the middle row and add to the top and then divide the top row which results by 3.

$$\begin{pmatrix} 1 & 0 & -5 & | & 3 \\ 0 & 1 & -10 & | & 0 \\ 0 & 0 & 0 & | & 0 \end{pmatrix}.$$

The equations corresponding to this augmented matrix are $y = 10z$ and $x = 3 + 5z$. Apparently z can equal any number. Lets call this number, t.[2] Therefore, the solution set of this system is $x = 3 + 5t, y = 10t$, and $z = t$ where t is completely arbitrary. The system has an infinite set of solutions which are given in the above simple way. This is what it is all about, giving a useful description of the solutions to the system.

In summary,

Definition 1.10. A **system of linear equations** is a list of equations

$$a_{11}x_1 + a_{12}x_2 + \cdots + a_{1n}x_n = b_1$$
$$a_{21}x_1 + a_{22}x_2 + \cdots + a_{2n}x_n = b_2$$

$$\vdots$$

$$a_{m1}x_1 + a_{m2}x_2 + \cdots + a_{mn}x_n = b_m$$

where a_{ij} are numbers, and b_j is a number. The above is a system of m equations in the n variables $x_1, x_2 \cdots, x_n$. Nothing is said about the relative size of m and n. Written more simply in terms of summation notation, the above can be written in the form

$$\sum_{j=1}^{n} a_{ij}x_j = f_j, \ i = 1, 2, 3, \cdots, m$$

It is desired to find $(x_1, \cdots, x_n)$ solving each of the equations listed.

[2]In this context t is called a **parameter.**

As illustrated above, such a system of linear equations may have a unique solution, no solution, or infinitely many solutions and these are the only three cases which can occur for any linear system. Furthermore, you do exactly the same things to solve any linear system. You write the augmented matrix and do row operations until you get a simpler system in which it is possible to see the solution. All is based on the observation that the row operations do not change the solution set. You can have more equations than variables fewer equations than variables etc. It does not matter. You always set up the augmented matrix and go to work on it.

Definition 1.11. A system of linear equations is called **consistent** if there exists a solution. It is called **inconsistent** if there is no solution.

These are reasonable words to describe the situations of having or not having a solution. If you think of each equation as a condition which must be satisfied by the variables consistent would mean there is some choice of variables which can satisfy all the conditions. Inconsistent means there is no choice of the variables which can satisfy each of the conditions.

1.11.2 Row Reduced Echelon Form

In doing row operations to find an augmented matrix which has an easy to see solution, it is helpful to work towards making the result in row reduced echelon form. This is described next. First of all, a **leading entry** of a row in a matrix is the first nonzero entry in reading from left to right. Now here is the description of row reduced echelon form.

Definition 1.12. An augmented matrix is in **row reduced echelon form** if

(1) All nonzero rows are above any rows of zeros.
(2) Each leading entry of a row is in a column to the right of the leading entries of any rows above it.
(3) All entries in a column above and below a leading entry are zero.
(4) Each leading entry is a 1, the only nonzero entry in its column.

Example 1.17. For example, the following matrices are in row reduced echelon form.

$$\begin{pmatrix} 1 & 2 & 0 & 4 \\ 0 & 0 & 1 & 1 \\ 0 & 0 & 0 & 0 \end{pmatrix}, \begin{pmatrix} 1 & 0 & 0 & -5 \\ 0 & 1 & 0 & 2 \\ 0 & 0 & 1 & 3 \end{pmatrix}$$

A lot more can be said about the row reduced echelon form than will be mentioned here. In particular, every augmented matrix has a unique row reduced echelon form, but this is a topic for linear algebra. Here it will just be a useful "target" when you do row operations. To illustrate, here is an example. Also, we usually do not bother to draw the separating line between the last column and the columns of the matrix which correspond to the variables. This is because the row reduced echelon form has far more general application than to augmented matrices.

Example 1.18. Find the solution(s) to the equations $x + 2y - z = 2$, $3x + 2y + z = 7$, $4x + y - z = 3$.

The augmented matrix is

$$\begin{pmatrix} 1 & 2 & -1 & 2 \\ 3 & 2 & 1 & 7 \\ 4 & 1 & -1 & 3 \end{pmatrix}$$

First take -3 times the first row and add to the second. Next take -4 times the first row and add to the third. This yields

$$\begin{pmatrix} 1 & 2 & -1 & 2 \\ 0 & -4 & 4 & 1 \\ 0 & -7 & 3 & -5 \end{pmatrix}$$

Next multiply the second row by $-1/4$ and after doing this, take 7 times it and add to the bottom row.

$$\begin{pmatrix} 1 & 2 & -1 & 2 \\ 0 & 1 & -1 & -1/4 \\ 0 & 0 & -4 & -\frac{27}{4} \end{pmatrix}$$

Now multiply the bottom row by $-1/4$ and then add to the second row and then add it to the top row.

$$\begin{pmatrix} 1 & 2 & 0 & \frac{59}{16} \\ 0 & 1 & 0 & \frac{23}{16} \\ 0 & 0 & 1 & \frac{27}{16} \end{pmatrix}$$

Finally, take -2 times the second row and add to the top.

$$\begin{pmatrix} 1 & 0 & 0 & \frac{13}{16} \\ 0 & 1 & 0 & \frac{23}{16} \\ 0 & 0 & 1 & \frac{27}{16} \end{pmatrix}$$

Then the system $x = 13/16$, $y = 23/16$, and $z = 27/16$ has the same set of solutions as the original system. This last one gives the answer.

It always works this way in the sense that when a row reduced echelon form is obtained it will be maximally easy to describe the solution to the original system of equations.

1.12 Exercises

(1) Find the general solution of the system whose augmented matrix is

$$\begin{pmatrix} 1 & 2 & 0 & | & 2 \\ 1 & 3 & 4 & | & 2 \\ 1 & 0 & 2 & | & 1 \end{pmatrix}.$$

(2) Find the general solution of the system whose augmented matrix is

$$\begin{pmatrix} 1 & 2 & 0 & | & 2 \\ 2 & 0 & 1 & | & 1 \\ 3 & 2 & 1 & | & 3 \end{pmatrix}.$$

(3) Find the general solution of the system whose augmented matrix is

$$\begin{pmatrix} 1 & 1 & 0 & | & 1 \\ 1 & 0 & 4 & | & 2 \end{pmatrix}.$$

(4) Solve the system

$$x + 2y + z - w = 2$$
$$x - y + z + w = 1$$
$$2x + y - z = 1$$
$$4x + 2y + z = 5$$

(5) Solve the system

$$x + 2y + z - w = 2$$
$$x - y + z + w - 0$$
$$2x + y - z = 1$$
$$4x + 2y + z = 3$$

(6) Find the general solution of the system whose augmented matrix is

$$\begin{pmatrix} 1 & 0 & 2 & 1 & 1 & | & 2 \\ 0 & 1 & 0 & 1 & 2 & | & 1 \\ 1 & 2 & 0 & 0 & 1 & | & 3 \\ 1 & 0 & 1 & 0 & 2 & | & 2 \end{pmatrix}.$$

(7) Find the general solution of the system whose augmented matrix is

$$\begin{pmatrix} 1 & 0 & 2 & 1 & 1 & | & 2 \\ 0 & 1 & 0 & 1 & 2 & | & 1 \\ 0 & 2 & 0 & 0 & 1 & | & 3 \\ 1 & -1 & 2 & 2 & 2 & | & 0 \end{pmatrix}.$$

(8) Give the complete solution to the system of equations, $7x + 14y + 15z = 22$, $2x + 4y + 3z = 5$, and $3x + 6y + 10z = 13$.

(9) Give the complete solution to the system of equations, $3x - y + 4z = 6$, $y + 8z = 0$, and $-2x + y = -4$.

(10) Give the complete solution to the system of equations, $9x - 2y + 4z = -17$, $13x - 3y + 6z = -25$, and $-2x - z = 3$.

(11) Give the complete solution to the system of equations, $8x + 2y + 3z = -3$, $8x + 3y + 3z = -1$, and $4x + y + 3z = -9$.

(12) Give the complete solution to the system of equations, $-8x + 2y + 5z = 18$, $-8x + 3y + 5z = 13$, and $-4x + y + 5z = 19$.

(13) Give the complete solution to the system of equations, $3x-y-2z = 3$, $y-4z = 0$, and $-2x + y = -2$.

(14) Give the complete solution to the system of equations, $-9x + 15y = 66$, $-11x + 18y = 79$, $-x + y = 4$, and $z = 3$.

(15) Consider the system $-5x + 2y - z = 0$ and $-5x - 2y - z = 0$. Both equations equal zero and so $-5x + 2y - z = -5x - 2y - z$ which is equivalent to $y = 0$. Thus x and z can equal anything. But when $x = 1$, $z = -4$, and $y = 0$ are plugged in to the equations, it does not work. Why?

(16) Four times the weight of Gaston is 150 pounds more than the weight of Ichabod. Four times the weight of Ichabod is 660 pounds less than seventeen times the weight of Gaston. Four times the weight of Gaston plus the weight of Siegfried equals 290 pounds. Brunhilde would balance all three of the others. Find the weights of the four sisters.

(17) Suppose a system of equations has fewer equations than variables. Must such a system be consistent? If so, explain why and if not, give an example which is not consistent.

(18) If a system of equations has more equations than variables, can it have a solution? If so, give an example and if not, tell why not.

(19) Find h such that

$$\begin{pmatrix} 2 & h & | & 4 \\ 3 & 6 & | & 7 \end{pmatrix}$$

is the augmented matrix of an inconsistent matrix.

(20) Find h such that

$$\begin{pmatrix} 1 & h & | & 3 \\ 2 & 4 & | & 6 \end{pmatrix}$$

is the augmented matrix of a consistent matrix.

(21) Find h such that

$$\begin{pmatrix} 1 & 1 & | & 4 \\ 3 & h & | & 12 \end{pmatrix}$$

is the augmented matrix of a consistent matrix.

(22) Choose h and k such that the augmented matrix shown has one solution. Then choose h and k such that the system has no solutions. Finally, choose h and k such that the system has infinitely many solutions.

$$\begin{pmatrix} 1 & h & | & 2 \\ 2 & 4 & | & k \end{pmatrix}.$$

Chapter 2

Functions

2.1 Functions And Sequences

By this time, you have seen several examples of functions such as the trig. functions. It is a good idea to formalize this concept before proceeding further. The concept of a function is that of something which gives a unique output for a given input.

Definition 2.1. Consider two sets D and R along with a rule which assigns a unique element of R to every element of D. This rule is called a function and it is denoted by a letter such as f. The symbol $D(f) = D$ is called the domain of f. The set R, also written $R(f)$, is called the range of f. It is also sometimes called the codomain. The set of all elements of R which are of the form $f(x)$ for some $x \in D$ is often denoted by $f(D)$. When $R = f(D)$, the function f, is said to be onto. It is common notation to write $f : D(f) \to R$ to denote the situation just described in this definition, where f is a function defined on D having values in R.

Example 2.1. Consider the list of numbers $\{1, 2, 3, 4, 5, 6, 7\} \equiv D$. Define a function which assigns an element of D to $R = \{2, 3, 4, 5, 6, 7, 8\}$ by $f(x) \equiv x + 1$ for each $x \in D$.

In this example there was a clearly defined procedure which determined the function. However, sometimes there is no discernible procedure which yields a particular function.

Example 2.2. Consider the ordered pairs $(1, 2), (2, -2), (8, 3), (7, 6)$ and let

$$D \equiv \{1, 2, 8, 7\},$$

the set of first entries in the given set of ordered pairs $R \equiv \{2, -2, 3, 6\}$, the set of second entries, and let $f(1) = 2$, $f(2) = -2$, $f(8) = 3$, and $f(7) = 6$.

Sometimes functions which are defined on all the real numbers do not come from a formula. For example, consider the following function defined on the positive real numbers having the following definition.

Example 2.3. For $x \in \mathbb{R}$ define

$$f(x) = \begin{cases} \frac{1}{n} \text{ if } x = \frac{m}{n} \text{ in lowest terms for } m, n \text{ integers} \\ 0 \text{ if } x \text{ is not rational} \end{cases} \tag{2.1}$$

This is a very interesting function called the Dirichlet function. Note that it is not defined in a simple way from a formula.

Example 2.4. Let D consist of the set of people who have lived on the earth except for Adam and for $d \in D$, let $f(d) \equiv$ the biological father of d. Then f is a function.

This function is not the sort of thing studied in calculus but it is a function just the same. The next functions are studied in calculus.

Example 2.5. Consider a weight which is suspended at one end of a spring which is attached at the other end to the ceiling. Suppose the weight has extended the spring so that the force exerted by the spring exactly balances the force resulting from the weight on the spring. Measure the displacement of the mass x from this point with the positive direction being up, and define a function as follows: $x(t)$ will equal the displacement of the spring at time t given knowledge of the velocity of the weight and the displacement of the weight at some particular time.

Example 2.6. Certain chemicals decay with time. Suppose A_0 is the amount of chemical at some given time. Then you could let $A(t)$ denote the amount of the chemical at time t.

These last two examples show how physical problems can result in functions. Examples 2.5 and 2.6 are considered later in the book and techniques for finding $x(t)$ and $A(t)$ from the given conditions are presented.

In this chapter the functions are defined on some subset of $\mathbb{R}$ having values in $\mathbb{R}$. Later this will be generalized. When $D(f)$ is not specified, it is understood to consist of everything for which f makes sense. The following definition gives several ways to make new functions from old ones.

Definition 2.2. Let f, g be functions with values in $\mathbb{R}$. Let a, b be points of $\mathbb{R}$. Then $af + bg$ is the name of a function whose domain is $D(f) \cap D(g)$ which is defined as

$$(af + bg)(x) = af(x) + bg(x).$$

The function fg is the name of a function which is defined on $D(f) \cap D(g)$ given by

$$(fg)(x) = f(x)g(x).$$

Similarly for k an integer, f^k is the name of a function defined as

$$f^k(x) = (f(x))^k$$

The function f/g is the name of a function whose domain is

$$D(f) \cap \{x \in D(g) : g(x) \neq 0\}$$

defined as

$$(f/g)(x) = f(x)/g(x).$$

If $f : D(f) \to X$ and $g : D(g) \to Y$, then $g \circ f$ is the name of a function whose domain is

$$\{x \in D(f) : f(x) \in D(g)\}$$

which is defined as

$$g \circ f(x) \equiv g(f(x)).$$

This is called the composition of the two functions.

You should note that $f(x)$ is not a function. It is the value of the function at the point x. The name of the function is f. Nevertheless, people often write $f(x)$ to denote a function and it does not cause too many problems in beginning courses. When this is done, the variable x should be considered as a generic variable free to be anything in $D(f)$. I will use this slightly sloppy abuse of notation whenever convenient. Thus, $x^2 + 4$ may mean the function f, given by $f(x) = x^2 + 4$.

Sometimes people get hung up on formulas and think that the only functions of importance are those which are given by some simple formula. It is a mistake to think this way. Functions involve a domain and a range and a function is determined by what it does. Functions are well described by a well known scripture in the Bible, Matthew 7:20, Wherefore by their fruits ye shall know them. When you have specified what it does to something in its domain, you have told what the function is.

Example 2.7. Let $f(t) = t$ and $g(t) - 1 + t$. Then $fg : \mathbb{R} \to \mathbb{R}$ is given by

$$fg(t) = t(1+t) = t + t^2.$$

Example 2.8. Let $f(t) = 2t + 1$ and $g(t) - \sqrt{1+t}$. Then

$$g \circ f(t) = \sqrt{1 + (2t+1)} = \sqrt{2t+2}$$

for $t \geq -1$. If $t < -1$ the inside of the square root sign is negative so makes no sense. Therefore $g \circ f : \{t \in \mathbb{R} : t \geq -1\} \to \mathbb{R}$.

Note that in this last example, it was necessary to fuss about the domain of $g \circ f$ because g is only defined for certain values of t.

The concept of a one to one function is very important. This is discussed in the following definition.

Definition 2.3. For any function $f : D(f) \subseteq X \to Y$, define the following set known as the inverse image of y.

$$f^{-1}(y) \equiv \{x \in D(f) : f(x) = y\}.$$

There may be many elements in this set, but when there is always at most one element in this set for all $y \in f(D(f))$, the function f is one to one, sometimes written $1 - 1$. Thus f is one to one $1 - 1$, if whenever $f(x) = f(x_1)$, then $x = x_1$. If f is one to one, the inverse function f^{-1} is defined on $f(D(f))$ and $f^{-1}(y) = x$ where $f(x) = y$. Thus from the definition, $f^{-1}(f(x)) = x$ for all $x \in D(f)$ and $f(f^{-1}(y)) = y$ for all $y \in f(D(f))$. Defining id by id$(z) \equiv z$ this says $f \circ f^{-1} = $ id and $f^{-1} \circ f = $ id.

Polynomials and rational functions are particularly easy functions to understand because they do come from a simple formula.

Definition 2.4. A function f is a polynomial if
$$f(x) = a_n x^n + a_{n-1} x^{n-1} + \cdots + a_1 x + a_0$$
where the a_i are real numbers and n is a nonnegative integer. In this case the degree of the polynomial $f(x)$ is n. Thus the degree of a polynomial is the largest exponent appearing on the variable.

f is a rational function if
$$f(x) = \frac{h(x)}{g(x)}$$
where h and g are polynomials.

For example, $f(x) = 3x^5 + 9x^2 + 7x + 5$ is a polynomial of degree 5 and
$$\frac{3x^5 + 9x^2 + 7x + 5}{x^4 + 3x + x + 1}$$
is a rational function.

Note that in the case of a rational function, the domain of the function might not be all of $\mathbb{R}$. For example, if
$$f(x) = \frac{x^2 + 8}{x + 1},$$
the domain of f would be all real numbers not equal to -1.

Closely related to the definition of a function is the concept of the graph of a function.

Definition 2.5. Given two sets, X and Y, the Cartesian product of the two sets, written as $X \times Y$, is assumed to be a set described as follows.
$$X \times Y = \{(x, y) : x \in X \text{ and } y \in Y\}.$$
$\mathbb{R}^2$ denotes the Cartesian product of $\mathbb{R}$ with $\mathbb{R}$.

The notion of Cartesian product is just an abstraction of the concept of identifying a point in the plane with an ordered pair of numbers.

Definition 2.6. Let $f : D(f) \to R(f)$ be a function. The graph of f consists of the set,
$$\{(x, y) : y = f(x) \text{ for } x \in D(f)\}.$$

Note that knowledge of the graph of a function is equivalent to knowledge of the function. To find $f(x)$, simply observe the ordered pair which has x as its first entry and the second entry equals $f(x)$. The graph of f can be represented by drawing a picture. For example, consider the picture of a part of the graph of the function $f(x) = 2x - 1$.

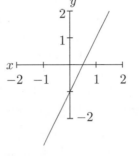

Here is part of the graph of the function $f(x) = x^2 - 2$

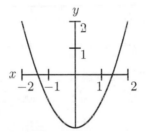

Definition 2.7. A function whose domain is defined as a set of the form

$$\{k, k+1, k+2, \cdots\}$$

for k an integer is known as a sequence. Thus you can consider $f(k), f(k+1), f(k+2)$, etc. Usually the domain of the sequence is either $\mathbb{N}$, the natural numbers consisting of $\{1, 2, 3, \cdots\}$, or the nonnegative integers, $\{0, 1, 2, 3, \cdots\}$. Also, it is traditional to write f_1, f_2, etc. instead of $f(1), f(2), f(3)$ etc. when referring to sequences. In the above context, f_k is called the first term, f_{k+1} the second and so forth. It is also common to write the sequence, not as f but as $\{f_i\}_{i=k}^{\infty}$ or just $\{f_i\}$ for short.

Example 2.9. Let $\{a_k\}_{k=1}^{\infty}$ be defined by $a_k \equiv k^2 + 1$.

This gives a sequence. In fact, $a_7 = a(7) = 7^2 + 1 = 50$ just from using the formula for the kth term of the sequence.

It is nice when sequences come in this way from a formula for the kth term. However, this is often not the case. Sometimes sequences are defined **recursively**.

This happens, when the first several terms of the sequence are given, and then a rule is specified which determines a_{n+1} from knowledge of $a_1, \cdots, a_n$. This rule which specifies a_{n+1} from knowledge of a_k for $k \leq n$ is known as a **recurrence relation**.

Example 2.10. Let $a_1 = 1$ and $a_2 = 1$. Assuming $a_1, \cdots, a_{n+1}$ are known, $a_{n+2} \equiv a_n + a_{n+1}$.

Thus the first several terms of this sequence, listed in order, are 1, 1, 2, 3, 5, 8,$\cdots$. This particular sequence is called the Fibonacci sequence and is important in the study of reproducing rabbits. Note this defines a function without giving a formula for it. Such sequences occur naturally in the solution of differential equations using power series methods and in many other situations of great importance.

For sequences, it is very important to consider something called a subsequence.

Definition 2.8. Let $\{a_n\}$ be a sequence and let $n_1 < n_2 < n_3, \cdots$ be any strictly increasing list of integers such that n_1 is at least as large as the first number in the domain of the function defining the sequence. Then if $b_k \equiv a_{n_k}$, $\{b_k\}$ is called a subsequence of $\{a_n\}$.

For example, suppose $a_n = (n^2 + 1)$. Thus $a_1 = 2$, $a_3 = 10$, etc. If

$$n_1 = 1, n_2 = 3, n_3 = 5, \cdots, n_k = 2k - 1,$$

then letting $b_k = a_{n_k}$, it follows

$$b_k = \left((2k - 1)^2 + 1\right) = 4k^2 - 4k + 2.$$

2.2   Exercises

(1) Let $g(t) \equiv \sqrt{2 - t}$ and let $f(t) = \frac{1}{t}$. Find $g \circ f$. Include the domain of $g \circ f$.

(2) Give the domains of the following functions.

(a) $f(x) = \frac{x+3}{3x-2}$

(b) $f(x) = \sqrt{x^2 - 4}$

(c) $f(x) = \sqrt{4 - x^2}$

(d) $f(x) = \sqrt{\frac{x-4}{3x+5}}$

(e) $f(x) = \sqrt{\frac{x^2-4}{x+1}}$

(3) Let $f : \mathbb{R} \to \mathbb{R}$ be defined by $f(t) \equiv t^3 + 1$. Is f one to one? Can you find a formula for f^{-1}?

(4) Suppose $a_1 = 1, a_2 = 3$, and $a_3 = -1$. Suppose also that for $n \geq 4$, it is known that $a_n = a_{n-1} + 2a_{n-2} + 3a_{n-3}$. Find a_4. Are you able to guess a formula for the kth term of this sequence?

(5) Let $f : \{t \in \mathbb{R} : t \neq -1\} \to \mathbb{R}$ be defined by $f(t) \equiv \frac{t}{t+1}$. Find f^{-1} if possible.

(6) A function $f : \mathbb{R} \to \mathbb{R}$ is a strictly increasing function if whenever $x < y$, it follows that $f(x) < f(y)$. If f is a strictly increasing function, does f^{-1} always exist? Explain your answer.

(7) Let $f(t)$ be defined by

$$f(t) = \begin{cases} 2t + 1 \text{ if } t \leq 1 \\ t \text{ if } t > 1 \end{cases}.$$

Find f^{-1} if possible.

(8) Suppose $f : D(f) \to R(f)$ is one to one, $R(f) \subseteq D(g)$, and $g : D(g) \to R(g)$ is one to one. Does it follow that $g \circ f$ is one to one? Explain your answer.

(9) If $f : \mathbb{R} \to \mathbb{R}$ and $g : \mathbb{R} \to \mathbb{R}$ are two one to one functions, which of the following are necessarily one to one on their domains? Explain why or why not by giving a proof or an example.

    (a) $f + g$

    (b) fg

    (c) f^3

    (d) f/g

(10) Draw the graph of the function $f(x) = x^3 + 1$.

(11) Draw the graph of the function $f(x) = x^2 + 2x + 2$.

(12) Draw the graph of the function $f(x) = \frac{x}{1+x}$.

(13) The function sin has domain equal to $\mathbb{R}$ and range $[-1, 1]$. However, this function is not one to one because $\sin(x + 2\pi) = \sin x$. Show that if the domain of the function is restricted to be $\left[-\frac{\pi}{2}, \frac{\pi}{2}\right]$, then sin still maps onto $[-1, 1]$ but is now also one to one on this restricted domain. Therefore, there is an inverse function, called arcsin which is defined by $\arcsin(x) \equiv$ the angle whose sin is x which is in the interval $\left[-\frac{\pi}{2}, \frac{\pi}{2}\right]$. Thus $\arcsin\left(\frac{1}{2}\right)$ is the angle whose sin is $\frac{1}{2}$ which is in $\left[-\frac{\pi}{2}, \frac{\pi}{2}\right]$. This angle is $\frac{\pi}{6}$. Suppose you wanted to find $\tan(\arcsin(x))$. How would you do it? Consider the following picture which corresponds to the case where $x > 0$.

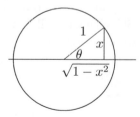

Then letting $\theta = \arcsin(x)$, the thing which is wanted is $\tan \theta$. Now from the picture, you see this is $\frac{x}{\sqrt{1-x^2}}$. If x were negative, you would have the little triangle pointing down rather than up as in the picture. The result would be the same for $\tan \theta$. Find the following:

    (a) $\cot(\arcsin(x))$

(b) $\sec\left(\arcsin\left(x\right)\right)$

(c) $\csc\left(\arcsin\left(x\right)\right)$

(d) $\cos\left(\arcsin\left(x\right)\right)$

(14) Using Problem 13 and the formulas for the trig functions of a sum of angles, find the following.

(a) $\cot\left(\arcsin\left(2x\right)\right)$

(b) $\sec\left(\arcsin\left(x+y\right)\right)$

(c) $\csc\left(\arcsin\left(x^2\right)\right)$

(d) $\cos\left(2\arcsin\left(x\right)\right)$

(e) $\tan\left(\arcsin\left(x\right)+\arcsin\left(y\right)\right)$

(15) The function cos, is onto $[-1,1]$ but fails to be one to one. Show that if the domain of cos is restricted to be $[0,\pi]$, then cos is one to one on this restricted domain and still is onto $[-1,1]$. Define $\arccos\left(x\right) \equiv$ the angle whose cosine is x which is in $[0,\pi]$. Find the following.

(a) $\tan\left(\arccos\left(x\right)\right)$

(b) $\cot\left(\arccos\left(x\right)\right)$

(c) $\sin\left(\arccos\left(x\right)\right)$

(d) $\csc\left(\arccos\left(x\right)\right)$

(e) $\sec\left(\arccos\left(x\right)\right)$

(16) Using Problem 15 and the formulas for the trig functions of a sum of angles, find the following. **Hint:** On some of these, pay attention to whether x,y are positive.

(a) $\cot\left(\arccos\left(2x\right)\right)$

(b) $\sec\left(\arccos\left(x+y\right)\right)$

(c) $\csc\left(\arccos\left(x^2\right)\right)$

(d) $\cos\left(\arcsin\left(x\right)+\arccos\left(y\right)\right)$

(e) $\tan\left(\arcsin\left(x\right)+\arccos\left(y\right)\right)$

(17) The function arctan is defined as $\arctan\left(x\right) \equiv$ the angle whose tangent is x which is in $\left(-\frac{\pi}{2},\frac{\pi}{2}\right)$. Show this is well defined and is the inverse function for tan if the domain of tan is restricted to be $\left(-\frac{\pi}{2},\frac{\pi}{2}\right)$. Find

(a) $\cos\left(\arctan\left(x\right)\right)$

(b) $\cot\left(\arctan\left(x\right)\right)$

(c) $\sin\left(\arctan\left(x\right)\right)$

(d) $\csc\left(\arctan\left(x\right)\right)$

(e) $\sec\left(\arctan\left(x\right)\right)$

(18) Using Problem 17 and the formulas for the trig functions of a sum of angles, find the following.

(a) $\cot\left(\arctan\left(2x\right)\right)$

(b) $\sec\left(\arctan\left(x+y\right)\right)$

(c) $\csc\left(\arccos\left(x^2\right)\right)$

(d) $\cos\left(2\arctan\left(x\right)+\arcsin\left(y\right)\right)$

(19) Suppose $a_n = \frac{1}{n}$ and let $n_k = 2^k$. Find b_k where $b_k = a_{n_k}$.

(20) For f a function defined on an interval $[a, b]$, we say f is an **increasing** function if whenever x, y are points of $[a, b]$ such that $x < y$, it follows $f\left(x\right) \le f\left(y\right)$. The function is **decreasing** if whenever x, y are as just described, $f\left(x\right) \ge f\left(y\right)$. Here are graphs of functions. Identify the intervals on which these functions are increasing.

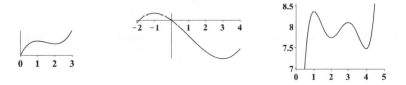

2.3 Continuous Functions

The concept of function is far too general to be useful in calculus. There are various ways to restrict the concept in order to study something interesting and the types of restrictions considered depend very much on what you find interesting. In Calculus, the most fundamental restriction made is to assume the functions are continuous. Continuous functions are those in which a sufficiently small change in x results in a small change in $f\left(x\right)$. They rule out things which could never happen physically. For example, it is not possible for a car to jump from one point to another instantly. Making this restriction precise turns out to be surprisingly difficult although many of the most important theorems about continuous functions seem intuitively clear.

Before giving the careful mathematical definitions, here are examples of graphs of functions which are not continuous at the point x_0.

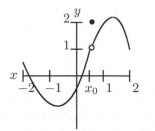

You see, there is a hole in the picture of the graph of this function and instead of filling in the hole with the appropriate value, $f\left(x_0\right)$ is too large. This is called a removable discontinuity because the problem can be fixed by redefining the function at the point x_0. Here is another example.

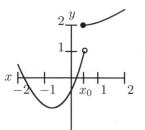

You see from this picture that there is no way to get rid of the jump in the graph of this function by simply redefining the value of the function at x_0. That is why it is called a nonremovable discontinuity or jump discontinuity. Now that pictures have been given of what it is desired to eliminate, it is time to give the precise definition.

The definition which follows, due to Cauchy[1] and Weierstrass[2] is the precise way to exclude the sort of behavior described above and all statements about continuous functions must ultimately rest on this definition from now on.

Definition 2.9. A function $f : D(f) \subseteq \mathbb{R} \to \mathbb{R}$ is continuous at $x \in D(f)$ if for each $\varepsilon > 0$ there exists $\delta > 0$ such that whenever $y \in D(f)$ and

$$|y - x| < \delta$$

it follows that

$$|f(x) - f(y)| < \varepsilon.$$

A function f is continuous if it is continuous at every point of $D(f)$.

---

[1] Augustin Louis Cauchy 1789-1857 was the son of a lawyer who was married to an aristocrat. He was born in France just after the fall of the Bastille and his family fled the reign of terror and hid in the countryside till it was over. Cauchy was educated at first by his father who taught him Greek and Latin. Eventually Cauchy learned many languages. He was also a good Catholic.

After the reign of terror, the family returned to Paris and Cauchy studied at the university to be an engineer but became a mathematician although he made fundamental contributions to physics and engineering. Cauchy was one of the most prolific mathematicians who ever lived. He wrote several hundred papers which fill 24 volumes. He also did research on many topics in mechanics and physics including elasticity, optics and astronomy. More than anyone else, Cauchy invented the subject of complex analysis. He is also credited with giving the first rigorous definition of continuity.

He married in 1818 and lived for 12 years with his wife and two daughters in Paris till the revolution of 1830. Cauchy refused to take the oath of allegiance to the new ruler and ended up leaving his family and going into exile for 8 years.

Notwithstanding his great achievements he was not known as a popular teacher.

[2] Wilhelm Theodor Weierstrass 1815-1897 brought calculus to essentially the state it is in now. When he was a secondary school teacher, he wrote a paper which was so profound that he was granted a doctor's degree. He made fundamental contributions to partial differential equations, complex analysis, calculus of variations, and many other topics. He also discovered some pathological examples such as space filling curves. Cauchy gave the definition in words and Weierstrass, somewhat later produced the totally rigorous ε δ definition presented here. The need for rigor in the subject of calculus was only realized over a long period of time.

In sloppy English this definition says roughly the following: A function f is continuous at x when it is possible to make $f(y)$ as close as desired to $f(x)$ provided y is taken close enough to x. In fact this statement in words is pretty much the way Cauchy described it. The completely rigorous definition above is due to Weierstrass. This definition does indeed rule out the sorts of graphs drawn above. Consider the second nonremovable discontinuity. The removable discontinuity case is similar.

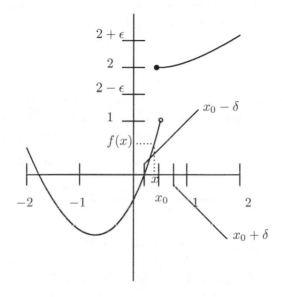

For the ε shown, you can see from the picture that no matter how small you take δ, there will be points x, between $x_0 - \delta$ and x_0 where $f(x) < 2 - \varepsilon$. In particular, for these values of x, $|f(x) - f(x_0)| > \varepsilon$. Therefore, the definition of continuity given above excludes the situation in which there is a jump in the function. Similar reasoning shows it excludes the removable discontinuity case as well. There are many ways a function can fail to be continuous and it is impossible to list them all by drawing pictures. This is why it is so important to use the definition. The other thing to notice is that the concept of continuity as described in the definition is a point property. That is to say, it is a property which a function may or may not have at a single point. Here is an example.

Example 2.11. Let

$$f(x) = \begin{cases} x & \text{if } x \text{ is rational} \\ 0 & \text{if } x \text{ is irrational} \end{cases}.$$

This function is continuous at $x = 0$, and nowhere else.

To verify the assertion about the above function, first show it is not continuous at x if $x \neq 0$. Take such an x and let $\varepsilon = |x|/2$. Now let $\delta > 0$ be completely arbitrary. In the interval $(x - \delta, x + \delta)$ there are rational numbers, y_1 such that

$|y_1| > |x|$ and irrational numbers y_2. Thus $|f(y_1) - f(y_2)| = |y_1| > |x|$. If f were continuous at x, there would exist $\delta > 0$ such that for every point $y \in (x - \delta, x + \delta)$, $|f(y) - f(x)| < \varepsilon$. But then, letting y_1 and y_2 be as just described,

$$|x| < |y_1| = |f(y_1) - f(y_2)|$$
$$\leq |f(y_1) - f(x)| + |f(x) - f(y_2)| < 2\varepsilon = |x|,$$

which is a contradiction. Since a contradiction is obtained by assuming that f is continuous at x, it must be concluded that f is not continuous there. To see f is continuous at 0, let $\varepsilon > 0$ be given and let $\delta = \varepsilon$. Then if $|y - 0| < \delta = \varepsilon$, then

$$|f(y) - f(0)| = 0 \text{ if } y \text{ is irrational}$$

$$|f(y) - f(0)| = |y| < \varepsilon \text{ if } y \text{ is rational.}$$

Either way, whenever $|y - 0| < \delta$, it follows $|f(y) - f(0)| < \varepsilon$ and so f is continuous at $x = 0$. How did I know to let $\delta = \varepsilon$? That is a very good question. The choice of δ for a particular ε is usually arrived at by using intuition, the actual ε δ argument reduces to a verification that the intuition was correct. Here is another example.

Example 2.12. Show the function $f(x) = -5x + 10$ is continuous at $x = -3$.

To do this, note first that $f(-3) = 25$ and it is desired to verify the conditions for continuity. Consider the following.

$$|-5x + 10 - (25)| = 5|x - (-3)|.$$

This allows one to find a suitable δ. If $\varepsilon > 0$ is given, let $0 < \delta \leq \frac{1}{5}\varepsilon$. Then if $0 < |x - (-3)| < \delta$, it follows from this inequality that

$$|-5x + 10 - (25)| = 5|x - (-3)| < 5\frac{1}{5}\varepsilon = \varepsilon.$$

Sometimes the determination of δ in the verification of continuity can be a little more involved. Here is another example.

Example 2.13. Show the function $f(x) = \sqrt{2x + 12}$ is continuous at $x = 5$.

First note $f(5) = \sqrt{22}$. Now consider:

$$\left|\sqrt{2x + 12} - \sqrt{22}\right| = \left|\frac{2x + 12 - 22}{\sqrt{2x + 12} + \sqrt{22}}\right|$$

$$= \frac{2}{\sqrt{2x + 12} + \sqrt{22}}|x - 5| \leq \frac{1}{11}\sqrt{22}|x - 5|$$

whenever $|x - 5| < 1$ because for such x, $\sqrt{2x + 12} > 0$. Now let $\varepsilon > 0$ be given. Choose δ such that $0 < \delta \leq \min\left(1, \frac{\varepsilon\sqrt{22}}{2}\right)$. Then if $|x - 5| < \delta$, all the inequalities above hold and

$$\left|\sqrt{2x + 12} - \sqrt{22}\right| \leq \frac{2}{\sqrt{22}}|x - 5| < \frac{2}{\sqrt{22}}\frac{\varepsilon\sqrt{22}}{2} = \varepsilon.$$

Example 2.14. Show $f(x) = -3x^2 + 7$ is continuous at $x = 7$.

First observe $f(7) = -140$. Now

$$\left|-3x^2 + 7 - (-140)\right| = 3\left|x + 7\right|\left|x - 7\right| \leq 3\left(\left|x\right| + 7\right)\left|x - 7\right|.$$

If $|x - 7| < 1$, it follows from the version of the triangle inequality which states that $||s| - |t|| \leq |s - t|$, that $|x| < 1 + 7$. Therefore, if $|x - 7| < 1$, it follows

$$\left|-3x^2 + 7 - (-140)\right| \leq 3\left((1 + 7) + 7\right)\left|x - 7\right|$$

$$= 3\left(1 + 27\right)\left|x - 7\right| = 84\left|x - 7\right|.$$

Now let $\varepsilon > 0$ be given. Choose δ such that $0 < \delta \leq \min\left(1, \frac{\varepsilon}{84}\right)$. Then for $|x - 7| < \delta$, it follows

$$\left|-3x^2 + 7 - (-140)\right| \leq 84\left|x - 7\right| < 84\left(\frac{\varepsilon}{84}\right) = \varepsilon.$$

These ε δ proofs will not be emphasized any more than necessary. However, you should try a few of them because, until you master this concept, you will not understand calculus as it has been understood since the last half of the nineteenth century. The best you can do without this definition is to gain an understanding of the subject as it was understood by people like Lagrange or Laplace[3] in the 1700's, before the need for rigor was realized.

2.4 Sufficient Conditions For Continuity

The next theorem is a fundamental result which is convenient for avoiding the ε δ definition of continuity.

Theorem 2.1. *The following assertions are valid for f, g functions and a, b numbers.*

(1) *The function $af + bg$ is continuous at x when f, g are continuous at $x \in D(f) \cap D(g)$ and $a, b \in \mathbb{R}$.*

(2) *If f and g are each real valued functions continuous at x, then fg is continuous at x. If, in addition to this, $g(x) \neq 0$, then f/g is continuous at x.*

---

[3]Lagrange and Laplace were two great mathematicians of the 1700's who, like most mathematicians of this time, were also interested in physics as well as many other subjects. They made fundamental contributions to the calculus of variations and to mechanics and astronomy. Lagrange is considered one of the inventors of calculus of variations. He also did work in many areas of mathematics and at one time was interested in botany.

(3) *If f is continuous at x, $f(x) \in D(g) \subseteq \mathbb{R}$, and g is continuous at $f(x)$, then $g \circ f$ is continuous at x.*

(4) *The function $f : \mathbb{R} \to \mathbb{R}$, given by $f(x) = |x|$ is continuous.*

The proof of this theorem is in the last section of this chapter but its conclusions are not surprising. For example the first claim says that $(af + bg)(y)$ is close to $(af + bg)(x)$ when y is close to x provided, the same can be said about f and g. For the third claim, continuity of f indicates that if y is close enough to x then $f(x)$ is close to $f(y)$ and so by continuity of g at $f(x)$, $g(f(y))$ is close to $g(f(x))$. The fourth claim is verified as follows. The triangle inequality implies

$$||x| - |y|| \le |x - y|.$$

It follows that if $\varepsilon > 0$ is given, one can take $\delta = \varepsilon$ and obtain that for $|x - y| < \delta = \varepsilon$,

$$||x| - |y|| \le |x - y| < \delta = \varepsilon,$$

which shows continuity of the function $f(x) = |x|$.

2.5   Continuity Of Circular Functions

The functions $\sin x$ and $\cos x$ are often called the circular functions. This is because for each $x \in \mathbb{R}$, $(\cos x, \sin x)$ is a point on the unit circle. You can skip the proof of the following theorem if you wish. This continuity is easily seen without the proof by simply referring to the geometric description of these functions as the coordinates of a point on the unit circle. The graphs of these functions are as shown below. The first is the graph of sin and the second is the graph of cos

sin(x)   cos(x)

Theorem 2.2. *The functions* cos *and* sin *are continuous.*

Proof: First it will be shown that cos and sin are continuous at 0. By Corollary 1.1 on Page 9 the following inequality is valid for small positive values of θ.

$$1 - \cos\theta + \sin\theta \ge \theta \ge \sin\theta.$$

It follows that for θ small and positive, $|\theta| \ge |\sin\theta| = \sin\theta$. If $\theta < 0$, then $-\theta = |\theta| > 0$ and $-\theta \ge \sin(-\theta)$. But then this means $|\sin\theta| = -\sin\theta = \sin(-\theta) \le -\theta = |\theta|$ in this case also. Therefore, whenever $|\theta|$ is small enough,

$$|\theta| \ge |\sin\theta|.$$

Now let $\varepsilon > 0$ be given and take $\delta = \varepsilon$. Then if $|\theta| < \delta$, it follows

$$|\sin\theta - 0| = |\sin\theta - \sin 0| = |\sin\theta| \le |\theta| < \delta = \varepsilon,$$

showing sin is continuous at 0.

Next, note that for $|\theta| < \pi/2$, $\cos\theta \geq 0$ and so for such θ,

$$\sin^2\theta \geq \frac{\sin^2\theta}{1+\cos\theta} = \frac{1-\cos^2\theta}{1+\cos\theta} = 1-\cos\theta \geq 0. \tag{2.2}$$

From the first part of this argument for sin, given $\varepsilon > 0$ there exists $\delta > 0$ such that if $|\theta| < \delta$, then $|\sin\theta| < \sqrt{\varepsilon}$. It follows from 2.2, that if $|\theta| < \delta$, then $\varepsilon > 1-\cos\theta \geq 0$. This proves these functions are continuous at 0. Now $y = (y-x) + x$ and so

$$\cos y = \cos(y-x)\cos x - \sin(x-y)\sin x.$$

Therefore,

$$\cos y - \cos x = \cos(y-x)\cos x - \sin(x-y)\sin x - \cos x$$

and so, since $|\cos x|, |\sin x| \leq 1$,

$$\begin{aligned}
|\cos y - \cos x| &\leq |\cos x (\cos(y-x)-1)| + |\sin x| |\sin(y-x)| \\
&\leq |\cos(y-x)-1| + |\sin(y-x)|.
\end{aligned}$$

From the first part of this theorem, if $|y-x|$ is sufficiently small, both of these last two terms are less than $\varepsilon/2$ and this proves cos is continuous at x. The proof that sin is continuous is left for you to verify. ■

2.6 Exercises

(1) Let $f(x) = 2x + 7$. Show f is continuous at every point x. **Hint:** You need to let $\varepsilon > 0$ be given. In this case, you should try $\delta \leq \varepsilon/2$. Note that if one δ works in the definition, then so does any smaller positive δ.

(2) Let $f(x) = x^2 + 1$. Show f is continuous at $x = 3$. **Hint:**

$$|f(x) - f(3)| = |x^2 + 1 - (9+1)| = |x+3||x-3|.$$

Thus if $|x-3| < 1$, it follows from the triangle inequality, $|x| < 1 + 3 = 4$ and so $|f(x) - f(3)| < 4|x-3|$. Now try to complete the argument by letting $\delta \leq \min(1, \varepsilon/4)$. The symbol min means to take the minimum of the two numbers in the parenthesis.

(3) Let $f(x) = x^2 + 1$. Show f is continuous at $x = 4$.

(4) Let $f(x) = 2x^2 + 1$. Show f is continuous at $x = 1$.

(5) Let $f(x) = x^2 + 2x$. Show f is continuous at $x = 2$. Then show it is continuous at every point.

(6) Let $f(x) = |2x+3|$. Show f is continuous at every point. **Hint:** Review the two versions of the triangle inequality for absolute values.

(7) Let $f(x) = \frac{1}{x^2+1}$. Show f is continuous at every value of x.

(8) Show sin is continuous.

(9) Let $f(x) = \sqrt{x}$ show f is continuous at every value of x in its domain. **Hint:** You might want to make use of the identity $\sqrt{x} - \sqrt{y} = \frac{x-y}{\sqrt{x}+\sqrt{y}}$ at some point in your argument.

(10) Using Theorem 2.1, show all polynomials are continuous and that a rational function is continuous at every point of its domain. **Hint:** First show the function given as $f(x) = x$ is continuous and then use the Theorem 2.1.

(11) Let $f(x) = \begin{cases} 1 \text{ if } x \in \mathbb{Q} \\ 0 \text{ if } x \notin \mathbb{Q} \end{cases}$ and consider $g(x) = f(x)\sin x$. Determine where g is continuous and explain your answer.

(12) Suppose f is any function whose domain is the integers. Thus $D(f) = \mathbb{Z}$, the set of whole numbers, $\cdots, -3, -2, -1, 0, 1, 2, 3, \cdots$. Then f is continuous. Why? **Hint:** In the definition of continuity, what if you let $\delta = \frac{1}{4}$? Would this δ work for a given $\varepsilon > 0$? This shows that the idea that a continuous function is one for which you can draw the graph without taking the pencil off the paper is a lot of nonsense.

(13) Describe the points where $\tan, \cot, \sec,$ and $\csc$ are continuous.

(14) Give an example of a function f which is not continuous at some point but $|f|$ is continuous at that point.

(15) Find two functions which fail to be continuous but whose product is continuous.

(16) Find two functions which fail to be continuous but whose sum is continuous.

(17) Find two functions which fail to be continuous but whose quotient is continuous.

(18) Where is the function $\sin(\tan(x))$ continuous? What is its domain?

(19) Where is the function $\tan(\sin(x))$ continuous? What is its domain?

(20) Let $f(x) = \begin{cases} \frac{x^2-4}{x-2} \text{ if } x \neq 2 \\ k \text{ if } x = 2 \end{cases}$. Find k such that the function is continuous for every $x \in \mathbb{R}$.

2.7 Properties Of Continuous Functions

Continuous functions have many important properties which are consequences of the completeness axiom. Proofs of these theorems are in the last section at the end of this chapter. The next theorem is called the intermediate value theorem and the following picture illustrates its conclusion. It gives the existence of a certain point.

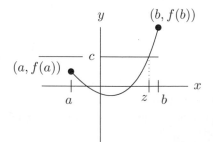

You see in the picture there is a horizontal line, $y = c$ and a continuous function which starts off less than c at the point a and ends up greater than c at point b. The intermediate value theorem says there is some point between a and b shown in the picture as z such that the value of the function at this point equals c. It may seem this is obvious but without completeness the conclusion of the theorem cannot be drawn. Nevertheless, the above picture makes this theorem very easy to believe.

Theorem 2.3. *Suppose $f : [a, b] \to \mathbb{R}$ is continuous and suppose $f(a) < c < f(b)$. Then there exists $z \in (a, b)$ such that $f(z) = c$.*

Example 2.15. Does there exist a solution to the equation $\sqrt{x^4 + 7} - x^3 \sin x = 0$?

By Theorem 2.1 and Problem 9 on Page 53 it follows easily that the function f, given by $f(x) = \sqrt{x^4 + 7} - x^3 \sin x$ is continuous. Also, $f(0) = \sqrt{7} > 0$ while

$$f\left(\frac{5\pi}{2}\right) = \sqrt{\left(\frac{5\pi}{2}\right)^4 + 7} - \left(\frac{5\pi}{2}\right)^3 \sin\left(\frac{5\pi}{2}\right)$$

which is approximately equal to $-422.731\,331\,831\,631\,6 < 0$. Therefore, by the intermediate value theorem there must exist $x \in \left(0, \frac{5\pi}{2}\right)$ such that $f(x) = 0$.

This example illustrates the use of this major theorem very well. It says something exists but it does not tell how to find it.

Definition 2.10. A function f, defined on some interval is strictly increasing if whenever $x < y$, it follows $f(x) < f(y)$. The function is strictly decreasing if whenever $x < y$, it follows $f(x) > f(y)$.

You should draw a picture of the graph of a strictly increasing or decreasing function from the definition.

Lemma 2.1. *Let $\phi : [a, b] \to \mathbb{R}$ be a one to one continuous function. Then ϕ is either strictly increasing or strictly decreasing.*

This lemma is not real easy to prove but it is one of those things which seems obvious. To say a function is one to one is to say that every horizontal line intersects the graph of the function in no more than one point. (This is called the horizontal line test.) Now if your function is continuous (having no jumps) and is one to one, try to imagine how this could happen without it being either strictly increasing or decreasing and you will soon see this is highly believable, and in fact, for it to fail would be incredible. The proof of this lemma is in the last section of this chapter in case you are interested. See Lemma 2.4.

Corollary 2.1. *Let $\phi : (a, b) \to \mathbb{R}$ be a one to one continuous function. Then ϕ is either strictly increasing or strictly decreasing.*

The proof of this corollary is the same as the proof of the lemma. The next corollary follows from the above.

Corollary 2.2. *Let $f : (a, b) \to \mathbb{R}$ be one to one and continuous. Then $f(a, b)$ is an open interval (c, d) and $f^{-1} : (c, d) \to (a, b)$ is continuous. Also, if $f : [a, b] \to \mathbb{R}$ is one to one and continuous, then $f([a, b])$ is a closed interval $[c, d]$ and $f^{-1} : [c, d] \to [a, b]$ is continuous.*

This corollary is not too surprising either. To view the graph of the inverse function, simply turn things on the side and switch x and y. If the original graph has no jumps in it, neither will the new graph. Of course, the concept of continuity is tied to a rigorous definition, not to the drawing of pictures. This is why there is a proof in the last section of this chapter.

The next theorem is known as the max min theorem or extreme value theorem. An easy proof is on Page 81.

Theorem 2.4. *Let $I = [a, b]$ and let $f : I \to \mathbb{R}$ be continuous. Then f achieves its maximum and its minimum on I. This means there exist $x_1, x_2 \in I$ such that for all $x \in I$,*

$$f(x_1) \leq f(x) \leq f(x_2).$$

The proof of this theorem is based on the following lemma called the nested interval lemma which depends on the completeness axiom.

Lemma 2.2. *Let $I_k = \left[a^k, b^k \right]$ and suppose that for all $k = 1, 2, \cdots$,*

$$I_k \supseteq I_{k+1}.$$

Then there exists a point which is in each of these intervals.

2.8 Exercises

(1) Give an example of a continuous function defined on $(0, 1)$ which does not achieve its maximum on $(0, 1)$. It suffices to draw a graph of this function.

(2) Give an example of a continuous function defined on $(0, 1)$ which is bounded but which does not achieve either its maximum or its minimum. It suffices to draw a graph of this function.

(3) Give an example of a discontinuous function defined on $[0, 1]$ which is bounded but does not achieve either its maximum or its minimum. It suffices to draw a graph of this function.

(4) Give an example of a continuous function defined on $[0, 1) \cup (1, 2]$ which is positive at 2, negative at 0 but is not equal to zero for any value of x. It suffices to draw a graph of this function.

(5) Let $f(x) = x^5 + ax^4 + bx^3 + cx^2 + dx + e$ where a, b, c, d, and e are numbers. Show there exists x such that $f(x) = 0$.

(6) Give an example of a function which is one to one but neither strictly increasing nor strictly decreasing. **Hint:** Look for discontinuous functions satisfying the horizontal line test. It suffices to draw a graph of this function.

(7) Do you believe in $\sqrt[7]{8}$? That is, does there exist a number which multiplied by itself seven times yields 8? Before you jump to any conclusions, the number you get on your calculator is wrong. In fact, your calculator does not even know about $\sqrt[7]{8}$. All it can do is try to approximate it and what it gives you

is this approximation. Why does it exist? **Hint:** Use the intermediate value theorem on the function $f(x) = x^7 - 8$.

(8) It has been known since the time of Pythagoras that $\sqrt{2}$ is irrational. If you throw out all the irrational numbers, show that the conclusion of the intermediate value theorem could no longer be obtained. That is, show there exists a function which starts off less than zero and ends up larger than zero and yet there is no number where the function equals zero. **Hint:** Try $f(x) = x^2 - 2$. You supply the details.

(9) Let f be a continuous function defined on a closed interval $I_1 \equiv [a, b]$ such that $f(a) < 0$ and $f(b) > 0$. Consider $\frac{a+b}{2}$. If $f\left(\frac{a+b}{2}\right) \geq 0$, let $I_2 = \left[a, \frac{a+b}{2}\right]$ and if $f\left(\frac{a+b}{2}\right) < 0$, let $I_2 \equiv \left[\frac{a+b}{2}, c\right]$. Thus $I_1 \supset I_2$ and the interval I_2 has exactly the same property that I_1 had in terms of f being negative at the left endpoint and nonnegative at the right endpoint. Continue this way obtaining a sequence of nested closed and bounded intervals $\{I_k\}$. From the nested interval lemma on Page 56, there exists a unique point x in all these intervals. Show $f(x) = 0$ This is called the method of bisection and can be used to find a solution to the equation $f(x) = 0$.

(10) Apply the method of bisection described in Problem 9 to find $\sqrt[7]{8}$. Use a calculator to raise things to the seventh power. It will be much easier than doing it by hand.

(11) A circular hula hoop lies partly in the shade and partly in the hot sun. Show there exist two points on the hula hoop which are at opposite sides of the hoop which have the same temperature. **Hint:** Imagine this is a circle and points are located by specifying their angle θ from a fixed diameter. Then letting $T(\theta)$ be the temperature in the hoop, $T(\theta + 2\pi) = T(\theta)$. You need to have $T(\theta) = T(\theta + \pi)$ for some θ. Assume T is a continuous function of θ.

(12) A car starts off on a long trip with a full tank of gas. The driver intends to drive the car till it runs out of gas. Show that at some time, the number of miles the car has gone, exactly equals the number of gallons of gas in the tank.

2.9 Limit Of A Function

A concept closely related to continuity is that of the limit of a function.

Definition 2.11. Let $f : D(f) \subseteq \mathbb{R} \to \mathbb{R}$ be a function where $D(f) \supseteq (x - r, x) \cup (x, x + r)$ for some $r > 0$. Note that f is not necessarily defined at x. Then

$$\lim_{y \to x} f(y) = L$$

if and only if the following condition holds. For all $\varepsilon > 0$ there exists $\delta > 0$ such that if

$$0 < |y - x| < \delta,$$

then,

$$|L - f(y)| < \varepsilon.$$

If everything is the same as the above, except y is required to be larger than x and f is only required to be defined on $(x, x + r)$, then the notation is

$$\lim_{y \to x+} f(y) = L.$$

If f is only required to be defined on $(x - r, x)$ and y is required to be less than x with the same conditions above, we write

$$\lim_{y \to x-} f(y) = L.$$

Limits are also taken as a variable "approaches" infinity. Of course nothing is "close" to infinity and so this requires a slightly different definition.

$$\lim_{x \to \infty} f(x) = L$$

if for every $\varepsilon > 0$ there exists l such that whenever $x > l$,

$$|f(x) - L| < \varepsilon \tag{2.3}$$

and

$$\lim_{x \to -\infty} f(x) = L$$

if for every $\varepsilon > 0$ there exists l such that whenever $x < l$, 2.3 holds.

The following pictures illustrate some of these definitions.

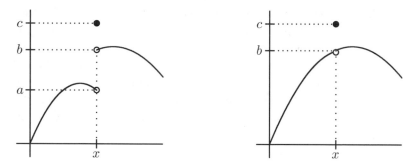

In the left picture is shown the graph of a function. Note the value of the function at x equals c while $\lim_{y \to x+} f(y) = b$ and $\lim_{y \to x-} f(y) = a$. In the second picture, $\lim_{y \to x} f(y) = b$. Note that the value of the function at the point x has nothing to do with the limit of the function in any of these cases. **The value of a function at x has nothing to do with the value of the limit at x!** This must always be kept in mind. You do not evaluate interesting limits by computing $f(x)$! In the above picture, $f(x)$ is always wrong! It may be the case that $f(x)$ is right but this is merely a happy coincidence when it occurs and as explained below in Theorem 2.8. It is often equivalent to f being continuous at x.

Theorem 2.5. *If $\lim_{y \to x} f(y) = L$ and $\lim_{y \to x} f(y) = L_1$, then $L = L_1$.*

Proof: Let $\varepsilon > 0$ be given. There exists $\delta > 0$ such that if $0 < |y - x| < \delta$, then

$$|f(y) - L| < \varepsilon, \ |f(y) - L_1| < \varepsilon.$$

Therefore, for such y,

$$|L - L_1| \leq |L - f(y)| + |f(y) - L_1| < \varepsilon + \varepsilon = 2\varepsilon.$$

Since $\varepsilon > 0$ was arbitrary, this shows $L = L_1$. ∎

The above theorem holds for any of the kinds of limits presented in the above definition.

Another concept is that of a function having either ∞ or $-\infty$ as a limit. In this case, the values of the function do not ever get close to their target because nothing can be close to $\pm\infty$. Roughly speaking, the limit of the function equals ∞ if the values of the function are ultimately larger than any given number. More precisely:

Definition 2.12. If $f(x) \in \mathbb{R}$, then $\lim_{y \to x} f(x) = \infty$ if for every number l, there exists $\delta > 0$ such that whenever $|y - x| < \delta$, then $f(x) > l$. $\lim_{x \to \infty} f(x) = \infty$ if for all k, there exists l such that $f(x) > k$ whenever $x > l$. One sided limits and limits as the variable approaches $-\infty$ are defined similarly.

It may seem there is a lot to memorize here. In fact, this is not so because all the definitions are intuitive when you understand them.

The proof of the following theorem is on Page 75. This is one of those messy theorems which is intuitively fairly easy to believe.

Theorem 2.6. *In this theorem, the symbol $\lim_{y \to x}$ denotes any of the limits described above. Suppose $\lim_{y \to x} f(y) = L$ and $\lim_{y \to x} g(y) = K$ where K and L are real numbers in $\mathbb{R}$. Then if $a, b \in \mathbb{R}$,*

$$\lim_{y \to x} (af(y) + bg(y)) = aL + bK, \tag{2.4}$$

$$\lim_{y \to x} fg(y) = LK \tag{2.5}$$

and if $K \neq 0$,

$$\lim_{y \to x} \frac{f(y)}{g(y)} = \frac{L}{K}. \tag{2.6}$$

Also, if h is a continuous function defined near L, then

$$\lim_{y \to x} h \circ f(y) = h(L). \tag{2.7}$$

Suppose $\lim_{y \to x} f(y) = L$. If $f(y) \leq a$ for all y of interest, then $L \leq a$ and if $f(y) \geq a$ then $L \geq a$.

A very useful theorem for finding limits is called the squeezing theorem. Imagine two men are walking toward a light post and a third man is right between them. Then the third man is also walking toward the light post. That is the idea of this theorem.

Theorem 2.7. *Suppose* $\lim_{x \to a} f(x) = L = \lim_{x \to a} g(x)$ *and for all* x *near* a,

$$f(x) \le h(x) \le g(x).$$

Then

$$\lim_{x \to a} h(x) = L.$$

Proof: If $L \ge h(x)$, then

$$|h(x) - L| \le |f(x) - L|.$$

If $L < h(x)$, then

$$|h(x) - L| \le |g(x) - L|.$$

Therefore,

$$|h(x) - L| \le |f(x) - L| + |g(x) - L|.$$

Now let $\varepsilon > 0$ be given. There exists δ_1 such that if $0 < |x - a| < \delta_1$,

$$|f(x) - L| < \varepsilon/2$$

and there exists δ_2 such that if $0 < |x - a| < \delta_2$, then

$$|g(x) - L| < \varepsilon/2.$$

Letting $0 < \delta \le \min(\delta_1, \delta_2)$, if $0 < |x - a| < \delta$, then

$$|h(x) - L| \le |f(x) - L| + |g(x) - L|$$
$$< \varepsilon/2 + \varepsilon/2 = \varepsilon. \qquad \blacksquare$$

Theorem 2.8. *For* $f : I \to \mathbb{R}$, *and* I *is an interval of the form* $(a, b), [a, b), (a, b]$, *or* $[a, b]$, *then* f *is continuous at* $x \in I$ *if and only if* $\lim_{y \to x} f(y) = f(x)$.

Proof: You fill in the details. Compare the definition of continuous and the definition of the limit just given. $\blacksquare$

Example 2.16. Find $\lim_{x \to 3} \frac{x^2 - 9}{x - 3}$.

Note that $\frac{x^2 - 9}{x - 3} = x + 3$ whenever $x \ne 3$. Therefore, if $0 < |x - 3| < \varepsilon$,

$$\left| \frac{x^2 - 9}{x - 3} - 6 \right| = |x + 3 - 6| = |x - 3| < \varepsilon.$$

It follows from the definition that this limit equals 6.

You should be careful to note that in the definition of limit, the variable **never equals the thing it is getting close to**. In this example, x is never equal to 3. This is very significant because, in interesting limits, the function whose limit is being taken will usually not be defined at the point of interest.

Example 2.17. Let

$$f(x) = \frac{x^2 - 9}{x - 3} \text{ if } x \neq 3.$$

How should f be defined at $x = 3$ so that the resulting function will be continuous there?

In the previous example, the limit of this function equals 6. Therefore, by Theorem 2.8 it is necessary to define $f(3) = 6$.

Example 2.18. Find $\lim_{x \to \infty} \frac{x}{1+x}$.

Write $\frac{x}{1+x} = \frac{1}{1+(1/x)}$. Now it seems clear that $\lim_{x \to \infty} 1 + (1/x) = 1 \neq 0$. Therefore, Theorem 2.6 implies

$$\lim_{x \to \infty} \frac{x}{1+x} = \lim_{x \to \infty} \frac{1}{1+(1/x)} = \frac{1}{1} = 1.$$

Example 2.19. Show $\lim_{x \to a} \sqrt{x} = \sqrt{a}$ whenever $a \geq 0$. In the case that $a = 0$, take the limit from the right.

There are two cases. First consider the case when $a > 0$. Let $\varepsilon > 0$ be given. Multiply and divide by $\sqrt{x} + \sqrt{a}$. This yields

$$\left| \sqrt{x} - \sqrt{a} \right| = \left| \frac{x - a}{\sqrt{x} + \sqrt{a}} \right|.$$

Now let $0 < \delta_1 < a/2$. Then if $|x - a| < \delta_1, x > a/2$ and so

$$\left| \sqrt{x} - \sqrt{a} \right| = \left| \frac{x - a}{\sqrt{x} + \sqrt{a}} \right| \leq \frac{|x - a|}{(\sqrt{a}/\sqrt{2}) + \sqrt{a}}$$

$$\leq \frac{2\sqrt{2}}{\sqrt{a}} |x - a|.$$

Now let $0 < \delta \leq \min\left(\delta_1, \frac{\varepsilon\sqrt{a}}{2\sqrt{2}}\right)$. Then for $0 < |x - a| < \delta$,

$$\left| \sqrt{x} - \sqrt{a} \right| \leq \frac{2\sqrt{2}}{\sqrt{a}} |x - a| < \frac{2\sqrt{2}}{\sqrt{a}} \frac{\varepsilon\sqrt{a}}{2\sqrt{2}} = \varepsilon.$$

Next consider the case where $a = 0$. In this case, let $\varepsilon > 0$ and let $\delta = \varepsilon^2$. Then if $0 < x - 0 < \delta = \varepsilon^2$, it follows that $0 \leq \sqrt{x} < \left(\varepsilon^2\right)^{1/2} = \varepsilon$.

2.10   Exercises

(1) Find the following limits if possible

    (a) $\lim_{x\to 0+} \frac{|x|}{x}$

    (b) $\lim_{x\to 0+} \frac{x}{|x|}$

    (c) $\lim_{x\to 0-} \frac{|x|}{x}$

    (d) $\lim_{x\to 4} \frac{x^2-16}{x+4}$

    (e) $\lim_{x\to 3} \frac{x^2-9}{x+3}$

    (f) $\lim_{x\to -2} \frac{x^2-4}{x-2}$

    (g) $\lim_{x\to\infty} \frac{x}{1+x^2}$

    (h) $\lim_{x\to\infty} -2\frac{x}{1+x^2}$

(2) Find $\lim_{h\to 0} \frac{\frac{1}{(x+h)^3} - \frac{1}{x^3}}{h}$.

(3) Find $\lim_{x\to 4} \frac{\sqrt[4]{x}-\sqrt{2}}{\sqrt{x}-2}$.

(4) Find $\lim_{x\to\infty} \frac{\sqrt[5]{3x}+\sqrt[4]{x}+7\sqrt{x}}{\sqrt{3x+1}}$.

(5) Find $\lim_{x\to\infty} \frac{(x-3)^{20}(2x+1)^{30}}{(2x^2+7)^{25}}$.

(6) Find $\lim_{x\to 2} \frac{x^2-4}{x^3+3x^2-9x-2}$.

(7) Find

$$\lim_{x\to\infty} \left(\sqrt{1-7x+x^2} - \sqrt{1+7x+x^2} \right).$$

(8) Prove Theorem 2.5 for right, left and limits as $y\to\infty$.

(9) Prove from the definition that $\lim_{x\to a} \sqrt[3]{x} = \sqrt[3]{a}$ for all $a \in \mathbb{R}$. **Hint:** You might want to use the formula for the difference of two cubes,

$$a^3 - b^3 = (a-b)\left(a^2+ab+b^2\right).$$

(10) Find $\lim_{h\to 0} \frac{(x+h)^2-x^2}{h}$.

(11) Prove Theorem 2.8 from the definitions of limit and continuity.

(12) Find $\lim_{h\to 0} \frac{(x+h)^3-x^3}{h}$

(13) Find $\lim_{h\to 0} \frac{1/(x+h)-1/x}{h}$

(14) Find $\lim_{x\to -3} \frac{x^3+27}{x+3}$

(15) Find $\lim_{h\to 0} \frac{\sqrt{(3+h)^2}-3}{h}$ if it exists.

(16) Find the values of x for which $\lim_{h\to 0} \frac{\sqrt{(x+h)^2}-x}{h}$ exists and find the limit.

(17) Find $\lim_{h\to 0} \frac{\sqrt[3]{(x+h)}-\sqrt[3]{x}}{h}$ if it exists. Here $x\neq 0$.

(18) Suppose $\lim_{y\to x+} f(y) = L_1 \neq L_2 = \lim_{y\to x-} f(y)$. Show $\lim_{y\to x} f(x)$ does not exist. **Hint:** Roughly, the argument goes as follows: For $|y_1 - x|$ small and $y_1 > x$, $|f(y_1) - L_1|$ is small. Also, for $|y_2 - x|$ small and $y_2 < x$, $|f(y_2) - L_2|$ is small. However, if a limit existed, then $f(y_2)$ and $f(y_1)$ would both need

to be close to some number and so both L_1 and L_2 would need to be close to some number. However, this is impossible because they are different.

(19) Show $\lim_{x\to 0}\frac{\sin x}{x}=1$. **Hint:** You might consider Theorem 2.2 on Page 52 to write the inequality $|\sin x|+1-\cos x \geq |x| \geq |\sin x|$ whenever $|x|$ is small. Then divide both sides by $|\sin x|$ and use some trig. identities to write $\frac{\sin^2 x}{|\sin x|(1+\cos x)}+1\geq \frac{|x|}{|\sin x|}\geq 1$ and then use squeezing theorem.

(20) Let $f(x,y)=\frac{x^2-y^2}{x^2+y^2}$. Find

$$\lim_{x\to 0}\left(\lim_{y\to 0} f(x,y)\right), \ \lim_{y\to 0}\left(\lim_{x\to 0} f(x,y)\right).$$

If you did it right you got -1 for one answer and 1 for the other. What does this tell you about interchanging limits?

2.11   The Limit Of A Sequence

A closely related concept is the limit of a sequence. This was defined precisely by Bolzano[4] a little before the definition of the limit. The following is the precise definition of what is meant by the limit of a sequence.

Definition 2.13. A sequence $\{a_n\}_{n=1}^{\infty}$ converges to a,

$$\lim_{n\to\infty} a_n = a \text{ or } a_n \to a$$

if and only if for every $\varepsilon > 0$ there exists n_ε such that whenever $n \geq n_\varepsilon$,

$$|a_n - a| < \varepsilon.$$

In words the definition says that given any measure of closeness ε, the terms of the sequence are eventually all this close to a. Note the similarity with the concept of limit. Here, the word "eventually" refers to n being sufficiently large. Earlier, it referred to y being sufficiently close to x on one side or another or else x being sufficiently large in either the positive or negative directions. The above definition is always the definition of what is meant by the limit of a sequence. If the a_n are complex numbers or later on, vectors the definition remains the same. If $a_n = x_n + iy_n$ and $a = x + iy$,

$$|a_n - a| = \sqrt{(x_n - x)^2 + (y_n - y)^2}.$$

Recall the way you measure distance between two complex numbers.

The next theorem says you can speak of **the** limit.

Theorem 2.9. *If* $\lim_{n\to\infty} a_n = a$ *and* $\lim_{n\to\infty} a_n = a_1$ *then* $a_1 = a$.

---

[4]Bernhard Bolzano lived from 1781 to 1848. He was a Catholic priest and held a position in philosophy at the University of Prague. He had strong views about the absurdity of war, educational reform, and the need for individual conscience. His convictions got him in trouble with Franz I of Austria and when he refused to recant, was forced out of the university. He understood the need for absolute rigor in mathematics. He also did work on physics.

Proof: Suppose $a_1 \neq a$. Then let $0 < \varepsilon < |a_1 - a|/2$ in the definition of the limit. It follows there exists n_ε such that if $n \geq n_\varepsilon$, then $|a_n - a| < \varepsilon$ and $|a_n - a_1| < \varepsilon$. Therefore, for such n,

$$|a_1 - a| \leq |a_1 - a_n| + |a_n - a|$$
$$< \varepsilon + \varepsilon < |a_1 - a|/2 + |a_1 - a|/2 = |a_1 - a|,$$

a contradiction. ∎

Example 2.20. Let $a_n = \frac{1}{n^2+1}$.

Then it seems clear that

$$\lim_{n \to \infty} \frac{1}{n^2 + 1} = 0.$$

In fact, this is true from the definition. Let $\varepsilon > 0$ be given. Let $n_\varepsilon \geq \sqrt{\varepsilon^{-1}}$. Then if

$$n > n_\varepsilon \geq \sqrt{\varepsilon^{-1}},$$

it follows that $n^2 + 1 > \varepsilon^{-1}$ and so

$$0 < \frac{1}{n^2 + 1} = a_n < \varepsilon$$

Thus $|a_n - 0| < \varepsilon$ whenever n is this large.

Note that **the definition was of no use in finding a candidate for the limit.** This had to be produced based on other considerations. The definition is for verifying beyond any doubt that something is the limit. It is also what must be referred to in establishing theorems which are good for finding limits.

Example 2.21. Let $a_n = n^2$

Then in this case $\lim_{n\to\infty} a_n$ does not exist. In this case, people sometimes will say $\lim_{n\to\infty} a_n = \infty$.

Example 2.22. Let $a_n = (-1)^n$.

In this case, $\lim_{n\to\infty} (-1)^n$ does not exist. This follows from the definition. Let $\varepsilon = 1/2$. If there exists a limit, l, then eventually, for all n large enough, $|a_n - l| < 1/2$. However, $|a_n - a_{n+1}| = 2$ and so,

$$2 = |a_n - a_{n+1}| \leq |a_n - l| + |l - a_{n+1}| < 1/2 + 1/2 = 1$$

which cannot hold. Therefore, there can be no limit for this sequence. This is all that can be said for this particular sequence.

The following is another of those theorems which is intuitively fairly clear. A proof is on Page 76.

Theorem 2.10. *Suppose $\{a_n\}$ and $\{b_n\}$ are sequences and that*

$$\lim_{n \to \infty} a_n = a \text{ and } \lim_{n \to \infty} b_n = b.$$

Also suppose x and y are real numbers. Then

$$\lim_{n \to \infty} x a_n + y b_n = xa + yb \tag{2.8}$$

$$\lim_{n \to \infty} a_n b_n = ab \tag{2.9}$$

If $b \neq 0$,

$$\lim_{n \to \infty} \frac{a_n}{b_n} = \frac{a}{b}. \tag{2.10}$$

Another very useful theorem for finding limits is the squeezing theorem.

Theorem 2.11. *Suppose $\lim_{n \to \infty} a_n = a = \lim_{n \to \infty} b_n$ and $a_n \leq c_n \leq b_n$ for all n large enough. Then $\lim_{n \to \infty} c_n = a$.*

Proof: Let $\varepsilon > 0$ be given and let n_1 be large enough that if $n \geq n_1$,

$$|a_n - a| < \varepsilon/2 \text{ and } |b_n - a| < \varepsilon/2.$$

Then for such n,

$$|c_n - a| \leq |a_n - a| + |b_n - a| < \varepsilon.$$

The reason for this is that if $c_n \geq a$, then

$$|c_n - a| = c_n - a \leq b_n - a \leq |a_n - a| + |b_n - a|$$

because $b_n \geq c_n$. On the other hand, if $c_n \leq a$, then

$$|c_n - a| = a - c_n \leq a - a_n \leq |a - a_n| + |b - b_n|. \qquad \blacksquare$$

As an example, consider the following.

Example 2.23. Let

$$c_n \equiv (-1)^n \frac{1}{n}$$

and let $b_n = \frac{1}{n}$, and $a_n = -\frac{1}{n}$. Then you may easily show that

$$\lim_{n \to \infty} a_n = \lim_{n \to \infty} b_n = 0.$$

Since $a_n \leq c_n \leq b_n$, it follows $\lim_{n \to \infty} c_n = 0$ also.

Theorem 2.12. $\lim_{n \to \infty} r^n = 0$ *whenever $|r| < 1$.*

Proof: If $0 < r < 1$ if follows $r^{-1} > 1$. Why? Letting $\alpha = \frac{1}{r} - 1$, it follows

$$r = \frac{1}{1 + \alpha}.$$

Therefore, by the Example 1.2,

$$0 < r^n = \frac{1}{(1 + \alpha)^n} \leq \frac{1}{1 + \alpha n}.$$

Therefore, $\lim_{n\to\infty} r^n = 0$ if $0 < r < 1$. Now in general, if $|r| < 1$, $|r^n| = |r|^n \to 0$ by the first part. $\blacksquare$

An important theorem is the one which states that if a sequence converges, then so does every subsequence. You should review Definition 2.8 on Page 44 at this point. Recall that if $\{x_n\}$ is a sequence and $n_1 < n_2 < \cdots$ is an increasing sequence of integers with $n_1 \geq 1$, then $\{x_{n_k}\}_{k=1}^{\infty}$ is called a subsequence.

Theorem 2.13. *Let $\{x_n\}$ be a sequence with $\lim_{n\to\infty} x_n = x$ and let $\{x_{n_k}\}$ be a subsequence. Then $\lim_{k\to\infty} x_{n_k} = x$. Also, if each $x_n \leq y$ $(x \geq y)$, then $x \leq y$ $(x \geq y)$.*

Proof: Let $\varepsilon > 0$ be given. Then there exists n_ε such that if $n > n_\varepsilon$, then $|x_n - x| < \varepsilon$. Suppose $k > n_\varepsilon$. Then $n_k \geq k > n_\varepsilon$ and so $|x_{n_k} - x| < \varepsilon$, showing $\lim_{k\to\infty} x_{n_k} = x$ as claimed.

If each $x_n \leq y$, and yet $x > y$, then from the definition of the limit of a sequence, it follows that for all n large enough, $x_n \geq (x + y)/2 > y$ which is a contradiction. The case where $x \geq y$ is similar. $\blacksquare$

2.11.1 Sequences And Completeness

You recall the definition of completeness which stated that every nonempty set of real numbers which is bounded above has a least upper bound and that every nonempty set of real numbers which is bounded below has a greatest lower bound and this is a property of the real line known as the completeness axiom. Geometrically, this involved filling in the holes. There is another way of describing completeness in terms of sequences which I believe is more useful than the least upper bound and greatest lower bound property.

Definition 2.14. $\{a_n\}$ is a **Cauchy sequence** if for all $\varepsilon > 0$, there exists n_ε such that whenever $n, m \geq n_\varepsilon$,

$$|a_n - a_m| < \varepsilon.$$

A sequence is Cauchy means the terms are "bunching up to each other" as m, n get large.

Theorem 2.14. *The set of terms in a Cauchy sequence in $\mathbb{R}$ is bounded above and below.*

Proof: Let $\varepsilon = 1$ in the definition of a Cauchy sequence and let $n > n_1$. Then from the definition, $|a_n - a_{n_1}| < 1$. It follows that for all $n > n_1, |a_n| < 1 + |a_{n_1}|$. Therefore, for all n, $|a_n| \leq 1 + |a_{n_1}| + \sum_{k=1}^{n_1} |a_k|$. $\blacksquare$

Theorem 2.15. *If a sequence $\{a_n\}$ in $\mathbb{R}$ converges, then the sequence is a Cauchy sequence. Hence every subsequence converges. Also, in contrast to an arbitrary* *sequence, if any subsequence of a Cauchy sequence converges, then the Cauchy sequence converges.*

Proof: Let $\varepsilon > 0$ be given and suppose $a_n \to a$. Then from the definition of convergence, there exists n_ε such that if $n > n_\varepsilon$, it follows that $|a_n - a| < \frac{\varepsilon}{2}$. Therefore, if $m, n \geq n_\varepsilon + 1$, it follows that

$$|a_n - a_m| \leq |a_n - a| + |a - a_m| < \frac{\varepsilon}{2} + \frac{\varepsilon}{2} = \varepsilon$$

showing that, since $\varepsilon > 0$ is arbitrary, $\{a_n\}$ is a Cauchy sequence.

Now suppose $\lim_{k \to \infty} a_{n_k} = a$ and $\{a_n\}$ is a Cauchy sequence. Let $\varepsilon > 0$ be given. There exists N_1 such that if $m, n \geq N_1$ then $|a_n - a_m| < \varepsilon/2$. There also exists N_2 such that if $k \geq N_2$, then $|a - a_{n_k}| < \varepsilon/2$. Pick

$$k > \max(N_1, N_2) \equiv N$$

then if $n \geq N$

$$|a - a_n| \leq |a - a_{n_k}| + |a_{n_k} - a_n|$$

Since the list of n_k is strictly increasing, it follows that $n_k > N$ and so the above is smaller than $\varepsilon/2 + \varepsilon/2$. ∎

Definition 2.15. The sequence $\{a_n\}$, is monotone increasing if for all n, $a_n \leq a_{n+1}$. The sequence is monotone decreasing if for all n, $a_n \geq a_{n+1}$.

If someone says a sequence is monotone, it usually means monotone increasing.

There exist different descriptions of the completeness axiom. If you like you can simply add the three new criteria in the following theorem to the list of things which you mean when you say $\mathbb{R}$ is complete and skip the proof. All versions of completeness involve the notion of filling in holes and they are really just different ways of expressing this idea.

In practice, it is often more convenient to use the first of the three equivalent versions of completeness in the following theorem which states that every Cauchy sequence converges. In fact, this version of completeness, although it is equivalent to the completeness axiom for the real line, also makes sense in many situations where Definition 1.1 on Page 2 does not make sense. For example, the concept of completeness is often needed in settings where there is no order. This happens as soon as one does multivariable calculus. From now on completeness will mean any of the three conditions in the following theorem.

It is the concept of completeness and the notion of limits which sets analysis apart from algebra. You will find that every existence theorem, a theorem which asserts the existence of something, depends in calculus on the assumption that something is complete. Here is the theorem. It is proved on Page 82.

Theorem 2.16. *The following conditions are equivalent to completeness.*

(1) Every Cauchy sequence converges.

(2) Every monotone increasing sequence which is bounded above converges.

(3) Every monotone decreasing sequence which is bounded below converges.

Theorem 2.17. *Let $\{a_n\}$ be a monotone increasing sequence which is bounded above. Then $\lim_{n\to\infty} a_n = \sup \{a_n : n \geq 1\}$. Similarly, if $\{a_n\}$ is a monotone decreasing sequence which is bounded below, then $\lim_{n\to\infty} a_n = \inf \{a_n : n \geq 1\}$.*

Proof: Let $a = \sup \{a_n : n \geq 1\}$ and let $\varepsilon > 0$ be given. Then from Proposition 10.1 on Page 238 there exists m such that $a - \varepsilon < a_m \leq a$. Since the sequence is increasing, it follows that for all $n \geq m$, $a - \varepsilon < a_n \leq a$. Thus $a = \lim_{n\to\infty} a_n$. The other case is entirely similar. You could also consider the sequence $\{-a_n\}$ to prove it directly from the first case. ∎

2.11.2 Decimals

You are all familiar with decimals. In the United States these are written in the form $.a_1a_2a_3 \cdots$ where the a_i are integers between 0 and 9.[5] Thus 0.23417432 is a number written as a decimal. You also recall the meaning of such notation in the case of a terminating decimal. For example, 0.234 is defined as $\frac{2}{10} + \frac{3}{10^2} + \frac{4}{10^3}$. Now what is meant by a nonterminating decimal?

Definition 2.16. Let $.a_1a_2 \cdots$ be a decimal. Define

$$.a_1a_2 \cdots \equiv \lim_{n\to\infty} \sum_{k=1}^{n} \frac{a_k}{10^k}.$$

Proposition 2.1. The above definition makes sense.

Proof: Note the sequence $\left\{ \sum_{k=1}^{n} \frac{a_k}{10^k} \right\}_{n=1}^{\infty}$ is an increasing sequence. Therefore, if there exists an upper bound, it follows from Theorem 2.17 that this sequence converges and so the definition is well defined.

$$\sum_{k=1}^{n} \frac{a_k}{10^k} \leq \sum_{k=1}^{n} \frac{9}{10^k} = 9 \sum_{k=1}^{n} \frac{1}{10^k}.$$

Now

$$\frac{9}{10} \left(\sum_{k=1}^{n} \frac{1}{10^k} \right) = \sum_{k=1}^{n} \frac{1}{10^k} - \frac{1}{10} \sum_{k=1}^{n} \frac{1}{10^k} = \sum_{k=1}^{n} \frac{1}{10^k} - \sum_{k=2}^{n+1} \frac{1}{10^k}$$

$$= \frac{1}{10} - \frac{1}{10^{n+1}}$$

and so

$$\sum_{k=1}^{n} \frac{1}{10^k} \leq \frac{10}{9} \left(\frac{1}{10} - \frac{1}{10^{n+1}} \right) \leq \frac{10}{9} \left(\frac{1}{10} \right) = \frac{1}{9}.$$

Therefore, since this holds for all n, it follows the above sequence is bounded above. It follows the limit exists. ∎

---

[5] In France and Russia they use a comma instead of a period. This looks very strange but that is just the way they do it.

2.11.3 *Continuity And The Limit Of A Sequence*

There is a very useful way of thinking of continuity in terms of limits of sequences found in the following theorem. In words, it says a function is continuous if it takes convergent sequences to convergent sequences whenever possible.

Theorem 2.18. *A function $f : D(f) \to \mathbb{R}$ is continuous at $x \in D(f)$ if and only if, whenever $x_n \to x$ with $x_n \in D(f)$, it follows $f(x_n) \to f(x)$.*

Proof: Suppose first that f is continuous at x and let $x_n \to x$. Let $\varepsilon > 0$ be given. By continuity, there exists $\delta > 0$ such that if $|y - x| < \delta$, then $|f(x) - f(y)| < \varepsilon$. However, there exists n_δ such that if $n \geq n_\delta$, then $|x_n - x| < \delta$ and so for all n this large,

$$|f(x) - f(x_n)| < \varepsilon$$

which shows $f(x_n) \to f(x)$.

Now suppose the condition about taking convergent sequences to convergent sequences holds at x. Suppose f fails to be continuous at x. Then there exists $\varepsilon > 0$ and $x_n \subset D(f)$ such that $|x - x_n| < \frac{1}{n}$, yet

$$|f(x) - f(x_n)| \geq \varepsilon.$$

But this is clearly a contradiction because, although $x_n \to x$, $f(x_n)$ fails to converge to $f(x)$. It follows f must be continuous after all. ■

2.12 Exercises

(1) Find $\lim_{n\to\infty} \frac{n}{3n+4}$.

(2) Find $\lim_{n\to\infty} \frac{n^3 - 100n}{2n^3 + 79n^2} = \frac{1}{2}$.

(3) Find $\lim_{n\to\infty} \frac{3n^4 + 7n + 1000}{n^4 + 1} = 3$.

(4) For $a_k, b_k \neq 0$, find $\lim_{n\to\infty} \frac{a_k n^k + a_{k-1} n^{k-1} + \cdots + a_1 n + a_0}{b_k n^k + b_{k-1} n^{k-1} + \cdots + b_1 n + b_0} = \frac{a_k}{b_k}$.

(5) For $b_k \neq 0$, find $\lim_{n\to\infty} \frac{a_{k-1} n^{k-1} + \cdots + a_1 n + a_0}{b_k n^k + b_{k-1} n^{k-1} + \cdots + b_1 n + b_0} = 0$.

(6) Find $\lim_{n\to\infty} \frac{2^n + 7(5^n)}{4^n + 2(5^n)}$.

(7) Find $\lim_{n\to\infty} n \tan \frac{1}{n}$. **Hint:** See Problem 19 on Page 63.

(8) Find $\lim_{n\to\infty} n \sin \frac{2}{n}$. **Hint:** See Problem 19 on Page 63.

(9) Find $\lim_{n\to\infty} \sqrt{\left(n \sin \frac{9}{n}\right)}$. **Hint:** See Problem 19 on Page 63.

(10) Find $\lim_{n\to\infty} \sqrt{(n^2 + 6n)} - n$. **Hint:** Multiply and divide by $\sqrt{(n^2 + 6n)} + n$.

(11) Find $\lim_{n\to\infty} \sum_{k=1}^{n} \frac{1}{10^k}$.

(12) Suppose $\{x_n + iy_n\}$ is a sequence of complex numbers which converges to the complex number $x + iy$. Show this happens if and only if $x_n \to x$ and $y_n \to y$.

(13) For $|r| < 1$, find $\lim_{n\to\infty} \sum_{k=0}^{n} r^k$. **Hint:** First show $\sum_{k=0}^{n} r^k = \frac{r^{n+1}}{r-1} - \frac{1}{r-1}$. Then recall Theorem 2.12.

(14) Suppose $x = .3434343\overline{34}$ where the bar over the last 34 signifies that this repeats forever. In elementary school you were probably given the following procedure for finding the number x as a quotient of integers. First multiply by 100 to get $100x = 34.3434343\overline{34}$ and then subtract to get $99x = 34$. From this you conclude that $x = 34/99$. Fully justify this procedure. **Hint:** $.3434343\overline{34} = \lim_{n\to\infty} 34 \sum_{k=1}^{n} \left(\frac{1}{100}\right)^k$ now use Problem 13.

(15) Suppose $D(f) = [0,1] \cup \{9\}$ and $f(x) = x$ on $[0,1]$ while $f(9) = 5$. Is f continuous at the point 9? Use whichever definition of continuity you like.

(16) Suppose $x_n \to x$ and $x_n \leq c$. Show that $x \leq c$. Also show that if $x_n \to x$ and $x_n \geq c$, then $x \geq c$. This was proved in the text, but you should have a try at doing it on your own. **Hint:** If this is not true, argue that for all n large enough $x_n > c$.

(17) Let $a \in [0,1]$. Show $a = .a_1 a_2 a_3 \cdots$ for a unique choice of integers, $a_1, a_2, \cdots$ if it is possible to do this. Otherwise, give an example.

(18) Find $\lim_{n\to\infty} n \sin n$ if it exists. If it does not exist, explain why it does not.

(19) Recall the axiom of completeness states that a set which is bounded above has a least upper bound and a set which is bounded below has a greatest lower bound. Show that a monotone decreasing sequence which is bounded below converges to its greatest lower bound. **Hint:** Let a denote the greatest lower bound and recall that because of this, it follows that for all $\varepsilon > 0$ there exist points of $\{a_n\}$ in $[a, a + \varepsilon)$.

(20) In Theorem 2.15 it was shown that if a subsequence of a Cauchy sequence converges to a, then so does the Cauchy sequence. Give an example of a sequence which has a convergent subsequence but which does not converge.

2.13   Uniform Continuity, Sequential Compactness*

There is a theorem about the integral of a continuous function which requires the notion of uniform continuity. This is discussed in this section. Consider the function $f(x) = \frac{1}{x}$ for $x \in (0,1)$. This is a continuous function because, by Theorem 2.1, it is continuous at every point of $(0,1)$. However, for a given $\varepsilon > 0$, the δ needed in the ε, δ definition of continuity becomes very small as x gets close to 0. The notion of uniform continuity involves being able to choose a single δ which works on the whole domain of f. Here is the definition.

Definition 2.17. Let $f : D \subseteq \mathbb{R} \to \mathbb{R}$ be a function. Then f is uniformly continuous if for every $\varepsilon > 0$, there exists a δ **depending only on** ε such that if $|x - y| < \delta$ then $|f(x) - f(y)| < \varepsilon$.

It is an amazing fact that under certain conditions continuity implies uniform continuity.

Definition 2.18. A set, $K \subseteq \mathbb{R}$ is sequentially compact if whenever $\{a_n\} \subseteq K$ is a sequence, there exists a subsequence, $\{a_{n_k}\}$ such that this subsequence converges to a point of K.

The following theorem is part of the Heine Borel theorem. Its proof is on Page 81.

Theorem 2.19. *Every closed interval $[a, b]$ is sequentially compact.*

Theorem 2.20. *Let $f : K \to \mathbb{R}$ be continuous where K is a sequentially compact set in $\mathbb{R}$. Then f is uniformly continuous on K.*

Proof: If this is not true, there exists $\varepsilon > 0$ such that for every $\delta > 0$ there exists a pair of points x_δ and y_δ such that even though $|x_\delta - y_\delta| < \delta$, $|f(x_\delta) - f(y_\delta)| \geq \varepsilon$. Taking a succession of values for δ equal to $1, 1/2, 1/3, \cdots$, and letting the exceptional pair of points for $\delta = 1/n$ be denoted by x_n and y_n,

$$|x_n - y_n| < \frac{1}{n}, |f(x_n) - f(y_n)| \geq \varepsilon.$$

Now since K is sequentially compact, there exists a subsequence $\{x_{n_k}\}$ such that $x_{n_k} \to z \in K$. Now $n_k \geq k$ and so

$$|x_{n_k} - y_{n_k}| < \frac{1}{k}.$$

Consequently, $y_{n_k} \to z$ also. (x_{n_k} is like a person walking toward a certain point and y_{n_k} is like a dog on a leash which is constantly getting shorter. Obviously y_{n_k} must also move toward the point also. You should give a precise proof of what is needed here.) By continuity of f and Problem 16 on Page 70,

$$0 = |f(z) - f(z)| = \lim_{k \to \infty} |f(x_{n_k}) - f(y_{n_k})| \geq \varepsilon,$$

an obvious contradiction. Therefore, the theorem must be true. ∎

The following corollary follows from this theorem and Theorem 2.19.

Corollary 2.3. *Suppose I is a closed interval $I = [a, b]$ and $f : I \to \mathbb{R}$ is continuous. Then f is uniformly continuous.*

2.14 Exercises

(1) A function $f : D \subseteq \mathbb{R} \to \mathbb{R}$ is Lipschitz continuous or just Lipschitz for short if there exists a constant K such that

$$|f(x) - f(y)| \leq K|x - y|$$

for all $x, y \in D$. Show every Lipschitz function is uniformly continuous.

(2) If $|x_n - y_n| \to 0$ and $x_n \to z$, show that $y_n \to z$ also.

(3) Consider $f : (1, \infty) \to \mathbb{R}$ given by $f(x) = \frac{1}{x}$. Show f is uniformly continuous even though the set on which f is defined is not sequentially compact.

(4) If f is uniformly continuous, does it follow that $|f|$ is also uniformly continuous? If $|f|$ is uniformly continuous does it follow that f is uniformly continuous? Answer the same questions with "uniformly continuous" replaced with "continuous". Explain why.

(5) Suppose K is a sequentially compact set and $f : K \to \mathbb{R}$. Show that f achieves both its maximum and its minimum on K. **Hint:** Let $M \equiv \sup \{f(x) : x \in K\}$. Argue there exists a sequence $\{x_n\} \subseteq K$ such that $f(x_n) \to M$. Now use sequential compactness to get a subsequence, $\{x_{n_k}\}$ such that $\lim_{k \to \infty} x_{n_k} = x \in K$ and use the continuity of f to verify that $f(x) = M$. Incidentally, this shows f is bounded on K as well. A similar argument works to give the part about achieving the minimum. This is called the extreme value theorem.

(6) Show the following functions are uniformly continuous on the specified sets.

    (a) $f(x) = \frac{x^2}{1+x^2}$, $x \in \mathbb{R}$

    (b) $f(x) = \frac{x}{1+x}$, $x \in [0, \infty)$

    (c) $f(x) = \sin(x)$, $x \in \mathbb{R}$

    (d) $f(x) = \cos(x)$, $x \in \mathbb{R}$

    (e) $f(x) = \sqrt{1 + x^2}$, $x \in \mathbb{R}$

(7) Show that if a function f is uniformly continuous on $[a, \infty), a > 0$, and continuous on $[0, a]$, then it must be uniformly continuous on $[0, \infty)$.

(8) Show that $f(x) = \sqrt{x}$ is uniformly continuous on $[0, \infty)$.

(9) A function f defined on $\mathbb{R}$ is called Holder continuous if there exists $\alpha > 0$ and C such that for all x, y,

$$|f(x) - f(y)| \le C |x - y|^\alpha.$$

Show every Holder continuous function is uniformly continuous.

2.15 Fundamental Theory*

In this section, proofs of some theorems which have not been proved yet are given.

2.15.1 *Combinations Of Functions And Sequences*

Theorem 2.21. *The following assertions are valid*

(1) The function $af + bg$ is continuous at x when f, g are continuous at $x \in D(f) \cap D(g)$ and $a, b \in \mathbb{R}$.

(2) If and f and g are each real valued functions continuous at x, then fg is continuous at x. If, in addition to this, $g(x) \ne 0$, then f/g is continuous at x.

(3) If f is continuous at x, $f(x) \in D(g) \subseteq \mathbb{R}$, and g is continuous at $f(x)$, then $g \circ f$ is continuous at x.

(4) The function $f : \mathbb{R} \to \mathbb{R}$, given by $f(x) = |x|$ is continuous.

Proof: First consider (1). Let $\varepsilon > 0$ be given. By assumption, there exist $\delta_1 > 0$ such that whenever $|x - y| < \delta_1$, it follows $|f(x) - f(y)| < \frac{\varepsilon}{2(|a|+|b|+1)}$ and there exists $\delta_2 > 0$ such that whenever $|x - y| < \delta_2$, it follows that $|g(x) - g(y)| < \frac{\varepsilon}{2(|a|+|b|+1)}$. Then let $0 < \delta \leq \min(\delta_1, \delta_2)$. If $|x - y| < \delta$, then everything happens at once. Therefore, using the triangle inequality

$$|af(x) + bf(x) - (ag(y) + bg(y))| \leq |a|\,|f(x) - f(y)| + |b|\,|g(x) - g(y)|$$

$$< |a| \left(\frac{\varepsilon}{2(|a| + |b| + 1)} \right) + |b| \left(\frac{\varepsilon}{2(|a| + |b| + 1)} \right) < \varepsilon.$$

Now consider (2). There exists $\delta_1 > 0$ such that if $|y - x| < \delta_1$, then $|f(x) - f(y)| < 1$. Therefore, for such y,

$$|f(y)| < 1 + |f(x)|.$$

It follows that for such y,

$$|fg(x) - fy(y)| \leq |f(x)g(x) - g(x)f(y)| + |g(x)f(y) - f(y)g(y)|$$

$$< |g(x)|\,|f(x) - f(y)| + |f(y)|\,|g(x) - g(y)|$$
$$\leq (1 + |g(x)| + |f(y)|)\,[|g(x) - g(y)| + |f(x) - f(y)|]$$
$$\leq (2 + |g(x)| + |f(x)|)\,[|g(x) - g(y)| + |f(x) - f(y)|]$$

Now let $\varepsilon > 0$ be given. There exists δ_2 such that if $|x - y| < \delta_2$, then

$$|g(x) - g(y)| < \frac{\varepsilon}{2(2 + |g(x)| + |f(x)|)},$$

and there exists δ_3 such that if $|x - y| < \delta_3$, then

$$|f(x) - f(y)| < \frac{\varepsilon}{2(2 + |g(x)| + |f(x)|)}$$

Now let $0 < \delta \leq \min(\delta_1, \delta_2, \delta_3)$. Then if $|x - y| < \delta$, all the above hold at once and so

$$|fg(x) - fg(y)| \leq (2 + |g(x)| + |f(x)|)\,[|g(x) - g(y)| + |f(x) - f(y)|]$$

$$< (2 + |g(x)| + |f(x)|) \left(\frac{\varepsilon}{2(2 + |g(x)| + |f(x)|)} + \frac{\varepsilon}{2(2 + |g(x)| + |f(x)|)} \right) = \varepsilon.$$

This proves the first part of (2). To obtain the second part, let δ_1 be as described above and let $\delta_0 > 0$ be such that for $|x - y| < \delta_0$,

$$|g(x) - g(y)| < |g(x)|/2$$

and so by the triangle inequality,

$$-|g(x)|/2 \leq |g(y)| - |g(x)| \leq |g(x)|/2$$

which implies $|g(y)| \geq |g(x)|/2$, and $|g(y)| < 3|g(x)|/2$.

Then if $|x-y| < \min(\delta_0, \delta_1)$,

$$\left| \frac{f(x)}{g(x)} - \frac{f(y)}{g(y)} \right| = \left| \frac{f(x)g(y) - f(y)g(x)}{g(x)g(y)} \right| \leq \frac{|f(x)g(y) - f(y)g(x)|}{\left(\frac{|g(x)|^2}{2} \right)}$$

$$= \frac{2|f(x)g(y) - f(y)g(x)|}{|g(x)|^2}$$

$$\leq \frac{2}{|g(x)|^2} \left[|f(x)g(y) - f(y)g(y) + f(y)g(y) - f(y)g(x)| \right]$$

$$\leq \frac{2}{|g(x)|^2} \left[|g(y)||f(x) - f(y)| + |f(y)||g(y) - g(x)| \right]$$

$$\leq \frac{2}{|g(x)|^2} \left[\frac{3}{2}|g(x)||f(x) - f(y)| + (1 + |f(x)|)|g(y) - g(x)| \right]$$

$$\leq \frac{2}{|g(x)|^2} (1 + 2|f(x)| + 2|g(x)|) \left[|f(x) - f(y)| + |g(y) - g(x)| \right]$$

$$\equiv M \left[|f(x) - f(y)| + |g(y) - g(x)| \right]$$

where M is defined by

$$M \equiv \frac{2}{|g(x)|^2} (1 + 2|f(x)| + 2|g(x)|)$$

Now let δ_2 be such that if $|x-y| < \delta_2$, then

$$|f(x) - f(y)| < \frac{\varepsilon}{2}M^{-1}$$

and let δ_3 be such that if $|x-y| < \delta_3$, then

$$|g(y) - g(x)| < \frac{\varepsilon}{2}M^{-1}.$$

Then if $0 < \delta \leq \min(\delta_0, \delta_1, \delta_2, \delta_3)$, and $|x-y| < \delta$, everything holds and

$$\left| \frac{f(x)}{g(x)} - \frac{f(y)}{g(y)} \right| \leq M \left[|f(x) - f(y)| + |g(y) - g(x)| \right]$$

$$< M \left[\frac{\varepsilon}{2}M^{-1} + \frac{\varepsilon}{2}M^{-1} \right] = \varepsilon.$$

This completes the proof of the second part of (2).

Note that in these proofs no effort is made to find some sort of "best" δ. The problem is one which has a yes or a no answer. Either is it or it is not continuous.

Now consider (3). If f is continuous at x, $f(x) \in D(g) \subseteq \mathbb{R}^p$, and g is continuous at $f(x)$, then $g \circ f$ is continuous at x. Let $\varepsilon > 0$ be given. Then there exists $\eta > 0$ such that if $|y-f(x)| < \eta$ and $y \in D(g)$, it follows that $|g(y) - g(f(x))| < \varepsilon$. From continuity of f at x, there exists $\delta > 0$ such that if $|x-z| < \delta$ and $z \in D(f)$, then

$|f(z) - f(x)| < \eta$. Then if $|x-z| < \delta$ and $z \in D(g \circ f) \subseteq D(f)$, all the above hold and so

$$|g(f(z)) - g(f(x))| < \varepsilon.$$

This proves part (3).

To verify part (4), let $\varepsilon > 0$ be given and let $\delta = \varepsilon$. Then if $|x-y| < \delta$, the triangle inequality implies

$$|f(x) - f(y)| = |\,|x| - |y|\,| \leq |x-y| < \delta = \varepsilon$$

This proves part (4) and completes the proof of the theorem. ■

Theorem 2.22. *In this theorem, the symbol $\lim_{y \to x}$ denotes limits from the right or left as well as limits at ∞ or $-\infty$. Suppose $\lim_{y \to x} f(y) = L$ and $\lim_{y \to x} g(y) = K$ where K and L are real numbers in $\mathbb{R}$. Then if $a, b \in \mathbb{R}$,*

$$\lim_{y \to x} (af(y) + bg(y)) = aL + bK, \tag{2.11}$$

$$\lim_{y \to x} fg(y) = LK \tag{2.12}$$

and if $K \neq 0$,

$$\lim_{y \to x} \frac{f(y)}{g(y)} = \frac{L}{K}. \tag{2.13}$$

Also, if h is a continuous function defined near L, then

$$\lim_{y \to x} h \circ f(y) = h(L). \tag{2.14}$$

Suppose $\lim_{y \to x} f(y) = L$. If $f(y) \leq a$ all y of interest, then $L \leq a$ and if $f(y) \geq a$ then $L \geq a$.

Proof: The proof of (2.11) is left for you. It is like a corresponding theorem for continuous functions. Next consider (2.12). Let $\varepsilon > 0$ be given. Then by the triangle inequality,

$$|fg(y) - LK| \leq |fg(y) - f(y)K| + |f(y)K - LK|$$
$$\leq |f(y)|\,|g(y) - K| + |K|\,|f(y) - L|. \tag{2.15}$$

There exists δ_1 such that if $0 < |y - x| < \delta_1$, then $|f(y) - L| < 1$, and so for such y, and the triangle inequality, $|f(y)| < 1 + |L|$. Therefore, for $0 < |y - x| < \delta_1$,

$$|fg(y) - LK| \leq (1 + |K| + |L|)\,[|g(y) - K| + |f(y) - L|]. \tag{2.16}$$

Now let $0 < \delta_2$ be such that for $0 < |x - y| < \delta_2$,

$$|f(y) - L| < \frac{\varepsilon}{2(1 + |K| + |L|)}, \quad |g(y) - K| < \frac{\varepsilon}{2(1 + |K| + |L|)}.$$

Then letting $0 < \delta \leq \min(\delta_1, \delta_2)$, it follows from (2.16) that $|fg(y) - LK| < \varepsilon$ and this proves (2.12). Limits as $x \to \pm\infty$ and one sided limits are handled similarly.

The proof of (2.13) is left to you. It is just like the theorem about the quotient of continuous functions being continuous provided the function in the denominator is nonzero at the point of interest.

Consider (2.14). Since h is continuous near L, it follows that for $\varepsilon > 0$ given, there exists $\eta > 0$ such that if $|y-L| < \eta$, then

$$|h(y) - h(L)| < \varepsilon$$

Now since $\lim_{y \to x} f(y) = L$, there exists $\delta > 0$ such that if $0 < |y-x| < \delta$, then

$$|f(y) - L| < \eta.$$

Therefore, if $0 < |y-x| < \delta$,

$$|h(f(y)) - h(L)| < \varepsilon.$$

The same theorem holds for one sided limits and limits as the variable moves toward $\pm\infty$. The proofs are left to you. They are minor modifications of the above.

It only remains to verify the last assertion. Assume $f(y) \leq a$. It is required to show that $L \leq a$. If this is not true, then $L > a$. Letting ε be small enough that $a < L - \varepsilon$, it follows that ultimately, for y close enough to x, $f(y) \in (L - \varepsilon, L + \varepsilon)$ which requires $f(y) > a$ contrary to assumption. ∎

Theorem 2.23. *Suppose $\{a_n\}$ and $\{b_n\}$ are sequences and that*

$$\lim_{n \to \infty} a_n = a \text{ and } \lim_{n \to \infty} b_n = b.$$

Also suppose x and y are real numbers. Then

$$\lim_{n \to \infty} xa_n + yb_n = xa + yb \tag{2.17}$$

$$\lim_{n \to \infty} a_n b_n = ab \tag{2.18}$$

If $b \neq 0$,

$$\lim_{n \to \infty} \frac{a_n}{b_n} = \frac{a}{b}. \tag{2.19}$$

Proof: The first of these claims is left for you to do. To do the second, let $\varepsilon > 0$ be given and choose n_1 such that if $n \geq n_1$ then

$$|a_n - a| < 1.$$

Then for such n, the triangle inequality implies

$$\begin{aligned} |a_n b_n - ab| &\leq |a_n b_n - a_n b| + |a_n b - ab| \\ &\leq |a_n||b_n - b| + |b||a_n - a| \\ &\leq (|a| + 1)|b_n - b| + |b||a_n - a|. \end{aligned}$$

Now let n_2 be large enough that for $n \geq n_2$,

$$|b_n - b| < \frac{\varepsilon}{2(|a| + 1)}, \text{ and } |a_n - a| < \frac{\varepsilon}{2(|b| + 1)}$$

Such a number exists because of the definition of limit. Therefore, let

$$n_\varepsilon > \max\left(n_1, n_2\right)$$

For $n \geq n_\varepsilon$,

$$
\begin{aligned}
|a_n b_n - ab| &\leq (|a| + 1)|b_n - b| + |b||a_n - a| \\
&< (|a| + 1)\frac{\varepsilon}{2(|a| + 1)} + |b|\frac{\varepsilon}{2(|b| + 1)} \leq \varepsilon
\end{aligned}
$$

This proves (2.18). Next consider (2.19).

Let $\varepsilon > 0$ be given and let n_1 be so large that whenever $n \geq n_1$, $|b_n - b| < \frac{|b|}{2}$. Thus for such n,

$$
\left|\frac{a_n}{b_n} - \frac{a}{b}\right| = \left|\frac{a_n b - ab_n}{bb_n}\right| \leq \frac{2}{|b|^2}\left[|a_n b - ab| + |ab - ab_n|\right] \leq \frac{2}{|b|}|a_n - a| + \frac{2|a|}{|b|^2}|b_n - b|
$$

Now choose n_2 so large that if $n > n_2$, then

$$
|a_n - a| < \frac{\varepsilon|b|}{4}, \text{ and } |b_n - b| < \frac{\varepsilon|b|^2}{4(|a| + 1)}
$$

Letting $n_\varepsilon > \max\left(n_1, n_2\right)$, it follows that for $n \geq n_\varepsilon$,

$$
\left|\frac{a_n}{b_n} - \frac{a}{b}\right| \leq \frac{2}{|b|}|a_n - a| + \frac{2|a|}{|b|^2}|b_n - b| < \frac{2}{|b|}\frac{\varepsilon|b|}{4} + \frac{2|a|}{|b|^2}\frac{\varepsilon|b|^2}{4(|a| + 1)} < \varepsilon.
$$

∎

2.15.2 The Intermediate Value Theorem

The following lemma says that if f is a continuous function defined on an interval and if $f(x) > 0$ then this situation persists near x and if $f(x) < 0$, then this also persists near x.

Lemma 2.3. *Let* $f : [a, b] \to \mathbb{R}$ *be continuous and suppose for some* $x \in [a, b]$, $f(x) > 0$ (< 0) *Then there exists* $\delta > 0$ *such that if* $y \in [a, b] \cap (x - \delta, x + \delta)$, *then* $f(y) > 0$. (< 0)

Proof: Suppose $f(x) \neq 0$. By continuity, there exists $\delta > 0$ such that if $|y - x| < \delta$, then

$$|f(x) - f(y)| < |f(x)|/2$$

If $f(x) > 0$ then for such y,

$$-f(x)/2 < f(y) - f(x)$$

and so $f(y) > f(x)/2 > 0$. If $f(x) < 0$, then for such y,

$$f(y) - f(x) < -f(x)/2$$

and so $f(y) < f(x)/2 < 0$. ∎

Next here is a proof of the intermediate value theorem. This theorem is a little like the assertion that if a chicken crosses the road, then it must pass over the center line somewhere.

Theorem 2.24. *Suppose* $f : [a, b] \to \mathbb{R}$ *is continuous and suppose* $f(a) < c < f(b)$. *Then there exists* $x \in (a, b)$ *such that* $f(x) = c$.

Proof: Since $f(a) < c$, the set S defined as
$$S \equiv \{y \in [a,b] : f(t) - c < 0 \text{ for all } t \le y\}$$
is nonempty. In particular $a \in S$. Also, since $S \subseteq [a,b]$ and S is bounded above by b, there exists a least upper bound $x \le b$. If $f(x) - c < 0$, then by Lemma 2.3, x is not an upper bound. If $f(x) - c > 0$ then by Lemma 2.3, x fails to be the **least** upper bound. Therefore, $f(x) - c = 0$ and consequently $x \in (a,b)$. ∎

The following is one of those theorems which seems obvious but is not at all obvious when you try to prove it. However, it is easy if you use the above intermediate value theorem.

Lemma 2.4. *Let $\phi : [a,b] \to \mathbb{R}$ be a continuous function and suppose ϕ is $1 - 1$ on (a,b). Then ϕ is either strictly increasing or strictly decreasing on $[a,b]$.*

Proof: First it is shown that ϕ is either strictly increasing or strictly decreasing on (a,b).

If ϕ is not strictly decreasing on (a,b), then there exists $x_1 < y_1$, $x_1, y_1 \in (a,b)$ such that
$$(\phi(y_1) - \phi(x_1))(y_1 - x_1) > 0.$$
If for some other pair of points $x_2 < y_2$ with $x_2, y_2 \in (a,b)$, the above inequality does not hold, then since ϕ is $1 - 1$,
$$(\phi(y_2) - \phi(x_2))(y_2 - x_2) < 0.$$
Let $x_t \equiv tx_1 + (1-t)x_2$ and $y_t \equiv ty_1 + (1-t)y_2$. Then $x_t < y_t$ for all $t \in [0,1]$ because
$$tx_1 \le ty_1 \text{ and } (1-t)x_2 \le (1-t)y_2$$
with strict inequality holding for at least one of these inequalities since not both t and $(1-t)$ can equal zero. Now define
$$h(t) \equiv (\phi(y_t) - \phi(x_t))(y_t - x_t).$$
Since h is continuous and $h(0) < 0$, while $h(1) > 0$, there exists $t \in (0,1)$ such that $h(t) = 0$. Therefore, both x_t and y_t are points of (a,b) and $\phi(y_t) - \phi(x_t) = 0$ contradicting the assumption that ϕ is one to one. It follows ϕ is either strictly increasing or strictly decreasing on (a,b).

This property of being either strictly increasing or strictly decreasing on (a,b) carries over to $[a,b]$ by the continuity of ϕ. Suppose ϕ is strictly increasing on (a,b), a similar argument holding for ϕ strictly decreasing on (a,b). If $x > a$, then pick $y \in (a,x)$ and from the above, $\phi(y) < \phi(x)$. Now by continuity of ϕ at a,
$$\phi(a) = \lim_{x \to a+} \phi(z) \le \phi(y) < \phi(x).$$
Therefore, $\phi(a) < \phi(x)$ whenever $x \in (a,b)$. Similarly $\phi(b) > \phi(x)$ for all $x \in (a,b)$. ∎

The following is an open mapping theorem.

Corollary 2.4. *Let $f : (a,b) \to \mathbb{R}$ be one to one and continuous. Then $f(a,b)$ is an open interval (c,d) and $f^{-1} : (c,d) \to (a,b)$ is continuous.*

Proof: Since f is either strictly increasing or strictly decreasing, it follows that $f(a, b)$ is an open interval (c, d). Assume f is decreasing. Now let $x \in (a, b)$. Why is f^{-1} is continuous at $f(x)$? Since f is decreasing, if $f(x) < f(y)$, then $y \equiv f^{-1}(f(y)) < x \equiv f^{-1}(f(x))$ and so f^{-1} is also decreasing. Let $\varepsilon > 0$ be given. Let $\varepsilon > \eta > 0$ and $(x - \eta, x + \eta) \subseteq (a, b)$. Then $f(x) \in (f(x + \eta), f(x - \eta))$. Let

$$\delta = \min\left(f(x) - f(x + \eta), f(x - \eta) - f(x)\right).$$

Then if $|f(z) - f(x)| < \delta$, it follows

$$z \equiv f^{-1}(f(z)) \in (x - \eta, x + \eta) \subseteq (x - \varepsilon, x + \varepsilon)$$

so

$$\left|f^{-1}(f(z)) - x\right| - \left|f^{-1}(f(z)) - f^{-1}(f(x))\right| < \varepsilon.$$

This proves the theorem in the case where f is strictly decreasing. The case where f is increasing is similar. ∎

2.15.3 *Extreme Value Theorem, Nested Intervals*

Now consider the extreme value theorem. This is based on the nested interval lemma.

In Russia there is a kind of doll called a matrushka doll. You pick it up and notice it comes apart in the center. Separating the two halves you find an identical doll inside. Then you notice this inside doll also comes apart in the center. Separating the two halves, you find yet another identical doll inside. This goes on quite a while until the final doll is in one piece. The nested interval lemma is like a matrushka doll except the process never stops. It involves a sequence of intervals, the first containing the second, the second containing the third, the third containing the fourth and so on. The fundamental question is whether there exists a point in all the intervals.

Lemma 2.5. *Let $I_k = [a_k, b_k]$ and suppose that for all $k = 1, 2, \cdots$,*

$$I_k \supseteq I_{k+1}.$$

Then there exists a point $c \in \mathbb{R}$ which is an element of every I_k.

Proof: Since $I_k \supseteq I_{k+1}$, this implies

$$a_k \leq a_{k+1}, \ b_k \geq b_{k+1}. \tag{2.20}$$

Consequently, if $k \leq l$,

$$a_l \leq b_l \leq b_k. \tag{2.21}$$

Now define

$$c \equiv \sup\{a_l : l = 1, 2, \cdots\}$$

By the first inequality in (2.20), and (2.21)

$$a_k \leq c = \sup\{a_l : l = k, k+1, \cdots\} \leq b_k \qquad (2.22)$$

for each $k = 1, 2, \cdots$. Thus $c \in I_k$ for every k. ∎

If this went too fast, the reason for the last inequality in (2.22) is that from (2.21), b_k is an upper bound to $\{a_l : l = k, k+1, \cdots\}$. Therefore, it is at least as large as the least upper bound.

This is a remarkable result and may not seem so obvious. Consider the intervals $I_k \equiv (0, 1/k)$. Then there is no point which lies in all these intervals because no negative number can be in all the intervals and $1/k$ is smaller than a given positive number whenever k is large enough. Thus the only candidate for being in all the intervals is 0, and 0 has been left out of them all. The problem here is that the endpoints of the intervals were not included contrary to the hypotheses of the above lemma in which all the intervals included the endpoints.

With the nested interval lemma, it becomes possible to prove the following lemma which shows a function continuous on a closed interval in $\mathbb{R}$ is bounded.

Lemma 2.6. *Let $I = [a, b]$ and let $f : I \to \mathbb{R}$ be continuous. Then f is bounded. That is, there exist numbers m and M such that for all $x \in [a, b]$, $m \leq f(x) \leq M$.*

Proof: Let $I \equiv I_0$ and suppose f is not bounded on I_0. Consider the two sets $\left[a, \frac{a+b}{2}\right]$ and $\left[\frac{a+b}{2}, b\right]$. Since f is not bounded on I_0, it follows that f must fail to be bounded on at least one of these sets. Let I_1 be one of these on which f is not bounded. Now do to I_1 what was done to I_0 to obtain $I_2 \subseteq I_1$ and for any two points $x, y \in I_2$

$$|x - y| \leq 2^{-1}\frac{b-a}{2} \leq 2^{-2}(b-a).$$

Continue in this way obtaining sets, I_k such that $I_k \supseteq I_{k+1}$ and for any two points in $I_k, x, y, |x-y| \leq 2^{-k}(b-a)$. By the nested interval lemma, there exists a point c which is contained in each I_k. Also, by continuity, there exists a $\delta > 0$ such that if $|c - y| < \delta$, then

$$|f(c) - f(y)| < 1. \qquad (2.23)$$

Let k be so large that $2^{-k}(b-a) < \delta$. Then for every $y \in I_k$, $|c-y| < \delta$ and so (2.23) holds for all such y. But this implies that for all $y \in I_k$,

$$|f(y)| \leq |f(c)| + 1$$

which shows that f is bounded on I_k contrary to the way I_k was chosen. This contradiction proves the lemma. ∎

Example 2.24. Let $f(x) = 1/x$ for $x \in (0,1)$.

Clearly, f is not bounded. Does this violate the conclusion of the nested interval lemma? It does not because the end points of the interval involved are not in the interval. The same function defined on $[.000001, 1)$ would have been bounded although in this case the boundedness of the function would not follow from the above lemma because it fails to include the right endpoint.

Theorem 2.25. *Let $I = [a,b]$ and let $f : I \to \mathbb{R}$ be continuous. Then f achieves its maximum and its minimum on I. This means there exist $x_1, x_2 \in I$ such that for all $x \in I$, $f(x_1) \leq f(x) \leq f(x_2)$.*

Proof: By completeness of $\mathbb{R}$ and Lemma 2.6, $f(I)$ has a least upper bound M. If for all $x \in I$, $f(x) \neq M$, then by Theorem 2.1, the function $g(x) = (M - f(x))^{-1} = \frac{1}{M - f(x)}$ is continuous on I. Since M is the least upper bound of $f(I)$ there exist points $x \in I$ such that $(M - f(x))$ is as small as desired. Consequently, g is not bounded above, contrary to Lemma 2.6. Therefore, there must exist some $x \in I$ such that $f(x) = M$. This proves f achieves its maximum. The argument for the minimum is similar. Alternatively, you could consider the function $h(x) = M - f(x)$. Then use what was just proved to conclude h achieves its maximum at some point x_1. Thus $h(x_1) \geq h(x)$ for all $x \in I$ and so $M - f(x_1) \geq M - f(x)$ for all $x \in I$ which implies $f(x_1) \leq f(x)$ for all $x \in I$. ∎

2.15.4 *Sequential Compactness Of Closed Intervals*

Theorem 2.26. *Every closed interval $[a,b]$ is sequentially compact.*

Proof: Let $\{x_n\} \subseteq [a,b] \equiv I_0$. Consider the two intervals $\left[a, \frac{a+b}{2}\right]$ and $\left[\frac{a+b}{2}, b\right]$, each of which has length $(b-a)/2$. At least one of these intervals contains x_n for infinitely many values of n. Call this interval I_1. Now do for I_1 what was done for I_0. Split it in half and let I_2 be an interval which contains x_n for infinitely many values of n. Continue this way, obtaining a sequence of nested intervals $I_0 \supseteq I_1 \supseteq I_2 \supseteq I_3 \cdots$, where the length of I_n is $(b-a)/2^n$. Now pick n_1 such that $x_{n_1} \in I_1$, n_2 such that $n_2 > n_1$ and $x_{n_2} \in I_2$, n_3 such that $n_3 > n_2$ and $x_{n_3} \in I_3$, etc. (This can be done because in each case the intervals contained x_n for infinitely many values of n.) By the nested interval lemma there exists a point c contained in all these intervals. Furthermore, $|x_{n_k} - c| < (b-a) 2^{-k}$, and so $\lim_{k \to \infty} x_{n_k} = c \in [a,b]$. ∎

2.15.5 *Different Versions Of Completeness*

If you want, you can use any of the following criteria in the next theorem as a definition of completeness. This is what the theorem says.

Theorem 2.27. *The following are equivalent to completeness of $\mathbb{R}$.*

(1) Every Cauchy sequence converges.
(2) Every increasing sequence which is bounded above converges.
(3) Every decreasing sequence which is bounded below converges.

Proof: Suppose 1 and suppose S is a nonempty set which is bounded above. Why does there exist a least upper bound? Consider for each n all points of the form $k2^{-n}$ where k is an integer. Thus the union of the half open intervals $[(k-1)2^{-n}, k2^{-n})$ includes all of $\mathbb{R}$. In particular, one of these $k2^{-n}$ must be an upper bound of S. Let k_n be the smallest integer at which this occurs. Let $b_n = k_n 2^{-n}$. Thus b_n is an upper bound but $b_n - 2^{-n}$ is not. Also, it follows from the construction and this observation, that $b_n - 2^{-n} \leq b_{n+1} \leq b_n$ and so $|b_n - b_{n+1}| \leq 2^{-n}$ and $\{b_n\}$ is a decreasing sequence. Therefore, if $m > n$,

$$|b_m - b_n| \leq \sum_{k=n+1}^{m} |b_k - b_{k-1}| \leq \sum_{k=n+1}^{m} 2^{-(k-1)} \leq 2^{-(n-1)}$$

See the exercises, Problem 26, for the last step. Therefore, $\{b_n\}$ is a Cauchy sequence converging to some $b \in \mathbb{R}$. Each b_n is an upper bound for S and so it follows from Theorem 2.13 that b is also an upper bound for S. If $c \leq b$ is an upper bound it follows from the construction, that there exists $x \in S$ such that $b_n - 2^{-n} < x \leq c \leq b \leq b_n$. Hence $|c - b| \leq 2^{-n}$ for every n and so b is the least upper bound.

Why does completeness of $\mathbb{R}$ imply that every Cauchy sequence converges? By Theorem 2.14, there exists an interval $[a, b]$ containing the values of the Cauchy sequence. Now from Theorem 2.26 there exists a convergent subsequence. By Theorem 2.15, the Cauchy sequence converges. This proves the equivalence of 1 with completeness.

By Theorem 2.17 completeness implies increasing sequences which are bounded above converge, and decreasing sequences which are bounded below also converge. Furthermore, by considering $\{-a_n\}$ it follows that the last two conditions above are equivalent. Suppose now that every increasing (decreasing) sequence which is bounded above (below) must converge. Why does this imply completeness? Let $\{a_n\}$ be a Cauchy sequence and let

$$A_n \equiv \sup\{a_k : k \geq n\}, \ B_n \equiv \inf\{a_k : k \geq n\}$$

Thus by Theorem 2.14, the values of the sequence are contained in some interval $[a, b]$. Hence A_n is bounded below by a and decreasing while B_n is bounded above by b and increasing. Since $\{a_n\}$ is a Cauchy sequence, it follows that $\lim_{n\to\infty}(A_n - B_n) = 0$, and by the assumption that $\{A_n\}$ converges, this implies each of $\{A_n\}$ and $\{B_n\}$ converge to some x. However, $A_n \geq a_n \geq B_n$ and so, by the squeezing theorem (Theorem 2.11), $\lim_{n\to\infty} a_n = x$ also. Thus every Cauchy sequence converges. ∎

2.16 Exercises

(1) If X_i are sets and for some j, $X_j = \emptyset$, the empty set. Verify carefully that $\prod_{i=1}^{n} X_i = \emptyset$.

(2) Suppose $f(x) + f\left(\frac{1}{x}\right) = 7x$ and f is a function defined on $\mathbb{R}\setminus\{0\}$, the nonzero real numbers. Does there exist any such function?

(3) Does there exist a function f, satisfying $f(x) - f\left(\frac{1}{x}\right) = 3x$ which has both x and $\frac{1}{x}$ in the domain of f?

(4) In the situation of the Fibonacci sequence show that the formula for the nth term can be found and is given by

$$a_n = \frac{\sqrt{5}}{5}\left(\frac{1+\sqrt{5}}{2}\right)^n - \frac{\sqrt{5}}{5}\left(\frac{1-\sqrt{5}}{2}\right)^n.$$

Hint: You might be able to do this by induction but a better way would be to look for a solution to the recurrence relation, $a_{n+2} \equiv a_n + a_{n+1}$ of the form r^n. You will be able to show that there are two values of r which work, one of which is $r - \frac{1+\sqrt{5}}{2}$. Next you can observe that if r_1^n and r_2^n both satisfy the recurrence relation then so does $cr_1^n + dr_2^n$ for any choice of constants c, d. Then you try to pick c and d such that the conditions, $a_1 = 1$ and $a_2 = 1$ both hold.

(5) In an ordinary annuity, you make constant payments P at the beginning of each payment period. These accrue interest at the rate of r per payment period. This means at the start of the first payment period, there is the payment $P \equiv A_1$. Then this produces an amount rP in interest so at the beginning of the second payment period, you would have $rP + P + P = A_2$. Thus $A_2 = A_1(1+r) + P$. Then at the beginning of the third payment period you would have $A_2(1+r) + P = A_3$. Continuing in this way, you see that the amount in the bank at the beginning of the nth payment period would be A_n given by $A_n = A_{n-1}(1+r) + P$ and $A_1 = P$. Thus A is a function defined on the positive integers given recursively as just described and A_n is the amount at the beginning of the nth payment period. Now if you wanted to find out A_n for large n, how would you do it? One way would be to use the recurrence relation n times. A better way would be to find a formula for A_n. Look for one in the form $A_n = Cz^n + s$ where C, z and s are to be determined. Show that $C = \frac{P}{r}$, $z = (1+r)$, and $s = -\frac{P}{r}$.

(6) A well known puzzle consists of three pegs and several disks each of a different diameter, each having a hole in the center which allows it to be slid down each of the pegs. These disks are piled one on top of the other on one of the pegs, in order of decreasing diameter, the larger disks always being below the smaller disks. The problem is to move the whole pile of disks to another peg such that you never place a disk on a smaller disk. If you have n disks, how many moves will it take? Of course this depends on n. If $n = 1$, you can do it in one move. If $n = 2$, you would need 3. Let A_n be the number required for n disks. Then in solving the puzzle, you must first obtain the top $n-1$ disks arranged in order

on another peg before you can move the bottom disk of the original pile. This takes A_{n-1} moves. Explain why $A_n = 2A_{n-1} + 1$, $A_1 = 1$ and give a formula for A_n. Look for one in the form $A_n = Cr^n + s$. This puzzle is called the tower of Hanoi. When you have found a formula for A_n, explain why it is not possible to do this puzzle if n is very large.

(7) Suppose f is a function defined on $\mathbb{R}$ and f is continuous at 0. Suppose also that $f(x + y) = f(x) + f(y)$. Show that if this is so, then f must be continuous at every value of $x \in \mathbb{R}$. Next show that for every rational number r, $f(r) = rf(1)$. Finally explain why $f(r) = rf(1)$ for every r a real number. **Hint:** To do this last part, you need to use the density of the rational numbers and continuity of f.

(8) Suppose f is a continuous function defined on $[0, 1]$ which maps $[0, 1]$ into $[0, 1]$. Show there exists $x \in [0, 1]$ such that $x = f(x)$. **Hint:** Consider $h(x) \equiv x - f(x)$ and the intermediate value theorem.

(9) Using the binomial theorem, Problem 2 on Page 13, prove that for all $n \in \mathbb{N}$, $\left(1 + \frac{1}{n}\right)^n \leq \left(1 + \frac{1}{n+1}\right)^{n+1}$. **Hint:** Show first that $\binom{n}{k} = \frac{n \cdot (n-1) \cdots (n-k+1)}{k!}$. By the binomial theorem,

$$\left(1 + \frac{1}{n}\right)^n = \sum_{k=0}^{n} \binom{n}{k} \left(\frac{1}{n}\right)^k = \sum_{k=0}^{n} \frac{\overbrace{n \cdot (n-1) \cdots (n-k+1)}^{k \text{ factors}}}{k! n^k}.$$

Now consider the term $\frac{n \cdot (n-1) \cdots (n-k+1)}{k! n^k}$ and note that a similar term occurs in the binomial expansion for $\left(1 + \frac{1}{n+1}\right)^{n+1}$ except you replace n with $n+1$ wherever this occurs. Argue the term got bigger and then note that in the binomial expansion for $\left(1 + \frac{1}{n+1}\right)^{n+1}$, there are more terms.

(10) Prove by induction that for all $k \geq 4$, $2^k \leq k!$

(11) First show that $2^k < k!$ for all $k > 4$. Now verify for all $n \in \mathbb{N}$, $\left(1 + \frac{1}{n}\right)^n \leq 3$.

(12) Prove $\lim_{n\to\infty} \left(1 + \frac{1}{n}\right)^n$ exists and equals a number less than 3.

(13) Using Problem 11, prove $n^{n+1} \geq (n+1)^n$ for all integers, $n \geq 3$.

(14) Let $A_n = \sum_{k=2}^{n} \frac{1}{k(k-1)}$ for $n \geq 2$. Show $\lim_{n\to\infty} A_n$ exists. **Hint:** Show there exists an upper bound to the A_n as follows.

$$\sum_{k=2}^{n} \frac{1}{k(k-1)} = \sum_{k=2}^{n} \left(\frac{1}{k-1} - \frac{1}{k}\right) = \frac{1}{2} - \frac{1}{n-1} \leq \frac{1}{2}$$

(15) Let $H_n = \sum_{k=1}^{n} \frac{1}{k^2}$ for $n \geq 2$. Show $\lim_{n\to\infty} H_n$ exists. **Hint:** Use the above problem to obtain the existence of an upper bound.

(16) Let a be a positive number and let $x_1 = b > 0$ where $b^2 > a$. Explain why there exists such a number b. Now having defined x_n, define $x_{n+1} \equiv \frac{1}{2}\left(x_n + \frac{a}{x_n}\right)$. Verify that $\{x_n\}$ is a decreasing sequence and that it satisfies $x_n^2 \geq a$ for all n and is therefore, bounded below. Explain why $\lim_{n\to\infty} x_n$ exists. If x is this limit, show that $x^2 = a$. Explain how this shows that every positive real number

has a square root. This is an example of a recursively defined sequence. Note this does not give a formula for x_n, just a rule which tells how to define x_{n+1} if x_n is known.

(17) Let $a_1 = 0$ and suppose that $a_{n+1} = \frac{9}{9-a_n}$. Write a_2, a_3, a_4. Now prove that for all n, it follows that $a_n \leq \frac{9}{2} - \frac{3}{2}\sqrt{5}$ (By Problem 7 on Page 56 there is no problem with the existence of various roots of positive numbers.) and so the sequence is bounded above. Next show that the sequence is increasing and so it converges. Find the limit of the sequence. **Hint:** You should prove these things by induction. Finally, to find the limit, let $n \to \infty$ in both sides and argue that the limit a, must satisfy $a = \frac{9}{9-a}$.

(18) If $x \in \mathbb{R}$, show there exists a sequence of rational numbers $\{x_n\}$ such that $x_n \to x$ and a sequence of irrational numbers $\{x'_n\}$ such that $x'_n \to x$. Now consider the following function.

$$f(x) = \begin{cases} 1 \text{ if } x \text{ is rational} \\ 0 \text{ if } x \text{ is irrational} \end{cases}$$

Show using the sequential version of continuity in Theorem 2.18 that f is discontinuous at every point.

(19) If $x \in \mathbb{R}$, show there exists a sequence of rational numbers $\{x_n\}$ such that $x_n \to x$ and a sequence of irrational numbers $\{x'_n\}$ such that $x'_n \to x$. Now consider the following function.

$$f(x) = \begin{cases} x \text{ if } x \text{ is rational} \\ 0 \text{ if } x \text{ is irrational} \end{cases}$$

Show using the sequential version of continuity in Theorem 2.18 that f is continuous at 0 and nowhere else.

(20) The nested interval lemma, Lemma 2.2 and Theorem 2.18 can be used to give an easy proof of the intermediate value theorem. Suppose $f(a) > 0$ and $f(b) < 0$ for f a continuous function defined on $[a, b]$. The intermediate value theorem states that under these conditions, there exists $x \in (a, b)$ such that $f(x) = 0$. Prove this theorem as follows: Let $c = \frac{a+b}{2}$ and consider the intervals $[a, c]$ and $[c, b]$. Show that on one of these intervals, f is nonnegative at one end and nonpositive at the other. Now consider that interval, divide it in half as was done for the original interval and argue that on one of these smaller intervals, the function has different signs at the two endpoints. Continue in this way. Next apply the nested interval lemma to get x in all these intervals and argue there exist sequences, $x_n \to x$ and $y_n \to x$ such that $f(x_n) < 0$ and $f(y_n) > 0$. By continuity, you can assume $f(x_n) \to f(x)$ and $f(y_n) \to f(x)$. Show this requires that $f(x) = 0$. See Problem 16 on Page 70.

(21) If $\lim_{n \to \infty} a_n = a$, does it follow that $\lim_{n \to \infty} |a_n| = |a|$? Prove or else give a counterexample.

(22) Show the following converge to 0.

(a) $\frac{n^5}{1.01^n}$

(b) $\frac{10^n}{n!}$

(23) * Suppose $\lim_{n\to\infty} x_n = x$. Show that then $\lim_{n\to\infty} \frac{1}{n}\sum_{k=1}^{n} x_k = x$. Give an example where $\lim_{n\to\infty} x_n$ does not exist but $\lim_{n\to\infty} \frac{1}{n}\sum_{k=1}^{n} x_k$ does.

(24) Prove $\lim_{n\to\infty} \sqrt[n]{n} = 1$. **Hint:** Let $e_n \equiv \sqrt[n]{n} - 1$ so that $(1 + e_n)^n = n$. Now observe that $e_n > 0$ and use the binomial theorem or Example 1.2 to conclude $1 + ne_n + \frac{n(n-1)}{2}e_n^2 \le n$. This nice approach to establishing this limit using only elementary algebra is in Rudin [23].

(25) Find $\lim_{n\to\infty} (x^n + 5)^{1/n}$ for $x \ge 0$. There are two cases here, $x < 1$ and $x \ge 1$. Show that if $x \ge 1$, the limit is x while if $x < 1$ the limit equals 1. **Hint:** Use the argument of Problem 24. This interesting example is in [8].

(26) Show that if $n < m$ are positive integers and $|r| < 1$, then

$$\sum_{k=n+1}^{m} r^{(k-1)} \le \frac{r^n}{1-r}$$

Hint: Let $S_m = \sum_{k=n+1}^{m} r^{(k-1)}$ and simplify $S_m - rS_m$ to obtain that this equals $r^n - r^m$. Now solve for S_m.

Chapter 3

Derivatives

3.1 Velocity

Imagine an object which is moving along the real line in the positive direction and that at time $t > 0$, the position of the object is $r(t) = -10 + 30t + t^2$ where distance is measured in kilometers and t in hours. Thus at $t - 0$, the object is at the point -10 kilometers and when $t = 1$, the object is at 21 kilometers. The average velocity during this time is the distance traveled divided by the elapsed time. Thus the average velocity would be $\frac{21-(-10)}{1} = 31$ kilometers per hour. It came out positive because the object moved in the positive direction along the real line, from -10 to 21. Suppose it was desired to find something which deserves to be referred to as the instantaneous velocity when $t - 1/2$? If the object were a car, it is reasonable to suppose that the magnitude of the average velocity of the object over a very small interval of time would be very close to the number that would appear on the speedometer. For example, if considering the average velocity of the object on the interval $[.5, .5 + .0001]$, this average velocity would be pretty close to the thing which deserves to be called the instantaneous velocity at $t = .5$ hours. Thus the velocity at $t = .5$ would be close to

$$(r(.5 + .0001) - r(.5))/.0001$$

$$= \left(30(.5 + .01) + (.5 + .0001)^2 - 30(.5) - (.5)^2\right)/.0001 = 31.0001$$

Of course, you would expect to be even closer using a time interval of length .000001 instead of just .0001. In general, consider a time interval of length h and then define the instantaneous velocity to be the number which all these average velocities get close to as h gets smaller and smaller. Thus in this case form the average velocity on the interval $[.5, .5 + h]$ to get

$$\left(30(.5 + h) + (.5 + h)^2 - \left(30(.5) + (.5)^2\right)\right)/h = 30 + 2(.5) + h.$$

What number does this average get close to as h gets smaller and smaller? Clearly it gets close to 31 and for this reason, the velocity at time .5 is defined as 31. It is positive because the object is moving in the positive direction. If the object were moving in the negative direction, the number would be negative. The notion just described of finding an instantaneous velocity has a geometrical application to finding the slope of a line tangent to a curve.

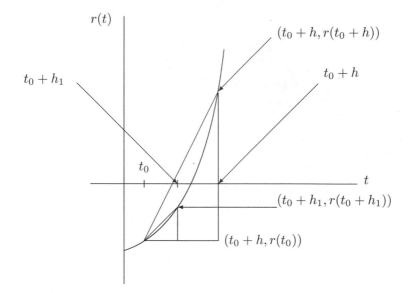

In the above picture, you see the slope of the line joining the two points $(t_0, r(t_0))$ and $(t_0 + h, r(t_0 + h))$ is given by

$$\frac{r(t_0 + h) - r(t_0)}{h}$$

which equals the average velocity on the time interval $[t_0, t_0 + h]$. You can also see the effect of making h closer and closer to zero as illustrated by changing h to the smaller h_1 in the picture. The slope of the resulting line segment appears to get closer and closer to what ought to be considered the slope of the line tangent to the curve at the point $(t_0, r(t_0))$.

It is time to make this heuristic material much more precise.

3.2   The Derivative

The derivative of a function of one variable is a function given by the following definition.

Definition 3.1. The derivative of a function $f'(x)$, is defined as the following limit whenever the limit exists. If the limit does not exist, then neither does $f'(x)$.

$$\lim_{h \to 0} \frac{f(x + h) - f(x)}{h} \equiv f'(x) \tag{3.1}$$

The function of h on the left is called the difference quotient.

Note that the difference quotient on the left of the equation is a function of h which is not defined at $h = 0$. This is why, in the definition of limit, $|h| > 0$. **It is not necessary to have the function defined at the point in order to consider its limit.** The distinction between the limit of a function and its value is very important and must be kept in mind. Also it is clear from setting $y = x + h$ that

$$f'(x) = \lim_{y \to x} \frac{f(y) - f(x)}{y - x}. \tag{3.2}$$

Theorem 3.1. *If $f'(x)$ exists, then f is continuous at x.*

Proof: Suppose $\varepsilon > 0$ is given and choose $\delta_1 > 0$ such that if $|h| < \delta_1$,

$$\left| \frac{f(x+h) - f(x)}{h} - f'(x) \right| < 1,$$

then for such h, the triangle inequality implies

$$|f(x+h) - f(x)| < |h| + |f'(x)| |h|.$$

Now letting $\delta < \min\left(\delta_1, \frac{\varepsilon}{1+|f'(x)|}\right)$, it follows if $|h| < \delta$, then

$$|f(x+h) - f(x)| < \varepsilon.$$

Letting $y - h + x$, this shows that if $|y - x| < \delta$,

$$|f(y) - f(x)| < \varepsilon$$

which proves f is continuous at x. ∎

It is very important to remember that just because f is continuous, does not mean f has a derivative. The following picture describes the situation.

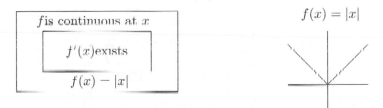

As indicated in the above picture the function $f(x) = |x|$ does not have a derivative at $x = 0$. To see this,

$$\lim_{h \to 0+} \frac{f(h) - f(0)}{h} = \lim_{h \to 0+} \frac{h}{h} = 1$$

while

$$\lim_{h \to 0-} \frac{f(h) - f(0)}{h} = \lim_{h \to 0-} \frac{-h}{h} = -1.$$

Thus the two limits, one from the right and one from the left do not agree as they would have to do if the function had a derivative at $x = 0$. See Problem 18 on Page 62. Geometrically, this lack of differentiability is manifested by the pointy place in the graph of $y = |x|$ at $x = 0$. In short, the function fails to have a derivative at values of x which correspond to pointy places.

Example 3.1. Let $f(x) = c$ where c is a constant. Find $f'(x)$.

Set up the difference quotient,

$$\frac{f(x+h) - f(x)}{h} = \frac{c - c}{h} = 0.$$

Therefore,

$$\lim_{h \to 0} \frac{f(x+h) - f(x)}{h} = \lim_{h \to 0} 0 = 0.$$

Example 3.2. Let $f(x) = cx$ where c is a constant. Find $f'(x)$.

Set up the difference quotient,

$$\frac{f(x+h) - f(x)}{h} = \frac{c(x+h) - cx}{h} = \frac{ch}{h} = c.$$

Therefore,

$$\lim_{h \to 0} \frac{f(x+h) - f(x)}{h} = \lim_{h \to 0} c = c.$$

Example 3.3. Let $f(x) = \sqrt{x}$ for $x > 0$. Find $f'(x)$.

Set up the difference quotient,

$$\frac{f(x+h) - f(x)}{h} = \frac{\sqrt{x+h} - \sqrt{x}}{h} = \frac{x+h-x}{h\left(\sqrt{x+h} + \sqrt{x}\right)}$$

$$= \frac{1}{\sqrt{x+h} + \sqrt{x}}$$

and so

$$\lim_{h \to 0} \frac{f(x+h) - f(x)}{h} = \lim_{h \to 0} \frac{1}{\sqrt{x+h} + \sqrt{x}} = \frac{1}{2\sqrt{x}}.$$

There are rules of derivatives which make finding the derivative very easy.

Theorem 3.2. *Let $a, b \in \mathbb{R}$ and suppose $f'(t)$ and $g'(t)$ exist. Then the following formulas are obtained.*

$$(af + bg)'(t) = af'(t) + bg'(t). \tag{3.3}$$

$$(fg)'(t) = f'(t)g(t) + f(t)g'(t). \tag{3.4}$$

The formula (3.4) is referred to as the product rule.
    If $g(t) \neq 0$,

$$\left(\frac{f}{g}\right)'(t) = \frac{f'(t)g(t) - g'(t)f(t)}{g^2(t)}. \tag{3.5}$$

Formula (3.5) is referred to as the quotient rule.
    If f is differentiable at ct where $c \neq 0$, Then letting $g(t) \equiv f(ct)$,

$$g'(t) = cf'(ct). \tag{3.6}$$

Written with a slight abuse of notation,

$$(f(ct))' = cf'(ct). \tag{3.7}$$

If f is differentiable on (a, b) and if $g(t) \equiv f(t + c)$, then g is differentiable on $(a - c, b - c)$ and

$$g'(t) = f'(t + c). \tag{3.8}$$

Written with a slight abuse of notation,

$$(f(t + c))' = f'(t + c) \tag{3.9}$$

For p an integer and $f'(t)$ exists, let $g_p(t) \equiv f(t)^p$. Then

$$(g_p)'(t) = pf(t)^{p-1} f'(t). \tag{3.10}$$

(In the case where $p < 0$, assume $f(t) \neq 0$.)
Written with a slight abuse of notation, an easy to remember version of 3.10 says

$$(f(t)^p)' = pf(t)^{p-1} f'(t).$$

Proof: The first formula is left for you to prove. Consider the second, (3.4).

$$\frac{fg(t + h) - fg(t)}{h} = \frac{f(t + h)g(t + h) - f(t + h)g(t)}{h} + \frac{f(t + h)g(t) - f(t)g(t)}{h}$$

$$= f(t + h)\frac{(g(t + h) - g(t))}{h} + \frac{(f(t + h) - f(t))}{h}g(t)$$

Taking the limit as $h \to 0$ and using Theorem 3.1 to conclude $\lim_{h \to 0} f(t + h) = f(t)$, it follows from Theorem 2.6 that (3.4) follows. Next consider the quotient rule.

$$h^{-1}\left(\frac{f}{g}(t + h) - \frac{f}{g}(t)\right) = \frac{f(t + h)g(t) - g(t + h)f(t)}{hg(t)g(t + h)}$$

$$= \frac{f(t + h)(g(t) - g(t + h)) + g(t + h)(f(t + h) - f(t))}{hg(t)g(t + h)}$$

$$= \frac{-f(t + h)}{g(t)g(t + h)}\frac{(g(t + h) - g(t))}{h} + \frac{g(t + h)}{g(t)g(t + h)}\frac{(f(t + h) - f(t))}{h}$$

and from Theorem 2.6 on Page 59,

$$\left(\frac{f}{g}\right)'(t) = \frac{g(t)f'(t) - g'(t)f(t)}{g^2(t)}.$$

Now consider Formula (3.6).

$$(g(t + h) - g(t))h^{-1} = h^{-1}(f(ct + ch) - f(ct))$$

$$= c\frac{f(ct + ch) - f(ct)}{ch}$$

$$= c\frac{f(ct + h_1) - f(ct)}{h_1}$$

where $h_1 = ch$. Then $h_1 \to 0$ if and only if $h \to 0$ and so taking the limit as $h \to 0$ yields

$$g'(t) = cf'(ct)$$

as claimed. Formulas (3.7) and (3.8) are left as exercises.

First consider (3.10) in the case where p equals a nonnegative integer. If $p = 0$, (3.10) holds because $g_0(t) = 1$ and so by Example 3.1,

$$g_0'(t) = 0 = 0 \left(f(t)\right)^{-1} f'(t).$$

Next suppose (3.10) holds for p an integer. Then

$$(g_{p+1}(t)) = f(t) g_p(t),$$

and so by the product rule,

$$g_{p+1}'(t) = f'(t) g_p(t) + f(t) g_p'(t)$$
$$= f'(t) (f(t))^p + f(t) \left(pf(t)^{p-1} f'(t)\right)$$
$$= (p+1) f(t)^p f'(t).$$

If the formula holds for some integer p then it holds for $-p$. Here is why.

$$g_{-p}(t) = g_p(t)^{-1}$$

and so

$$\frac{g_{-p}(t+h) - g_{-p}(t)}{h} = \left(\frac{g_p(t) - g_p(t+h)}{h}\right)\left(\frac{1}{g_p(t) g_p(t+h)}\right).$$

Taking the limit as $h \to 0$ and using the formula for p,

$$g_{-p}'(t) = -pf(t)^{p-1} f'(t) (f(t))^{-2p} = -p(f(t))^{-p-1} f'(t).$$

$\blacksquare$

Example 3.4. Let $p(x) = 3 + 5x + 6x^2 - 7x^3$. Find $p'(x)$.

From the above theorem, and abusing the notation,

$$p'(x) = \left(3 + 5x + 6x^2 - 7x^3\right)'$$
$$= 3' + (5x)' + \left(6x^2\right)' + \left(-7x^3\right)'$$
$$= 0 + 5 + (6)(2)(x)(x)' + (-7)(3)\left(x^2\right)(x)'$$
$$= 5 + 12x - 21x^2.$$

Note the process is to take the exponent and multiply by the coefficient and then make the new exponent one less in each term of the polynomial in order to arrive at the answer. This is the general procedure for differentiating a polynomial as shown in the next example.

Example 3.5. Let a_k be a number for $k = 0, 1, \cdots, n$ and let $p(x) = \sum_{k=0}^{n} a_k x^k$. Find $p'(x)$.

Use Theorem 3.2

$$p'(x) = \left(\sum_{k=0}^{n} a_k x^k \right)' = \sum_{k=0}^{n} a_k \left(x^k \right)' = \sum_{k=0}^{n} a_k k x^{k-1} \left(x \right)' = \sum_{k=0}^{n} a_k k x^{k-1}$$

Example 3.6. Find the derivative of the function $f(x) = \frac{x^2+1}{x^3}$.

Use the quotient rule

$$f'(x) = \frac{2x \left(x^3 \right) - 3x^2 \left(x^2 + 1 \right)}{x^6} = -\frac{1}{x^4} \left(x^2 + 3 \right)$$

Example 3.7. Let $f(x) = \left(x^2 + 1 \right)^4 \left(x^3 \right)$. Find $f'(x)$.

Use the product rule and (3.10). Abusing the notation for the sake of convenience,

$$\left(\left(x^2 + 1 \right)^4 \left(x^3 \right) \right)' = \left(\left(x^2 + 1 \right)^4 \right)' \left(x^3 \right) + \left(x^3 \right)' \left(\left(x^2 + 1 \right)^4 \right)$$

$$= 4 \left(x^2 + 1 \right)^3 (2x) \left(x^3 \right) + 3x^2 \left(x^2 + 1 \right)^4$$

$$= 4x^4 \left(x^2 + 1 \right)^3 + 3x^2 \left(x^2 + 1 \right)^4$$

Example 3.8. Let $f(x) = x^3 / \left(x^2 + 1 \right)^2$. Find $f'(x)$.

Use the quotient rule to obtain

$$f'(x) = \frac{3x^2 \left(x^2 + 1 \right)^2 - 2 \left(x^2 + 1 \right) (2x) x^3}{\left(x^2 + 1 \right)^4} = \frac{3x^2 \left(x^2 + 1 \right)^2 - 4x^4 \left(x^2 + 1 \right)}{\left(x^2 + 1 \right)^4}$$

Obviously, one could consider taking the derivative of the derivative and then the derivative of that and so forth. The main thing to consider about this is the notation. The second derivative is denoted with two primes, the third derivative by three and so forth.

Example 3.9. Let $f(x) = x^3 + 2x^2 + 1$. Find $f''(x)$ and $f'''(x)$.

To find $f''(x)$ take the derivative of the derivative. Thus $f'(x) = 3x^2 + 4x$ and so $f''(x) = 6x + 4$. Then $f'''(x) = 6$.

When high derivatives are taken, say the 5th derivative, it is customary to write $f^{(5)}(t)$ putting the number of derivatives in parentheses.

3.3 Exercises

(1) From the derivation of the derivative, you observe that when the derivative is positive, this corresponds to the slope of the tangent line being positive and suggests that the function is increasing near the point of interest. This will be made more precise later but suggest that intervals on which a function is

increasing may be identified as those on which the derivative is positive. The
following are some functions. Find from examining the derivative, intervals on
which the function is increasing.

(a) $f(x) = \frac{x^3}{3} - \frac{3}{2}x^2 + 2x$

(b) $f(x) = \frac{x^3}{3} - x^2 - 3x$

(c) $f(x) = \frac{1}{5}x^5 - \frac{5}{2}x^4 + \frac{35}{3}x^3 - 25x^2 + 24x$

(2) Find derivatives of the following functions.

(a) $3x^2 + 4x^3 - 7x + 11$

(b) $\frac{1}{x^3+1}$

(c) $\left(x^4 + x + 5\right)\left(x^5 + 7x\right)$

(d) $\frac{x^3+x}{x^2+1}$

(e) $\left(x^2 + 2\right)^4$

(f) $\left(x^3 + 2x + 1\right)^{-1}$

(3) For $f(x) = -3x^7 + 4x^5 + 2x^3 + x^2 - 5x$, find $f^{(3)}(x)$.

(4) For $f(x) = -x^7 + x^5 + x^3 - 2x$, find $f^{(3)}(x)$.

(5) Find $f'(x)$ for the given functions

(a) $f(x) = \frac{3x^3-x-1}{3x^2+1}$

(b) $f(x) = \frac{3x^3+3x-1}{x^2+1}$

(c) $f(x) = \frac{\left(x^2+5\right)^4}{\left(x^2+1\right)^2}$

(d) $f(x) = \left(4x + 7x^3\right)^6$

(e) $f(x) = \left(-2x^3 + x + 3\right)^3 \left(-2x^2 + 3x - 1\right)$

(6) For $f(x) = \sqrt{4x^2 + 1}$, find $f'(x)$ from the definition of the derivative.

(7) For $f(x) = \sqrt[3]{3x^2 + 1}$, find $f'(x)$ from the definition of the derivative. **Hint:**
You might use

$$b^3 - a^3 = \left(b^2 + ab + a^2\right)(b - a) \text{ for } b = \sqrt[3]{3(x+h)^2 + 1} \text{ and } a = \sqrt[3]{3x^2 + 1}.$$

(8) For $f(x) = (-3x + 5)^5$, find $f'(x)$ from the definition of the derivative. **Hint:**
You might use the formula

$$b^n - a^n = (b - a)\left(b^{n-1} + b^{n-2}a + \cdots + a^{n-2}b + b^{n-1}\right)$$

(9) Let

$$f(x) = (x + 2)^2 \sin\left(1/(x + 2)\right) + 6(x - 5)(x + 2)$$

for $x \neq -2$ and define $f(-2) \equiv 0$. Find $f'(-2)$ from the definition of the
derivative if this is possible. **Hint:** Note that $|\sin(z)| \leq 1$ for any real value
of z.

(10) Let $f(x) = (x + 5)\sin\left(1/(x + 5)\right)$ for $x \neq -5$ and define $f(-5) \equiv 0$. Show
$f'(-5)$ does not exist. **Hint:** Verify that $\lim_{h \to 0} \sin(1/h)$ does not exist and
then explain why this shows $f'(-5)$ does not exist.

(11) Suppose f is a continuous function and $y = mx + b$ is the equation of a straight line with slope m which intersects the graph of f only at the point $(x, f(x))$. Does it follow that $f'(x) = m$?

(12) Give an example of a function f which is not continuous but for which

$$\lim_{n \to \infty} \frac{f(x + (1/n)) - f(x - (1/n))}{2(1/n)}$$

exists for every x.

(13) Suppose f is defined on (a, b) and $f'(x)$ exists for some $x \in (a, b)$. Show $f'(x) = \lim_{h \to 0} \frac{f(x+h) - f(x-h)}{2h}$.

(14) Let $f(x) = x^2$ if x is rational and $f(x) = 0$ if x is irrational. Show that $f'(0) = 0$ but that f fails to be continuous at every other point.

(15) Suppose $|f(x)| \le x^2$ for all $x \in (-1, 1)$. Let $h(x) = x + f(x)$. Find $h'(0)$.

(16) Find a formula for the derivative of a product of three functions. Generalize to a product of n functions where n is a positive integer.

(17) Suppose f is a function defined on $\mathbb{R}$ and it satisfies the functional equation $f(a + b) = f(a) + f(b)$. Suppose also that $f'(0) = k$. Find $f'(x)$, $f(x)$.

(18) Suppose f is a function defined on $\mathbb{R}$ and it satisfies the functional equation $f(a + b) = f(a) + f(b) + ab$. Suppose also that $\lim_{h \to 0} \frac{f(h)}{h} = 7$. Find $f'(x)$, $f(x)$.

(19) From Corollary 1.1 on Page 9 the following inequality holds.

$$\sin x \mid (1 - \cos x) \ge x > \sin x. \tag{3.11}$$

Using this inequality, show

$$\lim_{x \to 0} \frac{\sin x}{x} - 1.$$

Hint: For $x > 0$, divide both sides of the inequality by $\sin x$. This yields

$$1 + \frac{1 - \cos x}{x} \ge \frac{x}{\sin x} \ge 1.$$

Now

$$0 \le \frac{1 - \cos x}{x} = \frac{1 - \cos^2 x}{x(1 + \cos x)} = \frac{\sin^2 x}{x(1 + \cos x)} \le \frac{\sin x}{(1 + \cos x)}. \tag{3.12}$$

If $x < 0$, $\frac{\sin x}{x} = \frac{\sin(-x)}{(-x)}$ and $-x > 0$.

(20) Show that $\lim_{h \to 0} \left(\frac{1 - \cos(h)}{h} \right) = 0$. **Hint:** For $h > 0$, consider the inequality (3.12) with x replaced with h.

(21) Show using Problems 19 and 20 and the definition of the derivative that $\sin'(x) = \cos x$. Also show $\cos'(x) = -\sin(x)$. **Hint:** Just write down the difference quotient and use the limits in these problems.

$$\frac{\sin(x + h) - \sin x}{h} = \frac{\sin(x)(\cos(h) - 1)}{h} + \cos(x) \frac{\sin(h)}{h}, \text{ etc.}$$

(22) Now that you know $\sin'(x) = \cos(x)$, use the definition of derivative to find $f'(x)$ where $f(x) = \sin(3x)$. What about $g'(x)$ where $g(x) = \cos(3x)$? Now show that both $\cos(\omega t)$ and $\sin(\omega t)$ are solutions to the differential equation $y'' + \omega^2 y = 0$. This is the equation for undamped oscillations and it will be discussed more later.

(23) Find the equation of the tangent line to the graph of the function $f(x) = x^2 + 4x - 3$ at the point $(1, 2)$.

3.4   Local Extrema

When you are on top of a hill, you are at a local maximum although there may be other hills higher than the one on which you are standing. Similarly, when you are at the bottom of a valley, you are at a local minimum even though there may be other valleys deeper than the one you are in. The word, "local" is applied to the situation because if you confine your attention only to points close to your location, you are indeed at either the top or the bottom.

Definition 3.2. Let $f : D(f) \to \mathbb{R}$ where here $D(f)$ is only assumed to be some subset of $\mathbb{R}$. Then $x \in D(f)$ is a local minimum (maximum) if there exists $\delta > 0$ such that whenever $y \in (x - \delta, x + \delta) \cap D(f)$, it follows $f(y) \geq (\leq) f(x)$. The plural of minimum is minima and the plural of maximum is maxima.

Derivatives can be used to locate local maxima and local minima. The following picture suggests how to do this. This picture is of the graph of a function having a local maximum and the tangent line to it.

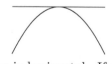

Note how the tangent line is horizontal. If you were not at a local maximum or local minimum, the function would be falling or climbing and the tangent line would not be horizontal.

Theorem 3.3. *Suppose $f : (a, b) \to \mathbb{R}$, and suppose $x \in (a, b)$ is a local maximum or minimum. Then $f'(x) = 0$.*

Proof: Suppose x is a local maximum. If $h > 0$ and is sufficiently small, then $f(x + h) \leq f(x)$ and so from Theorem 2.6 on Page 59,

$$f'(x) = \lim_{h \to 0+} \frac{f(x + h) - f(x)}{h} \leq 0.$$

Similarly,

$$f'(x) = \lim_{h \to 0-} \frac{f(x + h) - f(x)}{h} \geq 0.$$

The case when x is a local minimum is similar. ∎

Definition 3.3. Points where the derivative of a function equals zero are called critical points. It is also customary to refer to points where the derivative of a function does not exist as critical points.

Example 3.10. It is desired to find two positive numbers whose sum equals 16 and whose product is to be a large as possible.

The numbers are x and $16 - x$ and $f(x) = x(16 - x)$ is to be made as large as possible. The value of x which will do this would be a local maximum so by Theorem 3.3 the procedure is to take the derivative of f and find values of x where it equals zero. Thus $16 - 2x = 0$ and the only place this occurs is when $x = 8$. Therefore, the two numbers are 8 and 8.

Example 3.11. A farmer wants to fence a rectangular piece of land next to a straight river. What are the dimensions of the largest rectangle if there are exactly 600 meters of fencing available.

The two sides perpendicular to the river have length x and the third side has length $y - (600 - 2x)$. Thus the function to be maximized is $f(x) = 2x(600 - x) = 1200x - 2x^2$. Taking the derivative and setting it equal to zero gives

$$f'(x) = 1200 - 4x = 0$$

and so $x = 300$. Therefore, the desired dimensions are 300×600.

Example 3.12. A rectangular playground is to be enclosed by a fence and divided in 5 pieces by 4 fences parallel to one side of the playground. 1704 feet of fencing is used. Find dimensions of the playground which will have the largest total area.

Let x denote the length of one of these dividing fences and let y denote the length of the playground as shown in the following picture.

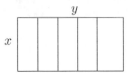

Thus $6x + 2y = 1704$ so $y = \frac{1704 - 6x}{2}$ and the function to maximize is

$$f(x) = x\left(\frac{1704 - 6x}{2}\right) = x(852 - 3x).$$

Therefore, to locate the value of x which will make $f(x)$ as large as possible, take $f'(x)$ and set it equal to zero.

$$852 - 6x = 0$$

and so $x = 142$ feet and $y = \frac{1704 - 6 \times 142}{2} = 426$ feet.

Revenue is defined to be the amount of money obtained in some transaction. Profit is defined as the revenue minus the costs.

Example 3.13. Sam, the owner of Spider Sam's Tarantulas and Creepy Critters finds he can sell 6 tarantulas every day at the regular price of $30 each. At his last spider celebration sale he reduced the price to $24 and was able to sell 12 tarantulas every day. He has to pay $.05 per day to exhibit a tarantula and his fixed costs are $30 per day, mainly to maintain the thousands of tarantulas he keeps on his tarantula breeding farm in the basement. What price should he charge to maximize his profit.

He assumes the demand for tarantulas is a linear function of price. Thus if y is the number of tarantulas demanded at price x, it follows $y = 36 - x$. Therefore, the revenue for price x equals $R(x) = (36 - x)x$. Now you have to subtract off the costs to get the profit. Thus

$$P(x) = (36 - x)x - (36 - x)(.05) - 30.$$

It follows the profit is maximized when $P'(x) = 0$ so $-2.0x + 36.05 = 0$ which occurs when $x = \$18.025$. Thus Sam should charge about $18 per tarantula.

Example 3.14. Lisa, the owner of Lisa's gags and gadgets sells 500 whoopee cushions per year. It costs $.25 per year to store a whoopee cushion. To order whoopee cushions it costs $8 plus $.90 per cushion. How many times a year and in what lot size should whoopee cushions be ordered to minimize inventory costs?

Let x be the times per year an order is sent for a lot size of $\frac{500}{x}$. If the demand is constant, it is reasonable to suppose there are about $\frac{500}{2x}$ whoopee cushions which have to be stored. Thus the cost to store whoopee cushions is $\frac{125.0}{2x} = .25\left(\frac{500}{2x}\right)$. Each time an order is made for a lot size of $\frac{500}{x}$ it costs $8 + \frac{450.0}{x} = 8 + .9\left(\frac{500}{x}\right)$ and this is done x times a year. Therefore, the total inventory cost is $x\left(8 + \frac{450.0}{x}\right) + \frac{125.0}{2x} = C(x)$. The problem is to minimize $C(x) = x\left(8 + \frac{450.0}{x}\right) + \frac{125.0}{2x}$. Taking the derivative yields

$$C'(x) = \frac{16x^2 - 125}{2x^2}$$

and so the value of x which will minimize $C(x)$ is $\frac{5}{4}\sqrt{5} = 2.79\cdots$ and the lot size is $\frac{500}{\frac{5}{4}\sqrt{5}} = 178..88$. Of course you would round these numbers off. Order 179 whoopee cushions 3 times a year.

3.5 Exercises

(1) If $f'(x) = 0$, is it necessary that x is either a local minimum or local maximum?
   Hint: Consider $f(x) = x^3$.

(2) Two positive numbers add to 32. Find the numbers if their product is to be as large as possible.

(3) The product of two positive numbers equals 16. Find the numbers if their sum is to be as small as possible.

(4) The product of two positive numbers equals 16. Find the numbers if twice the first plus three times the second is to be as small as possible.

(5) Emily, the owner of Slithery Serpent Emporium finds she can sell 6 boa constrictors at the regular price of $20 each. At the last serpent celebration sale she reduced the price to $14 and was able to sell 14. She has to pay $.05 per day to maintain a boa constrictor and her fixed costs are $30 per day. What price should she charge to maximize her profit.

(6) Eric, the owner of Shop of Dreadful Disguises, sells 500 clown masks per year. It costs $.25 per year to store a clown mask. To order clown masks, it costs $2 plus $.25 per mask. How many times a year and in what lot size should clown masks be ordered to minimize inventory costs?

(7) A continuous function f defined on $[a, b]$ is to be maximized. It was shown above in Theorem 3.3 that if the maximum value of f occurs at $x \in (a, b)$, and if f is differentiable there, then $f'(x) = 0$. However, this theorem does not say anything about the case where the maximum of f occurs at either a or b. Describe how to find the point of $[a, b]$ where f achieves its maximum. Does f have a maximum? Explain.

(8) Find the maximum and minimum values and the values of x where these are achieved for the function $f(x) = x + \sqrt{25 - x^2}$.

(9) A piece of wire of length L is to be cut in two pieces. One piece is bent into the shape of an equilateral triangle and the other piece is bent to form a square. How should the wire be cut to maximize the sum of the areas of the two shapes? How should the wire be bent to minimize the sum of the areas of the two shapes? **Hint:** Be sure to consider the case where all the wire is devoted to one of the shapes separately. This is a possible solution even though the derivative is not zero there.

(10) A cylindrical can is to be constructed of material which costs 3 cents per square inch for the top and bottom and only 2 cents per square inch for the sides. The can needs to hold 90π cubic inches. Find the dimensions of the cheapest can. **Hint:** The volume of a cylinder is $\pi r^2 h$ where r is the radius of the base and h is the height. The area of the cylinder is $2\pi r^2 + 2\pi r h$.

(11) A rectangular sheet of tin has dimensions 10 cm. by 20 cm. It is desired to make a topless box by cutting out squares from each corner of the rectangular sheet and then folding the rectangular tabs which remain. Find the volume of the largest box which can be made in this way.

(12) Let $f(x) = \frac{1}{3}x^3 - x^2 - 8x$ on the interval $[-1, 10]$. Find the point of $[-1, 10]$ at which f achieves its minimum.

(13) Lets find the point on the graph of $y = \frac{x^2}{4}$ which is closest to $(0, 1)$. One way to

do it is to observe that a typical point on the graph is of the form $\left(x, \frac{x^2}{4}\right)$ and then to minimize the function $f(x) = x^2 + \left(\frac{x^2}{4} - 1\right)^2$. Taking the derivative of f yields $x + \frac{1}{4}x^3$ and setting this equal to 0 leads to the solution, $x = 0$. Therefore, the point closest to $(0, 1)$ is $(0, 0)$. Now lets do it another way. Let us use $y = \frac{x^2}{4}$ to write $x^2 = 4y$. Now for (x, y) on the graph, it follows it is of the form $\left(\sqrt{4y}, y\right)$. Therefore, minimize $f(y) = 4y + (y - 1)^2$. Take the derivative to obtain $2 + 2y$ which requires $y = -1$. However, on this graph, y is never negative. What on earth is the problem?

(14) A rectangular garden 200 square feet in area is to be fenced off against rabbits. Find the least possible length of fencing if one side of the garden is already protected by a barn.

(15) A feed lot is to be enclosed by a fence and divided in 5 pieces by 4 fences parallel to one side. 1272 feet of fencing is used. Find dimensions of the feed lot which will have the largest total area.

(16) Find the dimensions of the largest rectangle that can be inscribed in a semi-circle of radius 8.

(17) Find the dimensions of the largest rectangle that can be inscribed in the ellipse $\frac{x^2}{9} + \frac{y^2}{4} = 1$.

(18) Find the largest volume for a cylinder inscribed in a ball of radius eight cm. The volume of a cylinder equals $\pi r^2 h$ where h is the height and r is the radius.

(19) A function f, is said to be odd if $f(-x) = -f(x)$ and a function f is said to be even if $f(-x) = f(x)$. Show that if f is even, then f' is odd and if f is odd, then f' is even. Sketch the graph of a typical odd function and a typical even function.

(20) Recall sin is an odd function and cos is an even function. Determine whether each of the trig functions is odd, even or neither.

(21) Find the x values of the critical points of the function $f(x) = 3x^2 - 5x^3$.

(22) Find the x values of the critical points of the function $f(x) = \sqrt{3x^2 - 6x + 8}$.

(23) Find the extreme points of the function $f(x) = x + \frac{25}{x}$ and tell whether the extreme point is a local maximum or a local minimum or neither.

(24) A piece of property is to be fenced on the front and two sides. Fencing for the sides costs $3.50 per foot and fencing for the front costs $5.60 per foot. What are the dimensions of the largest such rectangular lot if the available money is $1400?

(25) In a particular apartment complex of 200 units, it is found that all units remain occupied when the rent is $400 per month. For each $40 increase in the rent, 5 units become vacant, on the average. Occupied units require $80 per month for maintenance, while vacant units require none. Fixed costs for the buildings are $20 000 per month. What rent should be charged for maximum profit and what is the maximum profit?

(26) Find the point on the curve, $y = \sqrt{25 - 2x}$ which is closest to $(0, 0)$.

(27) A street is 200 feet long and there are two lights located at the ends of the street. One of the lights is $\frac{1}{8}$ times as bright as the other. Assuming the brightness of light from one of these street lights is proportional to the brightness of the light and the reciprocal of the square of the distance from the light, locate the darkest point on the street.

(28) * Find the volume of the smallest right circular cone which can be circumscribed about a sphere of radius 4 inches.

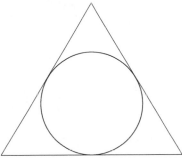

(29) Two cities are located on the same side of a straight river. One city is at a distance of 3 miles from the river and the other city is at a distance of 8 miles from the river. The distance between the two points on the river which are closest to the respective cities is 40 miles. Find the location of a pumping station which is to pump water to the two cities which will minimize the length of pipe used.

(30) A hungry spider is located on the wall four feet off the floor directly above a point four feet from the corner of the room. This is a Daring Jumping Spider[1], not a lazy web spinner who just sits in the web and waits for its prey. This kind of spider stalks its dinner like a lion hunting an impala. At a point on the floor which is 6 feet from the wall and 8 feet from the corner is a possible dinner, a plump juicy fly temporarily distracted as it slurps on a succulent morsel of rotting meat. What path should the hungry spider follow? Describe a way to do this problem geometrically without using any calculus.

(31) If Reid eats x pounds of spaghetti sauce, he will eat it with $4 - x^2$ pounds of noodles. How many pounds of spaghetti noodles and sauce can Reid consume?

3.6   Mean Value Theorem

The mean value theorem is one of the most important theorems about the derivative. The best versions of many other theorems depend on this fundamental result. The mean value theorem says that under suitable conditions, there exists a point in

---

[1]These are very beautiful spiders if you don't look too close. They are small furry and black with white markings. You sometimes see them running along walls and ceilings. Like many spiders they can climb even very slippery surfaces like glass with no difficulties.

(a, b), x, such that $f'(x)$ equals the slope of the secant line joining $(a, f(a))$ and $(b, f(b))$,

$$\frac{f(b) - f(a)}{b - a}.$$

The following picture is descriptive of this situation.

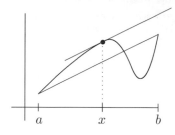

This theorem is an existence theorem and like the other existence theorems in analysis, it depends on the completeness axiom. The following is known as Rolle's[2] theorem.

Theorem 3.4. *Suppose $f : [a, b] \to \mathbb{R}$ is continuous,*

$$f(a) = f(b),$$

and $f : (a, b) \to \mathbb{R}$ has a derivative at every point of (a, b). Then there exists $x \in (a, b)$ such that $f'(x) = 0$.

Proof: Suppose first that $f(x) = f(a)$ for all $x \in [a, b]$. Then any $x \in (a, b)$ is a point such that $f'(x) = 0$. If f is not constant, either there exists $y \in (a, b)$ such that $f(y) > f(a)$ or there exists $y \in (a, b)$ such that $f(y) < f(b)$. In the first case, the maximum of f is achieved at some $x \in (a, b)$ and in the second case, the minimum of f is achieved at some $x \in (a, b)$. Either way, Theorem 3.3 on Page 96 implies $f'(x) = 0$. This proves Rolle's theorem. ∎

The next theorem is known as the Cauchy mean value theorem.

Theorem 3.5. *Suppose f, g are continuous on $[a, b]$ and differentiable on (a, b). Then there exists $x \in (a, b)$ such that*

$$f'(x)(g(b) - g(a)) = g'(x)(f(b) - f(a)).$$

Proof: Let

$$h(x) \equiv f(x)(g(b) - g(a)) - g(x)(f(b) - f(a)).$$

Then letting $x = a$ and then letting $x = b$, a short computation shows $h(a) = h(b)$. Also, h is continuous on $[a, b]$ and differentiable on (a, b). Therefore Rolle's theorem applies and there exists $x \in (a, b)$ such that

$$h'(x) = f'(x)(g(b) - g(a)) - g'(x)(f(b) - f(a)) = 0. \qquad ∎$$

---

[2]Rolle is remembered for Rolle's theorem and not for anything else he did. Ironically, he did not like calculus.

The usual mean value theorem, sometimes called the Lagrange mean value theorem, illustrated by the above picture is obtained by letting $g(x) = x$.

Corollary 3.1. *Let f be continuous on $[a, b]$ and differentiable on (a, b). Then there exists $x \in (a, b)$ such that $f(b) - f(a) = f'(x)(b - a)$.*

Corollary 3.2. *Suppose $f'(x) = 0$ for all $x \in (a, b)$ where $a \geq -\infty$ and $b \leq \infty$. Then $f(x) = f(y)$ for all $x, y \in (a, b)$. Thus f is a constant.*

Proof: If this is not true, there exists x_1 and x_2 such that $f(x_1) \neq f(x_2)$. Then by the mean value theorem,

$$0 \neq \frac{f(x_1) - f(x_2)}{x_1 - x_2} = f'(z)$$

for some z between x_1 and x_2. This contradicts the hypothesis that $f'(x) = 0$ for all x. ∎

Corollary 3.3. *Suppose $f'(x) > 0$ for all $x \in (a, b)$ where $a \geq -\infty$ and $b \leq \infty$. Then f is strictly increasing on (a, b). That is, if $x < y$, then $f(x) < f(y)$. If $f'(x) \geq 0$, then f is increasing in the sense that whenever $x < y$ it follows that $f(x) \leq f(y)$.*

Proof: Let $x < y$. Then by the mean value theorem, there exists $z \in (x, y)$ such that

$$0 < f'(z) = \frac{f(y) - f(x)}{y - x}.$$

Since $y > x$, it follows $f(y) > f(x)$ as claimed. Replacing $<$ by $\leq$ in the above equation and repeating the argument gives the second claim. ∎

Corollary 3.4. *Suppose $f'(x) < 0$ for all $x \in (a, b)$ where $a \geq -\infty$ and $b < \infty$. Then f is strictly decreasing on (a, b). That is, if $x < y$, then $f(x) > f(y)$. If $f'(x) \leq 0$, then f is decreasing in the sense that for $x < y$, it follows that $f(x) \geq f(y)$*

Proof: Let $x < y$. Then by the mean value theorem, there exists $z \in (x, y)$ such that

$$0 > f'(z) = \frac{f(y) - f(x)}{y - x}.$$

Since $y > x$, it follows $f(y) < f(x)$ as claimed. The second claim is similar except instead of a strict inequality in the above formula, you put $\geq$. ∎

3.7 Exercises

(1) Sally drives her Saturn over the 110 mile toll road in exactly 1.3 hours. The speed limit on this toll road is 70 miles per hour and the fine for speeding is 10 dollars per mile per hour over the speed limit. How much should Sally pay?

(2) Two cars are careening down a freeway weaving in and out of traffic. Car A passes car B and then car B passes car A as the driver makes obscene gestures. This infuriates the driver of car A who passes car B while firing his handgun at the driver of car B. Show there are at least two times when both cars have the same speed. Then show there exists at least one time when they have the same acceleration. The acceleration is the derivative of the velocity.

(3) A function f and an interval $[a, b]$ is given. Find $c \in (a, b)$ such that
$$f'(c) = \frac{f(b) - f(a)}{b - a}.$$
   (a) $f(x) = x^{1/2}, (0, 1)$ (see Example 3.3)
   (b) $f(x) = x^3 - 3x + 1, (-1, 1)$
   (c) $f(x) = \frac{x}{1-x}, (2, 3)$

(4) Let a, b be two positive numbers. Consider the function $f(x) = 1/x$ on $[a, b]$ and find the point at which the conclusion of the mean value theorem holds.

(5) Show the cubic function $f(x) = 5x^3 + 7x - 18$ has exactly one real zero.

(6) Suppose $f(x) = x^7 + |x| + x - 12$. How many solutions are there to the equation $f(x) = 0$?

(7) Let $f(x) = |x - 7| + (x - 7)^2 - 2$ on the interval $[6, 8]$. Then $f(6) = 0 = f(8)$. Does it follow from Rolle's theorem that there exists $c \in (6, 8)$ such that $f'(c) = 0$? Explain your answer.

(8) If $f(x) = ax^2 + bx + c$ where $a \neq 0$, show that there exists exactly one $z \in (p, q)$ such that
$$f'(z) = \frac{f(q) - f(p)}{q - p}.$$

(9) Consider the function $f(x) = (1 + x)^n - (1 + nx)$ where n is a positive integer. Explain why this function is no smaller than 0 for $x \geq 0$.

(10) For a, b positive numbers, use the mean value theorem to show that
$$\sqrt{ab} \leq \frac{a + b}{2}$$
   Hint: Fix positive b and consider the function
$$f(a) = \frac{a + b}{2} - \sqrt{ab}$$
   for $a \geq b$.

(11) Explain why the following equations have exactly one real solution.
   (a) $x^5 + 7x + \sin(x) = 0$
   (b) $x^3 + |x| x + 7 = 0$
   (c) $\frac{x}{1+x^2} + 3x = 0$

(12) Suppose f and g are differentiable functions defined on $\mathbb{R}$. Suppose also that it is known that $|f'(x)| > |g'(x)|$ for all x and that $|f'(t)| > 0$ for all t. Show that whenever $x \neq y$, it follows $|f(x) - f(y)| > |g(x) - g(y)|$. **Hint:** Use the Cauchy mean value theorem, Theorem 3.5.

(13) Show that, like continuous functions, functions which are derivatives have the intermediate value property. This means that if $f'(a) < 0 < f'(b)$ then there exists $x \in (a, b)$ such that $f'(x) = 0$. **Hint:** Argue the minimum value of f occurs at an interior point of $[a, b]$.

(14) ↑Consider the function

$$f(x) \equiv \begin{cases} 1 & \text{if } x \geq 0 \\ -1 & \text{if } x < 0 \end{cases}.$$

Is it possible that this function could be the derivative of some function? Why?

(15) Show by induction and Rolle's theorem that there are at most n real solutions to the equation

$$a_n x^n + a_{n-1} x^{n-1} + \cdots + a_1 x + a_0 = 0, a_n \neq 0, n \text{ is a positive integer.}$$

(16) *Suppose f, g are differentiable functions on the interval (a, b) and

$$\lim_{x \to a+} f(x) = \lim_{x \to a+} g(x) = 0$$

$$\lim_{x \to a+} g'(x) = q, \lim_{x \to a+} f'(x) = p.$$

Show that for x close enough to a, there exists $y \subset (a, x)$ such that

$$\frac{f(x)}{g(x)} = \frac{f'(y)}{g'(y)}$$

Next explain why

$$\lim_{x \to a+} \frac{f(x)}{g(x)} = \frac{p}{q}.$$

3.8 Curve Sketching

The theorems and corollaries given above can be used to aid in sketching the graphs of functions. The second derivative will also help in determining the shape of the function.

Definition 3.4. A differentiable function f, defined on an interval (a, b), is concave up if f' is an increasing function. A differentiable function defined on an interval (a, b), is concave down if f' is a decreasing function. A point where the graph of the function changes from being concave up to concave down or from concave down to concave up is called an inflection point.

From the geometric description of the derivative as the slope of a tangent line to the graph of the function, to say the derivative is an increasing function means that as you move from left to right, the slopes of the lines tangent to the graph of f become larger. Thus the graph of the function is bent up in the shape of a smile. It may also help to think of it as a cave when you view it from above, hence the term concave up. If the derivative is decreasing, it follows that as you move from left to right the slopes of the lines tangent to the graph of f become smaller. Thus the graph of the function is bent down in the form of a frown. It is concave down because it is like a cave when viewed from beneath. The following theorem will give a convenient criterion in terms of the second derivative for finding whether a function is concave up or concave down. The term, concavity, is used to refer to this property. Thus you determine the concavity of a function when you find whether it is concave up or concave down.

Theorem 3.6. *Suppose $f''(x) > 0$ for $x \in (a, b)$. Then f is concave up on (a, b). Suppose $f''(x) < 0$ on (a, b). Then f is concave down.*

Proof: This follows immediately from Corollaries 3.4 and 3.3 applied to the first derivative. ∎

The following picture may help in remembering this.

In this picture, the plus signs and the smile on the left correspond to the second derivative being positive. The smile gives the way in which the graph of the function is bent. In the second face, the minus signs correspond to the second derivative being negative. The frown gives the way in which the graph of the function is bent.

Example 3.15. Sketch the graph of the function $f(x) = (x^2 - 1)^2 = x^4 - 2x^2 + 1$.

Take the derivative of this function $f'(x) = 4x^3 - 4x = 4x(x - 1)(x + 1)$ which equals zero at $-1, 0$, and 1. It is positive on $(-1, 0)$, and $(1, \infty)$ and negative on $(0, 1)$ and $(-\infty, -1)$. Therefore, $x = 0$ corresponds to a local maximum and $x = -1$ and $x = 1$ correspond to local minima. The second derivative is $f''(x) = 12x^2 - 4$ and this equals zero only at the points $-1/\sqrt{3}$ and $1/\sqrt{3}$. The second derivative is positive on the intervals $(1/\sqrt{3}, \infty)$ and $(-\infty, -1/\sqrt{3})$ so the function f is smiling on these intervals. The second derivative is negative on the interval $(-1/\sqrt{3}, 1/\sqrt{3})$ and so the original function is frowning on this interval. This describes in words the qualitative shape of the function. It only remains to draw a picture which incorporates this description. Here is a graph of this function done by a computer algebra system.

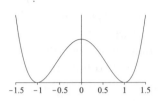

In general, if you are interested in getting a nice graph of a function, you should use a computer algebra system. An effective way to accomplish your graphing is to go to the help menu and copy and paste an example from this menu changing it as needed. Both mathematica and Maple have good help menus. Keep in mind there are certain conventions which must be followed. For example to write x raised to the second power you enter x ^2. In Maple, you also need to place an asterisk between quantities which are multiplied since otherwise it will not know you are multiplying and will not work. There are also easy to use versions of Maple available which involve essentially pointing and clicking. You won't learn any calculus from playing with a computer algebra system but you might have a lot of fun.

3.9   Exercises

(1) Here is a graph of a function. Identify intervals on which the function is concave up.

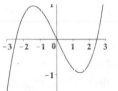

(2) A rock is thrown straight up. Sketch a graph of its distance above the ground as a function of t where t equals time. **Hint:** The height is increasing but at a slower and slower rate as it rises. Eventually, it goes as high as it will go and starts down when the speed of the rock is increasing till it hits the ground. The graph should reflect this information.

(3) Sketch the graph of the function $f(x) = x^3 - 3x + 1$ showing the intervals on which the function is concave up and down and identifying the intervals on which the function is increasing.

(4) Find intervals on which the function $f(x) = \sqrt{1 - x^2}$ is increasing and intervals on which it is concave up and concave down. Sketch a graph of the function.

(5) Sketch the graphs of $y = x^4, y = x^3$, and $y = -x^4$. What do these graphs tell you about the case when the second derivative equals zero?

(6) Sketch the graph of $f(x) = 1/(1 + x^2)$ showing the intervals on which the function is increasing or decreasing and the intervals on which the graph is

concave up and concave down.

(7) Sketch the graph of $f(x) = x/(1+x^2)$ showing the intervals on which the function is increasing or decreasing and the intervals on which the graph is concave up and concave down.

(8) Show that inflection points can be identified by looking at those points where the second derivative equals zero but that not every point where the second derivative equals zero is an inflection point. **Hint:** For the last part consider $y = x^3$ and $y = x^4$.

(9) Suppose $f''(x) = 0$ and $f'''(x) \neq 0$. Does it follow that x must be an inflection point?

(10) Find all inflection points for the function $f(x) = x^2/(1+x^2)$.

(11) Identify the intervals on which the following functions are concave up.

(a) $\sqrt{|x|}$

(b) $|x|\, x$

(c) $\frac{x}{1+x}$

(d) $x^3 - 2x$

(e) $\frac{1+x^2}{1-x^2}$

(f) $-x^3 + 3x$

(g) $x^4 + 2x^3 - x^2 + 3x$

(12) Suppose a function f satisfies the differential equation $f''(x) = (1+x)^2 (x-1)$ When is the graph of f concave up?

(13) Sketch the graph of the above function if it is known that $f(0) = 0$ and $f'(0) = 0$.

(14) Suppose $y' = 1 + y^2$. Show that y is concave up when $y > 0$ and concave down when $y < 0$.

(15) Suppose $y' + xy = x^2$ where y is a function of x and suppose it is known that $y(0) = 1$. Determine whether y is concave up when $x = 0$.

(16) Sketch the following rational functions.

(a) $\frac{x^2+1}{x-1}$

(b) $\frac{x+2}{x^2-1}$

(c) $\frac{x^2}{1-x^2}$

(17) A function L satisfies the differential equation $L'(x) = 1/x$, $L(1) = 0$. Sketch a graph of this function.

(18) A function $f(x)$ has a continuous second derivative on all of $\mathbb{R}$. The second derivative is positive for $x > 0$ and is negative for $x < 0$. Also $\lim_{x \to +\infty} f(x) = 1$. It also has a local maximum at $x = -2$. Sketch a possible graph for this function.

Chapter 4

Some Important Special Functions

4.1 The Circular Functions

The Trigonometric functions are also called the circular functions. Thus this section will be on the functions cos, sin, tan, sec, csc, and cot. The first thing to do is to give an important lemma. There are several approaches to this lemma. To see it done in terms of areas of a circular sector, see Apostol, [1], or almost any other calculus book. However, the book by Apostol has no loose ends in the presentation unlike most other books which use this approach. The proof given here is like that found in Tierney, [27] and Rose, [22] and is based on arc length.

Lemma 4.1. *The following limits hold.*

$$\lim_{x \to 0} \frac{\sin x}{x} = 1 \tag{4.1}$$

$$\lim_{x \to 0} \frac{1 - \cos x}{x} = 0 \tag{4.2}$$

Proof: First consider (4.1). In the following picture, it follows from Corollary 1.1 on Page 9 that for small positive x,

$$\sin x + (1 - \cos x) \geq x \geq \sin x. \tag{4.3}$$

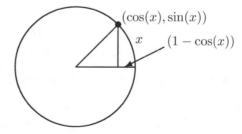

Now divide by $\sin x$ to get

$$1 + \frac{1 - \cos x}{|\sin x|} = 1 + \frac{1 - \cos x}{\sin x} \geq \frac{x}{\sin x} \geq 1.$$

For small negative values of x, it is also true that

$$1 + \frac{1 - \cos x}{|\sin x|} \geq \frac{x}{\sin x} \geq 1.$$

(Why?) From the trig. identities, it follows that for all small values of x,

$$1 + \frac{\sin^2 x}{|\sin x|\,(1 + \cos x)} = 1 + \frac{|\sin x|}{(1 + \cos x)} \geq \frac{x}{\sin x} \geq 1$$

and so from the squeezing theorem, Theorem 2.7 on Page 60,

$$\lim_{x \to 0} \frac{x}{\sin x} = 1,$$

and consequently, from the limit theorems,

$$\lim_{x \to 0} \frac{\sin x}{x} = \lim_{x \to 0} \frac{1}{\left(\frac{x}{\sin x}\right)} = 1.$$

Finally,

$$\frac{1 - \cos x}{x} = \frac{1 - \cos^2 x}{x\,(1 + \cos x)} = \sin x \frac{\sin x}{x} \frac{1}{1 + \cos x}.$$

Therefore, from Theorem 2.2 on Page 52 which says $\lim_{x \to 0} \sin(x) = 0$, and the limit theorems,

$$\lim_{x \to 0} \frac{1 - \cos x}{x} = 0. \qquad \blacksquare$$

With this, it is easy to find the derivative of sin. Using Lemma 4.1,

$$\lim_{h \to 0} \frac{\sin(x + h) - \sin x}{h} = \lim_{h \to 0} \frac{\sin(x)\cos(h) + \cos(x)\sin(h) - \sin x}{h}$$

$$= \lim_{h \to 0} \frac{(\sin x)\,(\cos(h) - 1)}{h} + \cos x \frac{\sin(h)}{h} = \cos x.$$

The derivative of cos can be found the same way. Alternatively, $\cos(x) = \sin(x + \pi/2)$ and so

$$\begin{aligned}
\cos'(x) &= \sin'(x + \pi/2) = \cos(x + \pi/2) \\
&= \cos x \cos(\pi/2) - \sin x \sin(\pi/2) = -\sin x.
\end{aligned}$$

The following theorem is now obvious and the proofs of the remaining parts are left for you.

Theorem 4.1. *The derivatives of the trig. functions are as follows.*

$$\begin{aligned}
\sin'(x) &= \cos x,\ \cos'(x) = -\sin x,\ \tan'(x) = \sec^2(x) \\
\cot'(x) &= -\csc^2(x),\ \sec'(x) = \sec x \tan x,\ \csc'(x) = -\csc x \cot x
\end{aligned}$$

Here are some examples of extremum problems which involve the use of the trig. functions.

Example 4.1. Two hallways intersect at a right angle. One is 5 feet wide and the other is 2 feet wide. What is the length of the longest thin rod which can be carried horizontally from one hallway to the other?

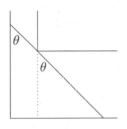

You must minimize the length of the rod which touches the inside corner of the two halls and extends to the outside walls. Letting θ be the angle between this rod and the outside wall for the hall having width 2, minimize

$$\overbrace{f(\theta) = 2\csc\theta + 5\sec\theta.}^{\text{length of rod}}$$

Therefore, using the rules of differentiation,

$$f'(\theta) = \frac{2\cos^3\theta - 5\sin\theta + 5\sin\theta\cos^2\theta}{(\cos^2\theta)(-1 + \cos^2\theta)} = 0$$

should be solved to get the angle where this length is as small as possible. Thus

$$2\cos^3\theta - 5\sin\theta + 5\sin\theta\cos^2\theta = 0.$$

and $2\cos^3\theta - 5\sin^3\theta = 0$ and so $\tan\theta = \frac{1}{5}\sqrt[3]{2}\left(\sqrt[3]{5}\right)^2$. Drawing a triangle, you see that at this value of θ, you have $\sec\theta = \frac{\sqrt{\left(\sqrt[3]{5}\right)^2 + \left(\sqrt[3]{2}\right)^2}}{\sqrt[3]{5}}$ and $\csc\theta = \frac{\sqrt{\left(\sqrt[3]{5}\right)^2 + \left(\sqrt[3]{2}\right)^2}}{\sqrt[3]{2}}$. Therefore, the minimum is obtained by substituting these values in to the equation for $f(\theta)$ yielding $\left(\sqrt{\left(\left(\sqrt[3]{5}\right)^2 + \left(\sqrt[3]{2}\right)^2\right)}\right)^3$.

Example 4.2. A fence 9 feet high is 2 feet from a building. What is the length of the shortest ladder which will lean against the top of the fence and touch the building?

Let θ be the angle of the ladder with the ground. Then the length of this ladder making this angle with the ground and leaning on the top of the fence while touching the building is $f(\theta) = \frac{9}{\sin\theta} + \frac{2}{\cos\theta}$. Then the final answer is $\left(\sqrt{\left(3\sqrt[3]{3} + \left(\sqrt[3]{2}\right)^2\right)}\right)^3$.
The details are similar to the problem of the two hallways.

4.2   Exercises

(1) Prove all parts of Theorem 4.1.

(2) Prove $\tan'(x) = 1 + \tan^2(x)$.

(3) Find and prove a formula for the derivative of $\sin^m(x)$ for m an integer.

(4) Find the derivative of the function $\sin^6(5x)$.

(5) Find the derivative of the function $\tan^7(4x)$.

(6) Find the derivative of the function $\frac{\sec^3(2x)}{\tan^3(3x)}$.

(7) Find derivatives of the following functions.

    (a) $\csc^3(x)$

    (b) $\tan^2(x)$

    (c) $\sin^2(x)\cos^2(x)$

    (d) $\cos^2(x) - \sin^2(x)$

    (e) $\sec^4(x)\tan(x)$

    (f) $x^2\tan^2(x)$

(8) Find all intervals where $\sin(2x)$ is concave down.

(9) Find the intervals where $\cos(3x)$ is increasing.

(10) Two hallways intersect at a right angle. One is 3 feet wide and the other is 4 feet wide. What is the length of the longest thin rod which can be carried horizontally from one hallway to the other?

(11) A fence 5 feet high is 2 feet from a building. What is the length of the shortest ladder which will lean against the top of the fence and touch the building?

(12) Suppose $f(x) = A\cos\omega x + B\sin\omega x$. Show there exists an angle ϕ such that $f(x) = \sqrt{A^2 + B^2}\sin(\omega x + \phi)$. The number $\sqrt{A^2 + B^2}$ gives the "amplitude" and ϕ is called the "phase shift" while ω is called the "frequency". This is very important because it is descriptive of what is going on. The amplitude gives the height of the periodic function f. **Hint:** Remember a point on the unit circle determines an angle. Write $f(x)$ in the form

$$\sqrt{A^2 + B^2}\left(\frac{A}{\sqrt{A^2 + B^2}}\cos\omega x + \frac{B}{\sqrt{A^2 + B^2}}\sin\omega x\right)$$

and note that $\left(\frac{B}{\sqrt{A^2+B^2}}, \frac{A}{\sqrt{A^2+B^2}}\right)$ is a point on the unit circle.

(13) Repeat Problem 12 but this time show $f(x) = \sqrt{A^2 + B^2}\cos(\omega x + \phi)$. How could you find ϕ?

(14) A square picture is 6 feet high and is fastened to the wall with its lowest edge one foot above the eye level of an observer. Where should he stand to maximize the angle subtended by the picture at his eye?

(15) A 100 foot tall lamp post with a light on top creates a shadow from a falling ball dropped from a height of 100 feet at a distance of 50 feet from the lamp post. Thus the distance the ball has fallen at time t is $16t^2$. Find the velocity of the shadow when the ball has dropped a distance of 64 feet.

4.3 The Exponential And Log Functions

4.3.1 *The Rules Of Exponents*

As mentioned earlier, b^m means to multiply b by itself m times assuming m is a positive integer. $b^0 \equiv 1$ provided $b \neq 0$. In the case where $b = 0$ the symbol is undefined. If $m < 0$, b^m is defined as $\frac{1}{b^{-m}}$. Then the following algebraic properties are obtained. Be sure you understand these properties for x and y integers.

$$b^{x+y} = b^x b^y, \ (ab)^x = a^x b^x \tag{4.4}$$

$$b^{xy} = (b^x)^y, \ b^{-1} = \frac{1}{b} \tag{4.5}$$

These properties are called the rules of exponents.

When x and y are not integers, the meaning of b^x is no longer clear. For example, suppose $b = -1$ and $x = 1/2$. What exactly is meant by $(-1)^{1/2}$? Even in the case where $b > 0$ there are difficulties. If x is a rational number m/n and $b > 0$ the symbol $b^{m/n}$ means $\sqrt[n]{b^m}$. That is its definition, and it is a useful exercise for you to verify (4.4) and (4.5) hold with this definition. There are no mathematical questions about the existence of this number. To see this, consider Problem 7 on Page 56. The problem is not one of theory but of practicality. Could you use this definition to find $2^{\frac{1234567812345}{1234567812344}}$? Consider what you would do. First find the number $2^{1234567812345}$ and then $\cdots$? Can you find this number? It is just too big. However, a calculator can find $2^{\frac{1234567812345}{1234567812344}}$. It yields $2^{\frac{1234567812345}{1234567812344}} = 2.000\,000\,000\,001\,123$ as an approximate answer. Clearly something else must be going on. To make matters even worse, what would you do with $2^{\sqrt{2}}$? As mentioned earlier, $\sqrt{2}$ is irrational and so cannot be written as the quotient of two integers. These are serious difficulties and must be dealt with.

4.3.2 *The Exponential Functions, A Wild Assumption*

Using your calculator or a computer you can obtain graphs of the functions $y = b^x$ for various choices of b. The following picture gives a few of these graphs.

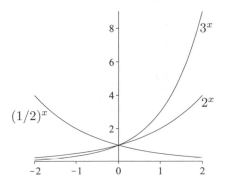

These graphs suggest that if $b < 1$ the function $y = b^x$ is decreasing while if $b > 1$, the function is increasing but just how was the calculator or computer able to draw those graphs? Also, do the laws of exponents continue to hold for all real values of x? The short answer is that they do and this is shown later but for now here is a wild assumption which glosses over these issues.

WILD
ASSUMPION!

Wild Assumption 4.2. For every $b > 0$ there exists a unique differentiable function $\exp_b (x) \equiv b^x$ valid for all real values of x such that (4.4) and (4.5) both hold for all $x, y \in \mathbb{R}$, $\exp_b (m/n) = \sqrt[n]{b^m}$ whenever m, n are integers, (In particular, $b^0 = 1$.) and $b^x > 0$ for all $x \in \mathbb{R}$. Furthermore, if $b \neq 1$ and $h \neq 0$, then $\exp_b (h) = b^h \neq 1$.

Instead of writing $\exp_b (x)$ I will often write b^x and I will also be somewhat sloppy and regard b^x as the name of a function and not just as $\exp_b (x)$, a given function defined at x. This is done to conform with standard usage. Also, the last claim in Wild Assumption 4.2 follows from the first part of this assumption. See Problem 1. I want it to be completely clear that the Wild Assumption is just that. No reason for believing in such an assumption has been given, notwithstanding the pretty pictures drawn by the calculator. Later in the book, the wild assumption will be completely justified. Based on Wild Assumption 4.2 one can easily find out all about b^x. In the following theorem, the number $\ln b$ defined there is the slope of the tangent line to the graph of $y = b^x$ at the point $(0, 1)$.

Theorem 4.3. *Let $\exp_b$ be defined in Wild Assumption 4.2 for $b > 0$. Then there exists a unique number, denoted by $\ln b$ for $b > 0$ satisfying*

$$\exp_b' (x) = (\ln b) \exp_b (x).\tag{4.6}$$

Furthermore,

$$\ln (ab) = \ln (a) + \ln (b), \ \ln 1 = 0,\tag{4.7}$$

and for all $y \in \mathbb{R}$,

$$\ln (b^y) = y \ln b.\tag{4.8}$$

$$\ln \left(\frac{a}{b}\right) = \ln (a) - \ln (b)\tag{4.9}$$

The function $x \to \ln x$ is differentiable and defined for all $x > 0$ and

$$\ln' (x) = \frac{1}{x}.\tag{4.10}$$

The function $x \to \ln x$ is one to one on $(0, \infty)$. Also, $\ln$ maps $(0, \infty)$ onto $(-\infty, \infty)$.

Proof: First consider (4.6).

$$\lim_{h \to 0} \frac{\exp_b (x+h) - \exp_b (x)}{h} = \lim_{h \to 0} \frac{b^{x+h} - b^x}{h} = \lim_{h \to 0} \left(\frac{b^h - 1}{h} \right) b^x.$$

The expression $\lim_{h \to 0} \left(\frac{b^h - 1}{h} \right)$ is assumed to exist thanks to Wild Assumption 4.2 and this is denoted by $\ln b$. This proves (4.6).

To verify (4.7), if $b = 1$ then $b^x = 1^x$ for all $x \in \mathbb{R}$. Now by (4.4) and (4.5),

$$1^x 1^x = 1^{x+x} = \left(1^2 \right)^x = 1^x$$

and so, dividing both sides by 1^x, an operation justified by Wild Assumption 4.2, $1^x = 1$ for all $x \in \mathbb{R}$. Therefore, $\exp_1 (x) = 1$ for all x and so $\exp_1' (x) = \ln 1 \exp_1 (x) = 0$. Thus $\ln 1 = 0$ as claimed. Next, by the product rule and Wild Assumption 4.2,

$$\begin{aligned} \ln (ab) (ab)^x &= ((ab)^x)' = (a^x b^x)' = (a^x)' b^x + a^x (b^x)' \\ &= (\ln a) a^x b^x + (\ln b) b^x a^x = [\ln a + \ln b] (ab)^x . \end{aligned}$$

Therefore, $\ln (ab) = \ln a + \ln b$ as claimed.

Next consider (4.8). Keeping y fixed, consider the function $x \to b^{xy} = (b^y)^x$. Then,

$$\ln (b^y) \exp_b (xy) = \ln (b^y) (b^y)^x = ((b^y)^x)' = g' (x)$$

where $g (x) \equiv b^{xy}$. Using (3.7) on Page 91, it is also the case that $g' (x) = y \exp_b' (xy)$ and so

$$y \ln (b) \exp_b (xy) = y \exp_b' (xy) = \ln (b^y) \exp_b (xy)$$

Now, dividing both sides by $\exp_b (xy)$ verifies (4.8).

To obtain (4.9) from this, note

$$\ln \left(\frac{a}{b} \right) = \ln \left(ab^{-1} \right) = \ln (a) + \ln \left(b^{-1} \right) = \ln (a) - \ln (b) .$$

It remains to verify (4.10). From (3.2) and the continuity of 2^x,

$$\ln' (1) = \lim_{h \to 0} \frac{\ln (2^h) - \ln 1}{2^h - 1} = \lim_{h \to 0} \frac{h \ln 2}{2^h - 1} = \frac{\ln 2}{\ln 2} = 1.$$

$\ln 2 \neq 0$ because if it were, then $(2^x)' - (\ln 2) 2^x = 0$ and by Corollary 3.2, this would imply 2^x is a constant function which it is not. Now the first part of this lemma implies

$$\ln' (x) = \lim_{y \to x} \frac{\ln y - \ln x}{y - x} = \lim_{y \to x} \frac{1}{x} \frac{\ln \left(\frac{y}{x} \right) - \ln 1}{\left(\frac{y}{x} \right) - 1} = \frac{1}{x} \ln' (1) = \frac{1}{x}.$$

It remains to verify $\ln$ is one to one. Suppose $\ln x = \ln y$. Then by the mean value theorem, there exists t between x and y such that $(1/t) (x - y) = \ln x - \ln y = 0$. Therefore, $x = y$ and this shows $\ln$ is one to one as claimed.

It only remains to verify that $\ln$ maps $(0, \infty)$ onto $(-\infty, \infty)$. By Wild Assumption 4.2 and the mean value theorem, Corollary 3.1, there exists $y \in (0, 1)$ such that

$$0 < \frac{2^1 - 1}{1} = \ln(2)\, 2^y$$

Since $2^y > 0$ it follows $\ln(2) > 0$ and so $x \to 2^x$ is strictly increasing. Therefore, by Corollary 3.1

$$\frac{2^1 - 1}{1} = \ln(2)\, 2^y \leq \ln(2)\, 2^1$$

and it follows that

$$\frac{1}{2} \leq \ln 2. \tag{4.11}$$

Also, from 4.7, $0 = \ln(2) + \ln\left(\frac{1}{2}\right) \geq \frac{1}{2} + \ln\left(\frac{1}{2}\right)$ which shows that

$$\ln\left(\frac{1}{2}\right) \leq -\frac{1}{2}. \tag{4.12}$$

It follows from 4.11 and 4.12 that $\ln$ achieves values which are arbitrarily large and arbitrarily large in the negative direction. Therefore, by the intermediate value theorem, $\ln$ achieves all values.

More precisely, let $y \in \mathbb{R}$. Then choose n large enough that $\frac{n}{2} > y$ and $-\frac{n}{2} < y$. Then from 4.11 and 4.12, $\ln\left(\left(\frac{1}{2}\right)^n\right) \leq \frac{-n}{2} < y < \frac{n}{2} < \ln(2^n)$. By the intermediate value theorem, there exists $x \in \left(\left(\frac{1}{2}\right)^n, 2^n\right)$ such that $\ln x = y$. ∎

Example 4.3. Find $f'(x)$ if $f(x) = 7^{2x}$.

$7^{2x} = (7^x)^2$ and so the derivative is $2(7^x)(7^x)' = 2(7^x)\ln 7\,(7^x) = 2\,(\ln 7)\,7^{2x}$.

Example 4.4. Find $f'(x)$ if $f(x) = 7^{(x^2)}$.

Set up the difference quotient,

$$\frac{7^{(x+h)^2} - 7^{x^2}}{h} = 7^{x^2}\left(\frac{7^{2xh+h^2} - 1}{h}\right) = 7^{x^2}\left(\frac{7^{2xh+h^2} - 1}{2xh + h^2}\right)\frac{2xh + h^2}{h}.$$

Taking the limit as $h \to 0$ and using the definition of $\ln 7$, this limit equals $7^{x^2}\ln(7)\,2x$.

4.3.3   The Special Number e

Since $\ln$ is one to one onto $\mathbb{R}$, it follows there exists a unique number e such that $\ln(e) = 1$. Therefore,

$$\exp'_e(x) \equiv (e^x)' = \ln(e)\,e^x \equiv \ln(e)\exp_e(x) = \exp_e(x)$$

showing that $\exp_e$ has the remarkable property that it equals its own derivative. This wonderful number is called Euler's number and it can be shown to equal approximately $2.718\,3$. It is customary to write $\exp(x)$ for $\exp_e(x)$. Thus

$$\exp'(x) = \exp(x). \tag{4.13}$$

4.3.4 The Function $\ln|x|$

The function ln is only defined on positive numbers. However, it is possible to write $\ln|x|$ whenever $x \neq 0$. What is the derivative of this function?

Corollary 4.1. *Let* $f(x) = \ln|x|$ *for* $x \neq 0$. *Then* $f'(x) = \frac{1}{x}$.

Proof: If $x > 0$ the formula is just (4.10). Suppose then that $x < 0$. Then $\ln|x| = \ln(-x)$ so by (3.7) on Page 91,

$$(\ln|x|)' = (\ln(-x))' = (\ln((-1)x))' = \frac{1}{-x}(-1) = \frac{1}{x}. \qquad \blacksquare$$

4.3.5 Logarithm Functions

Next a new function called $\log_b$ will be defined.

Definition 4.1. *For all* $b > 0$ *and* $b \neq 1$

$$\log_b(x) \equiv \frac{\ln x}{\ln b}. \tag{4.14}$$

Notice this definition implies (4.7) - (4.9) all hold with ln replaced with $\log_b$. Also, $\log_b : (0, \infty) \to \mathbb{R}$ is one to one and onto.

The fundamental relationship between the exponential function b^x and $\log_b x$ is in the following proposition. This proposition shows this new function is $\log_b$ which you may have studied in high school.

Proposition 4.1. *Let* $b > 0$ *and* $b \neq 1$. *Then for all* $x > 0$,

$$b^{\log_b x} = x, \tag{4.15}$$

and for all $y \in \mathbb{R}$,

$$\log_b b^y = y, \tag{4.16}$$

Also,

$$\log_b'(x) = \frac{1}{\ln b}\frac{1}{x}. \tag{4.17}$$

Proof: Formula (4.15) follows from (4.8).

$$\ln\left(b^{\log_b x}\right) = \log_b(x)\ln(b) = \ln(x)$$

and so, since ln is one to one, it follows (4.15) holds.

$$\log_b b^y \equiv \frac{\ln(b^y)}{\ln b} = \frac{y\ln b}{\ln b} = y$$

and this verifies (4.16). Formula (4.17) is obvious from (4.14). $\blacksquare$

The functions $\log_b$ are only defined on positive numbers. However, it is possible to write $\log_b|x|$ whenever $x \neq 0$. What is the derivative of these functions?

Corollary 4.2. *Let* $f(x) = \log_b|x|$ *for* $x \neq 0$. *Then* $f'(x) = \frac{1}{(\ln b)x}$.

Proof: If $x > 0$ the formula is just (4.17). Suppose then that $x < 0$. Then $\log_b |x| = \log_b (-x)$ so by (3.7) on Page 91,

$$(\log_b |x|)' = (\log_b (-x))' = (\log_b ((-1) x))' = \frac{1}{-x \ln b} (-1) = \frac{1}{x \ln b}. \qquad \blacksquare$$

Example 4.5. Using properties of logarithms, simplify the expression $\log_3 \left(\frac{1}{9}x\right)$.

From (4.7) - (4.9),

$$\log_3 \left(\frac{1}{9}x\right) = \log_3 \left(\frac{1}{9}\right) + \log_3 (x) = \log_3 (3^{-2}) + \log_3 (x) = -2 + \log_3 (x).$$

Example 4.6. Using properties of logarithms, solve $5^{x-1} = 3^{2x+2}$.

Take ln of both sides. Thus $(x - 1) \ln 5 = (2x + 2) \ln 3$. Then solving this for x yields $x = \frac{\ln 5 + 2 \ln 3}{\ln 5 - 2 \ln 3}$.

Example 4.7. Solve $\log_3 (x) + 2 = \log_9 (x + 3)$.

From the given equation,

$$3^{\log_3 (x) + 2} = 3^{\log_9 (x+3)} = 9^{\frac{1}{2}(\log_9 (x+3))} = 9^{\log_9 \sqrt{(x+3)}}$$

and so $9x = \sqrt{x + 3}$. Therefore, $x = \frac{1 + \sqrt{1 + 12 \times 81}}{2(81)} = \frac{1}{162} + \frac{1}{162}\sqrt{973}$. In the use of the quadratic formula, only one solution was possible. (Why?)

Example 4.8. Compare $\ln (x)$ and $\log_e (x)$.

Recall that

$$\log_e (x) \equiv \frac{\ln x}{\ln e}.$$

Since $\ln e = 1$ from the definition of e, it follows $\log_e (x) = \ln x$. These logarithms are called natural logarithms.

Example 4.9. Find the derivative of the function $f (x) = \log_5 (x)$.

$$f' (x) = \left(\frac{\ln (x)}{\ln 5}\right)' = \frac{1}{\ln 5} \frac{1}{x}.$$

Example 4.10. Find the derivative of $f (x) = \ln (x^3)$.

$$f' (x) = (\ln (x^3))' = (3 \ln x)' = \frac{3}{x}.$$

4.4   Exercises

(1) Prove the last part of the Wild Assumption follows from the first part of this assumption. That is, show that if $b \neq 1$, then $\exp_b(h) \neq 1$ if $h \neq 0$ follows from the first part. **Hint:** If $b^h = 1$ for $h \neq 0$, show $b^x = 1$ for all $x \in \mathbb{R}$.

(2) Simplify

  (a) $\log_4(16x)$

  (b) $\log_3(27x^3)$

  (c) $(\log_b a)(\log_a b)$ for a, b positive real numbers not equal to 1.

(3) Simplify

  (a) $\log_3(\log_3(27) - \log_3(9))$

  (b) $\log_2(3)\log_3(2)$

  (c) $\log_b(x)\log_a(b)$

  (d) $\log_{10}(100000^{1/3})$.

(4) Find the derivatives of the following functions. You may want to do this by looking at the definition of the derivative in some cases.

  (a) $\log_5(x^2)$

  (b) $\log_3(5x + 1)$

  (c) $\log_6(\sqrt{x})$

  (d) $\log_3(\sqrt{x^2 + 1})$.

(5) Find the derivatives of the following functions. You may want to do this by looking at the definition of the derivative in some cases.

  (a) 3^{x^2}

  (b) 2^{3x+1}

  (c) 5^{2x+7}

  (d) 7^{x^3}

  (e) $3^{\sqrt{x^2+1}}$

(6) Explain why the function $6^x(1 - x\ln 6)$ is never larger than 1. **Hint:** Consider $f(x) = 6^x(1 - x\ln 6)$ and find its maximum value.

(7) Solve $\log_2(x) + 3 = \log_2(3x + 8)$.

(8) Solve $\log_4(x) + 3 = \log_2(x + 8)$.

(9) Solve the equation $5^{2x+9} = 7^x$ in terms of logarithms.

(10) Using properties of logarithms, simplify the expression $\log_4\left(\frac{1}{64}x\right)$.

(11) Using properties of logarithms, solve $4^{x-1} = 3^{2x+2}$.

(12) The Wild Assumption gave the existence of a function b^x satisfying certain properties. Show there can be no more than one such function. **Hint:** Recall the rational numbers were dense in $\mathbb{R}$ and so one can obtain a rational number arbitrarily close to a given real number. Exploit this and the assumed continuity of $\exp_b$ to obtain uniqueness.

(13) Prove the function b^x is concave up and $\log_b(x)$ is concave down whenever $b > 1$.

(14) Using properties of logarithms and exponentials, solve $\log_3(27) + \ln(-3x) = 4 + \ln(3x^2)$.

(15) Let f be a differentiable function and suppose $f(a) \geq 0$ and that $f'(x) \geq 0$ for $x \geq a$. Show that $f(x) \geq f(a)$ for all $x \geq a$. **Hint:** Use the mean value theorem.

(16) Let e be defined by $\ln(e) = 1$ and suppose $e < x < y$. Find a relationship between x^y and y^x. **Hint:** Use Problem 15 and at some point consider the function $h(x) = \frac{\ln x}{x}$.

(17) Suppose f is any function defined on the positive real numbers and $f'(x) = g(x)$ where g is an odd function. ($g(-x) = -g(x)$.) Show $(f(|x|))' = g(x)$.

(18) You know $(e^x)' = e^x$. Use the definition of the derivative to verify that $(e^{ax})' = ae^{ax}$. Show that Ae^{ax} solves the differential equation $y' = ay$ along with the initial condition $y(0) = A$.

(19) Suppose $y' = ay$ and $y(0) = A$. Does it follow that $y = Ae^{ax}$? You know from Problem 18 that Ae^{ax} is one possibility. Is it the only possibility? **Hint:** Consider $y(x)e^{-ax}$ and use the product rule and Problem 18 to show $(y(x)e^{-ax})' = 0$. Then review the mean value theorem. What does the mean value theorem say about functions whose derivatives are always equal to zero?

(20) Show from the definition of the derivative that
$$(\sin(ax))' = a\cos(ax), \quad (\cos(ax))' = -a\sin(ax).$$
With this and Problem 18 find the derivatives of the following functions using the rules of derivatives. You will need to use the product and maybe the quotient rule along with properties of exponents.

(a) $\sin(3x)5^{6x}$

(b) $\tan(2x)3^{3x+1}$

(c) $\sin(2x)\cos(3x)\tan(4x)2^{3x}$

(21) Using Problems 18 and 20 try to determine real numbers b and a such that $y(t) = e^{bt}\cos at$ and $y(t) = e^{bt}\sin at$ both are solutions of the differential equation $y'' + 2\alpha y' + \beta^2 y = 0$ given that $\alpha^2 - \beta^2 < 0$. This is the equation of damped oscillation and will be discussed more carefully later. Typically y measures some sort of displacement from an equilibrium position and t represents time. Here is a picture of the graph of $y(t) = e^{-.3t}\cos(3t)$. This is called damped oscillation.

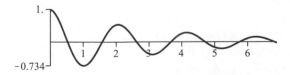

Chapter 5

Properties Of Derivatives

5.1 The Chain Rule And Derivatives Of Inverse Functions

5.1.1 The Chain Rule

The chain rule is one of the most important of differentiation rules. Special cases of it are in Theorem 3.2. Now it is time to consider the theorem in full generality.

Theorem 5.1. *Suppose* $f : (a, b) \to (c, d)$ *and* $g : (c, d) \to \mathbb{R}$. *Also suppose that* $f'(x)$ *exists and that* $g'(f(x))$ *exists. Then* $(g \circ f)'(x)$ *exists and*
$$(g \circ f)'(x) = g'(f(x)) f'(x).$$

Proof: Define
$$H(h) \equiv \begin{cases} \frac{g(f(x+h)) - g(f(x))}{f(x+h) - f(x)} & \text{if } f(x+h) - f(x) \neq 0 \\ g'(f(x)) & \text{if } f(x+h) - f(x) = 0 \end{cases}$$

Then for $h \neq 0$,
$$\frac{g(f(x+h)) - g(f(x))}{h} = H(h) \frac{f(x+h) - f(x)}{h}.$$

Note that $\lim_{h \to 0} H(h) = g'(f(x))$ due to Theorems 2.8 on Page 60 and 3.1 on Page 89. Therefore, taking the limit and using Theorem 2.6,
$$\lim_{h \to 0} \frac{g(f(x+h)) - g(f(x))}{h} = g'(f(x)) f'(x). \qquad \blacksquare$$

Example 5.1. Let $f(x) = \ln \left| \ln \left(x^4 + 1 \right) \right|$. Find $f'(x)$.

From the chain rule,
$$f'(x) = \ln' \left(\ln \left(x^4 + 1 \right) \right) \left(\ln \left(x^4 + 1 \right) \right)' = \frac{1}{\ln \left(x^4 + 1 \right)} \frac{1}{x^4 + 1} \left(x^4 \right)'$$
$$= \left(\frac{1}{\ln \left(x^4 + 1 \right)} \right) \left(\frac{1}{x^4 + 1} \right) \left(4x^3 \right).$$

Example 5.2. Let $f(x) = (2 + \ln |x|)^3$. Find $f'(x)$.

Use the chain rule again. Thus
$$f'(x) = 3 \left(2 + \ln |x| \right)^2 \left(2 + \ln |x| \right)' = \frac{3}{x} \left(2 + \ln |x| \right)^2.$$

5.1.2 *Implicit Differentiation*

Sometimes a function is not given explicitly in terms of a formula. For example, you might have $x^2 + y^2 = 4$. This relation defines y as a function of x near a given point such as $(0, 1)$. Near this point, $y = \sqrt{4 - x^2}$. Near the point $(0, -1)$, you have $y = -\sqrt{4 - x^2}$. Near the point $(1, 0)$, you cannot solve for y in terms of x but you can solve for x in terms of y. Thus near $(1, 0)$, $x = \sqrt{4 - y^2}$. This was a simple example but in general, you can't use algebra to solve for one of the variables in terms of the others even if the relation defines that variable as a function of the others. Here is an example in which, even though it is impossible to find $y(x)$ you can still find the derivative of y. The procedure by which this is accomplished is nothing more than the chain rule and other rules of differentiation.

Example 5.3. Suppose y is a differentiable function of x and $y^3 + 2yx = x^3 + 7 +$ $\ln |y|$. Find $y'(x)$.

This illustrates the technique of implicit differentiation. If you believe y is some differentiable function of x, then you can differentiate both sides with respect to x and write, using the chain rule and product rule.

$$3y^2 y' + 2xy' + 2y = 3x^2 + \frac{y'}{y}.$$

Now you can solve for y' and obtain $y' = -\frac{2y - 3x^2}{3y^3 + 2xy - 1} y.$

Of course there are significant mathematical considerations which are being ignored when it is assumed y is a differentiable function of x. It turns out that for problems like this, the equation relating x and y actually does define y as a differentiable function of x near points where it makes sense to formally solve for y' as just done. The theorems which give this justification are called the implicit and inverse function theorems. They are some of the most profound theorems in mathematics and are topics for advanced calculus. The interested reader should consult the book by Rudin, [23] for this and generalizations of all the hard theorems given in this book. I have also given a simpler proof of the Implicit function theorem in an appendix of this book (Volume 2). One case is of special interest in which $y = f(x)$ and it is desired to find $\frac{dx}{dy}$ or in other words, the derivative of the inverse function.

It happens that if f is a differentiable one to one function defined on an interval $[a, b]$, and $f'(x)$ exists and is nonzero then the inverse function f^{-1} has a derivative at the point $f(x)$. Recall that f^{-1} is defined according to the formula

$$f^{-1}(f(x)) = x.$$

Definition 5.1. Let $f : [a, b] \to \mathbb{R}$ be a continuous function. Define

$$f'(a) \equiv \lim_{x \to a+} \frac{f(x) - f(a)}{x - a}, \ f'(b) \equiv \lim_{x \to b-} \frac{f(x) - f(b)}{x - b}.$$

Recall the notation $x \rightarrow a+$ means that only $x > a$ are considered in the definition of limit. The notation $x \rightarrow b-$ is defined similarly. Thus, this definition includes the derivative of f at the endpoints of the interval and to save notation,

$$f'(x_1) \equiv \lim_{x \to x_1} \frac{f(x) - f(x_1)}{x - x_1}$$

where it is understood that x is always in $[a, b]$.

Theorem 5.2. *Let $f : [a, b] \rightarrow \mathbb{R}$ be continuous and one to one. Suppose $f'(x_1)$ exists for some $x_1 \in [a, b]$ and $f'(x_1) \neq 0$. Then $(f^{-1})'(f(x_1))$ exists and is given by the formula $(f^{-1})'(f(x_1)) = \frac{1}{f'(x_1)}$.*

Proof: By Lemma 2.1, and Corollary 2.2 on Page 55 f is either strictly increasing or strictly decreasing and f^{-1} is continuous. Therefore there exists $\eta > 0$ such that if $0 < |f(x_1) - f(x)| < \eta$, then

$$0 < |x_1 - x| = |f^{-1}(f(x_1)) - f^{-1}(f(x))| < \delta$$

where δ is small enough that for $0 < |x_1 - x| < \delta$,

$$\left| \frac{x - x_1}{f(x) - f(x_1)} - \frac{1}{f'(x_1)} \right| < \varepsilon.$$

It follows that if $0 < |f(x_1) - f(x)| < \eta$,

$$\left| \frac{f^{-1}(f(x)) - f^{-1}(f(x_1))}{f(x) - f(x_1)} - \frac{1}{f'(x_1)} \right| = \left| \frac{x - x_1}{f(x) - f(x_1)} - \frac{1}{f'(x_1)} \right| < \varepsilon$$

Therefore, since $\varepsilon > 0$ is arbitrary,

$$\lim_{y \to f(x_1)} \frac{f^{-1}(y) - f^{-1}(f(x_1))}{y - f(x_1)} = \frac{1}{f'(x_1)} \qquad \blacksquare$$

The following obvious corollary comes from the above by not bothering with end points.

Corollary 5.1. *Let $f : (a, b) \rightarrow \mathbb{R}$ be continuous and one to one. Suppose $f'(x_1)$ exists for some $x_1 \in (a, b)$ and $f'(x_1) \neq 0$. Then $(f^{-1})'(f(x_1))$ exists and is given by the formula $(f^{-1})'(f(x_1)) = \frac{1}{f'(x_1)}$.*

This is one of those theorems which is very easy to remember if you neglect the difficult questions and instead focus on formal manipulations. Consider the following.

$$f^{-1}(f(x)) = x.$$

Now use the chain rule on both sides to write

$$(f^{-1})'(f(x)) f'(x) = 1,$$

and then divide both sides by $f'(x)$ to obtain

$$\left(f^{-1}\right)'\left(f\left(x\right)\right) = \frac{1}{f'\left(x\right)}.$$

Of course this gives the conclusion of the above theorem rather effortlessly and it is formal manipulations like this which aid in remembering formulas such as the one given in the theorem.

Example 5.4. Let $f(x) = \ln\left(1 + x^2\right) + x^3 + 7$. Show that f has an inverse and find $\left(f^{-1}\right)'(7)$.

I am not able to find a formula for the inverse function. This is typical in useful applications so you need to get used to this idea. The methods of algebra are insufficient to solve hard problems in analysis. You need something more. The question is to determine whether f has an inverse. To do this,

$$f'(x) = \frac{2x}{1 + x^2} + 3x^2 + 7 > -1 + 3x^2 + 7 > 0$$

By Corollary 3.3 on Page 103, this function is strictly increasing on $\mathbb{R}$ and so it has an inverse function although I have no idea how to find an explicit formula for this inverse function. However, I can see that $f(0) = 7$ and so by the formula for the derivative of an inverse function,

$$\left(f^{-1}\right)'(7) = \left(f^{-1}\right)'\left(f\left(0\right)\right) = \frac{1}{f'\left(0\right)} = \frac{1}{7}.$$

Example 5.5. Suppose $f(a) = 0$ and $f'(x) = \sqrt{1 + x^4 + \ln\left(1 + x^2\right)}$. Find $\left(f^{-1}\right)'(0)$.

The function f is one to one because it is strictly increasing, due to the fact that its derivative is positive for all x. As in the last example, I have no idea how to find a formula for f^{-1} but I do see that $f(a) = 0$ and so

$$\left(f^{-1}\right)'(0) = \left(f^{-1}\right)'\left(f\left(a\right)\right) = \frac{1}{f'\left(a\right)} = \frac{1}{\sqrt{1 + a^4 + \ln\left(1 + a^2\right)}}.$$

The chain rule has a particularly attractive form in Leibniz's notation. Suppose $y = g(u)$ and $u = f(x)$. Thus $\mathbf{y} = g \circ f(x)$. Then from the above theorem

$$\left(g \circ f\right)'(x) = g'\left(f\left(x\right)\right) f'\left(x\right) = g'\left(u\right) f'\left(x\right)$$

or in other words,

$$\frac{dy}{dx} = \frac{dy}{du}\frac{du}{dx}.$$

Notice how the du's cancel. This particular form is a very useful crutch and is used extensively in applications.

5.2 Exercises

(1) In each of the following, find $\frac{dy}{dx}$.

(a) $y = e^{\sin x}$

(b) $y = \sqrt{7 + x^2 + \sin x}$

(c) $y = \ln\left(x^2 + 1\right)$

(d) $y = \sin\left(\ln\left(x^2 + 1\right)\right)$

(e) $y = \ln\left(\sin\left(x\right) + 3\right)$

(f) $y = \ln\left(\tan\left(x\right)\right)$

(g) $y = \sin^2\left(\ln\left(\tan\left(x\right)\right)\right)$

(h) $y = \sin\left(\left(x + \tan x\right)^6\right)$

(i) $y = \ln\left(\sin\left(x^2 + 7\right)\right)$

(j) $y = \tan\left(\cos\left(x^2\right)\right)$

(k) $y = \log_2\left(\sin\left(x\right) + 6\right)$

(l) $y - \sin\left(\log_3\left(x^2 + 1\right)\right)$

(m) $y = \frac{\sqrt{x^3 + 7}}{\sqrt{\sin(x) + 4}}$

(n) $y = 3^{\tan(\sin(x))}$

(o) $y = \left(\frac{x^2 + 2x}{\tan(x^2 + 1)}\right)^6$

(2) In each of the following, assume the relation defines y as a function of x for values of x and y of interest and use the process of implicit differentiation to find $y'(x)$.

(a) $xy^2 + \sin\left(y\right) = x^3 + 1$

(b) $y^3 + x\cos\left(y^2\right) = x^4$

(c) $y\cos\left(x\right) = \tan\left(y\right)\cos\left(x^2\right) + 2$

(d) $\left(x^2 + y^2\right)^6 = x^3 y + 3$

(e) $\frac{xy^2 + y}{y^5 + x} + \cos\left(y\right) = 7$

(f) $\sqrt{x^2 + y^4}\sin\left(y\right) = 3x$

(g) $y^3 \sin\left(x\right) + y^2 x^2 = 2^{x^2} y + \ln|y|$

(h) $y^2 \sin\left(y\right) x + \log_3\left(xy\right)$
$= y^2 + 11$

(i) $\sin\left(\tan\left(xy^2\right)\right) + y^3 = 16$

(j) $\cos\left(\sec\left(\tan\left(y\right)\right)\right) = 3x$

(3) In each of the following, assume the relation defines y as a function of x for values of x and y of interest and use the process of implicit differentiation to show y satisfies the given differential equation.

(a) $x^2 y + \sin y = 7$, $\left(x^2 + \cos y\right) y' + 2xy = 0$.

(b) $x^2 y^3 + \sin\left(y^2\right) = 5$, $2xy^3 + \left(3x^2 y^2 + 2\left(\cos\left(y^2\right)\right) y\right) y' = 0$.

(c) $y^2 \sin\left(y\right) + xy = 6$,

$$\left(2y\left(\sin\left(y\right)\right)+y^{2}\left(\cos\left(y\right)\right)+x\right)\cdot$$
$$y'+y=0.$$

(d) $\tan\left(x^{2}+y\right)+x^{x}=7,$
$$\left(1+\tan^{2}\left(x^{2}+y\right)\right)y'+$$
$$\left(1+\tan^{2}\left(x^{2}+y\right)\right)2x+$$
$$x^{x}\left(\ln x+1\right)=0.$$

(4) Show that if $D\left(g\right)\subseteq U\subseteq D\left(f\right)$, and if f and g are both one to one, then $f\circ g$ is also one to one.

(5) Using Problem 4 show that the following functions are one to one and find the derivative of the inverse function at the indicated point.

(a) $y=e^{x^{3}+1},e^{2}$

(b) $y=\left(x^{3}+x+2\right),0$

(c) $y=\tan\left(x+\frac{\pi}{4}\right),1$

(d) $y=\tan\left(-x+\frac{\pi}{4}\right),1$

5.3   The Function x^{r} For r A Real Number

Theorem 4.3 on Page 114 says that for $x>0$, and for r a real number,

$$\ln\left(x^{r}\right)=r\ln\left(x\right). \tag{5.1}$$

By this theorem, it also follows that $\ln^{-1}:\mathbb{R}\rightarrow\left(0,\infty\right)$ exists. Then by Corollary 5.1, $\ln^{-1}$ is differentiable. Also, for all $x\in\mathbb{R}$,

$$\ln\left(\ln^{-1}\left(x\right)\right)=x,\ \ln\left(\exp_{e}\left(x\right)\right)=\ln\left(e^{x}\right)=x\ln e=x.$$

Since $\ln$ is one to one, $\ln^{-1}\left(x\right)=\exp_{e}\left(x\right)=\exp\left(x\right)$. Thus

$$\exp\left(\ln x\right)=\ln^{-1}\left(\ln\left(x\right)\right)=x,\ \ln\left(\exp\left(y\right)\right)=y. \tag{5.2}$$

From (4.13) on Page 116, $\exp'\left(x\right)=\exp\left(x\right)$. Recall why this was. From the definition of $\ln$ and the fact $\ln e=1$,

$$\exp'\left(x\right)=\exp'_{e}\left(x\right)\equiv\ln\left(e\right)\exp_{e}\left(x\right)=\exp\left(x\right).$$

With this understanding, it becomes possible to find derivatives of functions raised to arbitrary real powers. First, note that upon taking exp of both sides of (5.1) and using (5.2),

$$x^{r}=\exp\left(r\ln x\right). \tag{5.3}$$

Theorem 5.3. *For $x>0$, $\left(x^{r}\right)'=rx^{r-1}$.*

Proof: Differentiate both sides of (5.3) using the chain rule. From the Wild Assumption on Page 114, in particular, the part about the validity of the laws of exponents,

$$\left(x^{r}\right)'=\exp'\left(r\ln x\right)\frac{r}{x}=\exp\left(r\ln x\right)\frac{r}{x}=\frac{r}{x}x^{r}=rx^{r-1}. \tag{5.4}$$

This shows from (5.4), that $(x^r)' = rx^{r-1}$ as claimed. ∎

Example 5.6. Suppose $f(x)$ is a nonzero differentiable function. Find the derivative of $|f(x)|^r$.

From (5.3),

$$|f(x)|^r = \exp\left(r\ln|f(x)|\right).$$

Therefore,

$$\left(|f(x)|^r\right)' = \exp\left(r\ln|f(x)|\right)\left(r\ln|f(x)|\right)'$$

$$= |f(x)|^r \, r\frac{f'(x)}{f(x)} = r\,|f(x)|^{r-2} \, f(x)\,f'(x).$$

5.3.1 Logarithmic Differentiation

Example 5.7. Let $f(x) = \left(1+x^2\right)^x$. Find $f'(x)$.

One way to do this is to take ln of both sides and use the chain rule to differentiate both sides with respect to x. Thus

$$\ln\left(f(x)\right) = x\ln\left(1+x^2\right)$$

and so, taking the derivative of both sides, using the chain and product rules,

$$\frac{f'(x)}{f(x)} = \frac{2x^2}{1+x^2} + \ln\left(1+x^2\right).$$

Then solve for $f'(x)$ to obtain

$$f'(x) - \left(1+x^2\right)^x \left(\frac{2x^2}{1+x^2} + \ln\left(1+x^2\right)\right)$$

This process is called logarithmic differentiation.

Example 5.8. Let $f(x) = \frac{\sqrt[3]{x^3 + \sin(x)}}{\sqrt[6]{x^4 + 2x}}$. Find $f'(x)$.

You could use the quotient and chain rules but it is easier to use logarithmic differentiation.

$$\ln\left(f(x)\right) = \frac{1}{3}\ln\left(x^3 + \sin(x)\right) - \frac{1}{6}\ln\left(x^4 + 2x\right)$$

Differentiating both sides,

$$\frac{f'(x)}{f(x)} = \frac{1}{3}\left(\frac{3x^2 + \cos x}{x^3 + \sin x}\right) - \frac{1}{6}\frac{4x^3 + 2}{x^4 + 2x}.$$

Therefore, the answer is

$$f'(x) = \frac{\sqrt[3]{x^3 + \sin(x)}}{\sqrt[6]{x^4 + 2x}}\left(\frac{1}{3}\left(\frac{3x^2 + \cos x}{x^3 + \sin x}\right) - \frac{1}{6}\frac{4x^3 + 2}{x^4 + 2x}\right)$$

I think you can see the advantage of doing it this way over using the quotient rule.

5.4 Exercises

(1) Let $f(x) \equiv x^3 + 1$. Find $f^{-1}(y)$. Now find $\left(f^{-1}\right)'(2)$.

(2) Let $f(x) \equiv x^3 + 7x + 3$. Explain why f has an inverse. Find $\left(f^{-1}\right)'(3)$.

(3) Derive the quotient rule from the product rule and the chain rule. This shows you don't need to remember the wretched quotient rule if you don't want to. It follows from two rules which you cannot survive without.

(4) What is wrong with the following "proof" of the chain rule? Here $g'(f(x))$ exists and $f'(x)$ exists.

$$\lim_{h \to 0} \frac{g(f(x+h)) - g(f(x))}{h}$$

$$= \lim_{h \to 0} \frac{g(f(x+h)) - g(f(x))}{f(x+h) - f(x)} \frac{f(x+h) - f(x)}{h}$$

$$= g'(f(x)) f'(x).$$

(5) Is the derivative of a function always continuous? **Hint:** Consider a differentiable function f which is periodic of period 1 and nonconstant. (Periodic of period 1 means $f(x+1) = f(x)$ for all $x \in \mathbb{R}$.) Now consider

$$h(x) = \begin{cases} x^2 f\left(\frac{1}{x}\right) & \text{if } x \neq 0 \\ 0 & \text{if } x = 0 \end{cases}.$$

Show $h'(0) = 0$. What is $h'(x)$ for $x \neq 0$? Is h' also periodic of period 1?

(6) Let $f(x) = x^3 + 1$. Find an explicit formula for f^{-1} and use it to compute $\left(f^{-1}\right)'(9)$. Then use the formula given in the theorem of this section to see you get the same answer.

(7) Find the derivatives of the following functions.

(a) $\sin(x^2) \ln(x^2 + 1)$

(b) $\ln(1 + x^2)$

(c) $(x^3 + 1)^6 \sin(x^2 + 7)$

(d) $\ln\left((x^3 + 1)^6 \sin(x^2 + 7)\right)$

(e) $\tan(\sec(\sin(x^2 + 1)))$

(f) $(\sin^2(x^2 + 5))^{\sqrt{7}}$

(8) Use (5.3) or logarithmic differentiation to differentiate the following functions.

(a) $(2 + \sin(x^2 + 6))^{\tan x}$

(b) x^x

(c) $(x^x)^x$

(d) $(\tan^2(x^4 + 4) + 1)^{\cos x}$

(e) $(\sin^2(x))^{\tan x}$

(9) A search light at a prison revolves five times every minute and is located 200 yards from a long straight wall. How fast in yards per minute is the light moving along the wall at a distance of 100 yards from the point of the wall closest to the light? If you are an escaping convict, where is the most dangerous location on this wall?

(10) A circular disk of paper has a circular sector removed and then the edges are joined to form a cone. What is the angle of the removed circular sector which will create the cone of largest volume?

(11) A light pole with a light on top of it is intended to illuminate the edge of a circle of radius 40 feet centered at the base of the light pole. The brightness of the illumination is of the form $k\frac{\cos\theta}{d^2}$ where θ and d are given in the following picture. Find the height x which will result in the brightest illumination at the edge of this circle.

40

(12) * Let θ be the angle between the two equal sides of an isosceles triangle. Suppose the sum of the lengths of the sides of this triangle is L. Find the value of θ which will maximize the area of the triangle.

(13) A window has total perimeter equal to L. It consists of a square surmounted by an equilateral triangle. (All sides equal) Find the dimensions which will make the largest window.

(14) * A car is proceeding down a road near Crystal Falls Michigan at 60 miles per hour. Fifty yards in front of the car there is a suicidal deer standing 50 feet from the road[1]. Consider the angle formed by the road and the line of sight to the deer. How fast is this angle changing? What happens to the rate of change of this angle as the driver gets closer to the deer? This is actually a related rates problem which will be discussed more later.

5.5   The Inverse Trigonometric Functions

It is desired to consider the inverse trigonometric functions. Graphing the function $y = \sin x$, it is clear sin is not one to one on $\mathbb{R}$ and so it is not possible to define an

---

[1]Crystal Falls Michigan is in the upper peninsula. It is a very beautiful place, especially in the fall when the leaves change colors but it is hazardous to drive there, especially at dusk, because of suicidal deer which jump out in front of cars unexpectedly. These are very large deer.

inverse function.

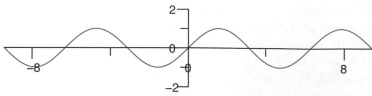

However, a little thing like this will not prevent the definition of useful inverse trig. functions. Observe the function sin, **is** one to one on the interval $\left[-\frac{\pi}{2}, \frac{\pi}{2}\right]$ shown in the above picture as the interval containing zero on which the function climbs from -1 to 1. Now the arcsin function is defined as the inverse of the sin when its domain is restricted to $\left[-\frac{\pi}{2}, \frac{\pi}{2}\right]$. In words, $\arcsin(x)$ is defined to be the angle whose sine is x which lies in $\left[-\frac{\pi}{2}, \frac{\pi}{2}\right]$. From Theorem 5.2 on Page 123 about the derivative of the inverse function, the derivative of $x \to \arcsin(x)$ exists for all $x \in [-1, 1]$. The formula in this theorem could be used to find the derivative of arcsin but it is more useful to simply use that theorem to resolve the existence question and apply the chain rule to find the formula. It is a mistake to memorize too many formulas. Let $y = \arcsin(x)$ so $\sin(y) = x$. Now taking the derivative of both sides,

$$\cos(y)\, y' = 1$$

and so

$$y' = \frac{1}{\cos y} = \frac{1}{\sqrt{1 - \sin^2(y)}} = \frac{1}{\sqrt{1 - x^2}}.$$

The positive value of the square root is used because for $y \in \left[-\frac{\pi}{2}, \frac{\pi}{2}\right]$, $\cos(y) \geq 0$. Thus

$$\frac{1}{\sqrt{1 - x^2}} = \arcsin'(x). \tag{5.5}$$

Next consider the inverse tangent function. Observe that tan is periodic of period π because

$$\tan(x + \pi) = \frac{\sin(x + \pi)}{\cos(x + \pi)} = \frac{\sin(x)\cos(\pi) + \cos(x)\sin(\pi)}{\cos(x)\cos(\pi) - \sin(x)\sin(\pi)}$$

$$= \frac{-\sin(x)}{-\cos(x)} = \tan(x).$$

Therefore, it is impossible to take the inverse of tan. However, tan is one to one on $\left(-\frac{\pi}{2}, \frac{\pi}{2}\right)$ and

$$\lim_{x \to \frac{\pi}{2}^-} \tan(x) = +\infty, \quad \lim_{x \to -\frac{\pi}{2}^+} \tan(x) = -\infty$$

as shown in the following graph of $y = \tan(x)$.

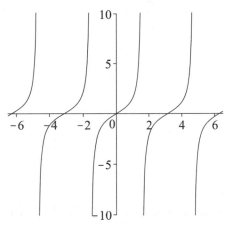

Therefore, arctan (x) for $x \in (-\infty, \infty)$ is defined according to the rule: arctan (x) is the angle whose tangent is x which is in $\left(-\frac{\pi}{2}, \frac{\pi}{2}\right)$. By Theorem 5.2 arctan has a derivative. Therefore, letting $y = \arctan(x)$, $\tan(y) = x$ and by the chain rule,

$$\sec^2(y) y' = 1.$$

Therefore,

$$y' = \frac{1}{\sec^2(y)} = \frac{1}{1 + \tan^2(y)} = \frac{1}{1 + x^2}.$$

and so

$$\frac{1}{1 + x^2} = \arctan'(x). \tag{5.6}$$

The inverse secant function can be defined similarly. There is no agreement on the best way to restrict the domain of sec. I will follow the book by Salas and Hille [25], recognizing that there are good reasons for doing it other ways also. The graph of sec is represented below. There is a vertical asymptote at $x = \frac{\pi}{2}$. Thus

$$\lim_{x \to \frac{\pi}{2}-} \sec(x) = +\infty, \quad \lim_{x \to \frac{\pi}{2}+} \sec(x) = -\infty$$

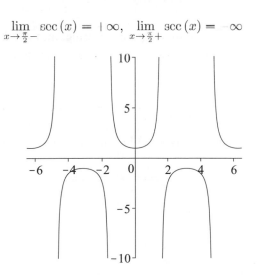

Similar to the case of arcsin and arctan, $\text{arcsec}(x)$ is the angle whose secant is x which lies in $[0, \frac{\pi}{2}) \cup (\frac{\pi}{2}, \pi]$. Let $y = \text{arcsec}(x)$ so $x = \sec(y)$ and using the chain rule,

$$1 = \sec(y)\tan(y)\, y'.$$

Now from the trig. identity $1 + \tan^2(y) = \sec^2(y)$,

$$y' = \frac{1}{\sec(y)\tan(y)}$$

$$= \frac{1}{x\left(\pm\sqrt{x^2 - 1}\right)}$$

and it is necessary to consider what to do with $\pm$. If $y \in [0, \frac{\pi}{2})$, both $x = \sec(y)$ and $\tan(y)$ are nonnegative and so in this case,

$$y' = \frac{1}{x\sqrt{x^2 - 1}} = \frac{1}{|x|\sqrt{x^2 - 1}}.$$

If $y \in (\frac{\pi}{2}, \pi]$, then $x = \sec(y) < 0$ and $\tan(y) \le 0$ so

$$y' = \frac{1}{x(-1)\sqrt{x^2 - 1}} = \frac{1}{(-x)\sqrt{x^2 - 1}} = \frac{1}{|x|\sqrt{x^2 - 1}}.$$

Thus either way,

$$y' = \frac{1}{|x|\sqrt{x^2 - 1}}.$$

This yields the formula

$$\frac{1}{|x|\sqrt{x^2 - 1}} = \text{arcsec}'(x). \tag{5.7}$$

As in the case of ln, there is an interesting and useful formula involving $\text{arcsec}(|x|)$. For $x < 0$, this function equals $\text{arcsec}(-x)$ and so by the chain rule, its derivative equals

$$\frac{1}{|-x|\sqrt{x^2 - 1}}(-1) = \frac{1}{x\sqrt{x^2 - 1}}$$

If $x > 0$, this function equals $\text{arcsec}(x)$ and its derivative equals

$$\frac{1}{|x|\sqrt{x^2 - 1}} = \frac{1}{x\sqrt{x^2 - 1}}$$

and so either way,

$$(\text{arcsec}(|x|))' = \frac{1}{x\sqrt{x^2 - 1}}. \tag{5.8}$$

5.6 The Hyperbolic And Inverse Hyperbolic Functions

The hyperbolic functions are given by

$$\sinh(x) \equiv \frac{e^x - e^{-x}}{2}, \quad \cosh(x) \equiv \frac{e^x + e^{-x}}{2}, \quad \tanh(x) \equiv \frac{\sinh(x)}{\cosh(x)}.$$

The first of these is called the hyperbolic sine and the second the hyperbolic cosine. I imagine you can guess what the third is called. If you guessed "hyperbolic tangent" you got it right. The other hyperbolic functions are defined by analogy to the circular functions.

The reason these are called hyperbolic functions is that

$$\cosh^2 t - \sinh^2 t = 1 \tag{5.9}$$

and so the point $(\cosh t, \sinh t)$ is a point on the hyperbola whose equation is $x^2 - y^2 - 1$. This is not important but is the source for the term hyperbolic. It follows directly from the definitions that

$$\cosh'(x) = \sinh(x), \quad \sinh'(x) = \cosh x.$$

Also, you see that $\sinh(0) = 0$, $\cosh(0) = 1$ and that $\sinh(x) < 0$ if $x < 0$ while $\sinh(x) > 0$ for $x > 0$, but $\cosh(x) > 0$ for all x. Therefore, sinh is an increasing function, concave down for $x < 0$ and concave up for $x > 0$ because $\sinh''(x) = \sinh(x)$ while cosh is decreasing for $x < 0$ and increasing for $x > 0$. Since $\cosh''(x) = \cosh(x)$, the graph of this function is concave up for all x. Thus the graphs of these functions are as follows.

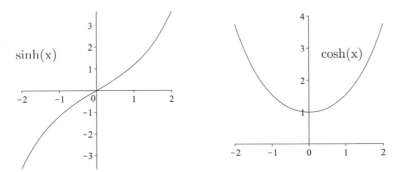

Also, a short computation shows $\tanh(x) = \left(1 - e^{-2x}\right) / \left(1 + e^{-2x}\right)$. Its graph is below.

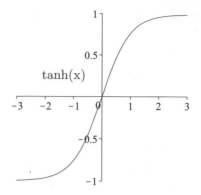

tanh(x)

Since $x \to \sinh(x)$ is strictly increasing, it has an inverse function $\sinh^{-1}(x)$. If $y = \sinh^{-1}(x)$, then $\sinh(y) = x$ and so using the chain rule and the theorem about the existence of the derivative of the inverse function, $y' \cosh(y) = 1$. From the identity (5.9) $\cosh(y) = \sqrt{1 + \sinh^2(y)} = \sqrt{1 + x^2}$. Therefore,

$$y' = \frac{1}{\sqrt{1 + x^2}}$$

which gives the formula

$$\frac{1}{\sqrt{1 + x^2}} = \left(\sinh^{-1}\right)'(x). \qquad (5.10)$$

The derivative of the hyperbolic tangent is also easy to find. This yields after a short computation

$$(\tanh x)' = 1 - \tanh^2 x. \qquad (5.11)$$

Another notation for the inverse hyperbolic functions which is sometimes used is arcsinh or arccosh or arctanh by analogy with the inverse trig. functions.

5.7 Exercises

(1) Verify (5.11).
(2) Simplify the following.
  (a) $\sin(\arctan(1))$
  (b) $\cos(\arctan\sqrt{3})$
  (c) $\tan(\arcsin(\sqrt{3}/2))$
  (d) $\sec(\arcsin(\sqrt{2}/2))$
  (e) $\tan(\arcsin(1/2))$
  (f) $\cos(\arctan(\sqrt{3}) - \arcsin(\sqrt{2}/2))$
  (g) $\sin(\arctan(x))$
  (h) $\cos(\arcsin(x))$
  (i) $\cos(2\arcsin(x))$
  (j) $\sec(2\arctan(x))$

(k) $\tan\left(2\arcsin\left(x\right)+\arctan\left(x^2\right)\right)$

(l) $\sec\left(\arcsin\left(x\right)+2\arccos\left(x\right)\right)$

(3) Find the derivatives and give the domains of the following functions.

(a) $\arcsin\left(x^2+3x\right)$

(b) $\arctan\left(3x+5\right)$

(c) $\operatorname{arcsec}\left(2x+1\right)$

(d) $\arctan\left(\sinh\left(x\right)\right)$

(e) $\sinh\left(\tan\left(3x\right)\right)$

(f) $\cosh\left(\csc\left(3x\right)\right)$

(g) $\tanh\left(\sec\left(3x\right)\right)$

(4) For a and b positive constants, find $\frac{d}{dx}\left(\frac{\arctan(bx/a)}{ab}\right)$.

(5) Find the derivative of

$$\left(\arctan\frac{2}{\sqrt{(9x^2-4)}}+\operatorname{arcsec}\left|\frac{3x}{2}\right|\right).$$

(6) Use the process of implicit differentiation to find y' in the following examples.

(a) * $\arcsin\left(x^2y\right)+\arctan\left(xy^2\right)=\sinh\left(y\right)+3$

(b) $x^2\tanh\left(yx\right)+\arctan\left(x^2\right)=5$

(c) $y^4\sinh\left(x\right)+\arccos\left(x^2+y^2\right)=1$

(d) $\sinh\left(x\tanh\left(x^2+y^2\right)\right)=2$

(7) Find the derivatives of the following functions.

(a) $\sinh^{-1}\left(x^2+7\right)$

(b) $\tanh^{-1}\left(x\right)$

(c) $\sin\left(\sinh^{-1}\left(x^2+2\right)\right)$

(d) $\sin\left(\tanh\left(x\right)\right)$

(e) $x^2\sinh\left(\sin\left(\cos\left(x\right)\right)\right)$

(f) $\left(\cosh\left(x^3\right)\right)^{\sqrt{6}}$

(g) $\left(1+x^4\right)^{\sin x}$

(8) Simplify for $x>0$, $\arcsin x+\arccos x$.

(9) A wonderful identity which was used to compute π for over 200 years[2] is the following.

$$\frac{\pi}{4}=4\arctan\left(\frac{1}{5}\right)-\arctan\left(\frac{1}{239}\right).$$

---

[2] John Machin computed π to 100 decimal places in 1706 through the use of this identity. Later in 1873 William Shanks did it to over 700 places using this identity. The next advance was in 1948, 808 decimal places. After this, computers began to be used and currently π is "known" to millions of decimal places. Many other schemes have been used besides this identity for computing π.

Establish this identity by taking the tangent of both sides and using an appropriate formula for the tangent of the difference of two angles. Use De Moivre's theorem to get some help in finding a formula for $\tan(4\theta)$.

(10) Find a formula for $\tanh^{-1}$ in terms of ln.

(11) Find a formula for $\sinh^{-1}$ in terms of ln.

(12) Prove $1 - \tanh^2 x = \operatorname{sech}^2 x$.

(13) Prove $\coth^2 x - 1 = \operatorname{csch}^2 x$.

(14) What about $\cosh^{-1}$? Define it by restricting the domain of $\cosh$ to be nonnegative numbers? What is $\cosh^{-1}$ in terms of ln?

(15) * Show $\arcsin x = \arctan\left(\frac{x}{\sqrt{1-x^2}}\right)$. It is possible to start with the arctan function and obtain all the other trig functions in terms of this one. If you knew the function arctan explain how to define sin and cos. This is interesting because there is a simple way to define arctan directly as a function of a real variable[15]. Approaches like these avoid all reference to plane geometry.

(16) A divided highway is separated by a median which is $1/10$ of a mile wide. Two cars pass each other, one going west and the other east. The west bound car is traveling at 70 miles per hour while the east bound car is traveling at 60 miles per hour. The situation is described in the following picture.

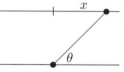

How fast is the distance between the two cars changing when x, shown in the picture, equals $1/10$ mile? How fast is the angle θ, shown in the picture, changing at this instant?

(17) Graph the functions $y = \arctan(x)$ and $y = \arcsin(x)$ on the intervals $[-5, 5]$ and $[-1, 1]$ respectively. Compare to the graphs of the functions $y = \tan(x)$ and $y = \sin(x)$ on the intervals $\left(-\frac{\pi}{2}, \frac{\pi}{2}\right)$ and $\left[-\frac{\pi}{2}, \frac{\pi}{2}\right]$ respectively. Next give the graph of $y = \operatorname{arcsec}(x)$ as defined above.

Chapter 6

Applications Of Derivatives

6.1 L'Hôpital's Rule

There is an interesting rule which is often useful for evaluating difficult limits, called L'Hôpital's[1] rule. The best versions of this rule are based on the Cauchy Mean value theorem, Theorem 3.5 on Page 102.

Theorem 6.1. *Let $[a, b] \subseteq [-\infty, \infty]$ and suppose f, g are functions which satisfy,*

$$\lim_{x \to b-} f(x) = \lim_{x \to b-} g(x) = 0, \qquad (6.1)$$

and f' and g' exist on (a, b) with $g'(x) \neq 0$ on (a, b). Suppose also that

$$\lim_{x \to b-} \frac{f'(x)}{g'(x)} = L. \qquad (6.2)$$

Then

$$\lim_{x \to b-} \frac{f(x)}{g(x)} = L. \qquad (6.3)$$

Proof: By the definition of limit and (6.2) there exists $c < b$ such that if $t > c$, then

$$\left| \frac{f'(t)}{g'(t)} - L \right| < \frac{\varepsilon}{2}.$$

Now pick x, y such that $c < x < y < b$. By the Cauchy mean value theorem, there exists $t \in (x, y)$ such that

$$g'(t)(f(x) - f(y)) = f'(t)(g(x) - g(y)).$$

Since $g'(s) \neq 0$ for all $s \in (a, b)$ it follows $g(x) - g(y) \neq 0$. Therefore,

$$\frac{f'(t)}{g'(t)} = \frac{f(x) - f(y)}{g(x) - g(y)}$$

---

[1]L'Hôpital published the first calculus book in 1696. This rule, named after him, appeared in this book. The rule was actually due to Bernoulli who had been L'Hôpital's teacher. L'Hôpital did not claim the rule as his own but Bernoulli accused him of plagiarism. Nevertheless, this rule has become known as L'Hôpital's rule ever since. The version of the rule presented here is superior to what was discovered by Bernoulli and depends on the Cauchy mean value theorem which was found over 100 years after the time of L'Hôpital.

and so, since $t > c$,

$$\left| \frac{f(x) - f(y)}{g(x) - g(y)} - L \right| < \frac{\varepsilon}{2}.$$

Now letting $y \to b-$,

$$\left| \frac{f(x)}{g(x)} - L \right| \leq \frac{\varepsilon}{2} < \varepsilon.$$

Since $\varepsilon > 0$ is arbitrary, this shows (6.3). ∎

The following corollary is proved in the same way.

Corollary 6.1. *Let $[a, b] \subseteq [-\infty, \infty]$ and suppose f, g are functions which satisfy,*

$$\lim_{x \to a+} f(x) = \lim_{x \to a+} g(x) = 0, \tag{6.4}$$

and f' and g' exist on (a, b) with $g'(x) \neq 0$ on (a, b). Suppose also that

$$\lim_{x \to a+} \frac{f'(x)}{g'(x)} = L. \tag{6.5}$$

Then

$$\lim_{x \to a+} \frac{f(x)}{g(x)} = L. \tag{6.6}$$

Here is a simple example which illustrates the use of this rule.

Example 6.1. Find $\lim_{x \to 0} \frac{5x + \sin 3x}{\tan 7x}$.

The conditions of L'Hôpital's rule are satisfied because the numerator and denominator both converge to 0 and the derivative of the denominator is nonzero for x close to 0. Therefore, if the limit of the quotient of the derivatives exists, it will equal the limit of the original function. Thus,

$$\lim_{x \to 0} \frac{5x + \sin 3x}{\tan 7x} = \lim_{x \to 0} \frac{5 + 3\cos 3x}{7 \sec^2 (7x)} = \frac{8}{7}.$$

Sometimes you have to use L'Hôpital's rule more than once.

Example 6.2. Find $\lim_{x \to 0} \frac{\sin x - x}{x^3}$.

Note that $\lim_{x \to 0} (\sin x - x) = 0$ and $\lim_{x \to 0} x^3 = 0$. Also, the derivative of the denominator is nonzero for x close to 0. Therefore, if $\lim_{x \to 0} \frac{\cos x - 1}{3x^2}$ exists and equals L, it will follow from L'Hôpital's rule that the original limit exists and equals L. However, $\lim_{x \to 0} (\cos x - 1) = 0$ and $\lim_{x \to 0} 3x^2 = 0$ so L'Hôpital's rule can be applied again to consider $\lim_{x \to 0} \frac{-\sin x}{6x}$. From L'Hôpital's rule, if this limit exists and equals L, it will follow that $\lim_{x \to 0} \frac{\cos x - 1}{3x^2} = L$ and consequently $\lim_{x \to 0} \frac{\sin x - x}{x^3} = L$. But from Lemma 4.1 on Page 109, $\lim_{x \to 0} \frac{-\sin x}{6x} = \frac{-1}{6}$. Therefore, by L'Hôpital's rule, $\lim_{x \to 0} \frac{\sin x - x}{x^3} = \frac{-1}{6}$.

Warning: Be sure to check the assumptions of L'Hôpital's rule before using it.

Example 6.3. Find $\lim_{x \to 0+} \frac{\cos 2x}{x}$.

The numerator becomes close to 1 and the denominator gets close to 0. Therefore, the assumptions of L'Hôpital's rule do not hold, and so it does not apply. In fact there is no limit unless you define the limit to equal $+\infty$. Now lets try to use the conclusion of L'Hôpital's rule even though the conditions for using this rule are not verified. Take the derivative of the numerator and the denominator which yields $\frac{-2\sin 2x}{1}$, an expression whose limit as $x \to 0+$ equals 0. This is a good illustration of the above warning.

Some people get the unfortunate idea that one can find limits by doing experiments with a calculator. If the limit is taken as x gets close to 0, these people think one can find the limit by evaluating the function at values of x which are closer and closer to 0. Theoretically, this should work although you have no way of knowing how small you need to take x to get a good estimate of the limit. In practice, the procedure may fail miserably.

Example 6.4. Find $\lim_{x \to 0} \frac{\ln|1+x^{10}|}{x^{10}}$.

This limit equals $\lim_{y \to 0} \frac{\ln|1+y|}{y} = \lim_{y \to 0} \frac{\left(\frac{1}{1+y}\right)}{1} = 1$ where L'Hôpital's rule has been used. This is an amusing example. You should plug .001 in to the function $\frac{\ln|1+x^{10}|}{x^{10}}$ and see what your calculator or computer gives you. If it is like mine, it will give the answer, 0 and will keep on returning the answer of 0 for smaller numbers than .001. This illustrates the folly of trying to compute limits through calculator or computer experiments. Indeed, you could say that a calculator is as useful for taking limits as a bicycle is for swimming.

There is another form of L'Hôpital's rule in which

$$\lim_{x \to b-} f(x) = \pm\infty, \quad \lim_{x \to b-} g(x) = \pm\infty.$$

Theorem 6.2. *Let $[a, b] \subseteq [-\infty, \infty]$ and suppose f, g are functions which satisfy,*

$$\lim_{x \to b-} f(x) = \pm\infty \text{ and } \lim_{x \to b-} g(x) = +\infty, \tag{6.7}$$

and f' and g' exist on (a, b) with $g'(x) \neq 0$ on (a, b). Suppose also

$$\lim_{x \to b-} \frac{f'(x)}{g'(x)} = L. \tag{6.8}$$

Then

$$\lim_{x \to b-} \frac{f(x)}{g(x)} = L. \tag{6.9}$$

Proof: By the definition of limit and (6.8) there exists $c < b$ such that if $t > c$, then

$$\left| \frac{f'(t)}{g'(t)} - L \right| < \frac{\varepsilon}{2}.$$

Now pick x, y such that $c < x < y < b$. By the Cauchy mean value theorem, there exists $t \in (x, y)$ such that

$$g'(t)(f(x) - f(y)) = f'(t)(g(x) - g(y)).$$

Since $g'(s) \neq 0$ on (a, b), it follows $g(x) - g(y) \neq 0$. Therefore,

$$\frac{f'(t)}{g'(t)} = \frac{f(x) - f(y)}{g(x) - g(y)}$$

and so, since $t > c$,

$$\left| \frac{f(x) - f(y)}{g(x) - g(y)} - L \right| < \frac{\varepsilon}{2}.$$

Now this implies

$$\left| \frac{f(y) \left(\frac{f(x)}{f(y)} - 1 \right)}{g(y) \left(\frac{g(x)}{g(y)} - 1 \right)} - L \right| < \frac{\varepsilon}{2}$$

where for all y large enough, both $\frac{f(x)}{f(y)} - 1$ and $\frac{g(x)}{g(y)} - 1$ are not equal to zero. Continuing to rewrite the above inequality yields

$$\left| \frac{f(y)}{g(y)} - L \frac{\left(\frac{g(x)}{g(y)} - 1 \right)}{\left(\frac{f(x)}{f(y)} - 1 \right)} \right| < \frac{\varepsilon}{2} \left| \frac{\left(\frac{g(x)}{g(y)} - 1 \right)}{\left(\frac{f(x)}{f(y)} - 1 \right)} \right|.$$

Therefore, for y large enough,

$$\left| \frac{f(y)}{g(y)} - L \right| \leq \left| L - L \frac{\left(\frac{g(x)}{g(y)} - 1 \right)}{\left(\frac{f(x)}{f(y)} - 1 \right)} \right| + \frac{\varepsilon}{2} \left| \frac{\left(\frac{g(x)}{g(y)} - 1 \right)}{\left(\frac{f(x)}{f(y)} - 1 \right)} \right| < \varepsilon$$

due to the assumption (6.7) which implies

$$\lim_{y \to b-} \frac{\left(\frac{g(x)}{g(y)} - 1 \right)}{\left(\frac{f(x)}{f(y)} - 1 \right)} = 1.$$

Therefore, whenever y is large enough,

$$\left| \frac{f(y)}{g(y)} - L \right| < \varepsilon$$

and this is what is meant by (6.9). ∎

As before, there is no essential difference between the proof in the case where $x \to b-$ and the proof when $x \to a+$. This observation is stated as the next corollary.

Corollary 6.2. *Let $[a, b] \subseteq [-\infty, \infty]$ and suppose f, g are functions which satisfy,*

$$\lim_{x \to a+} f(x) = \pm\infty \text{ and } \lim_{x \to a+} g(x) = \pm\infty, \tag{6.10}$$

and f' and g' exist on (a, b) with $g'(x) \neq 0$ on (a, b). Suppose also that

$$\lim_{x \to a+} \frac{f'(x)}{g'(x)} = L. \tag{6.11}$$

Then

$$\lim_{x \to a+} \frac{f(x)}{g(x)} = L. \tag{6.12}$$

Theorems 6.1, 6.2 and Corollaries 6.1 and 6.2 will be referred to as L'Hôpital's rule from now on. Theorem 6.1 and Corollary 6.1 involve the notion of indeterminate forms of the form $\frac{0}{0}$. Please do not think any meaning is being assigned to the nonsense expression $\frac{0}{0}$. It is just a symbol to help remember the sort of thing described by Theorem 6.1 and Corollary 6.1. Theorem 6.2 and Corollary 6.2 deal with indeterminate forms which are of the form $\frac{\pm\infty}{\infty}$. Again, this is just a symbol which is helpful in remembering the sort of thing being considered. There are other indeterminate forms which can be reduced to these forms just discussed. Do not ever try to assign meaning to such symbols.

Example 6.5. Find $\lim_{y\to\infty}\left(1+\frac{x}{y}\right)^y$.

It is good to first see why this is called an indeterminate form. One might think that as $y\to\infty$, it follows $x/y\to 0$ and so $1+\frac{x}{y}\to 1$. Now 1 raised to anything is 1 and so it would seem this limit should equal 1. On the other hand, if $x > 0$, $1+\frac{x}{y} > 1$ and a number raised to higher and higher powers should approach ∞. It really is not clear what this limit should be. It is an indeterminate form which can be described as 1^∞. By definition,

$$\left(1+\frac{x}{y}\right)^y = \exp\left(y\ln\left(1+\frac{x}{y}\right)\right).$$

Now using L'Hôpital's rule,

$$\lim_{y\to\infty} y\ln\left(1+\frac{x}{y}\right) = \lim_{y\to\infty} \frac{\ln\left(1+\frac{x}{y}\right)}{1/y}$$

$$= \lim_{y\to\infty} \frac{\frac{1}{1+(x/y)}\left(-x/y^2\right)}{(-1/y^2)}$$

$$= \lim_{y\to\infty} \frac{x}{1+(x/y)} = x$$

Therefore,

$$\lim_{y\to\infty} y\ln\left(1+\frac{x}{y}\right) = x$$

Since exp is continuous, it follows

$$\lim_{y\to\infty}\left(1+\frac{x}{y}\right)^y = \lim_{y\to\infty} \exp\left(y\ln\left(1+\frac{x}{y}\right)\right) = e^x.$$

6.1.1 *Interest Compounded Continuously*

Suppose you put money in the bank and it accrues interest at the rate of r per payment period. These terms need a little explanation. If the payment period is one month, and you started with $100 then the amount at the end of one month would equal $100(1+r) = 100+100r$. In this the second term is the interest and the first is called the principal. Now you have $100(1+r)$ in the bank. This becomes the new principal. How much will you have at the end of the second month? By analogy to what was just done it would equal

$$100\left(1+r\right)+100\left(1+r\right)r=100\left(1+r\right)^{2}.$$

In general, the amount you would have at the end of n months is $100\left(1+r\right)^{n}$.

When a bank says they offer 6% compounded monthly, this means r, the rate per payment period equals .06/12. Consider the problem of a rate of r per year and compounding the interest n times a year and letting n increase without bound. This is what is meant by compounding continuously. The interest rate per payment period is then r/n and the number of payment periods after time t years is approximately tn. From the above, the amount in the account after t years is

$$P\left(1+\frac{r}{n}\right)^{nt} \tag{6.13}$$

Recall from Example 6.5 that $\lim_{y\to\infty}\left(1+\frac{x}{y}\right)^{y}=e^{x}$. The expression in (6.13) can be written as

$$P\left[\left(1+\frac{r}{n}\right)^{n}\right]^{t}$$

and so, taking the limit as $n\to\infty$, you get

$$Pe^{rt}=A.$$

This shows how to compound interest continuously.

Example 6.6. Suppose you have $100 and you put it in a savings account which pays 6% compounded continuously. How much will you have at the end of 4 years?

From the above discussion, this would be $100e^{(.06)4}=127.12$. Thus, in 4 years, you would gain interest of about $27.

6.2 Exercises

(1) Find the limits.

(a) $\lim_{x\to0}\frac{3x-4\sin3x}{\tan3x}$

(b) $\lim_{x\to\frac{\pi}{2}-}\left(\tan x\right)^{x-(\pi/2)}$

(c) $\lim_{x\to1}\frac{\arctan(4x-4)}{\arcsin(4x-4)}$

(d) $\lim_{x\to0}\frac{\arctan3x-3x}{x^3}$

(e) $\lim_{x\to0+}\frac{9^{\sec x-1}-1}{3^{\sec x-1}-1}$

(f) $\lim_{x\to0}\frac{3x+\sin4x}{\tan2x}$

(g) $\lim_{x\to\pi/2}\frac{\ln(\sin x)}{x-(\pi/2)}$

(h) $\lim_{x\to0}\frac{\cosh2x-1}{x^2}$

(i) $\lim_{x\to0}\frac{-\arctan x+x}{x^3}$

(j) $\lim_{x\to0}\frac{x^8\sin\frac{1}{x}}{\sin3x}$

(k) $\lim_{x\to\infty} (1+5^x)^{\frac{2}{x}}$

(l) $\lim_{x\to 0} \frac{-2x+3\sin x}{x}$

(m) $\lim_{x\to 1} \frac{\ln(\cos(x-1))}{(x-1)^2}$

(n) $\lim_{x\to 0+} \sin^{\frac{1}{x}} x$

(o) $\lim_{x\to 0} (\csc 5x - \cot 5x)$

(p) $\lim_{x\to 0+} \frac{3^{\sin x}-1}{2^{\sin x}-1}$

(q) $\lim_{x\to 0+} (4x)^{x^2}$

(r) $\lim_{x\to\infty} \frac{x^{10}}{(1.01)^x}$

(s) $\lim_{x\to 0} (\cos 4x)^{(1/x^2)}$

(2) Find the following limits.

(a) $\lim_{x\to 0+} \frac{1-\sqrt{\cos 2x}}{\sin^4(4\sqrt{x})}$

(b) $\lim_{x\to 0} \frac{2^{x^2}-2^{5x}}{\sin\left(\frac{x^2}{5}\right)-\sin(3x)}$

(c) $\lim_{n\to\infty} n\left(\sqrt[n]{7}-1\right)$

(d) $\lim_{x\to\infty} \left(\frac{3x+2}{5x-9}\right)^{x^2}$

(e) $\lim_{x\to\infty} \left(\frac{3x+2}{5x-9}\right)^{1/x}$

(f) $\lim_{n\to\infty} \left(\cos\frac{2x}{\sqrt{n}}\right)^n$

(g) $\lim_{n\to\infty} \left(\cos\frac{2x}{\sqrt{5n}}\right)^n$

(h) $\lim_{x\to 3} \frac{x^x-27}{x-3}$

(i) $\lim_{n\to\infty} \cos\left(\pi\frac{\sqrt{4n^2+13n}}{n}\right)$

(j) $\lim_{x\to\infty} \left(\sqrt[3]{x^3+7x^2}-\sqrt{x^2-11x}\right)$

(k) $\lim_{x\to\infty} \left(\sqrt[5]{x^5+7x^4}-\sqrt[3]{x^3-11x^2}\right)$

(l) $\lim_{x\to\infty} \left(\frac{5x^2+7}{2x^2-11}\right)^{\frac{x}{1-x}}$

(m) $\lim_{x\to\infty} \left(\frac{5x^2+7}{2x^2-11}\right)^{\frac{x\ln x}{1-x}}$

(n) $\lim_{x\to 0+} \frac{\ln\left(e^{2x^2}+7\sqrt{x}\right)}{\sinh(\sqrt{x})}$

(o) $\lim_{x\to 0+} \frac{\sqrt[7]{x}-\sqrt[5]{x}}{\sqrt[9]{x}-\sqrt[11]{x}}$

(3) Find the following limits.

(a) $\lim_{x\to 0+} (1+3x)^{\cot 2x}$

(b) $\lim_{x\to 0} \frac{\sin x-x}{x^2} = 0$

(c) $\lim_{x\to 0} \frac{\sin x-x}{x^3}$

(d) $\lim_{x\to 0} \frac{\tan(\sin x)-\sin(\tan x)}{x^7}$

(e) $\lim_{x\to 0} \frac{\tan(\sin 2x)-\sin(\tan 2x)}{x^7}$

(f) $\lim_{x\to 0} \frac{\sin(x^2)-\sin^2(x)}{x^4}$

(g) $\lim_{x\to 0} \frac{e^{-\left(1/x^2\right)}}{x}$

(h) $\lim_{x\to 0} \left(\frac{1}{x} - \cot(x)\right)$

(i) $\lim_{x\to 0} \frac{\cos(\sin x)-1}{x^2}$

(j) $\lim_{x\to\infty} \left(x^2 \left(4x^4 + 7\right)^{1/2} - 2x^4\right)$

(k) $\lim_{x\to 0} \frac{\cos(x)-\cos(4x)}{\tan(x^2)}$

(l) $\lim_{x\to 0} \frac{\arctan(3x)}{x}$

(m) $\lim_{x\to\infty} \left[\left(x^9 + 5x^6\right)^{1/3} - x^3\right]$

(4) Suppose you want to have \$2000 saved at the end of 5 years. How much money should you place into an account which pays 7% per year compounded continuously?

(5) Using a good calculator, find $e^{.06} - \left(1 + \frac{.06}{360}\right)^{360}$. Explain why this gives a measure of the difference between compounding continuously and compounding daily.

(6) Let $a > 1$. Find

$$\lim_{x\to\infty} \left(\frac{1}{x}\frac{a^x - 1}{a - 1}\right)^{1/x}.$$

This is a case of a problem which appeared on the 1956 Putnam exam. **Hint:** Consider the ln of the function and split it up. It is not too bad if you do this.

6.3   Related Rates

Sometimes some variables are related by a formula and it is known how fast all are changing but one. The related rates problem asks for how fast the remaining variable is changing.

Example 6.7. A cube of ice is melting such that $\frac{dV}{dt} = -4\text{cm}^3/\sec$ where V is the volume. How fast are the sides changing when they are equal to 5 centimeters in length?

The volume is $V = x^3$ where x is the length of a side of the cube. Therefore, the chain rule implies

$$-4 = \frac{dV}{dt} = 3x^2\frac{dx}{dt}$$

and the problem is to find $\frac{dx}{dt}$ when $x = 5$. Therefore,

$$\frac{dx}{dt} = \frac{-4}{3\,(25)} = \frac{-4}{75}\text{cm/second}$$

at this time.

Note there is no way of knowing the volume or the sides as a functions of t.

Example 6.8. One car travels north at 70 miles per hour and the other travels east at 60 miles per hour toward an intersection. How fast is the distance between the two cars changing when this distance equals five miles and the car heading north is at a distance of three miles from the intersection?

Let l denote the distance between the cars. Thus if x is the distance from the intersection of the car traveling east and y is the distance from the intersection of the car traveling north, $l^2 = x^2 + y^2$. When $y = 3$, it follows that $x = 4$. Therefore, at the instant described

$$2ll' = 2xx' + 2yy'$$
$$10l' = 8(-60) + 6(-70)$$

and so $l' = -90$ miles per hour at this instant.

6.4 Exercises

(1) One car travels north at 70 miles per hour toward an intersection and the other travels east at 60 miles per hour away from the intersection. How fast is the distance between the two cars changing when this distance equals five miles and the car heading north is at a distance of three miles from the intersection?

(2) A trash compactor compacts some trash which is in the shape of a box having a square base and a height equal to twice the length of a side of the base. Suppose each side of the base is changing at the rate of -3 inches per second. How fast is the volume changing when the side of the base equals 10 inches?

(3) An isosceles triangle has two sides of equal length. Imagine such a triangle in which the two legs have length 8 inches and denote the included angle by θ and the area by A. Suppose $\frac{dA}{dt} = \sqrt{3}$ square inches per minute. How fast is θ changing when $\theta = \pi/6$ radians?

(4) A point having coordinates (x, y) moves over the ellipse $\frac{x^2}{4} + \frac{y^2}{9} = 2$. If $\frac{dy}{dt} = 2$, find $\frac{dx}{dt}$ at the point $(2, 3)$.

(5) A spectator at a tennis tournament sits 10 feet from the end of the net and on the line determined by the net. He watches the ball go back and forth, and will have a very sore neck when he wakes up the next morning. How fast is the angle between his line of sight and the line determined by the net changing when the ball crosses over the net at a point 12 feet from the end assuming the ball travels at a speed of 60 miles per hour?

(6) The surface area of a sphere of radius r equals $4\pi r^2$ and the volume of the ball of radius r equals $(4/3)\pi r^3$. A balloon in the shape of a ball is being inflated at the rate of 6 cubic inches per minute. How fast is the surface area changing when the volume of the ball equals 20π cubic inches?

(7) * A mother cheetah attempts to fix dinner, a Thompson gazelle, for her hungry children. She moves at 100 feet per second while dinner travels at 80 feet per second. How fast is the distance between her and dinner decreasing when she is located at the point $(0, 40)$ feet and dinner is moving in the direction of the positive x axis at the point $(30, 0)$ feet? **Hint:** Let (x, y) denote the coordinates of the cheetah and let $(z, 0)$ denote the coordinates of the gazelle. Then if l is the desired distance, $l^2 = (x - z)^2 + y^2$. At the instant described, assuming the cheetah moves toward the gazelle at all times, $y'/x' = -4/3$ (why?) and also $\sqrt{(x')^2 + (y')^2} = 100$.

(8) A six foot high man walks at a speed of 5 feet per second away from a light post which is 12 feet high that has the light right on the top. How fast is the end of his shadow moving when he is at a distance of 10 feet from the base of the light pole? How fast is the end of the shadow moving when he is 5 feet from the pole? (Assume he does not walk normally but instead oozes along like a giant amoeba so that his head is always exactly 6 feet above the ground.)

(9) The volume of a right circular cone is $\frac{1}{3}\pi r^2 h$. Grain comes off a conveyor belt and falls to the ground making a right circular cone. It is observed that $r'(t) = .5$ feet per minute and $h'(t) = .3$ feet per minute. It is also known that the rate at which the grain falls off the conveyor belt is 100π cubic feet per minute. When the radius of the cone is 10 feet what is the height of the cone?

(10) A hemispherical dish of radius 5 inches is sitting on a table. Soup is being poured in at the constant rate of 4 cubic inches per second. How fast is the level of soup rising when the radius of the top surface of the soup equals 3 inches? The volume of soup at depth y will be shown later to equal $V(y) = \pi\left(5y^2 - \frac{y^3}{3}\right)$.

(11) A vase of water is sitting on a table.

It will be shown later that if $V(y)$ is the total volume of the vase up to height y, then $\frac{dV}{dy} = A(y)$ where $A(y)$ is the surface area of the top surface of the water at this height. (To see this is very reasonable, note that a little chunk of volume of the vase between heights y and $y + dy$ would be $dV = A(y)\,dy$, area times height.) Also, the rate at which the water evaporates is proportional to the surface area of the exposed water. Thus $\frac{dV}{dt} = -kA(y)$. Show $\frac{dy}{dt}$ is a constant even though the surface area of the exposed water is constantly changing in a typical vase.

(12) A revolving search light at a prison is 1/4 mile from the nearest point of a long wall. It makes one revolution per minute. How fast is the light traveling along the wall at a point which is 1/4 mile from the nearest point? Give your answer in miles per hour.

(13) A painter is on top of a 13 foot ladder which leans against a house. The base of the ladder is moving away from the house at the rate of 2 feet per second causing the top of the ladder to move down the house. How fast is the painter descending when the base of the ladder is at a distance of 5 feet from the house?

(14) A rope fastened to the bow of a row boat has the other end wound around a windlass which is 4 feet above the level of the bow of the boat. The current pulls the boat away at the rate of 2 feet per second. How fast is the rope unwinding when the distance between the bow of the boat and the windlass is 10 feet?

(15) A kite 100 feet above the ground is being blown away from the person holding its string in a direction parallel to the ground at the rate of 10 feet per second. At what rate must the string be let out when the length of string already let out is 200 feet?

(16) A kite is moving horizontally at the rate of 10 feet per second and is 100 feet high. How fast is the angle of elevation of the kite string changing when 200 feet of string have been let out?

(17) A certain volume of an ideal gas satisfies $PV = kT$ where T is the absolute temperature, P is the pressure, V is the volume and k is a constant which depends on the amount of the gas and the sort of gas in the sample. Find a formula for $\frac{dV}{dt}$ in terms of k, P, V and their derivatives.

(18) A disposable cup is made in the shape of a right circular cone with height 5 inches and radius 2 inches. Water flows in to this conical cup at the rate of 4 cubic inches per minute. How fast is the water level rising when the water in the cone is three inches deep? The volume of a cone is $\frac{1}{3}\pi r^2 h$ where r is the radius and h is the height.

(19) The two equal sides of an isosceles triangle have length x inches and the third leg has length y inches. Suppose $\frac{dx}{dt} = 2$ inches per minute and that the length of the other side changes in such a way that the area of the triangle is always 10 square inches. For θ the angle between the two equal sides, find $\frac{d\theta}{dt}$ when $x = 5$ inches.

(20) A object is moving over the ellipse whose equation is $\frac{x^2}{9} + \frac{y^2}{4} = 1$. Near the point $\left(\frac{3\sqrt{2}}{2}, \sqrt{2}\right)$, it is observed that x is changing at the rate of 1 unit per second. How fast is y changing when the object is at this point?

(21) * Consider the following diagram.

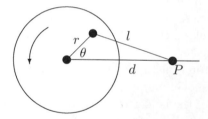

This represents a wheel which is spinning at a constant angular velocity equal to ω. Thus $\frac{d\theta}{dt} = \omega$. The circles represent axles and the two top lines joining the circles represent two rigid bars. The circle on the far right is allowed to slide on the indicated line. Thus, as the wheel turns, the point P moves back and forth on the horizontal line shown. Find the velocity of P when $\theta = \pi/4$ in terms of r, l, and ω. **Hint:** Remember the law of cosines.

6.5 The Derivative And Optimization

There are existence theorems such as Theorem 2.25 on Page 81 which ensure a maximum or minimum of a function exists, but in this section the goal is to give ways to find the maximum or minimum values of a function.

Suppose f is continuous on $[a, b]$. The minimum or maximum could occur at either end point or it could occur at a point in the open interval (a, b). If it occurs at a point of the open interval (a, b), say at x_0, and if $f'(x_0)$ exists, then from Theorem 3.3 on Page 96, $f'(x_0) = 0$. Therefore, the following simple procedure can be used to locate the maximum or minimum of a function f. Find all points x, in (a, b) where $f'(x) = 0$ and all points x, in (a, b) where $f'(x)$ does not exist. Then consider these points along with the end points of the interval. Evaluate f at the end points and at these points where the derivative is zero or does not exist. The largest must be the maximum value of f on the interval $[a, b]$, and the smallest must be the minimum value of f on the interval $[a, b]$. Typically, this involves checking only finitely many points.

Sometimes there are no end points. In this case, you do not necessarily know a maximum value or a minimum value of a continuous function even exists. However, if the function is differentiable and if a maximum or minimum exists, it can still be found by looking at the points where the derivative equals zero.

Example 6.9. Find the maximum and minimum values of the function $f(x) = x^3 - 3x + 1$ on the interval $[-2, 2]$.

The points where $f'(x) = 3x^2 - 3 = 0$ are $x = 1$ or -1. There are no points where the derivative does not exist. Therefore, evaluate the function at $-1, -2, 2$, and 1. Thus $f(-1) = 3, f(1) = -1, f(-2) = -1$, and $f(2) = 3$. Therefore, the maximum value of the function occurs at the point -1 and 2 and has the value of 3

while the minimum value of the function occurs at -1 and -2 and equals -1. The following is a graph of this function.

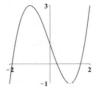

Example 6.10. Find the maximum and minimum values of the function $f(x) = |x^2 - 1|$ on the interval $[-.5, 2]$.

You should verify that this function fails to have a derivative at the point $x = 1$. For $x \in (-.5, 1)$ the function equals $1 - x^2$ and so its derivative equals zero at $x = 0$. For $x > 1$, the function equals $x^2 - 1$ and so there is no point larger than 1 where the derivative equals zero. Therefore, the points to look at are the end points $-.5, 2$, the points where the derivative fails to exist, 1 and the point where the derivative equals zero, $x = 0$. Now $f(-.5) = .75, f(0) = 1, f(1) = 0$, and $f(2) = 3$. It follows the function achieves its maximum at the end point $x = 2$ and its minimum at the point $x = 1$ where the derivative fails to exist. The following is a graph of this function.

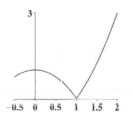

Example 6.11. Find the minimum value of the function $f(x) = x + \frac{1}{x}$ for $x \in (0, \infty)$.

The graph of this function is given below.

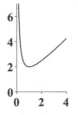

From the graph, it seems there should exist a minimum value at the bottom of the graph. To find it, take the derivative of f and set it equal to zero and then solve for the value of x. Thus $1 - \frac{1}{x^2} = 0$ and so $x = 1$. The solution to the equation $x = -1$ is of no interest because it is not greater than zero. Therefore, the minimum value of the function on the interval $(0, \infty)$ equals $f(1) = 2$ as suggested by the graph.

Example 6.12. An eight foot high wall stands one foot from a warehouse. What is the length of the shortest ladder which extends from the ground to the warehouse.

A diagram of this situation is the following picture.

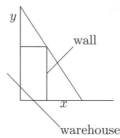

In this picture, the slanted line represents the ladder and x and y are as shown. By similar triangles, $y/1 = 8/x$. Therefore, $xy = 8$. From the Pythagorean theorem the length of the ladder equals $\sqrt{1 + y^2} + \sqrt{x^2 + 64}$. Now using the relation between x and y, the function of a single variable x, to minimize is

$$f(x) = \sqrt{1 + \frac{64}{x^2}} + \sqrt{x^2 + 64}.$$

Clearly $x > 0$ so there are no endpoints to worry about. (Why?) Also the function is differentiable and so it suffices to consider only points where the derivative equals zero. This is a little messy but finally

$$f'(x) = \frac{1}{\sqrt{(x^2 + 64)}} \frac{-64 + x^3}{x^2}$$

and the value where this equals zero is $x = 4$. It follows the shortest ladder is of length $\sqrt{1 + \frac{64}{4^2}} + \sqrt{4^2 + 64} = 5\sqrt{5}$ feet.

Example 6.13. Fermat's principle says that light travels on a path which will minimize the total time. Consider the following picture of light passing from (x_1, y_1) to (x_2, y_2) as shown. The angle θ_1 is called the angle of incidence while the angle θ_2 is called the angle of refraction. The picture indicates a situation in which $c_1 > c_2$.

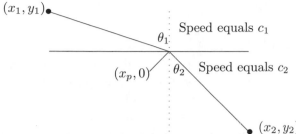

Then define by x the quantity $x_p - x_1$. What is the relation between θ_1 and θ_2?

The time it takes for the light to go from (x_1, y_1) to the point $(x_p, 0)$ equals

$$\sqrt{x^2 + y_1^2}/c_1$$

and the time it takes to go from $(x_p, 0)$ to (x_2, y_2) is $\sqrt{(x_2 - x_1 - x)^2 + y_2^2}/c_2$.

Therefore, the total time is $T = \dfrac{\sqrt{x^2 + y_1^2}}{c_1} + \dfrac{\sqrt{(x_2 - x_1 - x)^2 + y_2^2}}{c_2}$. Thus T is minimized if

$$\frac{dT}{dx} = \frac{d}{dx}\left(\frac{\sqrt{x^2 + y_1^2}}{c_1} + \frac{\sqrt{(x_2 - x_1 - x)^2 + y_2^2}}{c_2} \right)$$

$$= \frac{x}{c_1 \sqrt{x^2 + y_1^2}} - \frac{(x_2 - x_1 - x)}{c_2 \sqrt{(x_2 - x_1 - x)^2 + y_2^2}}$$

$$= \frac{\sin(\theta_1)}{c_1} - \frac{\sin(\theta_2)}{c_2} = 0$$

at this point. Therefore, this yields the desired relation between θ_1 and θ_2. This is called Snell's law.

6.6   Exercises

(1) Find the maximum and minimum values for the following functions defined on the given intervals.

(a) $x^3 - 3x^2 + x - 7$, $[0, 4]$

(b) $\ln(x^2 - x + 2)$, $[0, 2]$

(c) $x^3 + 3x$, $[-1, 10]$

(d) $\frac{x^2 + 1 + 3x^3}{3x^2 + 5}$, $[-1, 1]$

(e) $\sin(x^3 - x)$, $[-1, 1]$

(f) $x^2 - x\tan x$, $[-1, 1]$

(g) $1 - 2x^2 + x^4$, $[-2, 2]$

(h) $\ln(2 - 2x^2 + x^4)$, $[-1, 2]$

(i) $x^2 + 4x - 8$, $[-4, 2]$

(j) $x^2 - 3x + 6$, $[-2, 4]$

(k) $-x^2 + 3x$, $[-4, 2]$

(l) $x + \frac{1}{x}$, $(0, \infty)$

(2) A cylindrical can is to be constructed to hold 30 cubic inches. The top and bottom of the can are constructed of a material costing one cent per square inch and the sides are constructed of a material costing 2 cents per square inch. Find the minimum cost for such a can.

(3) Two positive numbers sum to 8. Find the numbers if their product is to be as large as possible.

(4) The ordered pair (x, y) is on the ellipse $x^2 + 4y^2 = 4$. Form the rectangle which has (x, y) as one end of a diagonal and $(0, 0)$ at the other end. Find the rectangle of this sort which has the largest possible area.

(5) A rectangle is inscribed in a circle of radius r. Find the formula for the rectangle of this sort which has the largest possible area.

(6) A point is picked on the ellipse $x^2 + 4y^2 = 4$ which is in the first quadrant. Then a line tangent to this point is drawn which intersects the x axis at a point x_1 and the y axis at the point y_1. The area of the triangle formed by the y axis, the x axis, and the line just drawn is thus $\frac{x_1 y_1}{2}$. Out of all possible triangles formed in this way, find the one with smallest area.

(7) Find maximum and minimum values if they exist for the function $f(x) = \frac{\ln x}{x}$ for $x > 0$.

(8) Describe how you would find the maximum value of the function $f(x) = \frac{\ln x}{2 + \sin x}$ for $x \in (0, 6)$ if it exists. **Hint:** You might want to use a calculator to graph this and get an idea what is going on.

(9) A rectangular beam of height h and width w is to be sawed from a circular log of radius 1 foot. Find the dimensions of the strongest such beam assuming the strength is of the form $kh^2 w$. Here k is some constant which depends on the type of wood used.

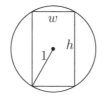

(10) A farmer has 600 feet of fence with which to enclose a rectangular piece of land that borders a river. If he can use the river as one side, what is the largest area that he can enclose.

(11) An open box is to be made by cutting out little squares at the corners of a rectangular piece of cardboard which is 20 inches wide and 40 inches long and then folding up the rectangular tabs which result. What is the largest possible volume which can be obtained?

(12) A feeding trough is to be made from a rectangular piece of metal which is 3 feet wide and 12 feet long by folding up two rectangular pieces of dimension one foot by 12 feet. What is the best angle for this fold?

(13) Find the dimensions of the right circular cone which has the smallest area given the volume is 30π cubic inches. The volume of the right circular cone is $(1/3)\pi r^2 h$ and the area of the cone is $\pi r \sqrt{h^2 + r^2}$.

(14) A wire of length 10 inches is cut into two pieces, one of length x and the other of length $10 - x$. One piece is bent into the shape of a square and the other piece is bent into the shape of a circle. Find the two lengths such that the sum of the areas of the circle and the square is as large as possible. What are the lengths if the sum of the two areas is to be as small as possible.

(15) A hiker begins to walk to a cabin in a dense forest. He is walking on a road which runs from East to West and the cabin is located exactly one mile north of a point two miles down the road. He walks 5 miles per hour on the road but only 3 miles per hour in the woods. Find the path which will minimize the time it takes for him to get to the cabin.

(16) A park ranger needs to get to a fire observation tower which is one mile from a long straight road in a dense forest. The point on the road closest to the observation tower is 10 miles down the road on which the park ranger is standing. Knowing that he can walk at 4 miles per hour on the road but only one mile per hour in the forest, how far down the road should he walk before entering the forest, in order to minimize the travel time?

(17) A refinery is on a straight shore line. Oil needs to flow from a mooring place for oil tankers to this refinery. Suppose the mooring place is two miles off shore from a point on the shore 8 miles away from the refinery which is also on the shore and that it costs five times as much to lay pipe under water than above the ground. Describe the most economical route for a pipeline from the mooring place to the refinery.

(18) Two hallways, one 5 feet wide and the other 6 feet wide meet. It is desired to carry a ladder horizontally around the corner. What is the longest ladder which can be carried in this way? **Hint:** Consider a line through the inside corner which extends to the opposite walls. The shortest such line will be the length of the longest ladder. You might also consider Example 4.1 on Page 111.

(19) A triangle is inscribed in a circle in such a way that one side of the triangle is always the same length, s. Show that out of all such triangles the maximum area is obtained when the triangle is an isosceles triangle. **Hint:** From theorems in plane geometry, the angle opposite the side having fixed length is a constant, no matter how you draw the triangle. (See Problem 32 on Page 15 to see why this is if the geometry interests you.) Use the law of sines and this fact. In the following picture, α is a constant.

(20) A window is to be constructed for the wall of a church which is to consist of a rectangle of height b surmounted by a half circle of radius a. Suppose the total perimeter of the window is to be no more than $4\pi + 8$ feet. Find the dimensions of the window which will admit the most light.

(21) You know $\lim_{x \to \infty} \ln x = \infty$. Show that if $a > 0$, then $\lim_{x \to \infty} \frac{\ln x}{x^\alpha} = 0$.

(22) * A parabola opens down. The vertex is at the point $(0, a)$ and the parabola intercepts the x axis at the points $(-b, 0)$ and $(b, 0)$. A tangent line to the parabola is drawn in the first quadrant which has the property that the triangle formed by this tangent line and the x and y axes has smallest possible area. Find a relationship between a and b such that the normal line to the point of tangency passes through $(0, 0)$. Also determine what kind of triangle this is.

(23) * Suppose p and q are two positive numbers larger than 1 which satisfy $\frac{1}{p} + \frac{1}{q} = 1$. Now let a and b be two positive numbers and consider $f(t) = \frac{1}{p}(at)^p + \frac{1}{q}\left(\frac{b}{t}\right)^q$ for $t > 0$. Show the minimum value of f is ab. Prove the important inequality,

$$ab \leq \frac{a^p}{p} + \frac{b^q}{q}.$$

(24) * Using Problem 23 establish the following magnificent inequality which is a case of Holder's inequality. For $\frac{1}{p} + \frac{1}{q} = 1$, and a_i, b_i nonnegative numbers,

$$\sum_{i=1}^n a_i b_i \leq \left(\sum_{i=1}^n a_i^p\right)^{1/p} \left(\sum_{i=1}^n b_i^q\right)^{1/q}.$$

Hint: You might consider using the inequality of Problem 23 on the following expression.

$$\sum_{i=1}^n \frac{a_i}{\left(\sum_{i=1}^n a_i^p\right)^{1/p}} \frac{b_i}{\left(\sum_{i=1}^n b_i^q\right)^{1/q}}$$

(25) * Using the inequality of Problem 24 establish the following inequality for $p > 1$ which is called the triangle inequality.

$$\left(\sum_{i=1}^n |a_i + b_i|^p\right)^{1/p} \leq$$

$$\left(\sum_{i=1}^n |a_i|^p\right)^{1/p} + \left(\sum_{i=1}^n |b_i|^p\right)^{1/p}.$$

The reason this inequality is important is that people sometimes define the length of a vector in $\mathbb{R}^n, \mathbf{a} \equiv (a_1, \cdots, a_n)$ in the following way.

$$||\mathbf{a}||_p \equiv \left(\sum_{k=1}^{n} |a_k|^p \right)^{1/p}$$

and the triangle inequality says

$$||\mathbf{a} + \mathbf{b}||_p \leq ||\mathbf{a}||_p + ||\mathbf{b}||_p$$

which turns out to be one of the axioms for a norm.

6.7 The Newton Raphson Method

The Newton Raphson method is a way to get approximations of solutions to various equations. For example, suppose you want to find $\sqrt{2}$. The existence of $\sqrt{2}$ is not difficult to establish by considering the continuous function $f(x) = x^2 - 2$ which is negative at $x = 0$ and positive at $x = 2$. Therefore, by the intermediate value theorem, there exists $x \in (0, 2)$ such that $f(x) = 0$ and this x must equal $\sqrt{2}$. The problem consists of how to find this number, not just to prove it exists. The following picture illustrates the procedure of the Newton Raphson method.

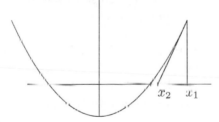

In this picture, a first approximation, denoted in the picture as x_1 is chosen and then the tangent line to the curve $y = f(x)$ at the point $(x_1, f(x_1))$ is obtained. The equation of this tangent line is

$$y - f(x_1) = f'(x_1)(x - x_1).$$

Then extend this tangent line to find where it intersects the x axis. In other words, set $y = 0$ and solve for x. This value of x is denoted by x_2. Thus

$$x_2 = x_1 - \frac{f(x_1)}{f'(x_1)}.$$

This second point x_2 is the second approximation and the same process is done for x_2 that was done for x_1 in order to get the third approximation x_3. Thus

$$x_3 = x_2 - \frac{f(x_2)}{f'(x_2)}.$$

Continuing this way, yields a sequence of points $\{x_n\}$ given by

$$x_{n+1} = x_n - \frac{f(x_n)}{f'(x_n)}. \tag{6.14}$$

which hopefully has the property that $\lim_{n \to \infty} x_n = x$ where $f(x) = 0$. You can see from the above picture that this must work out in the case of $f(x) = x^2 - 2$.

Now carry out the computations in the above case for $x_1 = 2$ and $f(x) = x^2 - 2$. From (6.14),

$$x_2 = 2 - \frac{2}{4} = 1.5, x_3 = 1.5 - \frac{(1.5)^2 - 2}{2(1.5)} \le 1.417,$$

$$x_4 = 1.417 - \frac{(1.417)^2 - 2}{2(1.417)} = 1.414\,216\,302\,046\,577.$$

What is the true value of $\sqrt{2}$? To several decimal places this is $\sqrt{2} = 1.414\,213\,562$ $373\,095$, showing that the Newton Raphson method has yielded a very good approximation after only a few iterations, even starting with an initial approximation 2, which was not very good.

This method does not always work. For example, suppose you wanted to find the solution to $f(x) = 0$ where $f(x) = x^{1/3}$. You should check that the sequence of iterates which results does not converge. This is because, starting with x_1 the above procedure yields $x_2 = -2x_1$ and so as the iteration continues, the sequence oscillates between positive and negative values as its absolute value gets larger and larger.

However, if $f(x_0) = 0$ and $f''(x) > 0$ for x near x_0, you can draw a picture to show that the method will yield a sequence which converges to x_0 provided the first approximation x_1 is taken sufficiently close to x_0. Similarly, if $f''(x) < 0$ for x near x_0, then the method produces a sequence which converges to x_0 provided x_1 is close enough to x_0.

6.8   Exercises

(1) By drawing representative pictures, show convergence of the Newton Raphson method in the cases described above where $f''(x) > 0$ near x_0 where $f(x_0) = 0$ or $f''(x) < 0$ near x_0.
(2) Draw some graphs to illustrate that the Newton Raphson method does not yield a convergent sequence in the case where $f(x) = x^{1/3}$.
(3) Use the Newton Raphson method to approximate the first positive solution of $x - \cos x = 0$.
(4) Use the Newton Raphson method to approximate the first positive solution of $x - \tan x = 0$. **Hint:** You may need to use a calculator to deal with $\tan x$.
(5) Use the Newton Raphson method to compute an approximation to $\sqrt{3}$ which is within 10^{-6} of the true value. Explain how you know you are this close.

(6) Using the Newton Raphson method and an appropriate picture, discuss the convergence of the recursively defined sequence $x_{n+1} = \left((p-1)\,x_n + cx_n^{1-p} \right)/p$ where $x_1, c > 0$ and $p > 1$.

6.9   Review Exercises

(1) Where is the function $\cot\left(\sin\left(x\right)\right)$ continuous?

(2) By completing the square, determine whether the following is an ellipse, hyperbola, or parabola and find its vertices. $-3x^2 - 24x - 46 + 3y^2 + 6y = 0$

(3) Sketch the graph of $f\left(x\right) = \frac{-3x+4}{4x-1}$.

(4) Find $\lim_{x\to 0} \frac{1}{5}\frac{\tan 3x}{x}$.

(5) Find $\lim_{x \to 0} \frac{\ln\left(1+27x^3\right)}{x^3}$.

(6) Where is the function $\cot\left(\cos\left(x\right)\right)$ continuous?

(7) By completing the square, determine whether the following is an ellipse, hyperbola, or parabola and find its vertices. $x^2 - 4x + 2 - y^2 + 2y = 0$

(8) Show the function $f\left(x\right) = x^2$ is continuous at every point x.

(9) Find $\lim_{x \to \infty} \left(\sqrt{\left(1 - x + x^2\right)} - \sqrt{\left(1 - 3x + x^2\right)} \right)$

(10) By completing the square, determine whether the following is an ellipse, hyperbola, or parabola and find its vertices. $-2x^2 + 8x - 11 + y = 0$

(11) Let $f\left(x\right) = -3x^5 + 3x^4 + 6x + \left|\sin 6x\right|$. Show there exists a point x where $f\left(x\right) = 0$.

(12) Find the inverse function $f\left(t\right) = -3t^3 + 3$.

(13) Find the cube roots of 1.

(14) Simplify the following involving complex numbers. $\frac{-3+3i}{1-4i} + -6 - i$

(15) Suppose θ is in $\left[-\pi/2, \pi/2\right]$ and $\sin\theta = -3x/\sqrt{\left(5 + 9x^2\right)}$. Find $\tan\theta, \sec\theta, \cos\theta, \csc\theta, \cot\theta$.

(16) By completing the square, determine whether the following is an ellipse, hyperbola, or parabola and find its vertex if it is a parabola, the lengths of its two axes and center if it is an ellipse and its two vertices if it is a hyperbola. $2x^2 - 12x + 33 + 4y^2 + 16y = 0$

(17) Find $\frac{d}{dx}\left(\arcsin\left(x^5 + 8x\right)\right)$.

(18) Find the intervals where $\sin 5x$ is concave down.

(19) State the Cauchy mean value theorem.

(20) Find y' at every point (x, y) which satisfies the relation $y^2 + y^3\left(3x^2 + 5x\right) = 7$.

(21) Two positive numbers add to 10. Find the numbers if their product is to be as large as possible.

(22) Find y' at every point (x, y) which satisfies the relation $y^2 \sin\left(4x\right) + y\ln\left(4 + x^2\right) = 7$.

(23) Find $f'\left(x\right)$ where $f\left(x\right) = -\frac{3\sin x^2 + x}{1 + 6x}$.

(24) A point having coordinates (x, y) moves over the ellipse $\frac{x^2}{4} + \frac{y^2}{100} = 2$. Find $\frac{dx}{dt}$ at the point $(2, 10)$ if $\frac{dy}{dt} = 3$.

(25) A rectangular garden 287 square feet is to be fenced off against kangaroos. Find the least possible length of fencing if one side of the garden is already protected by a barn.

(26) Find $\lim_{n \to \infty} \frac{-3n^3 + n^2 + 1}{n^3 + n + 7}$.

(27) Let $f(x) = \sin\left(\cos\left(\ln\left(1 + x^5\right)\right)\right)$. Find $f'(x)$.

(28) Find $f'(x)$ where $f(x) = 6^{x^2 + 6x}$.

(29) Suppose a function f satisfies the differential equation

$$f'(x) = (10 + x)^2 (ax - b).$$

Find the intervals on which f is increasing and then find the intervals on which f is concave up.

(30) Let $f(x) = \ln\left(\left(x^2 + 1\right)^5\right)$. Find $f'(x)$.

(31) Let $f(x) = \left(4x^3 + 8x + 3\right)$. Find $\left(f^{-1}\right)'(3)$.

(32) Find $\lim_{n \to \infty} \sum_{k=0}^{n} \left(\frac{1}{8}\right)^k$.

(33) Find $\frac{d}{dx}\left(\sinh^{-1}\left(5x^2 + 7\right)\right)$.

(34) Simplify $\log_4(64)$.

(35) If $\sin\theta = x$ for $|x| < 1$, find $\tan\theta$.

(36) Find the amplitude of the function $f(x) = 5\sin 2x + 3\cos 2x$.

(37) Find $\lim_{x \to 0} \left(\frac{1}{8} \frac{\sin 8x}{x}\right)^{\frac{8}{x^2}}$.

(38) Find $\frac{d}{dx}\left(x^2 + 1\right)^{\sin(3x)}$.

(39) Find $\lim_{x \to 0} \left(\cos(3x)\right)^{\frac{6}{x^2}}$.

(40) Solve $8^{5x+4} = 4^{x+5}$.

(41) Find $\lim_{x \to 0} \left(\cos\left(9x^2\right)\right)^{\frac{4}{x^4}}$.

(42) Find $f'(x)$ where $f(x) = -2x^2$.

(43) Find $\frac{d}{dx}\left(\sinh^{-1}\left(\sin\left(2x^4 + 3\right)\right)\right)$.

(44) Find $\lim_{x \to 0+} \left(1 + 4x\right)^{\cot(4x)}$.

(45) Let $f(x) = \frac{1}{1+x^2}$. Show f is uniformly continuous.

(46) Find $\frac{d}{dx}\left(\sin\left(\sinh^{-1}(2x)\right)\right)$.

(47) Find $\frac{d}{dx}\log_4\left(\sin(9x) + 8\right)$.

(48) Find $\lim_{x \to 0} \frac{6x + 9x^2 + 2\tan 6x}{\sin 2x + \tan 2x^2}$.

(49) Find $\frac{d}{dx}\left(\arcsin(4x + 7)\right)$.

(50) Find $\frac{d}{dx}\left(\arctan\left(\sinh(6x)\right)\right)$.

Chapter 7

Antiderivatives

A differential equation is an equation which involves an unknown function and its derivatives. Differential equations are the unifying idea in this chapter. Many interesting problems may be solved by formulating them as solutions of a suitable differential equation with initial condition called an initial value problem.

7.1 Initial Value Problems

The initial value problem is to find a function $y(x)$ for $x \in [a, b]$ such that
$$y'(x) = f(x, y(x)), \ y(a) = y_0.$$
Various assumptions are made on $f(x, y)$. At this time it is assumed that f does not depend on y and f is a given continuous defined $[a, b]$. Thus the initial value problem of interest here is one of the form
$$y'(x) = f(x), \ y(a) = y_0. \tag{7.1}$$

As an example of an application of an initial value problem, recall the discussion which led to the derivative on Page 87. There $r(t)$ was the x coordinate of a point moving on the x axis at time t and it was shown there that the velocity of this object was $r'(t)$. When this is positive, the object is moving to the right on the x axis and when it is negative, the object is moving to the left. Suppose $r'(t)$ was known, say $r'(t) = f(t)$ along with $r(0) = x_0$ and you wanted to find $r(t)$. Then you are really asking for the solution to an initial value problem of the form
$$r'(t) = f(t), \ r(0) = x_0.$$

Theorem 7.1. *There is at most one solution to the initial value problem 7.1 which is continuous on $[a, b]$.*

Proof: Suppose both $A(x)$ and $B(x)$ are solutions to this initial value problem for $x \in (a, b)$. Then letting $H(x) \equiv A(x) - B(x)$, it follows that $H(a) = 0$ and $H'(x) = A'(x) - B'(x) = f(x) - f(x) = 0$. Therefore, from Corollary 3.2 on Page 103, it follows $H(x)$ equals a constant on (a, b). By continuity of H, this constant must equal $H(a) = 0$. ∎

The main difficulty in solving these initial value problems like (7.1) is in finding a function whose derivative equals the given function. This is in general a very hard problem, although techniques for doing this are presented later which will cover many cases of interest. The functions whose derivatives equal a given function $f(x)$, are called antiderivatives and there is a special notation used to denote them.

Definition 7.1. Let f be a function. $\int f(x)\, dx$ denotes the set of antiderivatives of f. Thus $F \in \int f(x)\, dx$ means $F'(x) = f(x)$. It is customary to refer to $f(x)$ as the integrand. This symbol is also called the indefinite integral and sometimes is referred to as an integral although this last usage is not correct.

The reason this last usage is not correct is that the integral of a function is a single number not a whole set of functions. Nevertheless, you cannot escape the fact that it is common usage to call that symbol an integral.

Lemma 7.1. *Suppose $F, G \in \int f(x)\, dx$ for $x \in (a, b)$. Then there exists a constant C such that for all $x \in (a, b)$, $F(x) = G(x) + C$.*

Proof:

$$F'(x) - G'(x) = f(x) - f(x) = 0$$

for all $x \in (a, b)$. Consequently, by Corollary 3.2 on Page 103, $F(x) - G(x) = C$. ∎

There is another simple lemma about antiderivatives.

Lemma 7.2. *If a and b are nonzero real numbers, and if $\int f(x)\, dx$ and $\int g(x)\, dx$ are nonempty, then*

$$\int (af(x) + bg(x))\, dx = a \int f(x)\, dx + b \int g(x)\, dx$$

Proof: The symbols on the two sides of the equation denote sets of functions. It is necessary to verify the two sets of functions are the same. Suppose then that $F \in \int f(x)\, dx$ and $G \in \int g(x)\, dx$. Then $aF + bG$ is a typical function of the right side of the equation. Taking the derivative of this function, yields $af(x) + bg(x)$ and so this shows the set of functions on the right side is a subset of the set of functions on the left.

Now take $H \in \int (af(x) + bg(x))\, dx$ and pick $F \in \int f(x)\, dx$. Then $aF \in a \int f(x)\, dx$ and

$$(H(x) - aF(x))' = af(x) + bg(x) - af(x) = bg(x)$$

showing that

$$H - aF \in \int bg(x)\, dx = b \int g(x)\, dx$$

because $b \neq 0$. Therefore

$$H \in aF + b \int g(x)\,dx \subseteq a \int f(x)\,dx + b \int g(x)\,dx.$$

This has shown the two sets of functions are the same and proves the lemma. ∎

From Lemma 7.1 it follows that if $F(x) \in \int f(x)\,dx$, then every other function in $\int f(x)\,dx$ is of the form $F(x) + C$ for a suitable constant C. Thus it is customary to write

$$\int f(x)\,dx = F(x) + C$$

where it is understood that C is an arbitrary constant, called a constant of integration.

The formulas for derivatives presented earlier imply the following table of antiderivatives.

Table of antiderivatives

$f(x)$	$\int f(x)\,dx$
$x^n, n \neq -1$	$\dfrac{x^n}{n+1} + C$
x^{-1}	$\ln|x| + C$
$\cos(x)$	$\sin(x) + C$
$\sin(x)$	$-\cos(x) + C$
$\sec^2(x)$	$\tan(x) + C$
e^x	$e^x + C$
$\cosh(x)$	$\sinh(x) + C$
$\sinh(x)$	$\cosh(x) + C$
$\dfrac{1}{\sqrt{1-x^2}}$	$\arcsin(x) + C$
$\dfrac{1}{\sqrt{1+x^2}}$	$\operatorname{arcsinh} x + C$
$\dfrac{1}{x\sqrt{x^2-1}}$	$\operatorname{arcsec}|x| + C$
$\dfrac{1}{1+x^2}$	$\arctan(x) + C$

The above table is a good starting point for other antiderivatives. For example,

Proposition 7.1. Let $\sum_{k=0}^{n} a_k x^k$ be a polynomial. Then

$$\int \sum_{k=0}^{n} a_k x^k\,dx = \sum_{k=0}^{n} a_k \frac{x^{k+1}}{k+1} + C$$

Proof: This follows from the above table and Lemma 7.2. ∎

Example 7.1. Suppose the velocity of an object moving on the x axis is given by the function $\cos(t)$ and when $t = 0$, the object is at 0. Find the position of the object.

As explained above, you need to solve

$$r'(t) = \cos(t), \; r(0) = 0.$$

To do this, note that from the differential equation, $r(t) = \sin(t) + C$ and it only remains to find the constant C such that $r(0) = 0$. Thus $0 = r(0) = \sin(0) + C = C$ and so $r(t) = \sin(t)$.

Example 7.2. Consider the same problem as in Example 7.1 but this time let $r(1) = 5$. What is the answer in this case?

From the differential equation, it is still the case that $r(t) = \sin(t) + C$ but now C needs to be chosen such that $r(1) = 5$. Thus $5 = r(1) = \sin(1) + C$ and so $C = 5 - \sin(1)$. Therefore, the solution to the initial value problem

$$r'(t) = \cos(t), \; r(1) = 5$$

is $r(t) = \sin(t) + (5 - \sin(1))$.

They are all like this. You find an antiderivative. The answer you want is this antiderivative added to an appropriate constant chosen to satisfy the given initial condition. The problem is in finding the antiderivative. In general, this is a very hard problem but there are techniques for solving it in some cases. The next section is one such technique.

7.2 The Method Of Substitution

The method of substitution is based on the following formula which is merely a restatement of the chain rule.

$$\int f(g(x))\, g'(x)\, dx = F(g(x)) + C, \tag{7.2}$$

where $F'(y) = f(y)$. Here are some examples of the method of substitution.

Example 7.3. Find $\int \sin(x)\cos(x)\sqrt{3 + 2^{-1}\sin^2(x)}\, dx$

Note it is of the form given in (7.2) with $g(x) = 2^{-1}\sin^2(x)$ and $F(u) = \frac{2}{3}\left(\sqrt{(3+u)}\right)^3$. Therefore,

$$\int \sin(x)\cos(x)\sqrt{3 + 2^{-1}\sin^2(x)}\, dx = \frac{1}{12}\left(\sqrt{(12 + 2\sin^2 x)}\right)^3 + C.$$

Example 7.4. Find $\int \left(1 + x^2\right)^6 2x\, dx$.

This is a special case of (7.2) when $g(x) = 1 + x^2$ and $F(u) = u^7/7$. Therefore, the answer is

$$\int \left(1 + x^2\right)^6 2x \, dx = \left(1 + x^2\right)^7 / 7 + C.$$

Example 7.5. Find $\int \left(1 + x^2\right)^6 x \, dx$.

This equals

$$\frac{1}{2} \int \left(1 + x^2\right)^6 2x \, dx = \frac{1}{2} \frac{\left(1 + x^2\right)^7}{7} + C = \frac{\left(1 + x^2\right)^7}{14} + C.$$

Actually, it is not necessary to recall (7.2) and massage things to get them in that form. There is a trick based on the Leibnitz notation for the derivative which is very useful and illustrated in the following example.

Example 7.6. Find $\int \cos(2x) \sin^2(2x) \, dx$.

Let $u = \sin(2x)$. Then $\frac{du}{dx} = 2 \cos(2x)$. Now formally

$$\frac{du}{2} = \cos(2x) \, dx.$$

Thus

$$\int \cos(2x) \sin^2(2x) \, dx = \frac{1}{2} \int u^2 \, du = \frac{u^3}{6} + C$$

$$= \frac{(\sin(2x))^3}{6} + C$$

The expression $(1/2) \, du$ replaced the expression $\cos(2x) \, dx$ which occurs in the original problem and the resulting problem in terms of u was much easier. This was solved and finally the original variable was replaced. When using this method, it is a good idea to check your answer to be sure you have not made a mistake. Thus in this example, the chain rule implies $\left(\frac{(\sin(2x))^3}{6}\right)' = \cos(2x) \sin^2(2x)$ which verifies the answer is right. Here is another example.

Example 7.7. Find $\int \sqrt[3]{2x + 7} x \, dx$.

In this example $u = 2x + 7$ so that $du = 2dx$. Then

$$\int \sqrt[3]{2x + 7} x \, dx = \int \sqrt[3]{u} \overbrace{\frac{u - 7}{2}}^{x} \overbrace{\frac{1}{2}}^{dx} \, du$$

$$= \int \left(\frac{1}{4} u^{4/3} - \frac{7}{4} u^{1/3}\right) \, du$$

$$= \frac{3}{28} u^{7/3} - \frac{21}{16} u^{4/3} + C$$

$$= \frac{3}{28} (2x + 7)^{7/3} - \frac{21}{16} (2x + 7)^{4/3} + C$$

Example 7.8. Find $\int x 3^{x^2} \, dx$.

Let $u = 3^{x^2}$ so that $\frac{du}{dx} = 2x \ln(3) 3^{x^2}$ and $\frac{du}{2\ln(3)} = x3^{x^2} dx$. Thus

$$\int x3^{x^2} dx = \frac{1}{2\ln(3)} \int du = \frac{1}{2\ln(3)} [u + C]$$

$$= \frac{1}{2\ln(3)} 3^{x^2} + \left(\frac{1}{2\ln(3)}\right) C$$

Since the constant is an arbitrary constant, this is written as

$$\frac{1}{2\ln(3)} 3^{x^2} + C.$$

Example 7.9. Find $\int \cos^2(x) \, dx$.

Recall that $\cos(2x) = \cos^2(x) - \sin^2(x)$ and $1 = \cos^2(x) + \sin^2(x)$. Then subtracting and solving for $\cos^2(x)$,

$$\cos^2(x) = \frac{1 + \cos(2x)}{2}.$$

Therefore,

$$\int \cos^2(x) \, dx = \int \frac{1 + \cos(2x)}{2} \, dx$$

Now letting $u = 2x$, $du = 2dx$ and so

$$\int \cos^2(x) \, dx = \int \frac{1 + \cos(u)}{4} \, du$$

$$= \frac{1}{4} u + \frac{1}{4} \sin u + C$$

$$= \frac{1}{4} (2x + \sin(2x)) + C.$$

Also

$$\int \sin^2(x) \, dx = -\frac{1}{2} \cos x \sin x + \frac{1}{2} x + C$$

which is left as an exercise.

Example 7.10. Find $\int \tan(x) \, dx$.

Let $u = \cos x$ so that $du = -\sin(x) \, dx$. Then writing the antiderivative in terms of u, this becomes $\int \frac{-1}{u} \, du$. At this point, recall that $(\ln |u|)' = 1/u$. Thus this antiderivative is $-\ln |u| + C = \ln |u^{-1}| + C$ and so $\int \tan(x) \, dx = \ln |\sec x| + C$.

This illustrates a general procedure.

Procedure 7.2. $\int \frac{f'(x)}{f(x)} dx = \ln |f(x)| + C$.

This follows from the chain rule.

Example 7.11. Find $\int \sec(x) \, dx$.

This is usually done by a trick. You write as

$$\int \frac{\sec(x)(\sec(x) + \tan(x))}{(\sec(x) + \tan(x))} dx$$

and note that the numerator of the integrand is the derivative of the denominator. Thus

$$\int \sec(x)\, dx = \ln|\sec(x) + \tan(x)| + C.$$

Example 7.12. Find $\int \csc(x)\, dx$.

This is done like the antiderivatives for the secant.

$$\frac{d}{dx} \csc(x) = -\csc(x)\cot(x)$$

and $\frac{d}{dx} \cot(x) = -\csc^2(x)$. Write the integral as

$$-\int \frac{-\csc(x)(\cot(x) + \csc(x))}{(\cot(x) + \csc(x))} dx = -\ln|\cot(x) + \csc(x)| + C.$$

7.3 Exercises

(1) Find the indicated antiderivatives.
   (a) $\int \frac{x}{\sqrt{2x-3}}\, dx$
   (b) $\int x\left(3x^2 + 6\right)^5 dx$
   (c) $\int x \sin\left(x^2\right) dx$
   (d) $\int \sin^3(2x)\cos(2x)$
   (e) $\int \frac{1}{\sqrt{1+4x^2}}\, dx$ **Hint:** Remember the $\sinh^{-1}$ function and its derivative.

(2) Solve the initial value problems.
   (a) $\frac{dy}{dx} = \frac{x}{\sqrt{2x-3}}, y(2) = 1$
   (b) $\frac{dy}{dx} = 5x\left(3x^2 + 6\right)^5, y(0) = 3$
   (c) $\frac{dy}{dx} = 3x^2 \sin\left(2x^3\right), y(1) = 1$
   (d) $y'(x) = \frac{1}{\sqrt{1+3x^2}}.y(1) = 1$
   (e) $y'(x) = \sec(x), y(0) = 3$
   (f) $y'(x) = x\csc\left(x^2\right), y(1) = 1$

(3) An object moves on the x axis having velocity equal to $\frac{3t^3}{7+t^4}$. Find the position of the object given that at $t = 1$, it is at the point 2.

(4) An object moves on the x axis having velocity equal to $t\sin\left(2t^2\right)$. Find the position of the object given that at $t = 1$, it is at the point 1.

(5) An object moves on the x axis having velocity equal to $\sec(t)$. Find the position of the object given that at $t = 1$, it is at the point -2.

(6) Find the indicated antiderivatives.

(a) $\int \sec(3x)\, dx$

(b) $\int \sec^2(3x)\tan(3x)\, dx$

(c) $\int \frac{1}{3+5x^2}\, dx$

(d) $\int \frac{1}{\sqrt{5-4x^2}}\, dx$

(e) $\int \frac{3}{x\sqrt{4x^2-5}}\, dx$

(7) Find the indicated antiderivatives.

(a) $\int x\cosh(x^2+1)\, dx$

(b) $\int x^3 5^{x^4}\, dx$

(c) $\int \sin(x)\, 7^{\cos(x)}\, dx$

(d) $\int x\sin(x^2)\, dx$

(e) $\int x^5\sqrt{2x^2+1}\, dx$ **Hint:** Let $u = 2x^2 + 1$.

(8) Find $\int \sin^2(x)\, dx$. **Hint:** Derive and use $\sin^2(x) = \frac{1-\cos(2x)}{2}$.

(9) Find the indicated antiderivatives.

(a) $\int \frac{\ln x}{x}\, dx$

(b) $\int \frac{x^3}{3+x^4}\, dx$

(c) $\int \frac{1}{x^2+2x+2}\, dx$ **Hint:** Complete the square in the denominator and then let $u = x + 1$.

(d) $\int \frac{1}{\sqrt{4-x^2}}\, dx$

(e) $\int \frac{1}{x\sqrt{x^2-9}}\, dx$ **Hint:** Let $x = 3u$.

(f) $\int \frac{\ln(x^2)}{x}\, dx$

(g) Find $\int \frac{x^3}{\sqrt{6x^2+5}}\, dx$

(h) Find $\int x\sqrt[3]{6x+4}\, dx$

(10) Find the indicated antiderivatives.

(a) $\int x\sqrt{2x+4}\, dx$

(b) $\int x\sqrt{3x+2}\, dx$

(c) $\int \frac{1}{\sqrt{36-25x^2}}\, dx$

(d) $\int \frac{1}{\sqrt{9-4x^2}}\, dx$

(e) $\int \frac{1}{\sqrt{1+4x^2}}\, dx$

(f) $\int \frac{x}{\sqrt{(3x-1)}}\, dx$

(g) $\int \frac{x}{\sqrt{5x+1}}\, dx$

(h) $\int \frac{1}{x\sqrt{9x^2-4}}\, dx$

(i) $\int \frac{1}{\sqrt{9+4x^2}}\, dx$

(11) Find $\int \frac{1}{x^{1/3}+x^{1/2}}\, dx$. **Hint:** Try letting $x = u^6$ and use long division.

(12) Suppose f is a function defined on $\mathbb{R}$ and it satisfies the functional equation $f(a+b) = f(a) + f(b)$. Suppose also $f'(0) = k$. Find $f(x)$.

(13) Suppose f is a function defined on $\mathbb{R}$ having values in $(0, \infty)$ and it satisfies the functional equation $f(a+b) = f(a) f(b)$. Suppose also $f'(0) = k$. Find $f(x)$.

(14) Suppose f is a function defined on $(0, \infty)$ having values in $\mathbb{R}$ and it satisfies the functional equation $f(ab) = f(a) + f(b)$. Suppose also $f'(1) = k$. Find $f(x)$.

(15) Suppose f is a function defined on $\mathbb{R}$ and it satisfies the functional equation $f(a+b) = f(a) + f(b) + 3ab$. Suppose also that $\lim_{h \to 0} \frac{f(h)}{h} = 7$. Find $f(x)$ if possible.

7.4 Integration By Parts

Another technique for finding antiderivatives is called integration by parts and is based on the product rule. Recall the product rule. If u' and v' exist, then

$$(uv)'(x) = u'(x) v(x) + u(x) v'(x). \qquad (7.3)$$

Therefore,

$$(uv)'(x) - u'(x) v(x) = u(x) v'(x)$$

Proposition 7.2. Let u and v be differentiable functions for which $\int u(x) v'(x) \, dx$ and $\int u'(x) v(x) \, dx$ are nonempty. Then

$$uv - \int u'(x) v(x) \, dx = \int u(x) v'(x) \, dx. \qquad (7.4)$$

Proof: Let $F \in \int u'(x) v(x) \, dx$. Then

$$(uv - F)' = (uv)' - F' = (uv)' - u'v = uv'$$

by the chain rule. Therefore every function from the left in (7.4) is a function found in the right side of (7.4). Now let $G \in \int u(x) v'(x) \, dx$. Then $(uv - G)' = -uv' + (uv)' = u'v$ by the product rule. It follows that $uv - G \in \int u'(x) v(x) \, dx$ and so $G \in uv - \int u'(x) v(x) \, dx$. Thus every function from the right in (7.4) is a function from the left. ∎

Example 7.13. Find $\int x \sin(x) \, dx$.

Let $u(x) = x$ and $v'(x) = \sin(x)$. Then applying (7.4),

$$\int x \sin(x) \, dx = (-\cos(x))x - \int (-\cos(x)) \, dx$$

$$= -x \cos(x) + \sin(x) + C.$$

Example 7.14. Find $\int x \ln(x) \, dx$.

Let $u(x) = \ln(x)$ and $v'(x) = x$. Then from (7.4),

$$\int x \ln(x) \, dx = \frac{x^2}{2} \ln(x) - \int \frac{x^2}{2} \left(\frac{1}{x}\right)$$

$$= \frac{x^2}{2} \ln(x) - \int \frac{x}{2}$$

$$= \frac{x^2}{2} \ln(x) - \frac{1}{4}x^2 + C$$

Example 7.15. Find $\int \arctan(x) \, dx$.

Let $u(x) = \arctan(x)$ and $v'(x) = 1$. Then from (7.4),

$$\int \arctan(x) \, dx = x \arctan(x) - \int x \left(\frac{1}{1+x^2}\right) dx$$

$$= x \arctan(x) - \frac{1}{2} \int \frac{2x}{1+x^2} \, dx$$

$$= x \arctan(x) - \frac{1}{2} \ln\left(1+x^2\right) + C.$$

Sometimes you want to find antiderivatives for something like $\int f g \, dx$ where $f^{(m)} = 0$ for some positive integer m. For example, $\int x^5 \sin x \, dx$. If you do integration by parts repeatedly, what do you get? Let $G'_1 = g, G'_2 = G_1, G'_3 = G_2$ etc. Then the first application of integration by parts yields $f G_1 - \int G_1 f' dx$. The next application of integration by parts yields $f G_1 - G_2 f' + \int G_2 f'' dx$. Yet another application of integration by parts yields $f G_1 - G_2 f' + G_3 f'' - \int G_3 f''' dx$. Eventually the process will stop because a high enough derivative of f equals zero. This justifies the following procedure for finding antiderivatives in this case.

Procedure 7.3. Suppose $f^{(m)} = 0$ for some m a positive integer and let $G'_k = G_{k-1}$ for all k and $G_0 = g$. Then

$$\int fg \, dx = f G_1 - f' G_2 + f'' G_3 - f''' G_4 + \cdots$$

Just keep writing these terms, alternating signs until the process yields a zero. Then add on an arbitrary constant of integration and stop. Sometimes people remember this in the form of a table.

$$g$$

$$
\begin{array}{ccc}
f & \overset{+}{\to} & G_1 \\
f' & \overset{-}{\to} & G_2 \\
f'' & \overset{+}{\to} & G_3 \\
f''' & \overset{-}{\to} & G_4
\end{array}
$$

Thus you fill in the table until the left column ends in a 0 and then do the arrows, $f G_1 - f' G_2 + f'' G_3 \cdots$ till the process ends. Then add C, a constant of integration.

Example 7.16. Find $\int x^5 \sin x \, dx$.

From the above procedure, and letting $f(x) = x^5$, this equals

$$x^5 \left(-\cos(x)\right) - 5x^4 \left(-\sin(x)\right) + 20x^3 \left(\cos(x)\right) - 60x^2 \left(\sin(x)\right)$$
$$+120x \left(-\cos(x)\right) - 120 \left(-\sin(x)\right) + C.$$

7.5 Exercises

(1) Find the following antiderivatives.

  (a) $\int x^3 e^{-3x} \, dx$
  (b) $\int x^4 \cos x \, dx$
  (c) $\int x^5 e^x \, dx$
  (d) $\int x^6 \sin(2x) \, dx$
  (e) $\int x^3 \cos(x^2) \, dx$

(2) Find the following antiderivatives.

  (a) $\int xe^{-3x} \, dx$
  (b) $\int \frac{1}{x(\ln(|x|))^2} \, dx$
  (c) $\int x\sqrt{2-x} \, dx$
  (d) $\int (\ln|x|)^2 \, dx$ **Hint:** Let $u(x) = (\ln|x|)^2$ and $v'(x) = 1$.
  (e) $\int x^3 \cos(x^2) \, dx$

(3) Show that $\int \sec^3(x) \, dx =$

$$\frac{1}{2} \tan(x) \sec(x) + \frac{1}{2} \ln|\sec x + \tan x| + C.$$

(4) Find $\int \frac{xe^x}{(1+x)^2} \, dx$

(5) Consider the following argument. Integrate by parts, letting $u(x) = x$ and $v'(x) = \frac{1}{x^2}$ to get

$$\int \frac{1}{x} \, dx = \int x \left(\frac{1}{x^2}\right) \, dx = \left(-\frac{1}{x}\right) x + \int \frac{1}{x} \, dx$$

$$= -1 + \int \frac{1}{x} \, dx.$$

Now subtracting $\int \frac{1}{x} \, dx$ from both sides, $0 = -1$. Is there anything wrong here? If so, what?

(6) Find the following antiderivatives.

  (a) $\int x^3 \arctan(x) \, dx$
  (b) $\int x^3 \ln(x) \, dx$
  (c) $\int x^2 \sin(x) \, dx$
  (d) $\int x^2 \cos(x) \, dx$

(e) $\int x \arcsin(x) \, dx$

(f) $\int \cos(2x) \sin(3x) \, dx$

(g) $\int x^3 e^{x^2} \, dx$

(h) $\int x^3 \cos(x^2) \, dx$

(7) Find the antiderivatives

(a) $\int x^2 \sin x \, dx$

(b) $\int x^3 \sin x \, dx$

(c) $\int x^3 7^x \, dx$

(d) $\int x^2 \ln x \, dx$

(e) $\int (x+2)^2 e^x \, dx$

(f) $\int x^3 2^x \, dx$

(g) $\int \sec^3(2x) \tan(2x) \, dx$

(h) $\int x^2 7^x \, dx$

(8) Solve the initial value problem $y'(x) = f(x)$, $\lim_{x\to 0+} y(x) = 1$ where $f(x)$ is each of the integrands in Problem 7.

(9) Solve the initial value problem $y'(x) = f(x)$, $\lim_{x\to 0+} y(x) = 2$ where $f(x)$ is each of the integrands in Problem 6.

(10) Try doing $\int \sin^2 x \, dx$ the obvious way. If you do not make any mistakes, the process will go in circles. Now do it by taking

$$\int \sin^2 x \, dx = x \sin^2 x - 2 \int x \sin x \cos x \, dx$$

$$= x \sin^2 x - \int x \sin(2x) \, dx.$$

(11) An object moves on the x axis having velocity equal to $t \sin t$. Find the position of the object given that at $t = 1$, it is at the point 2.

(12) An object moves on the x axis having velocity equal to $\sec^3(t)$. Find the position of the object given that at $t = 0$, it is at the point 2. **Hint:** You might want to use Problem 3.

(13) Find the antiderivatives.

(a) $\int x \cos(x^2) \, dx$

(b) $\int \sin(\sqrt{x}) \, dx$

(c) $\int \ln(|\sin(x)|) \cos(x) \, dx$

(d) $\int \cos^4(x) \, dx$

(e) $\int \arcsin(x) \, dx$

(f) $\int \sec^3(x) \tan(x) \, dx$

(g) $\int \tan^2(x) \sec(x) \, dx$

(14) A car is moving at 14 feet per second when the driver applies the brake causing the car to slow down at the constant rate of 2 feet per second per second until it stops. How far does the car travel during the time the brake was applied?

7.6 Trig. Substitutions

Certain antiderivatives are easily obtained by making an auspicious substitution involving a trig. function. The technique will be illustrated by presenting examples.

Example 7.17. Find $\int \frac{1}{(x^2+2x+2)^2} \, dx$.

Complete the square as before and write

$$\int \frac{1}{(x^2 + 2x + 2)^2} \, dx = \int \frac{1}{\left((x+1)^2 + 1\right)^2} \, dx$$

Use the following substitution next.

$$x + 1 = \tan u \tag{7.5}$$

so $dx = \left(\sec^2 u\right) du$. Therefore, this last indefinite integral becomes

$$\int \frac{\sec^2 u}{(\tan^2 u + 1)^2} \, du \;=\; \int \left(\cos^2 u\right) du$$

$$= \int \frac{1 + \cos 2u}{2} \, du$$

$$= \frac{u}{2} + \frac{\sin 2u}{4} + C$$

$$= \frac{u}{2} + \frac{2 \sin u \cos u}{4} + C$$

Next write this in terms of x using the following device based on the following picture.

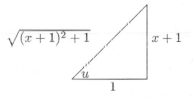

In this picture which is descriptive of (7.5), $\sin u = \frac{x+1}{\sqrt{(x+1)^2+1}}$ and

$$\cos u = \frac{1}{\sqrt{(x+1)^2 + 1}}.$$

Therefore, putting in this information to change back to the x variable,

$$\int \frac{1}{(x^2 + 2x + 2)^2} \, dx$$

$$= \frac{1}{2} \arctan(x + 1) + \frac{1}{2} \frac{x + 1}{\sqrt{(x + 1)^2 + 1} \sqrt{(x + 1)^2 + 1}} + C$$

$$= \frac{1}{2} \arctan(x + 1) + \frac{1}{2} \frac{x + 1}{(x + 1)^2 + 1} + C.$$

Example 7.18. Find $\int \frac{1}{\sqrt{x^2+7}} \, dx$.

Let $x = \sqrt{7}\tan u$ so $dx = \sqrt{7}\left(\sec^2 u\right)du$. Making the substitution, consider

$$\int \frac{1}{\sqrt{7}\sqrt{\tan^2 u + 1}}\sqrt{7}\left(\sec^2 u\right)du$$

$$= \int \left(\sec u\right)du = \ln\left|\sec u + \tan u\right| + C$$

Now the following diagram is descriptive of the above transformation.

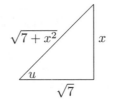

Using the above diagram, $\sec u = \frac{\sqrt{7+x^2}}{\sqrt{7}}$ and $\tan u = \frac{x}{\sqrt{7}}$. Therefore, restoring the x variable,

$$\int \frac{1}{\sqrt{x^2+3}}dx = \ln\left|\frac{\sqrt{7+x^2}}{\sqrt{7}} + \frac{x}{\sqrt{7}}\right| + C$$

$$= \ln\left|\sqrt{7+x^2} + x\right| + C.$$

Note the constant C changed in going from the top to the bottom line. It is $C - \ln\sqrt{7}$ but it is customary to simply write this as C because C is arbitrary.

Example 7.19. Find $\int \left(4x^2 + 3\right)^{1/2}dx$.

Let $2x = \sqrt{3}\tan u$ so $2dx = \sqrt{3}\sec^2\left(u\right)du$. Then making the substitution,

$$\sqrt{3}\int \left(\tan^2 u + 1\right)^{1/2}\frac{\sqrt{3}}{2}\sec^2\left(u\right)du$$

$$= \frac{3}{2}\int \sec^3\left(u\right)du. \tag{7.6}$$

Now use integration by parts to obtain

$$\int \sec^3\left(u\right)du = \int \sec^2\left(u\right)\sec\left(u\right)du$$

$$= \tan\left(u\right)\sec\left(u\right) - \int \tan^2\left(u\right)\sec\left(u\right)du$$

$$= \tan\left(u\right)\sec\left(u\right) - \int \left(\sec^2\left(u\right) - 1\right)\sec\left(u\right)du$$

$$= \tan\left(u\right)\sec\left(u\right) + \int \sec\left(u\right)du - \int \sec^3\left(u\right)du$$

$$= \tan\left(u\right)\sec\left(u\right) + \ln\left|\sec\left(u\right) + \tan\left(u\right)\right| - \int \sec^3\left(u\right)du$$

Therefore,

$$2 \int \sec^3 (u) \, du = \tan (u) \sec (u) + \ln |\sec (u) + \tan (u)| + C$$

and so

$$\int \sec^3 (u) \, du = \frac{1}{2} [\tan (u) \sec (u) + \ln |\sec (u) + \tan (u)|] + C. \qquad (7.7)$$

Now it follows from (7.6) that in terms of u the set of antiderivatives is given by

$$\frac{3}{4} [\tan (u) \sec (u) + \ln |\sec (u) + \tan (u)|] + C$$

Use the following diagram to change back to the variable x.

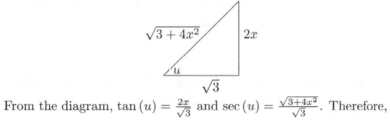

From the diagram, $\tan (u) = \frac{2x}{\sqrt{3}}$ and $\sec (u) = \frac{\sqrt{3+4x^2}}{\sqrt{3}}$. Therefore,

$$\int (4x^2 + 3)^{1/2} \, dx$$

$$= \frac{3}{4} \left[\frac{2x}{\sqrt{3}} \frac{\sqrt{3 + 4x^2}}{\sqrt{3}} + \ln \left| \frac{\sqrt{3 + 4x^2}}{\sqrt{3}} + \frac{2x}{\sqrt{3}} \right| \right] + C$$

$$= \frac{3}{4} \left[\frac{2x}{\sqrt{3}} \frac{\sqrt{3 + 4x^2}}{\sqrt{3}} + \ln \left| \frac{\sqrt{3 + 4x^2}}{\sqrt{3}} + \frac{2x}{\sqrt{3}} \right| \right] + C$$

$$= \frac{1}{2} x \sqrt{(3 + 4x^2)} + \frac{3}{4} \ln \left| \sqrt{3 + 4x^2} + 2x \right| + C$$

Note that these examples involved something of the form $\left(a^2 + (bx)^2 \right)$ and the trig substitution

$$bx = a \tan u$$

was the right one to use. This is the auspicious substitution which often simplifies these sorts of problems. However, there is a possibly better way to do these kinds.

Example 7.20. Find $\int (4x^2 + 3)^{1/2} \, dx$ another way.

Let $2x = \sqrt{3}\sinh u$ and so $2dx = \sqrt{3}\cosh(u)\,du$. Then substituting in the integral leads to

$$\int \sqrt{3}\sqrt{1 + \sinh^2(u)}\,\frac{\sqrt{3}}{2}\cosh(u)\,du$$

$$= \frac{3}{2}\int \cosh^2(u)\,du + C$$

$$= \frac{3}{4}\cosh(u)\sinh(u) + \frac{3}{4}u + C$$

$$= \frac{3}{4}\sqrt{1 + \sinh^2(u)}\sinh(u) + \frac{3}{4}u + C$$

$$= \frac{1}{2}x\sqrt{(3 + 4x^2)} + \frac{3}{4}\sinh^{-1}\left(\frac{2x}{\sqrt{3}}\right) + C$$

This other way is often used by computer algebra systems. If you solve for $\sinh^{-1}x$ in terms of $\ln$, you get the same set of antiderivatives. The function $\sinh^{-1}$ is also written as arcsinh by analogy to the trig. functions.

Example 7.21. Find $\int \sqrt{3 - 5x^2}\,dx$.

In this example, let $\sqrt{5}x = \sqrt{3}\sin(u)$ so $\sqrt{5}dx = \sqrt{3}\cos(u)\,du$. The reason this might be a good idea is that it will get rid of the square root sign as shown below. Making the substitution,

$$\int \sqrt{3 - 5x^2}\,dx = \sqrt{3}\int \sqrt{1 - \sin^2(u)}\,\frac{\sqrt{3}}{\sqrt{5}}\cos(u)\,du$$

$$= \frac{3}{\sqrt{5}}\int \cos^2(u)\,du = \frac{3}{\sqrt{5}}\int \frac{1 + \cos 2u}{2}\,du$$

$$= \frac{3}{\sqrt{5}}\left(\frac{u}{2} + \frac{\sin 2u}{4}\right) + C = \frac{3}{2\sqrt{5}}u + \frac{3}{2\sqrt{5}}\sin u \cos u + C$$

The appropriate diagram is the following.

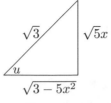

From the diagram, $\sin(u) = \frac{\sqrt{5}x}{\sqrt{3}}$ and $\cos(u) = \frac{\sqrt{3 - 5x^2}}{\sqrt{3}}$. Therefore, changing back to x,

$$\int \sqrt{3 - 5x^2}\,dx =$$

$$\frac{3}{2\sqrt{5}}\arcsin\left(\frac{\sqrt{5}x}{\sqrt{3}}\right) + \frac{3}{2\sqrt{5}}\frac{\sqrt{5}x}{\sqrt{3}}\frac{\sqrt{3-5x^2}}{\sqrt{3}} + C$$

$$= \frac{3}{10}\sqrt{5}\arcsin\left(\frac{1}{3}\sqrt{15}x\right) + \frac{1}{2}x\sqrt{(3-5x^2)} + C$$

Example 7.22. Find $\int \sqrt{5x^2 - 3}\,dx$.

In this example, let $\sqrt{5}x = \sqrt{3}\sec(u)$ so $\sqrt{5}dx = \sqrt{3}\sec(u)\tan(u)\,du$. Then changing the variable, consider

$$\sqrt{3}\int \sqrt{\sec^2(u)-1}\frac{\sqrt{3}}{\sqrt{5}}\sec(u)\tan(u)\,du = \frac{3}{\sqrt{5}}\int \tan^2(u)\sec(u)\,du$$

$$= \frac{3}{\sqrt{5}}\left[\int \sec^3(u)\,du - \int \sec(u)\,du\right]$$

Now from (7.7), this equals

$$\frac{3}{\sqrt{5}}\left[\frac{1}{2}\left[\tan(u)\sec(u) + \ln|\sec(u) + \tan(u)|\right] - \ln|\tan(u) + \sec(u)|\right] + C$$

$$= \frac{3}{2\sqrt{5}}\tan(u)\sec(u) - \frac{3}{2\sqrt{5}}\ln|\sec(u) + \tan(u)| + C.$$

Now it is necessary to change back to x. The diagram is as follows.

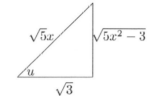

Therefore, $\tan(u) = \frac{\sqrt{5x^2-3}}{\sqrt{3}}$ and $\sec(u) = \frac{\sqrt{5}x}{\sqrt{3}}$ and so

$$\int \sqrt{5x^2 - 3}\,dx$$

$$= \frac{3}{2\sqrt{5}}\frac{\sqrt{5x^2-3}}{\sqrt{3}}\frac{\sqrt{5}x}{\sqrt{3}} - \frac{3}{2\sqrt{5}}\ln\left|\frac{\sqrt{5}x}{\sqrt{3}} + \frac{\sqrt{5x^2-3}}{\sqrt{3}}\right| + C$$

$$= \frac{1}{2}\left(\sqrt{5x^2-3}\right)x - \frac{3}{10}\sqrt{5}\ln\left|\sqrt{5}x + \sqrt{(-3+5x^2)}\right| + C$$

To summarize, here is a short table of auspicious substitutions corresponding to certain expressions.

Table Of Auspicious Substitutions

Expression	$a^2 + b^2x^2$	$a^2 - b^2x^2$	$a^2x^2 - b^2$
Trig. substitution	$bx = a\tan(u)$	$bx = a\sin(u)$	$ax = b\sec(u)$
Hyperbolic substitution	$bx = a\sinh(u)$		

Of course there are no "magic bullets" but these substitutions will often simplify an expression enough to allow you to find an antiderivative. These substitutions are often especially useful when the expression is enclosed in a square root.

7.7   Exercises

(1) Find the antiderivatives.

(a) $\int \frac{x}{\sqrt{4-x^2}} \, dx$

(b) $\int \frac{3}{\sqrt{36-25x^2}} \, dx$

(c) $\int \frac{3}{\sqrt{16-25x^2}} \, dx$

(d) $\int \frac{1}{\sqrt{4-9x^2}} \, dx$

(e) $\int \frac{1}{\sqrt{36-x^2}} \, dx$

(f) $\int \left(\sqrt{9-16x^2}\right)^3 \, dx$

(g) $\int \left(\sqrt{16-x^2}\right)^5 \, dx$

(h) $\int \sqrt{25-36x^2} \, dx$

(i) $\int \left(\sqrt{4-9x^2}\right)^3 \, dx$

(j) $\int \sqrt{1-9x^2} \, dx$

(2) Find the antiderivatives.

(a) $\int \sqrt{36x^2-25} \, dx$

(b) $\int \sqrt{x^2-4} \, dx$

(c) $\int \left(\sqrt{16x^2-9}\right)^3 \, dx$

(d) $\int \sqrt{25x^2-16} \, dx$

(3) Find the antiderivatives.

(a) $\int \frac{1}{26+x^2-2x} \, dx$ **Hint:** Complete the square.

(b) $\int \sqrt{x^2+9} \, dx$

(c) $\int \sqrt{4x^2+25} \, dx$

(d) $\int x\sqrt{4x^4+9} \, dx$

(e) $\int x^3 \sqrt{4x^4+9} \, dx$

(f) * $\int \frac{1}{\left(16+25(x-3)^2\right)^2} \, dx$

(g) $\int \frac{1}{261+25x^2-150x} \, dx$ **Hint:** Complete the square.

(h) $\int \left(\sqrt{25x^2+9}\right)^3 \, dx$

(i) $\int \frac{1}{25+16x^2} \, dx$

(4) Find the antiderivatives. **Hint:** Complete the square.

(a) $\int \sqrt{4x^2+16x+15} \, dx$

(b) $\int \sqrt{x^2+6x} \, dx$

(c) $\int \frac{3}{\sqrt{-32-9x^2-36x}} \, dx$

(d) $\int \frac{3}{\sqrt{-5-x^2-6x}}\, dx$

(e) $\int \frac{1}{\sqrt{9-16x^2-32x}}\, dx$

(f) $\int \sqrt{4x^2+16x+7}\, dx$

(5) Find $\int x^5\sqrt{1+x^4}dx.$

(6) Find $\int \frac{x}{\sqrt{1-x^4}}dx.$ **Hint:** Try $x^2 = \sin(u).$

7.8 Partial Fractions

The main technique for finding antiderivatives in the case $f(x) = \frac{p(x)}{q(x)}$ for p and q polynomials is the technique of partial fractions. Before presenting this technique, a few more examples are presented.

Example 7.23. Find $\int \frac{1}{x^2+2x+2}\, dx.$

To do this, complete the square in the denominator to write

$$\int \frac{1}{x^2+2x+2}\, dx = \int \frac{1}{(x+1)^2+1}\, dx$$

Now change the variable, letting $u = x+1$, so that $du = dx$. Then the last indefinite integral reduces to

$$\int \frac{1}{u^2+1}\, du = \arctan u + C$$

and so

$$\int \frac{1}{x^2+2x+2}\, dx = \arctan(x+1) + C.$$

Example 7.24. Find $\int \frac{1}{3x+5}\, dx.$

Let $u = 3x+5$ so $du = 3dx$ and changing the variable,

$$\frac{1}{3}\int \frac{1}{u}\, du = \frac{1}{3}\ln|u| + C.$$

Therefore,

$$\int \frac{1}{3x+5}\, dx = \frac{1}{3}\ln|3x+5| + C.$$

Example 7.25. Find $\int \frac{3x+2}{x^2+x+1}\, dx.$

First complete the square in the denominator.

$$\int \frac{3x+2}{x^2+x+1}\, dx = \int \frac{3x+2}{x^2+x+\frac{1}{4}+\frac{3}{4}}\, dx$$

$$= \int \frac{3x+2}{\left(x+\frac{1}{2}\right)^2+\frac{3}{4}}\, dx.$$

Now let

$$\left(x + \frac{1}{2}\right)^2 = \frac{3}{4}u^2$$

so that $x + \frac{1}{2} = \frac{\sqrt{3}}{2}u$. Therefore, $dx = \frac{\sqrt{3}}{2}du$ and changing the variable,

$$\frac{4}{3} \int \frac{3\left(\frac{\sqrt{3}}{2}u - \frac{1}{2}\right) + 2}{u^2 + 1} \frac{\sqrt{3}}{2} \, du$$

$$= \frac{\sqrt{3}}{2}\left(2\sqrt{3} \int \frac{u}{u^2 + 1} \, du - \frac{2}{3} \int \frac{1}{u^2 + 1} \, du\right)$$

$$= \frac{\sqrt{3}}{2}\left(\sqrt{3} \int \frac{2u}{u^2 + 1} \, du - \frac{2}{3} \int \frac{1}{u^2 + 1} \, du\right)$$

$$= \frac{3}{2} \ln\left(u^2 + 1\right) - \frac{\sqrt{3}}{3} \arctan u + C$$

Therefore,

$$\int \frac{3x + 2}{x^2 + x + 1} \, dx =$$

$$\frac{3}{2} \ln\left(\left(\frac{2}{\sqrt{3}}\left(x + \frac{1}{2}\right)\right)^2 + 1\right) - \frac{\sqrt{3}}{3} \arctan\left(\frac{2}{\sqrt{3}}\left(x + \frac{1}{2}\right)\right) + C.$$

The method of partial fractions splits rational functions into a sum of functions which are like those which were just done successfully. In using this method it is essential that in the rational function the degree of the numerator is smaller than the degree of the denominator. The following simple but important Lemma shows that you can always reduce to this case.

Lemma 7.3. *Let $f(x)$ and $g(x)$ be polynomials. Then there exists a polynomial, $q(x)$ such that*

$$f(x) = q(x)g(x) + r(x)$$

where the degree of $r(x)$ is less than the degree of $g(x)$ or $r(x) = 0$.

Proof: Consider the polynomials of the form $f(x) - g(x)l(x)$ and out of all these polynomials, pick one which has the smallest degree. This can be done because every nonempty set of nonnegative integers has a smallest member (Why? Try to prove this by induction.). Let this take place when $l(x) = q_1(x)$ and let

$$r(x) = f(x) - g(x)q_1(x).$$

It is required to show degree of $r(x)$ < degree of $g(x)$ or else $r(x) = 0$.

Suppose $f(x) - g(x)l(x)$ is never equal to zero for any $l(x)$. Then $r(x) \neq 0$. It is required to show the degree of $r(x)$ is smaller than the degree of $g(x)$. If this does not happen, then the degree of $r \geq$ the degree of g. Let

$$r(x) = b_m x^m + \cdots + b_1 x + b_0$$
$$g(x) = a_n x^n + \cdots + a_1 x + a_0$$

where $m \geq n$ and b_m and a_n are nonzero. Then let $r_1(x)$ be given by

$$r_1(x) = r(x) - \frac{x^{m-n}b_m}{a_n} g(x)$$

$$= (b_m x^m + \cdots + b_1 x + b_0) - \frac{x^{m-n}b_m}{a_n}(a_n x^n + \cdots + a_1 x + a_0)$$

which has smaller degree than m, the degree of $r(x)$. But

$$r_1(x) = \overbrace{f(x) - g(x) q_1(x)}^{r(x)} - \frac{x^{m-n}b_m}{a_n} g(x)$$

$$= f(x) - g(x)\left(q_1(x) + \frac{x^{m-n}b_m}{a_n}\right),$$

and this is not zero by the assumption that $f(x) - g(x) l(x)$ is never equal to zero for any $l(x)$ yet has smaller degree than $r(x)$ which is a contradiction to the choice of $r(x)$. ■

Corollary 7.1. *Let $f(x)$ and $g(x)$ be polynomials. Then there exists a polynomial, $r(x)$ such that the degree of $r(x) <$ degree of $g(x)$ and a polynomial, $q(x)$ such that*

$$\frac{f(x)}{g(x)} = q(x) + \frac{r(x)}{g(x)}.$$

Here is an example where the degree of the numerator exceeds the degree of the denominator.

Example 7.26. Find $\int \frac{3x^5+7}{x^2-1}\, dx$.

In this case the degree of the numerator is larger than the degree of the denominator and so long division must first be used. Thus

$$\frac{3x^5 + 7}{x^2 - 1} = 3x^3 + 3x + \frac{7 + 3x}{x^2 - 1}$$

Now look for a partial fractions expansion of the form

$$\frac{7 + 3x}{x^2 - 1} - \frac{a}{(x-1)} + \frac{b}{(x+1)}.$$

Therefore,

$$7 + 3x = a(x+1) + b(x-1).$$

Letting $x = 1$, $a = 5$. Then letting $x = -1$, it follows $b = -2$. Therefore,

$$\frac{7 + 3x}{x^2 - 1} = \frac{5}{x - 1} - \frac{2}{x + 1}$$

and so

$$\frac{3x^5 + 7}{x^2 - 1} = 3x^3 + 3x + \frac{5}{x - 1} - \frac{2}{x + 1}.$$

Therefore,

$$\int \frac{3x^5 + 7}{x^2 - 1}\, dx = \frac{3}{4}x^4 + \frac{3}{2}x^2 + 5\ln(x-1) - 2\ln(x+1) + C.$$

Here is another example.

Example 7.27. Find $\int \frac{-x^3+11x^2+24x+14}{(2x+3)(x+5)(x^2+x+1)}\, dx$.

In this problem, first check to see if the degree of the numerator in the integrand is less than the degree of the denominator. In this case, this is so. If it is not so, use long division to write the integrand as the sum of a polynomial with a rational function in which the degree of the numerator is less than the degree of the denominator. See the preceding corollary which guarantees this can be done. Now look for a partial fractions expansion for the integrand which is in the following form.

$$\frac{a}{2x+3} + \frac{b}{x+5} + \frac{cx+d}{x^2+x+1}$$

and try to find constants, a, b, c, and d so that the above rational functions sum to the integrand. The reason $cx + d$ is used in the numerator of the last expression is that $x^2 + x + 1$ cannot be factored using real polynomials. Thus the problem involves finding a, b, c, d, such that

$$\frac{-x^3 + 11x^2 + 24x + 14}{(2x+3)(x+5)(x^2+x+1)} = \frac{a}{2x+3} + \frac{b}{x+5} + \frac{cx+d}{x^2+x+1}$$

and so

$$-x^3 + 11x^2 + 24x + 14 = a\,(x+5)\,(x^2+x+1) +$$

$$b\,(2x+3)\,(x^2+x+1) + (cx+d)\,(2x+3)\,(x+5). \tag{7.8}$$

Now these are two polynomials which are supposed to be equal. Therefore, they have the same coefficients. Multiplying the right side out and collecting the terms,

$$-x^3 + 11x^2 + 24x + 14 =$$

$$(2b + 2c + a)\,x^3 + (6a + 5b + 13c + 2d)\,x^2 + (6a + 13d + 5b + 15c)\,x + 15d + 5a + 3b$$

and therefore, it is necessary to solve the equations

$$2b + 2c + a = -1$$
$$6a + 5b + 13c + 2d = 11$$
$$6a + 13d + 5b + 15c = 24$$
$$15d + 5a + 3b = 14$$

The solution is $c = 1, a = 1, b = -2, d = 1$. Therefore,

$$\frac{-x^3 + 11x^2 + 24x + 14}{(2x+3)(x+5)(x^2+x+1)} = \frac{1}{2x+3} - \frac{2}{x+5} + \frac{1+x}{x^2+x+1}.$$

This may look like a fairly formidable problem. In reality it is not that bad. First let $x = -5$ in (7.8) and obtain a simple equation for finding b. Next let $x = -3/2$ to get a simple equation for a. This reduces the above system to a more manageable size. Anyway, it is now possible to find an antiderivative of the given function.

$$\int \frac{-x^3 + 11x^2 + 24x + 14}{(2x+3)(x+5)(x^2+x+1)}\,dx =$$

$$\int \frac{1}{2x+3}\,dx - \int \frac{2}{x+5}\,dx + \int \frac{1+x}{x^2+x+1}\,dx.$$

Each of these indefinite integrals can be found using the techniques given above. Thus the antiderivative is

$$\frac{1}{2}\ln|2x+3| - 2\ln|x+5| +$$

$$\frac{1}{2}\ln\left(x^2+x+1\right) + \frac{1}{3}\sqrt{3}\arctan\left(\frac{\sqrt{3}}{3}\left(2x+1\right)\right) + C.$$

What is done when the factors are repeated?

Example 7.28. Find $\int \frac{3x+7}{(x+2)^2(x+3)}\,dx$.

First observe that the degree of the numerator is less than the degree of the denominator. In this case the correct form of the partial fraction expansion is

$$\frac{a}{(x+2)} + \frac{b}{(x+2)^2} + \frac{c}{(x+3)}.$$

The reason there are two terms devoted to $(x+2)$ is that this is squared. Computing the constants yields

$$\frac{3x+7}{(x+2)^2(x+3)} = \frac{1}{(x+2)^2} + \frac{2}{x+2} - \frac{2}{x+3}$$

and therefore,

$$\int \frac{3x+7}{(x+2)^2(x+3)}\,dx = -\frac{1}{x+2} + 2\ln(x+2) - 2\ln(x+3) + C.$$

Example 7.29. Find the proper form for the partial fractions expansion of

$$\frac{x^3+7x+9}{(x^2+2x+2)^3(x+2)^2(x+1)(x^2+1)}.$$

First check to see if the degree of the numerator is smaller than the degree of the denominator. Since this is the case, look for a partial fractions decomposition in the following form.

$$\frac{ax+b}{(x^2+2x+2)} + \frac{cx+d}{(x^2+2x+2)^2} + \frac{ex+f}{(x^2+2x+2)^3} +$$

$$\frac{A}{(x+2)} + \frac{B}{(x+2)^2} + \frac{D}{(x+1)} + \frac{gx+h}{x^2+1}.$$

These examples illustrate what to do when using the method of partial fractions. You first check to be sure the degree of the numerator is less than the degree of the denominator. If this is not so, do a long division. Then you factor the denominator into a product of factors, some linear of the form $ax+b$ and others quadratic, ax^2+bx+c which cannot be factored further. Next follow the procedure illustrated in the above examples and summarized below.

Warning: When you use partial fractions, **be sure you look for something which is of the right form.** Otherwise you may be looking for something which is not there. The rules are summarized next.

7.8.0.1 Rules For Finding Partial Fractions Expansion Of A Rational Function

(1) Check to see if the numerator has smaller degree than the denominator. If this is not so, correct the situation by doing long division.

(2) Factor the denominator into a product of linear factors, (Things like $(ax + b)$) and irreducible quadratic factors, (Things like $(ax^2 + bx + c)$ where $b^2 - 4ac < 0$.)[1]

(3) Let m, n be positive integers. Corresponding to $(ax + b)^m$ in the denominator, you should have a sum of the form $\sum_{i=1}^{m} \frac{c_i}{(ax+b)^i}$ in the partial fractions expansion. Here the c_i are the constants to be found. Corresponding to $(ax^2 + bx + c)^n$ in the denominator where $b^2 - 4ac < 0$, you should have a sum of the form $\sum_{i=1}^{m} \frac{p_i x + q_i}{(ax^2+bx+c)^i}$ in the partial fractions expansion. Here the p_i and q_i are to be found.

(4) Find the constants, c_i, p_i, and q_i. Use whatever method you like. You might see if you can make up new ways to do this if you like. If you have followed steps 1 - 3 correctly, it will work out.

The above technique for finding the coefficients is fine but some people like to do it other ways. It really does not matter how you do it. Here is another example.

Example 7.30. Find the partial fractions expansion for $\frac{15x^4+44x^3+71x^2+64x+28+2x^5}{(x+2)^2(x^2+2x+2)^2}$.

The degree of the top is 4 and the degree of the bottom is 6 so you do not need to do long division. You do have to look for the right thing however. The correct form for the partial fractions expansion is

$$\frac{a}{x+2} + \frac{b}{(x+2)^2} + \frac{cx+d}{x^2+2x+2} + \frac{ex+f}{(x^2+2x+2)^2}$$

$$= \frac{15x^4 + 44x^3 + 71x^2 + 64x + 28 + 2x^5}{(x+2)^2(x^2+2x+2)^2}$$

Multiply by the denominator of the right side. This yields

$$15x^4 + 44x^3 + 71x^2 + 64x + 28 + 2x^5$$
$$= a(x+2)(x^2+2x+2)^2 + b(x^2+2x+2)^2$$
$$+ (cx+d)(x+2)^2(x^2+2x+2) + (ex+f)(x+2)^2$$

You could multiply out the right side and match the coefficients but you could also do other things. First set $x = -2$ because this will cause many of the terms on the right to equal zero. The left side equals

$$15(-2)^4 + 44(-2)^3 + 71(-2)^2 + 64(-2) + 28 + 2(-2)^5 = 8.$$

---

[1]Of course this factoring of the denominator is easier said than done. In general you cannot do it at all. Of course there are big theorems which guarantee the existence of such a factorization but these theorems do not tell how to find it. This is an example of the gap between theory and practice which permeates mathematics.

the right side equals

$$b\left((-2)^2 + 2(-2) + 2\right)^2 = 4b$$

and so $b = 2$. Progress has been made. Now subtract the term $2\left(x^2 + 2x + 2\right)^2$ from both sides and then divide both sides by $(x+2)$. First the subtraction:

$$15x^4 + 44x^3 + 71x^2 + 64x + 28 + 2x^5 - 2\left(x^2 + 2x + 2\right)^2$$
$$= a(x+2)\left(x^2 + 2x + 2\right)^2 + (cx+d)(x+2)^2\left(x^2 + 2x + 2\right) + (ex+f)(x+2)^2.$$

Simplifying this yields

$$13x^4 + 36x^3 + 55x^2 + 48x + 20 + 2x^5$$
$$= a(x+2)\left(x^2 + 2x + 2\right)^2 + (cx+d)(x+2)^2\left(x^2 + 2x + 2\right) + (ex+f)(x+2)^2.$$

Now $(x+2)$ divides the right side so it must also divide the left. Do the division by whatever method you like best.

$$2x^4 + 9x^3 + 18x^2 + 19x + 10$$
$$= a\left(x^2 + 2x + 2\right)^2 + (cx+d)(x+2)\left(x^2 + 2x + 2\right) + (ex+f)(x+2). \quad (7.9)$$

Now let $x = -2$ again. This yields

$$2(-2)^4 + 9(-2)^3 + 18(-2)^2 + 19(-2) + 10 - a\left((-2)^2 + 2(-2) + 2\right)^2$$

and so $4 = a(4)$ which shows $a = 1$. Now use this information in (7.9). Thus

$$2x^4 + 9x^3 + 18x^2 + 19x + 10 - \left(x^2 + 2x + 2\right)^2$$
$$= (cx+d)(x+2)\left(x^2 + 2x + 2\right) + (ex+f)(x+2).$$

It looks like you can divide both sides by $(x+2)$ again. Lets do so.

$$x^3 + 3x^2 + 4x + 3 = (cx+d)\left(x^2 + 2x + 2\right) + (ex+f).$$

Now finally, lets expand both sides.

$$x^3 + 3x^2 + 4x + 3 = cx^3 + (2c+d)x^2 + (2d+2c+c)x + 2d+f. \quad (7.10)$$

At this point, you could match the coefficients and get a system of linear equations or you could be more clever. Lets be clever. Differentiate both sides. Then

$$3x^2 + 6x + 4 = 3cx^2 + 4cx + 2dx + 2d + 2c + e. \quad (7.11)$$

Now lets differentiate both sides again.

$$6x + 6 = 6cx + 4c + 2d. \quad (7.12)$$

Differentiate yet again. $6 = 6c$. Thus $c = 1$. Now let $x = 0$ (7.12) to get $6 = 4 + 2d$ so $d = 1$. Now let $x = 0$ in (7.11) to get $4 = 2 + 2 + e$ showing $e - 0$. Now it only remains to find f. Let $x = 0$ in (7.10) and obtain $3 = 2 + f$ so $f = 1$. It follows the partial fractions expansion is

$$\frac{1}{x+2} + \frac{2}{(x+2)^2} + \frac{x+1}{x^2+2x+2} + \frac{1}{(x^2+2x+2)^2}.$$

One other thing should be mentioned. Suppose you wanted to find the integral in this example. The first three terms are by now routine. How about the last one?

Example 7.31. Find $\int \frac{1}{(x^2+2x+2)^2} dx$.

First complete the square to write this as $\int \frac{1}{((x+1)^2+1)^2}dx$. Now do a trig. substitution. You should let $x + 1 = \tan\theta$. Then the integral becomes

$$\int \frac{1}{\sec^4\theta}\sec^2\theta d\theta = \int \cos^2\theta d\theta = \int \frac{1+\cos(2\theta)}{2}d\theta$$

$$= \frac{\theta}{2} + \frac{2\sin\theta\cos\theta}{4} + C.$$

Setting up a little triangle as in the section on trig. substitutions, you can restore the original variables to obtain

$$\frac{1}{2}\arctan(x+1) + \frac{1}{2}\left(\frac{x+1}{x^2+2x+2}\right) + C.$$

7.9 Rational Functions Of Trig. Functions

There is a technique which reduces certain kinds of integrals involving trig. functions to the technique of partial fractions. This is illustrated in the following example.

Example 7.32. Find $\int \frac{\cos\theta}{1+\cos\theta}d\theta$.

The integrand is an example of a rational function of cosines and sines. When such a thing occurs there is a substitution which will reduce the integrand to a rational function like those above which can then be integrated using partial fractions. The substitution is $u = \tan\left(\frac{\theta}{2}\right)$. Thus in this example, $du = \left(1+\tan^2\left(\frac{\theta}{2}\right)\right)\frac{1}{2}d\theta$ and so in terms of this new variable, the indefinite integral is

$$\int \frac{2\cos(2\arctan u)}{(1+\cos(2\arctan u))(1+u^2)}du.$$

You can evaluate $\cos(2\arctan u)$ exactly. This equals $2\cos^2(\arctan u) - 1$. Setting up a little triangle as above, $\cos(\arctan u)$ equals $1/\sqrt{1+u^2}$ and so the integrand reduces to

$$\frac{2\left(2\left(1/\sqrt{1+u^2}\right)^2 - 1\right)}{\left(1+\left(2\left(1/\sqrt{1+u^2}\right)^2 - 1\right)\right)(1+u^2)} = \frac{1-u^2}{1+u^2} = -1 + \frac{2}{1+u^2}$$

therefore, in terms of u the antiderivative equals $-u + 2\arctan u$. Now replace u to obtain

$$-\tan\left(\frac{\theta}{2}\right) + 2\arctan\left(\tan\left(\frac{\theta}{2}\right)\right) + C.$$

This procedure can be expected to work in general. Suppose you want to find

$$\int \frac{p(\cos\theta,\sin\theta)}{q(\cos\theta,\sin\theta)}d\theta$$

where p and q are polynomials in each argument. Make the substitution $u = \tan\frac{\theta}{2}$. As above this means

$$du = \left(1+\tan^2\left(\frac{\theta}{2}\right)\right)\frac{1}{2}d\theta = \frac{1}{2}\left(1+u^2\right)d\theta.$$

It remains to substitute for $\sin\theta$ and $\cos\theta$. Recall that $\sin\left(\frac{\theta}{2}\right) = \pm\sqrt{\frac{1-\cos\theta}{2}}$ and $\cos\left(\frac{\theta}{2}\right) = \pm\sqrt{\frac{1+\cos\theta}{2}}$. Thus,

$$\tan\left(\frac{\theta}{2}\right) = \frac{\pm\sqrt{1-\cos\theta}}{\sqrt{1+\cos\theta}}$$

and so

$$u^2 = \tan^2\left(\frac{\theta}{2}\right) = \frac{1-\cos\theta}{1+\cos\theta}$$

and solving this for $\cos\theta$ and $\sin\theta$ yields

$$\cos\theta = \frac{1-u^2}{1+u^2}, \quad \sin\theta = \pm\frac{2u}{1+u^2}.$$

It follows that in terms of u the integral becomes

$$\int \frac{p\left(\frac{1-u^2}{1+u^2}, \pm\frac{2u}{1+u^2}\right)}{q\left(\frac{1-u^2}{1+u^2}, \pm\frac{2u}{1+u^2}\right)} \frac{2du}{1+u^2}$$

which is a rational function of u and so in theory, you might be able to find the integral from the method of partial fractions.

7.10 Exercises

(1) Give a condition on $a, b,$ and c such that $ax^2 + bx + c$ cannot be factored as a product of two polynomials which have real coefficients.

(2) Find the partial fractions expansion of the following rational functions.
   (a) $\frac{2x+7}{(x+1)^2(x+2)}$
   (b) $\frac{5x+1}{(x^2+1)(2x+3)}$
   (c) $\frac{5x+1}{(x^2+1)^2(2x+3)}$
   (d) $\frac{5x^4+10x^2+3+4x^3+6x}{(x+1)(x^2+1)^2}$

(3) Find the antiderivatives
   (a) $\int \frac{x^5+4x^4+5x^3+2x^2+2x+7}{(x+1)^2(x+2)}\,dx$
   (b) $\int \frac{5x+1}{(x^2+1)(2x+3)}\,dx$
   (c) $\int \frac{5x+1}{(x^2+1)^2(2x+3)}$

(4) Each of $\cot\theta, \tan\theta, \sec\theta,$ and $\csc\theta$ is a rational function of $\cos\theta$ and $\sin\theta$. Use the technique of substituting $u = \tan\left(\frac{\theta}{2}\right)$ to find antiderivatives for each of these.

(5) Find $\int \frac{\sin\theta}{1+\sin\theta}\,d\theta$. **Hint:** Use the above procedure of letting $u = \tan\left(\frac{\theta}{2}\right)$ and then multiply both the top and the bottom by $(1-\sin\theta)$ to see another way of doing it.

(6) Find $\int \frac{\cos\theta+1}{\cos\theta+2} d\theta$ using the substitution $u = \tan\left(\frac{\theta}{2}\right)$.

(7) In finding $\int \sec(x)\, dx$, try the substitution $u = \sin(x)$.

(8) In finding $\int \csc(x)\, dx$ try the substitution $u = \cos(x)$.

(9) Find the antiderivatives.

   (a) $\int \frac{17x-3}{(6x+1)(x-1)}\, dx$

   (b) $\int \frac{50x^4-95x^3-20x^2-3x+7}{(5x+3)(x-2)(2x-1)}\, dx$ **Hint:** Notice the degree of the numerator is larger than the degree of the denominator.

   (c) $\int \frac{8x^2+x-5}{(3x+1)(x-1)(2x-1)}\, dx$

   (d) $\int \frac{3x+2}{(5x+3)(x+1)}\, dx$

(10) Find the antiderivatives

   (a) $\int \frac{52x^2+68x+46+15x^3}{(x+1)^2(5x^2+10x+8)}\, dx$

   (b) $\int \frac{9x^2-42x+38}{(3x+2)(3x^2-12x+14)}\, dx$

   (c) $\int \frac{9x^2-6x+19}{(3x+1)(3x^2-6x+5)}\, dx$

(11) Solve the initial value problem $y' = f(x)$, $y(0) = 1$ for $f(x)$ equal to each of the integrands in Problem 10.

(12) *Find the antiderivatives. You will need to complete the square and then make a trig. substitution.

   (a) $\int \frac{1}{(3x^2+12x+13)^2}\, dx =$

   (b) $\int \frac{1}{(5x^2+10x+7)^2}\, dx =$

   (c) $\int \frac{1}{(5x^2-20x+23)^2}\, dx =$

(13) Solve the initial value problem $y' = f(x), y(0) = 1$ for $f(x)$ equal to each of the integrands in Problem 12.

7.11   Practice Problems For Antiderivatives

The process of finding antiderivatives has absolutely nothing to do with mathematics. However, it is fun and it is good to become adept at doing it. This can only be accomplished through working lots of problems. Here are lots of practice problems for finding antiderivatives. Some of these are very hard but you do not have to do all of them if you do not want to. However, the more you do, the better you will be at taking antiderivatives. Most of these problems are modifications of problems I found in a Russian calculus book. This book had some which were even harder. For even more examples, consult an old table of integrals. I shall give answers to these problems so you can see whether you have it right. Beware that sometimes you may get it right even though it looks different than the answer given. Also, there is no guarantee that my answers are right.

(1) Find $\int \frac{\sqrt{2x+1}}{x}\, dx$.

Answer:

$\int \frac{\sqrt{2x+1}}{x} dx = 2\sqrt{2}\sqrt{x} + \ln x + C$

(2) Find $\int \frac{te^t}{(t+1)^2} dt$.

Hint: Write this as $\int \left(\frac{(1+t)e^t}{(t+1)^2} - \frac{e^t}{(1+t)^2} \right) dt = \int \left(\frac{e^t}{(t+1)} - \frac{e^t}{(1+t)^2} \right) dt$.

Answer: $\frac{e^t}{t+1} + C$

(3) Find $\int \frac{5-2x^2}{5+2x^2} dx$. **Hint:** $\frac{5-2x^2}{5+2x^2} = -1 + \frac{10}{5+2x^2}$.

Answer: $\int \frac{5-2x^2}{5+2x^2} dx = -x + \sqrt{10}\arctan \frac{1}{5}x\sqrt{10} + C$

(4) Find $\int \left(3 - x^5\right)^2 dx$.

Answer: $\int \left(3 - x^5\right)^2 dx = \frac{1}{11}x^{11} - x^6 + 9x + C$

(5) Find $\int (2x + 3)^{-25} dx$. **Hint:** Let $u = 2x + 3$

Answer: $\int (2x + 3)^{-25} dx = -\frac{1}{48(2x+3)^{24}} + C$

(6) Find $\int 5^x dx$.

Answer: $\int 5^x dx = \frac{1}{\ln 5}5^x + C$

(7) Find $\int \cosh^2 8x dx$. **Hint:** Try integration by parts.

Answer: $\int \cosh^2 8x dx = \frac{1}{16} \cosh 8x \sinh 8x + \frac{1}{2}x + C$

(8) Find $\int \tanh^2 2x dx$.

Answer:

$\int \tanh^2 2x dx = -\frac{1}{2}\tanh 2x - \frac{1}{4}\ln(-1 + \tanh 2x) + \frac{1}{4}\ln(1 + \tanh 2x) + C$

(9) Find $\int \cosh(3x + 3) dx$.

Answer: $\int \cosh(3x + 3) dx = \frac{1}{3}\sinh(3x + 3) + C$

(10) Find $\int 8^{1+x} dx$. **Hint:** This equals

$8 \int e^{x \ln 8} dx$.

Answer: $\int 8^{1+x} dx = \frac{8}{\ln 8}8^x + C$

(11) Find $\int \frac{\sqrt{(36+x^2)}+\sqrt{(36-x^2)}}{\sqrt{(1296-x^4)}} dx$. **Hint:** The integrand equals $\frac{\sqrt{(36+x^2)}+\sqrt{(36-x^2)}}{\sqrt{(36-x^2)(36+x^2)}}$

$= \frac{1}{\sqrt{36-x^2}} + \frac{1}{\sqrt{36+x^2}}$.

Answer:

$\int \frac{\sqrt{36+x^2}+\sqrt{36-x^2}}{\sqrt{1296-x^4}} dx = \int \frac{1}{\sqrt{36-x^2}} dx + \int \frac{1}{\sqrt{(36+x^2)}} dx - \arcsin \frac{1}{6}x +$

$\ln \left(x + \sqrt{36 + x^2}\right) + C$

(12) Find $\int (7x + 1)^{30} dx$.

Answer: $\int (7x + 1)^{30} dx = \frac{1}{217}(7x + 1)^{31} + C$

(13) Find $\int \sqrt{1 + \sin 3x} dx$. **Hint:** The integrand equals $\frac{\cos(3x)}{\sqrt{1-\sin(3x)}}$.

Answer: $\int \sqrt{1 + \sin 3x} dx = \frac{2}{3}(\sin 3x - 1)\frac{\sqrt{(1+\sin 3x)}}{\cos 3x} + C$

(14) Find $\int \left(\sqrt{3 + 2x}\right)^5 dx$.

Answer: $\int \left(\sqrt{3 + 2x}\right)^5 dx = \frac{1}{7}\left(\sqrt{3 + 2x}\right)^7 + C$

(15) Find $\int \frac{1}{49+4x^2} dx$.

Answer: $\int \frac{1}{49+4x^2} dx = \frac{1}{14}\arctan \frac{2}{7}x + C$

(16) Find $\int \frac{1}{\sqrt{4x^2-9}} dx$.

Answer: $\int \frac{1}{\sqrt{4x^2-9}}\, dx = \frac{1}{2}\ln\left(2x + \sqrt{4x^2-9}\right) + C$

(17) Find $\int \frac{1}{\sin^2(2x+3)}\, dx$.

Answer: $\int \frac{1}{\sin^2(2x+3)}\, dx = -\frac{1}{2\sin(2x+3)}\cos(2x+3) + C$

(18) Find $\int \frac{1}{1+\cos 6x}\, dx$. **Hint:** You could let $u = 6x$ and then use the technique for rational functions of $\cos x$ and $\sin x$.

Answer: $\int \frac{1}{1+\cos(6x)}\, dx = \frac{1}{6}\tan 3x + C$.

(19) Find $\int \frac{1}{1+\sin(3x)}\, dx$.

Answer: $\int \frac{1}{1+\sin 3x}\, dx = -\frac{2}{3\left(1+\tan \frac{3}{2}x\right)} + C$

(20) Find $\int x^2\sqrt{4x^3+2}\, dx$. **Hint:** Let $u = 4x^3 + 2$.

Answer: $\int x^2\sqrt{4x^3+2}\, dx = \frac{1}{18}\left(\sqrt{4x^3+2}\right)^3 + C$

(21) Find $\int x^8\left(\sqrt{3x^9+2}\right)^5 dx$.

Answer: $\int x^8\left(\sqrt{3x^9+2}\right)^5 dx = \frac{2}{189}\left(\sqrt{3x^9+2}\right)^7 + C$

(22) Find $\int \frac{x}{3+8x^4}\, dx$. **Hint:** Try $u = x^2$.

Answer: $\int \frac{x}{3+8x^4}\, dx = \frac{1}{24}\sqrt{6}\arctan\left(\frac{2}{3}x^2\sqrt{6}\right) + C$

(23) Find $\int \frac{1}{3(2+x)\sqrt{x}}\, dx$. **Hint:** Try $x = u^2$.

Answer: $\int \frac{1}{3(2+x)\sqrt{x}}\, dx = \frac{1}{9}\sqrt{3}\sqrt{6}\arctan\left(\frac{1}{6}\sqrt{3}\sqrt{x}\sqrt{6}\right) + C$

(24) Find $\int \frac{1}{x\sqrt{(25x^2-9)}}\, dx$.

Answer: You could let $3\sec u = 5x$ so $(3\sec u\tan u)\, du = 5\, dx$ and then $\int \frac{1}{x\sqrt{(25x^2-9)}}\, dx = \frac{1}{3}\int du = \frac{u}{3} + C$. Now restoring the original variables, this yields $\frac{1}{3}\operatorname{arcsec}\frac{5}{3}|x| + C$.

(25) Find $\int \frac{1}{\sqrt{(x(3x-5))}}\, dx$. **Hint:** You might try completing the square in $x(3x-5)$ and then changing the variable in an appropriate manner.

Answer: $\int \frac{dx}{\sqrt{(x(3x-5))}} = \frac{1}{3}\sqrt{3}\ln\left(\sqrt{3}\left(x - \frac{5}{6}\right) + \sqrt{(3x^2 - 5x)}\right) + C$

(26) Find $\int \frac{1}{x\cos(\ln 4x)}\, dx$. **Hint:** You might try letting $u = \ln(4x)$.

Answer: $\int \frac{1}{x\cos(\ln 4x)}\, dx = \ln\left(\sec(\ln 4x) + \tan(\ln 4x)\right) + C$

(27) Find $\int x^7 e^{x^8}\, dx$.

Answer: $\int x^7 e^{x^8}\, dx = \frac{1}{8}e^{x^8} + C$

(28) Find $\int \frac{\ln^4 x}{x}\, dx$.

Answer: $\int \frac{\ln^4 x}{x}\, dx = \frac{1}{5}\ln^5 x + C$

(29) Find $\int \frac{1}{x(2+\ln 2x)}\, dx$.

Answer: $\int \frac{1}{x(2+\ln 2x)}\, dx = \ln(2 + \ln 2x) + C$

(30) Find $\int \frac{\sin 7x+\cos 7x}{\sqrt{\sin 7x-\cos 7x}}\, dx$. **Hint:** Try $u = \sin 7x - \cos 7x$.

Answer: $\int \frac{\sin 7x+\cos 7x}{\sqrt{(\sin 7x-\cos 7x)}}\, dx = \frac{2}{7}\sqrt{(\sin 7x - \cos 7x)} + C$

(31) Find $\int \csc 4x\, dx$.

Answer: $\int \csc 4x\, dx = \frac{1}{4}\ln|\csc 4x - \cot 4x| + C$

(32) Find $\int \frac{\arctan 5x}{1+25x^2}\, dx$. **Hint:** Try $u = \arctan(5x)$.

Answer: $\int \frac{\arctan 5x}{1+25x^2}\, dx = \frac{1}{10}\arctan^2 5x + C$

(33) Find $\int \frac{18}{(6+2x)(6-x)} \cos\left(\ln \frac{6+2x}{6-x}\right) dx$.

Hint: It might help to first let $u = \ln \frac{6+2x}{6-x}$ and see if it simplifies.

Answer: $\int \frac{18}{(6+2x)(6-x)} \cos\left(\ln \frac{6+2x}{6-x}\right) dx = \sin\left(\ln \frac{6+2x}{6-x}\right) + C$

(34) Find $\int x^{23} \left(2 - 6x^{12}\right)^{10} dx$

Hint: Maybe let $u = 2 - 6x^{12}$.

Answer: $\int x^{23} \left(2 - 6x^{12}\right)^{10} dx = \frac{1}{5184}\left(2 - 6x^{12}\right)^{12} - \frac{1}{2376}\left(2 - 6x^{12}\right)^{11} + C$

(35) Find $\int \frac{x^5}{\sqrt{6-3x^2}} dx$.

Answer: $\int \frac{x^5}{\sqrt{(6-3x^2)}} dx = -\frac{1}{15}x^4\sqrt{(6-3x^2)} - \frac{8}{45}x^2\sqrt{(6-3x^2)} - \frac{32}{45}\sqrt{(6-3x^2)}$
$+ C$

(36) Find $\int \cos^3 (3x) \sin^{\frac{1}{2}} (3x) \, dx$.

Answer: $\int \cos^3 3x \sin^{\frac{1}{2}} 3x \, dx = \int \left(\cos 3x \left(1 - \sin^2 3x\right) \sin^{\frac{1}{2}} 3x\right) dx =$
$-\frac{2}{21} \sin^{\frac{7}{2}} 3x + \frac{2}{9} \sin^{\frac{3}{2}} 3x + C$

(37) Find $\int \frac{1}{e^{2x}+e^x} dx$. **Hint:** Try $u = e^x$.

Answer: $\int \frac{1}{e^{2x}+e^x} dx = \frac{-1-xe^x}{e^x} + \ln(e^x + 1) + C$

(38) Find $\int \frac{1}{\sqrt{e^x+1}} dx$.

Answer: Let $u^2 = e^x$ so $2u\,du = e^x dx = au^2 dx$. In terms of u this is $2\int \frac{1}{u\sqrt{1+u^2}} du$. Now let $u = \tan\theta$ so $du = (\sec^2\theta)\,d\theta$. Then the indefinite integral becomes $2\int \frac{\sec^2\theta}{\tan\theta\sec(\theta)} d\theta = 2\int \csc\theta\,d\theta = 2\ln|\csc\theta - \cot\theta| + C$. In terms of u this is $2\ln\left|\frac{\sqrt{u^2+1}}{u} - \frac{1}{u}\right| + C$ and in terms of x this is

$2\ln\left|\frac{\sqrt{(e^x+1)}}{e^{\frac{1}{2}x}} - e^{-\frac{1}{2}x}\right| + C$.

(39) Find $\int \frac{\arctan\sqrt{x}}{\sqrt{x}(1+x)} dx$.

Answer: $\int \frac{\arctan\sqrt{x}}{\sqrt{x}(1+x)} = \arctan^2\sqrt{x} + C$

(40) Find $\int \sqrt{\left(\frac{1+x}{1-x}\right)} dx$.

Answer: Multiply the fraction on the top and bottom by $1 + x$ to get $\int \frac{1+x}{\sqrt{1-x^2}} dx$. Now let $x = \sin\theta$ so $dx = \cos\theta\,d\theta$. Then this is $\int \frac{1+\sin\theta}{\sqrt{1-\sin^2\theta}} \cos\theta\,d\theta = \int (1+\sin\theta)\,d\theta = \theta - \cos\theta + C$. In terms of x this gives $\arcsin x - \sqrt{1-x^2} + C$.

(41) Find $\int \sqrt{\frac{x-2}{x+2}} dx$.

Answer: Multiply the fraction on the top and bottom by $x + 2$ to get $\int \frac{\sqrt{x^2-4}}{x+2} dx$. Now let $x = 2\sec\theta$ so $dx = 2\sec\theta\tan\theta\,d\theta$ and in terms of θ the indefinite integral is $2\int \frac{\sqrt{\sec^2(\theta)-1}}{\sec\theta+1} \sec\theta\tan\theta\,d\theta = 2\int \frac{\tan^2\theta\sec(\theta)}{1+\sec\theta} d\theta =$
$2\int \frac{\tan^2\theta}{1+\cos\theta} = 2\int \sec^2\theta\,d\theta - 2\int \sec\theta\,d\theta = 2\tan\theta - 2\ln(\sec\theta + \tan\theta) + C$. Now in terms of x this is $\sqrt{(x^2-4)} - 2\ln\left(\frac{1}{2}x + \frac{1}{2}\sqrt{(x^2-4)}\right) + C$.

(42) Find $\int \frac{x^2}{\sqrt{(4+9x^2)}} dx$.

Answer: $\int \frac{x^2}{\sqrt{4+9x^2}} \, dx = \frac{1}{18} x \sqrt{4 + 9x^2} - \frac{2}{27} \ln \left(3x + \sqrt{4 + 9x^2} \right) + C$

(43) Find $\int \frac{1}{\sqrt{x^2+49}} \, dx$.

Answer: $\int \frac{1}{\sqrt{x^2+49}} \, dx = \ln \left(x + \sqrt{x^2 + 49} \right) + C$

(44) Find $\int \frac{1}{\sqrt{x^2-36}} \, dx$.

Answer: $\int \frac{1}{\sqrt{x^2-36}} \, dx = \ln \left(x + \sqrt{x^2 - 36} \right) + C$

(45) Find $\int x \ln (3x) \, dx$.

Answer: $\int x \ln (3x) \, dx = \frac{1}{2} x^2 \ln 3x - \frac{1}{4} x^2 + C$

(46) Find $\int x \ln^2 (6x) \, dx$.

Answer: $\int x \ln^2 6x \, dx = \frac{1}{2} x^2 \ln^2 6x - \frac{1}{2} x^2 \ln 6x + \frac{1}{4} x^2 + C$

(47) Find $\int x^3 e^{x^2} \, dx$.

Answer: $\int x^3 e^{x^2} \, dx = \frac{1}{2} x^2 e^{x^2} - \frac{1}{2} e^{x^2} + C$

(48) Find $\int x^2 \sin 8x \, dx$.

Answer: $\int x^2 \sin 8x \, dx = -\frac{1}{8} x^2 \cos 8x + \frac{1}{256} \cos 8x + \frac{1}{32} x \sin 8x + C$

(49) Find $\int \arcsin x \, dx$.

Answer: $\int \arcsin x \, dx = x \arcsin x + \sqrt{(1 - x^2)} + C$

(50) Find $\int \arctan x \, dx$.

Answer: $\int \arctan x \, dx = x \arctan x - \frac{1}{2} \ln \left(1 + x^2 \right) + C$

(51) Find $\int \frac{\arctan 6x}{x^3} \, dx$. **Hint:** Do integration by parts on this one. This will get things started. Then recall partial fractions.

Answer: $\int \frac{\arctan 6x}{x^3} \, dx = -\frac{1}{2x^2} \arctan 6x - \frac{3}{x} - 18 \arctan 6x + C$

(52) Find $\int \sin (4x) \ln (\tan 4x) \, dx$.

Answer:

Integration by parts gives $-\frac{1}{4} \cos 4x \ln (\tan 4x) + \frac{1}{4} \ln (\csc 4x - \cot 4x) + C$

(53) Find $\int e^{2x} \sqrt{e^{4x} + 1} \, dx$. **Hint:** Try $u = e^{2x}$. This will yield something which will look a little different than the answer I have given below.

Answer: $\frac{1}{4} e^{2x} \sqrt{(e^{4x} + 1)} + \frac{1}{4} \text{arcsinh} \left(e^{2x} \right) + C$

(54) Find $\int \cos (\ln 7x) \, dx$.

Answer: $\int \cos (\ln 7x) \, dx = x \cos (\ln 7x) + \int x \sin (\ln (7x)) \left(\frac{1}{x} \right) dx = x \cos (\ln 7x) + \left[x \sin (\ln 7x) - \int \cos (\ln (7x)) \, dx \right]$ and so $\int \cos (\ln 7x) \, dx = \frac{1}{2} \left[x \cos (\ln 7x) + x \sin (\ln 7x) \right] + C$

(55) Find $\int \sin (\ln 7x)$.

Answer: $\int \sin (\ln 7x) \, dx = \frac{1}{2} \left[x \sin (\ln 7x) - x \cos (\ln 7x) \right] + C$

(56) Find $\int e^{5x} \sin 3x \, dx$.

Answer: $\int e^{5x} \sin 3x \, dx = -\frac{3}{34} e^{5x} \cos 3x + \frac{5}{34} e^{5x} \sin 3x + C$

(57) Find $\int e^{3x} \sin^2 2x \, dx$.

Answer: $\int e^{3x} \sin^2 2x \, dx = \frac{1}{6} e^{3x} - \frac{3}{25} e^{3x} (\cos 2) x - \frac{4}{25} e^{3x} (\sin 2) x + C$

(58) Find $\int \frac{x+4}{(x+3)^2} \, dx$.

Answer: $\int \frac{x+4}{(x+3)^2} \, dx = \ln (x + 3) - \frac{1}{x+3} + C$

(59) Find $\int \frac{7x^2+100x+347}{(x+2)(x+7)^2} \, dx$.

Answer: $\int \frac{7x^2+100x+347}{(x+2)(x+7)^2} \, dx = 7 \ln (x + 2) - \frac{2}{x+7} + C$

(60) Find $\int \frac{3x^2+7x+8}{(x+2)(x^2+2x+2)}\,dx$.

Answer: $\int \frac{3x^2+7x+8}{(x+2)(x^2+2x+2)}\,dx = 3\ln(x+2) + \arctan(1+x) + C$

(61) Find $\int \frac{4x^2+65x+266}{(x+2)(x^2+16x+66)}\,dx$.

Answer: $\int \frac{4x^2+65x+266}{(x+2)(x^2+16x+66)}\,dx = 4\ln(2+x) + \frac{1}{2}\sqrt{2}\arctan\frac{1}{4}(2x+16)\sqrt{2}$

(62) Find $\int \frac{1}{x^3+8}\,dx$. **Hint:** You need to first factor x^3+8 and then use partial fractions.

Answer:

$\int \frac{1}{x^3+8}\,dx = \frac{1}{12}\ln(x+2) - \frac{1}{24}\ln(x^2-2x+4) + \frac{1}{12}\sqrt{3}\arctan\frac{1}{6}(2x-2)\sqrt{3}+C$

(63) Find $\int \frac{1}{x^4+x^2+1}\,dx$.

Answer: $\int \frac{1}{x^4+x^2+1}\,dx = \frac{1}{4}\ln(x^2+x+1) + \frac{1}{6}\sqrt{3}\arctan\frac{1}{3}(2x+1)\sqrt{3} - \frac{1}{4}\ln(x^2-x+1) + \frac{1}{6}\sqrt{3}\arctan\frac{1}{3}(2x-1)\sqrt{3} + C$

(64) Find $\int \frac{1}{x^4+81}\,dx$. **Hint:**
$x^4+81 = \left(x^2-3x\sqrt{2}+9\right)\left(x^2+3x\sqrt{2}+9\right).$

Answer: Now you can use partial fractions to find $\int \frac{1}{x^4+81}\,dx =$

$\frac{1}{216}\sqrt{2}\ln\frac{x^2+3x\sqrt{2}+3}{x^2-3x\sqrt{2}+3} + \frac{1}{108}\sqrt{2}\arctan\left(\frac{1}{3}x\sqrt{2}+1\right) + \frac{1}{108}\sqrt{2}\arctan\left(\frac{1}{3}x\sqrt{2}-1\right)+C$

(65) Find $\int \frac{1}{x^6+64}\,dx$.

Answer:

First factor the denominator.

$x^6+64 = \left(x^2+4\right)\left(\left(x-\sqrt{3}\right)^2+1\right)\left(\left(x+\sqrt{3}\right)^2+1\right) \cdot \frac{1}{x^6+64} = \frac{Ax+B}{x^2+4} +$

$\frac{Cx+D}{\left(x-\sqrt{3}\right)^2+1}$

$+\frac{Ex+F}{\left(x+\sqrt{3}\right)^2+1}$ and so after wading through much affliction, the partial fractions

decomposition is $\frac{16}{3x^2+12} + \frac{-\frac{1}{192}\sqrt{3}x+\frac{1}{48}}{\left(x-\sqrt{3}\right)^2+1} + \frac{\frac{1}{192}\sqrt{3}x+\frac{1}{48}}{\left(x+\sqrt{3}\right)^2+1}$. Therefore, the indefinite

integral is $\int \left(\frac{16}{3x^2+12} + \frac{-\frac{1}{192}\sqrt{3}x+\frac{1}{48}}{\left(x-\sqrt{3}\right)^2+1} + \frac{\frac{1}{192}\sqrt{3}x+\frac{1}{48}}{\left(x+\sqrt{3}\right)^2+1}\right)dx =$

$\frac{1}{96}\arctan\frac{1}{2}x - \frac{1}{384}\sqrt{3}\ln\left(x^2-2\sqrt{3}x+4\right) + \frac{1}{192}\arctan\left(x-\sqrt{3}\right) + \frac{1}{384}\sqrt{3}\ln\left(x^2+2\sqrt{3}x+4\right) + \frac{1}{192}\arctan\left(x+\sqrt{3}\right)+C$

(66) Find $\int \frac{x^2}{(x-3)^{100}}\,dx$.
Hint: You ought to let $u = x-3$.

Answer:$\int \frac{x^2}{(x-3)^{100}}\,dx = \frac{1}{11(3-x)^{99}} - \frac{3}{49(3-x)^{98}} + \frac{1}{97(3-x)^{97}} + C$

(67) Find $\int 2\frac{4+4x+3x^2+x^3}{(x^2+1)(x^2+2x+5)}\,dx$.

Answer: $\int 2\frac{4+4x+3x^2+x^3}{(x^2+1)(x^2+2x+5)}\,dx = \frac{1}{2}\ln(x^2+1) + \arctan x + \frac{1}{2}\ln(x^2+2x+5) + \arctan\left(\frac{1}{2}+\frac{1}{2}x\right) + C$

(68) Find $\int \frac{1-x^7}{x(1+x^7)}\,dx$.

Answer: $\int \frac{1-x^7}{x(1+x^7)}\,dx = \ln x - \frac{2}{7}\ln\left(1+x^7\right) + C$

(69) Find $\int \frac{x^2+1}{x^4-2x^2+1}\,dx$.

Answer: $\int \frac{x^2+1}{x^4-2x^2+1}\,dx = -\frac{1}{2(x-1)} - \frac{1}{2(1+x)} + C$

(70) Find $\int \frac{x^5}{x^8+1}\,dx$.

Answer: Let $u = x^2$. $\int \frac{x^5}{x^8+1} dx = \int \frac{u^2}{2u^4+2} du$. Now use partial fractions. $\int \frac{x^5}{x^8+1} dx = \frac{1}{16}\sqrt{2}\ln\frac{x^4-x^2\sqrt{2}+1}{x^4+x^2\sqrt{2}+1} + \frac{1}{8}\sqrt{2}\arctan\left(x^2\sqrt{2}+1\right) + \frac{1}{8}\sqrt{2}\arctan\left(x^2\sqrt{2}-1\right) + C$

(71) Find $\int \frac{x^2+4}{x^6+64} dx$.

Hint: $x^6 + 64 = \left(x^2 + 4\right)\left(x^4 - 4x^2 + 16\right)$.

Answer:

$\int \frac{x^2+4}{x^6+64} dx = -\frac{1}{96}\sqrt{3}\ln\left(x^2 - 2\sqrt{3}x + 4\right) + \frac{1}{16}\arctan\left(x - \sqrt{3}\right) + \frac{1}{96}\sqrt{3}\ln\left(x^2 + 2\sqrt{3}x + 4\right) + \frac{1}{16}\arctan\left(x + \sqrt{3}\right) + C$

(72) Find $\int \frac{dx}{x\left(2+3\sqrt{x}+\sqrt[3]{x}\right)}$.

Answer: $\int \frac{dx}{x\left(2+3\sqrt{x}+\sqrt[3]{x}\right)} = \int \frac{6}{u\left(2+3u^3+u^2\right)} du = 3\ln u - \frac{6}{7}\ln\left(1+u\right) - \frac{15}{14}\ln\left(3u^2 - 2u + 2\right) - \frac{3}{35}\sqrt{5}\arctan\frac{1}{10}\left(6u-2\right)\sqrt{5} + C = 3\ln x^{1/6} - \frac{6}{7}\ln\left(1+x^{1/6}\right) - \frac{15}{14}\ln\left(3x^{1/3} - 2x^{1/6} + 2\right) - \frac{3}{35}\sqrt{5}\arctan\frac{1}{10}\left(6x^{1/6} - 2\right)\sqrt{5} + C$

(73) Find $\int \frac{\sqrt{x+3}-\sqrt{x-3}}{\sqrt{x+3}+\sqrt{x-3}} dx$.

Hint: Show the integrand equals $\frac{1}{3}x - \frac{1}{3}\sqrt{x^2-9}$.

Answer:

$\int \frac{\sqrt{x+3}-\sqrt{x-3}}{\sqrt{x+3}+\sqrt{x-3}} dx = \int \left(\frac{1}{3}x - \frac{1}{3}\sqrt{x^2-9}\right) dx = \frac{1}{6}x^2 - \frac{1}{6}x\sqrt{x^2-9} + \frac{3}{2}\ln\left(x + \sqrt{x^2-9}\right) + C$

(74) Find $\int \frac{x^2}{\sqrt{x^2+6x+13}} dx$. **Hint:** Complete the square in the denominator.

Answer: $\int \frac{x^2}{\sqrt{x^2+6x+13}} dx = \frac{1}{2}x\sqrt{x^2+6x+13} - \frac{9}{2}\sqrt{x^2+6x+13} + 7\ln\left(x+3+\sqrt{x^2+6x+13}\right) + C$

(75) Find $\int \frac{\sqrt{x^2+4x+5}}{x} dx$.

Answer: $\int \frac{\sqrt{x^2+4x+5}}{x} dx = \sqrt{x^2+4x+5} + 2\operatorname{arcsinh}\left(x+2\right) - \sqrt{5}\operatorname{arctanh}\frac{1}{5}\left(5+2x\right)\frac{\sqrt{5}}{\sqrt{x^2+4x+5}} + C$

You might try letting $u = \sinh^{-1}\left(x+2\right)$.

(76) Find $\int \frac{1}{x^3\sqrt{x^2+9}} dx$.

Answer: $\int \frac{1}{x^3\sqrt{x^2+9}} dx = -\frac{1}{18x^2}\sqrt{x^2+9} + \frac{1}{54}\operatorname{arctanh}\frac{3}{\sqrt{x^2+9}} + C$

You might try letting $u = \frac{1}{\sqrt{x^2+9}}$.

(77) Find $\int \frac{dx}{x^4\sqrt{x^2-9}}$.

Answer: $\int \frac{dx}{x^4\sqrt{x^2-9}} = \frac{1}{27x^3}\sqrt{\left(x^2-9\right)} + \frac{2}{243x}\sqrt{\left(x^2-9\right)} + C$

(78) Find $\int \sin^6\left(3x\right) dx$.

Answer: $\int \sin^6\left(3x\right) dx = -\frac{1}{2}\sin^5 x \cos x - \frac{5}{8}\sin^3 x \cos x - \frac{15}{16}\cos x \sin x + \frac{5}{16}x + C$

(79) Find $\int \frac{\sin^3(x)}{\cos^4(x)} dx$.

Answer: $\int \frac{\sin^3(x)}{\cos^4(x)} dx = \frac{1}{3}\frac{\sin^4 x}{\cos^3 x} - \frac{1}{3}\frac{\sin^4 x}{\cos x} - \frac{1}{3}\sin^2 x \cos x - \frac{2}{3}\cos x + C$

(80) Find $\int \tan^5\left(2x\right) dx$.

Answer: $\int \tan^5\left(2x\right) dx = -\frac{1}{4}\tan^2 2x + \frac{1}{8}\tan^4 2x + \frac{1}{4}\ln\left(2 + 2\tan^2 2x\right) + C$

(81) Find $\int \frac{dx}{\sqrt{\tan(2x)}}$.

Answer: $\int \frac{dx}{\sqrt{\tan(2x)}} = 2\frac{\tan^{\frac{1}{2}}x}{1+2\tan^2 x} + 2\frac{\tan^{\frac{5}{2}}x}{1+2\tan^2 x} - 2\tan^{\frac{1}{2}}x +$

$\frac{1}{4}\sqrt{2}\arctan 2\sqrt{2}\frac{\tan^{\frac{1}{2}}x}{1-2\tan x} + \frac{1}{4}\sqrt{2}\ln\frac{2\tan x+2\sqrt{2}\tan^{\frac{1}{2}}x+1}{\sqrt{(1+2\tan^2 x)}} + C$

You might try the substitution $u = \tan(x)$.

(82) Find $\int \frac{dx}{\cos x+3\sin x+4}$.

Answer: $\int \frac{dx}{\cos x+3\sin x+4} = \frac{1}{3}\sqrt{6}\arctan\frac{1}{12}\left(6\tan\frac{1}{2}x + 6\right)\sqrt{6} + C$

Try the substitution $u = \tan\left(\frac{x}{2}\right)$.

7.12 Computers And Antiderivatives

It should be fairly clear by now that there are standard techniques for finding antiderivatives which sometimes work. All the main methods have been presented but there are many functions which are considered important enough that they are given names even though they are not given names in the calculus books. Sometimes the desired antiderivative is expressible in terms of one of these functions. Computer algebra systems know about these functions even though you have not encountered them. Also, the routine mechanical process of searching for an antiderivative can often be done by a computer more quickly than you could do it by hand. For example to find $\int \frac{\sin x}{x}dx$ you use the maple input

$$> int(sin(x)/x, x);$$

and then press return. This gives $Si(x)$ for the answer. Of course you need to add in an arbitrary constant of integration for it to be correct. Maple doesn't bother to do this for you. This function Si is one not usually studied in calculus but it is known to the computer algebra system. Sometimes the integrals involve elementary functions but they are very hard to do. For example suppose you wanted to find $\int \frac{1}{(x^4+1)}dx$. In Maple this would involve the command

$$> int(1/(1+x\hat{\ }4), x);$$

You would then press return and it will give

$$\frac{1}{8}\sqrt{2}\ln\frac{x^2+x\sqrt{2}+1}{x^2-x\sqrt{2}+1} + \frac{1}{4}\sqrt{2}\arctan\left(x\sqrt{2}+1\right) + \frac{1}{4}\sqrt{2}\arctan\left(x\sqrt{2}-1\right).$$

Then you add in the arbitrary constant of integration and you are done with the wretched task of finding the antiderivative. There are many versions of computer algebra systems right now other than Maple. Scientific workplace, the software used to type this book, contains one very easy to use system which is not as good as the full version of Maple or Mathematica but which is entirely adequate for finding antiderivatives. In fact, this system is what actually computed the above answers. All that was necessary to find $\int \frac{x}{(x^4+4)}dx$ was to enter the preceding expression in math mode and then to press evaluate. This yields $\int \frac{x}{(x^4+4)}dx = \frac{1}{4}\arctan\frac{1}{2}x^2$. Then to make it right, you need to add in an arbitrary constant of integration.

It is not the purpose of this book to feature computer algebra systems at length but it is important that you realize these things exist and make finding antiderivatives fairly routine when they can be found. If you have access to a computer algebra system, you should experiment with using it to find antiderivatives. It is fun to have a machine do laborious computations for you. However, it is essential that you acquire some facility with the various techniques independent of a machine.

Chapter 8

Applications Of Antiderivatives

8.1 Areas

Consider the problem of finding the area between the graph of a function of one variable and the x axis as illustrated in the following picture.

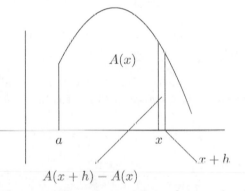

$$A(x+h) - A(x)$$

The curved line on the top represents the graph of the function $y = f(x)$ and the symbol $A(x)$ represents the area between this curve and the x axis between the point a and the point x as shown. The vertical line from the point $x + h$ up to the curve and the vertical line from x up to the curve define the area, $A(x+h) - A(x)$ as indicated. You can see that this area is between $hf(x)$ and $hf(x+h)$. This happens because the function is decreasing near x. In general, for continuous functions f, Theorem 2.25 on Page 81 implies there exists $x_M, x_m \in [x, x+h]$ with the properties $f(x_M) \equiv \max\{f(x) : x \in [x, x+h]\}$ and $f(x_m) \equiv \min\{f(x) : x \in [x, x+h]\}$. Then, $f(x_m) = \frac{hf(x_m)}{h} \leq \frac{A(x+h)-A(x)}{h} \leq \frac{hf(x_M)}{h} = f(x_M)$. Therefore, using the squeezing theorem, Theorem 2.7, and the continuity of f, $A'(x) \equiv \lim_{h\to 0} \frac{A(x+h)-A(x)}{h} = f(x)$. The consideration of $h < 0$ is also straightforward. This discussion implies the following theorem.

Theorem 8.1. *Let $a < b$ and let $f : [a, b] \to [0, \infty)$ be continuous. Then letting $A(x)$ denote the area between a, x, the graph of the function and the x axis,*

$$A'(x) = f(x) \text{ for } x \in (a, b), \ A(a) = 0. \tag{8.1}$$

Also, A is continuous on $[a, b]$.

The problem for A described in the above theorem is called an initial value problem and the equation $A'(x) = f(x)$ is a differential equation. It is called this because it is an equation for an unknown function $A(x)$ written in terms of the derivative of this unknown function. The assertion that A should be continuous on $[a, b]$ follows from the fact that it has to be continuous on (a, b) because of the existence of its derivative and the above argument can also be used to obtain one sided derivatives for A at the end points a and b, which yields continuity on $[a, b]$.

Example 8.1. Let $f(x) = x^2$ for $x \in [1, 2]$. Find the area between the graph of the function, the points 1 and 2, and the x axis.

The function $\frac{x^3}{3} + C$ has the property that its derivative gives x^2. This is true for any C. It only remains to choose C in such a way that the function equals zero at $x = 1$. Thus $C = \frac{-1}{3}$. It follows that $A(x) = \frac{x^3}{3} - \frac{1}{3}$. Therefore, the area described equals $A(2) = \frac{8}{3} - \frac{1}{3} = \frac{7}{3}$ square units.

Example 8.2. Find the area between the graph of the function $y = 1/x^2$ and the x axis for x between $1/2$ and 3.

The function $-\frac{1}{x} + C$ has the property that its derivative equals $1/x^2$. Letting $C = 2$, $A(x) = -\frac{1}{x} + 2$ satisfies the appropriate initial value problem and so the area equals $A(3) = \frac{5}{3}$.

8.2   Area Between Graphs

It is a minor generalization to consider the area between the graphs of two functions. Consider the following picture.

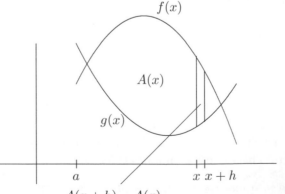

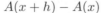

You see that sometimes the function $f(x)$ is on top and sometimes the function $g(x)$ is on top. It is the length of the vertical line joining the two graphs which is of importance and this length is always $|f(x) - g(x)|$ regardless of which function is larger. By Theorem 2.25 on Page 81 there exist x_M, $x_m \in [x, x+h]$ satisfying

$$|f(x_M) - g(x_M)| \equiv \max\{|f(x) - g(x)| : x \in [x, x+h]\}$$

and

$$|f(x_m) - g(x_m)| \equiv \min\{|f(x) - g(x)| : x \in [x, x+h]\}.$$

Then

$$\frac{|f(x_m) - g(x_m)| \, h}{h} \leq \frac{A(x+h) - A(x)}{h} \leq \frac{|f(x_M) - g(x_M)| \, h}{h},$$

and using the squeezing theorem, Theorem 2.7 on Page 60, as $h \to 0$

$$A'(x) = |f(x) - g(x)|$$

Also $A(a) = 0$ as before. This yields the following theorem which generalizes the one presented earlier because the x axis is the graph of the function $y = 0$.

Theorem 8.2. *Let $a < b$ and let $f, g : [a, b] \to \mathbb{R}$ be continuous. Then letting $A(t)$ denote the area between the graphs of the two functions for $x \in [a, t]$,*

$$A'(t) - |f(t) - g(t)| \ \text{for } t \in (a, b), \ A(a) = 0. \tag{8.2}$$

Also, A is continuous on $[a, b]$.

This theorem provides the justification for the following procedure for finding the area between the graphs of two functions.

Procedure 8.3. **To find the area between the graphs of the functions $y = f(x)$ and $y = g(x)$ for $x \in [a, b]$, split the interval into nonoverlapping subintervals, $I_1, I_2, \cdots, I_k$ which have the property that on $I_i, |f(x) - g(x)|$ equals either $f(x) - g(x)$ or $g(x) - f(x)$. If $I_i = [p_i, q_i]$, take an antiderivative of $|f(x) - g(x)|$ on $[p_i, q_i]$, H_i. (This might be a reasonable problem because on this interval, you will not need to write in absolute value signs.) Then the area between the curves for $x \in I_i$ is $H_i(q_i) - H_i(p_i)$. The desired area is the sum of these.**

Proof: Consider the area between the curves for $x \in [p_i, q_i]$. You need $A(q_i)$ where $A' = |f(x) - g(x)|$ and $A(p_i) = 0$. Let H be any antiderivative. Then from Lemma 7.1, $A(x) = H(x) + C$ where C is some constant. Thus $C = -H(p_i)$ and so $A(q_i) = H(q_i) - H(p_i)$. ∎

Notation 8.1. For a continuous function f defined on an interval $[a, b]$, if $F'(x) = f(x)$, the expression $F(b) - F(a)$ is denoted by $\int_a^b f(x) \, dx$. Thus to find the area between the two functions which are defined on an interval $[a, b]$, one can write it in the form $\int_a^b |f(x) - g(x)| \, dx$. This is also called a definite integral and it will be discussed more later. Note that the expression $F(b) - F(a)$ is independent of the antiderivative used because if G is another antiderivative, then G equals $F + C$ for a suitable constant C. Thus $G(b) - G(a) = F(b) - F(a)$. The constants cancel.

This also illustrates the following general principle.

Proposition 8.1. Consider the initial value problem $\frac{dy}{dx} = f(x)$, $y(a) = c$, where f is a continuous function. Then $y(x) = c + F(x) - F(a) \equiv c + \int_a^x f(t)\, dt$ where F is any antiderivative for $f(x)$.

Proof: This follows from the definition of $\int_a^x f(t)\, dt$ and the fact that if the differential equation holds, then $y(x) = F(x) + C$ for a suitable constant. Thus $c + F(x) - F(a) = c + y(x) - y(a) = y(x)$. ■

Example 8.3. Let $f(x) = 8 - \frac{x^2}{2}$ and $g(x) = \frac{x^2}{2} - 1$. Find the area between the graphs of the two functions for $x \in [-4, 3]$.

The following graph is of the absolute value of the difference of the two functions.

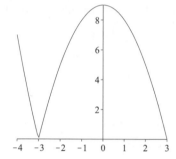

You should always draw such a graph to determine what is going on. $|f(x) - g(x)| = |9 - x^2| = \begin{cases} x^2 - 9 & \text{if } x \in [-4, -3] \\ 9 - x^2 & \text{if } x \in [-3, 3] \end{cases}$. It follows that on $(-4, -3)$, an antiderivative is $H(x) = \frac{x^3}{3} - 9x$. Therefore, the area between the curves for $x \in [-4, -3]$ is

$$\left(\frac{(-3)^3}{3} - 9(-3) \right) - \left(\frac{(-4)^3}{3} - 9(-4) \right) = \frac{10}{3}$$

Now consider $x \in [-3, 3]$. On this interval, an antiderivative is $H(x) = 9x - \frac{x^3}{3}$ and so the area between the curves for x in this interval is

$$\left(9(3) - \frac{(3)^3}{3} \right) - \left(9(-3) - \frac{(-3)^3}{3} \right) = 36$$

The total area is the sum of these. Thus the total area is $36 + \frac{10}{3} = \frac{118}{3}$.

Sometimes you have to worry about the difference of the functions changing sign at many points.

Example 8.4. Find the area between the graphs of $y = \sin 2x$ and $y = \cos(2x)$ for $x \in [0, \pi]$.

Here is a picture of the graphs of the two functions. In the picture, $y = \cos(2x)$ starts out on top. The dotted line is the graph of $\cos(2x)$ while the dashed line is the graph of $\sin(2x)$.

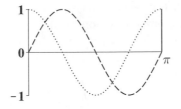

The points where things change are $2x = \pi/4$ so $x = \pi/8$, $2x = 5\pi/4$ so $x = 5\pi/8$, and $2x = 9\pi/4$ so $x = 9\pi/8$, but this last one is outside the interval so the area between these two curves for the indicated values of x is

$$\int_0^{\pi/8} (\cos(2x) - \sin(2x))\, dx + \int_{\pi/8}^{5\pi/8} (\sin(2x) - \cos(2x))\, dx$$

$$\left| \int_{5\pi/8}^{\pi} (\cos(2x) - \sin(2x))\, dx = 2\sqrt{2} \right.$$

A similar procedure holds for finding the area between two functions which are of the form $x = f(y)$ and $x = q(y)$ for $y \in [c, d]$. You just let y play the role of x in the above.

Example 8.5. Find the area between $x = 4 - y^2$ and $x = -3y$.

Here is a graph of these two curves.

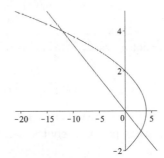

First find where the two graphs intersect. $4 - y^2 = -3y$. The solution is $y = -1$ and 4. For y in this interval, you can verify that $4 - y^2 > -3y$ and so $|4 - y^2 - (-3y)| = 4 - y^2 + 3y$. An antiderivative is $4y - \frac{y^3}{3} + \frac{3y^2}{2}$ and so the desired area is

$$\int_{-1}^{4} (4 - y^2 + 3y)\, dy = 4(4) - \frac{(4)^3}{3} + \frac{3(4)^2}{2} - \left(4(-1) - \frac{(-1)^3}{3} + \frac{3(-1)^2}{2}\right) = \frac{125}{6}.$$

Example 8.6. Find the area of the circular sector shown below.

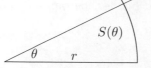

This is interesting because it is necessary to split the integral into two pieces corresponding to the two different functions which are on the top. The first of these is $y = x \tan(\theta)$ for $x \in [0, r\cos(\theta)]$ and the second is $y = \sqrt{r^2 - x^2}$ for $x \in [r\cos(\theta), r]$. Thus, from the above procedure, the area equals

$$\int_0^{r\cos\theta} x \tan(\theta)\, dx + \int_{r\cos\theta}^r \sqrt{r^2 - x^2}\, dx$$

An antiderivative for the first integrand is $\frac{x^2}{2}\tan\theta$. To find an antiderivative of the second, you need to use the trig. substitution $x = r\sin(u)$. When this is done, you find that an antiderivative of the second integral is

$$\frac{1}{2}\left(\arcsin\frac{x}{r}\right)r^2 + \frac{1}{2}x\sqrt{(r^2 - x^2)}$$

Therefore, using this information along with some laborious computations, it follows the area equals

$$\frac{1}{2}r^2\cos\theta\sin\theta + -\frac{1}{2}r^2\sin\theta\cos\theta + \frac{1}{2}r^2\theta = \frac{1}{2}r^2\theta$$

Of course this is entirely the wrong way to go about it! The sensible way is to consider little triangles rather than rectangles. Consider the following sector of a circle in which θ will be **small.**

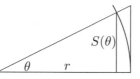

Then the area of this sector, $A(\theta)$ lies between the area of the two triangles. Thus

$$\frac{1}{2}r^2\sin(\theta) \le A(\theta) \le \frac{1}{2}r^2\tan(\theta)$$

Replacing θ by $d\theta$, a very small change in θ and $A(\theta)$ by dA and noting that as θ gets smaller $\sin(\theta)$ and $\tan(\theta)$ are approximately the same and equal to θ, $dA = \frac{1}{2}r^2 d\theta$, so $\frac{dA}{d\theta} = \frac{r^2}{2}$.

Example 8.7. Find the area of the circular sector illustrated in Example 8.6.

You need to have

$$\frac{dA}{d\theta} = \frac{r^2}{2}, \quad A(0) = 0.$$

The solution is now obviously $A(\theta) = \int_0^\theta \frac{r^2}{2}\, dt = \frac{1}{2}r^2\theta$.

8.3 Exercises

(1) Find the area between the graphs of the functions $y = x^2 + 1$ and $y = 3x + 5$.

(2) Find the area between the graphs of the functions $y = 2x^2$ and $y = 6x + 8$.

(3) Find the area between the graphs of $y = 5x + 14$ and $y = x^2$.

(4) Find the area between the graphs of the functions $y = x + 1$ and $y = 2x$ for $x \in [0, 3]$.

(5) Find the area between $y = |x|$ and the x axis for $x \in [-2, 2]$.

(6) Find the area between the graphs of $y = x$ and $y = \sin x$ for $x \in \left[-\frac{\pi}{2}, \pi\right]$.

(7) Find the area between the graphs of $x = y^2$ and $y = 2 - x$.

(8) Find the area between the x axis and the graph of the function $2x\sqrt{1 + x^2}$ for $x \in [0, 2]$. **Hint:** Recall the chain rule for derivatives.

(9) Show that the area of a right triangle equals one half the product of two sides which are not the hypotenuse.

(10) Let A denote the region between the x axis and the graph of the function $f(x) = x - x^2$. For $k \in (0, 1)$, the line $y = kx$ divides this region into two pieces. Explain why there exists a number k such that the area of these two pieces is exactly equal. **Hint:** This will likely involve the intermediate value theorem. Write an equation satisfied by k and then find an approximate value for k.

(11) Find the area between the graph of $f(x) = 1/x$ for $x \in [1, 2]$ and the x axis in terms of known functions.

(12) Find the area between the graph of $f(x) = 1/x^2$ for $x \in [1, 2]$ and the x axis in terms of known functions.

(13) Find the area between $y = \sin x$ and $y = \cos x$ for $x \in \left[0, \frac{\pi}{4}\right]$.

(14) Find the area between e^x and $\cos x$ for $x \in [0, 2\pi]$.

(15) Find the area between the graphs of $y = \sin(2x)$ and $y = \cos(2x)$ for $x \in [0, 2\pi]$.

(16) Find the area between the graphs of $y = \sin^2 x$, and the x axis, for $x \in [0, 2\pi]$.

(17) Find the area between the graphs of $y = \cos^2 x$ and $y = \sin^2 x$ for $x \in [0, \pi/2]$.

(18) Find the area between the graph of $f(x) = x^3 / (2 + 3x^4)$ and the x axis for $x \in [0, 4]$.

(19) Find the area between the graph of $f(x) = x^3 \sin(x^2) + 30$ and the x axis for $x \in [0, \pi]$.

(20) Find the area between the graph of $f(x) = x \sin 2x$ and $x \cos x - 4$ for $x \in [0, \pi]$.

(21) Find the area between $y = \dfrac{3x^3 + 9x^2 + 10x + 6}{(x+1)^2(x^2 + 2x + 2)}$ and the x axis for $x \subset [0, 4]$.

(22) Find the area between $y = \dfrac{5x^2 + 4 + 8x + 2x^4 + 3x^3}{(2x+3)(x^2+1)}$ and the x axis for $x \in [0, 2]$.

(23) Find the area between $\arctan(x)$ and $\ln x$ for $x \in [0, b]$ where $\arctan(b) - \ln(b) = 0$ and b is the first positive number for which this is so. $\arctan(b) - \ln(b) = 0$. You will need to use Newton's method or graphing on a calculator to find b.

(24) Find the area between $y = e^x$ and $y = 2x + 1$ for $x > 0$. In order to do this, you

have to find a solution to $e^x = 2x+1$ and this will require a numerical procedure such as Newton's method or graphing and zooming on your calculator.

(25) Find the area between $y = \ln x$ and $y = \sin x$ and $x \in [1, b]$ where b is such that the two functions are equal there. In order to do this, you have to find a solution to $\ln x = \sin x$ and this will require a numerical procedure such as Newton's method or graphing and zooming on your calculator.

(26) A ten foot ladder leans against a wall. The bottom of the ladder starts out six feet from the wall when the bottom of the ladder begins to move away from the wall eventually causing the ladder to crash to the ground. A point on the ladder which is two feet down the ladder from the top has coordinates $(x, f(x))$. Find $f(x)$. Next find the area between the resulting curve traced out by this point and the ground.

(27) Let $p > 1$. An inequality which is of major importance is

$$ab \le \frac{a^p}{p} + \frac{b^q}{q}$$

where here q is defined by $1/p + 1/q = 1$. Establish this inequality by adding up areas in the following picture.

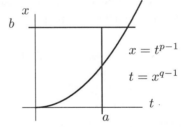

In the picture the right side of the inequality represents the sum of all the areas and the left side is the area of the rectangle determined by $(a, 0)$ and $(0, b)$.

(28) Refer to Example 8.7 and the discussion before it for this problem. Suppose θ is an angle measured from the positive x axis and you draw the ray from $(0, 0)$ out a distance of $r(\theta)$. Thus as θ varies, the distance of the point varies also.

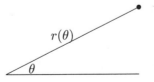

Suppose θ takes values in some interval $[a, b] \subseteq [0, 2\pi]$. Show the area between the curve traced out by this point and the two rays corresponding to $\theta = a$ and $\theta = b$ is given by

$$\frac{1}{2} \int_a^b (r(\theta))^2 \, d\theta$$

(29) In Problem 28 suppose $r(\theta) = 1 - \cos(\theta)$ where $\theta \in [0, 2\pi]$. Sketch a graph of the region described in this problem and find its area.

(30) Find the area between the graph of $y = 1/(1 + x^2)$ and the x axis for $x \in [-R, R]$. Now take a limit of what you just obtained as $R \to \infty$.

8.4 Volumes

8.4.1 Volumes Using Cross Sections

Imagine a line next to a three dimensional solid as shown in the next picture. For each y between a and b, let $A(y)$ denote the area of the cross section of the solid obtained by intersecting this solid with a plane through y perpendicular to the indicated line. Then

$$\frac{V(y + h) - V(y)}{h} \approx \frac{hA(y)}{h} = A(y)$$

the approximation getting better as h gets smaller. Thus in the limit, $V'(y) = A(y)$

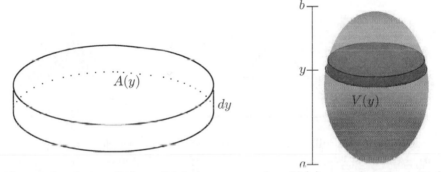

and so the total volume of the solid between a and y, $V(y)$, satisfies the initial value problem

$$\frac{dV}{dy} = A(y), \quad V(a) = 0.$$

The volume of the solid is $V(b)$. Here is a simple procedure which follows from this and Proposition 8.1.

Procedure 8.4. Given a three dimensional solid S and a line as shown in the above picture, the volume of S which lies between $y = a$ and $y = b$ is

$$\int_a^b A(y)\, dy.$$

Example 8.8. The parabola $y = 4 - x^2$ for $y \geq 0$ is revolved about the y axis to form a solid of revolution. Find the volume of this solid of revolution.

Below is a picture of the solid and a cross section of this solid.

As suggested above, one finds the area of a cross section at y, $A(y)$ and then solves $dV/dy = A(y)$, $V(0) = 0$. Equivalently one takes the definite integral $\int_0^4 A(y)\, dy$. At height y a cross section is a circle of radius $\sqrt{4-y}$, and so the area of this cross section is $\pi(4-y)$. Thus an antiderivative is $\pi\left(4y - \frac{1}{2}y^2\right)$, and so the volume is

$$\int_0^4 \pi(4-y)\, dy = \pi\left(4(4) - \frac{1}{2}(4)^2\right) = 8\pi$$

Example 8.9. Consider a pyramid which sits on a square base of length 500 feet and suppose the pyramid has height 300 feet. Find the volume of the pyramid in cubic feet.

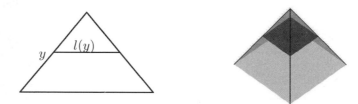

At height y, it follows from similar triangles that the length of one of the sides, $l(y)$, satisfies $\frac{l(y)}{300-y} = \frac{500}{300} = \frac{5}{3}$, and so $l(y) = \frac{5}{3}(300-y)$. Therefore, the cross section at y has area $\left(\frac{5}{3}(300-y)\right)^2$. An antiderivative is $-\frac{1}{5}\left(500 - \frac{5}{3}y\right)^3$ and so the total volume of this pyramid is

$$\int_0^{300} \left(\frac{5}{3}(300-y)\right)^2 dy = -\frac{1}{5}\left(500 - \frac{5}{3}(300)\right)^3 + \frac{1}{5}(500)^3 = 25,000,000$$

Example 8.10. The base of a solid is a circle of radius 10 meters in the xy plane. When this solid is cut with a plane which is perpendicular to the xy plane and the x axis, the result is a rectangle with height equal to one half the length of the side in the base. Find the volume of the resulting solid.

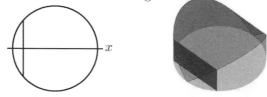

Note the cross section in the solid intersects the base in the vertical line shown.

Reasoning as above for the building, and letting $V(x)$ denote the volume between -10 and x, yields

$$\frac{dV}{dx} = A(x), \ V(-10) = 0$$

as an appropriate initial value problem for this volume. Here $A(x)$ is the area of the surface resulting from the intersection of the plane with the solid. The length of the side of this surface in the base is $2\sqrt{100 - x^2}$ because the equation of the circle which bounds the base is $x^2 + y^2 = 100$. The height of the rectangle is $\sqrt{100 - x^2}$. Therefore, $A(x) = 2(100 - x^2)$. An antiderivative is $V(x) = 200x - \frac{2}{3}x^3$. Therefore, the volume is

$$\int_{-10}^{10} 2(100 - x^2)\, dx = \left(200(10) - \frac{2}{3}(10)^3\right) - \left(200(-10) - \frac{2}{3}(-10)^3\right) = \frac{8000}{3}$$

Example 8.11. Find the volume of a sphere of radius R.

The sphere is obtained by revolving a disk of radius R about the y axis. Thus the radius of the cross section at height y would be $\sqrt{R^2 - y^2}$ and so the area of this cross section is $\pi(R^2 - y^2)$. An antiderivative is $V(y) = \pi\left(R^2 y - \frac{1}{3}y^3\right)$. Then the volume of the sphere is

$$\int_{R}^{R} \pi(R^2 - y^2)\, dy = \pi\left(R^2(R) - \frac{1}{3}(R)^3\right) - \left(\pi\left(R^2(-R) - \frac{1}{3}(-R)^3\right)\right) = \frac{4}{3}\pi R^3$$

Example 8.12. The bounded region in the first quadrant $(x, y > 0)$ determined by the graphs of $y = x^2$ and $y = \sqrt{x}$ is revolved about the x axis to form a solid in three dimensions. Find the volume of this solid.

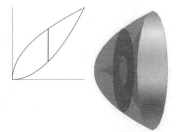

As illustrated in the above picture, for each x, the cross sections consist of the area between two concentric circles, the smaller one having radius x^2 and the larger having radius $\sqrt{x}$. Note that for $0 < x < 1, x^2 < \sqrt{x}$. The area of this "washer" is

$\pi \left(\sqrt{x}\right)^2 - \pi x^4 = \pi \left(x - x^4\right) = A(x)$. An antiderivative is $V(x) = \pi \left(\frac{1}{2}x^2 - \frac{1}{5}x^5\right)$.
Therefore, the volume is

$$\int_0^1 \left(\pi \left(x - x^4\right)\right) dx = \pi \left(\frac{1}{2} - \frac{1}{5}\right) - \left(\pi \left(\frac{1}{2}(0) - \frac{1}{5}(0)\right)\right) = \frac{3}{10}\pi$$

8.4.2   Volumes Using Shells

There is another way to find some volumes without using cross sections. This method involves the notion of shells. Consider the following picture of a circular shell.

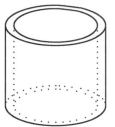

In this picture the radius of the inner circle will be r and the radius of the outer circle will be $r + \Delta r$ while the height of the shell is H. Therefore, the volume of the shell is the difference in the volumes of the two cylinders,

$$\pi(r + \Delta r)^2 H - \pi r^2 h = 2\pi H r (\Delta r) + \pi H (\Delta r)^2. \tag{8.3}$$

In the following picture, there is a 2 dimensional region in the xy plane which is revolved about a line. That which results is on the right. Notice the way the thin approximate rectangle on the left results in a shell in the solid of revolution on the right. Also note how the hole in the solid results from the gap between the line and the region in the plane.

To be more specific, consider the problem of revolving the region between $y = f(x)$ and $y = g(x)$ for $x \in [a, b]$ about the line $x = c$ for $c < a$. The following picture is descriptive of the situation.

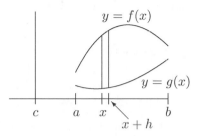

Let $V(x)$ denote the volume of the solid which results from revolving the region between the graphs of f and g above the interval $[a, x]$ about the line $x = c$. Thus $V(x + h) - V(x)$ equals the volume which results from revolving the region between the graphs of f and g which is also between the two vertical lines shown in the above picture. This results in a solid which is very nearly a circular shell like the one shown above and the approximation gets better as h decreases to zero. Therefore,

$$V'(x) = \lim_{h \to 0} \frac{V(x + h) - V(x)}{h} \tag{8.4}$$

$$= \lim_{h \to 0} \frac{2\pi |f(x) - g(x)|(x - c)h + \pi |f(x) - g(x)| h^2}{h}$$

$$= 2\pi |f(x) - g(x)|(x - c).$$

Also, $V(a) = 0$, and this means that to find the volume of revolution it suffices to solve the initial value problem

$$\frac{dV}{dx} = 2\pi |f(x) - g(x)|(x - c), \quad V(a) = 0 \tag{8.5}$$

and the volume of revolution will equal $V(b)$. Note that in the above formula, it is not necessary to worry about which is larger, $f(x)$ or $g(x)$ because it is expressed in terms of the absolute value of their difference. However, in doing the computations necessary to solve a given problem, you typically will have to worry about which is larger. Always draw a picture of the two dimensional region and the line to get some understanding for what is going on.

Procedure 8.5. To find the volume of the region between $y = f(x), y = g(x)$ for x between a and b which is obtained by revolving about the line $x = c$ which is either to the left or to the right of the entire two dimensional region, compute

$$\int_a^b 2\pi |f(x) - g(x)|(x - c)\, dx$$

Example 8.13. Find the volume of the solid formed by revolving the region between $y = \sin(x)$ and the x axis for $x \in [0, \pi]$ about the y axis.

In this example, $c = 0$ and since $\sin(x) \geq 0$ for $x \in [0, \pi]$, the initial value problem is

$$\frac{dV}{dx} = 2\pi x \sin(x), \quad V(0) = 0.$$

Then using integration by parts,

$$V(x) = 2\pi(\sin x - x \cos x) + C$$

and from the initial condition, $C = 0$. Therefore, the volume is $V(\pi) = 2\pi^2$. Alternatively, just find $H \in \int 2\pi x \sin(x)$ and compute $H(\pi) - H(0)$.

Example 8.14. The region between $y = \ln x$ and the x axis which lies between $x = 1$ and $x = 2$ is revolved about the line $x = 1/4$. Find the volume of the resulting solid of revolution.

Using the procedure described above, $\int 2\pi \left(x - \frac{1}{4}\right) \ln(x) \, dx = \pi x^2 \ln x - \frac{1}{2}\pi x^2 - \frac{1}{2}\pi x \ln x + \frac{1}{2}\pi x + C$. Then since $\ln 1 = 0$, this volume equals

$$\pi(2)^2 \ln 2 - \frac{1}{2}\pi 2^2 - \frac{1}{2}\pi 2 \ln(2) + \frac{1}{2}\pi 2 - \left(-\frac{1}{2}\pi + \frac{1}{2}\pi\right) = 3\pi \ln 2 - \pi$$

Example 8.15. Find the volume of the solid formed by revolving the region between $y = \sin x$, $y = \cos x$ for $x \in [0, \pi/4]$ about the line $x = -4$

In this example, $\cos x > \sin x$ for $x \in [0, \pi/4]$ and so using the above procedure, the volume is

$$\int_0^{\pi/4} 2\pi(x+4)(\cos x - \sin x) \, dx = 8\pi\sqrt{2} + \frac{1}{2}\pi^2\sqrt{2} - 10\pi$$

Example 8.16. Find the volume of a ball of radius R using the method of shells.

You draw the picture. In this case the ball is obtained by revolving the region between $y = \sqrt{R^2 - x^2}$, $y = -\sqrt{R^2 - x^2}$ for $x \in [0, R]$ about the y axis. Thus, from the procedure, this volume equals

$$\int_0^R 2\pi x \left(\sqrt{R^2 - x^2} - \left(-\sqrt{R^2 - x^2}\right)\right) dx = \frac{4}{3}R^3\pi,$$

the same formula obtained earlier using the method of cross sections.

8.5   Exercises

(1) The equation of an ellipse is $\left(\frac{x}{a}\right)^2 + \left(\frac{y}{b}\right)^2 = 1$. Sketch the graph of this in the case where $a = 2$ and $b = 3$. Now find the volume of the solid obtained by revolving the general ellipse about the y axis. What is the volume if it is revolved about the x axis? What is the volume of the solid obtained by revolving it about the line $x = -2a$?

(2) A ball of radius R has a hole drilled through a diameter which is centered at the diameter and of radius $r < R$. Find the volume of what is left after the hole has been drilled. What is the volume of the material which was taken out?

(3) Show the volume of a right circular cone is $(1/3) \times$ area of the base $\times$ height.

(4) Let R be a region in the xy plane of area A and consider the cone formed by fixing a point in space h units above R and taking the union of all lines starting at this point which end in R. Show that under these general conditions, the volume of the cone is $(1/3) \times A \times h$. **Hint:** The cross sections at height y look just like R but shrunk. Argue that the area at height y, denoted by $A(y)$ is simply $A(y) = A\frac{(h-y)^2}{h^2}$.

(5) Find the volume of the solid which results by revolving $y - \sin x$ about the x axis for $x \in [0, \pi]$.

(6) A circle of radius r in the xy plane is the base of a solid which has the property that cross sections perpendicular to the x axis are equilateral triangles. Find the volume of this solid.

(7) The region between $y = x^2$ and $y = x^3$ for $x \in [0, 1]$ is revolved about the y axis. Find the volume of the resulting solid using the method of cross sections. Now find it using the method of shells.

(8) A square having each side equal to r in the xy plane is the base of a solid which has the property that cross sections perpendicular to the x axis are equilateral triangles. Find the volume of this solid.

(9) The ellipse $\frac{x^2}{4} + \frac{y^2}{9} \leq 1$ is the base of a solid which has the property that cross sections perpendicular to the x axis are parabolas of height equal to the width of the ellipse. What is the volume of this solid?

(10) The area between $y = \sqrt{x}$ and $x = 4$ is the base of a solid which has the property that cross sections perpendicular to the x axis are parabolas of height equal to the width of the base. What is the volume of this solid?

(11) The circle $x^2 + y^2 \leq 4$ is the base of a solid which has the property that cross sections perpendicular to the y axis are isosceles right triangles having one leg equal to the width of the base. Find the volume of the solid.

(12) The circle $x^2 + y^2 \leq 4$ is the base of a solid which has the property that cross sections perpendicular to the y axis are isosceles right triangles having the hypotenuse equal to the width of the base. Find the volume of the solid.

(13) The part of the ellipse $\frac{x^2}{4} + \frac{y^2}{9} \leq 1$ which is in the first quadrant is the base of a solid which has the property that cross sections perpendicular to the x axis are parabolas of height equal to the width of the base. What is the volume of this solid?

(14) The ellipse $x^2 + \frac{y^2}{9} \leq 1$ is the base of a solid which has the property that cross sections perpendicular to the x axis are circles having a diameter perpendicular to the x axis and equal to the width of the ellipse. What is the volume of this solid? This looks like a football.

(15) The bounded region between $y = x^2$ and $y = x$ is revolved about the x axis. Find the volume of the solid which results.

(16) The region between $y = \ln x$, and the x axis for $x \in [1, 3]$ is revolved about the y axis. What is the volume of the resulting solid?

(17) The region between $y = x^2, y = 1$ for $x \in [0, 1]$ is revolved about the line, $x = -4$. Find the volume of the solid which results.

(18) The region between $y = \arctan(x)$, and the x axis for $x \in [0, 2]$ is revolved about the y axis. Find the volume of the resulting solid.

(19) The region between $y = \arctan(x)$, and the x axis for $x \in [0, 2]$ is revolved about the line $x = -1$. Find the volume of the resulting solid.

(20) The region between $y = \sin(x)$, and the x axis for $x \in [0, \pi]$ is revolved about the y axis. Find the volume of the resulting solid.

(21) The region between $y = \sin(x)$, and the x axis for $x \in [0, \pi]$ is revolved about the line $x = -1$. Find the volume of the resulting solid.

(22) The region between $y = \sin(x)$, and the x axis for $x \in [0, \pi]$ is revolved about the line $x = 5$. Find the volume of the resulting solid.

(23) The region between $y = \sin(x)$, and the x axis for $x \in [0, \pi]$ is revolved about the line $y = 2$. Find the volume of the resulting solid.

(24) The region between $y = 1 + \sin x$ and the x axis for $x \in [0, 2\pi]$ is revolved about the y axis. Find the volume of the solid which results.

(25) The region between $y = 2 + \sin 3x$ and the x axis for $x \in [0, \pi/3]$ is revolved about the line $x = -1$. Find the volume of the solid which results.

(26) The region between $y = x^3 - x$ and the x axis for $x > 0$ is revolved about the line $x - -1$. Find the volume of the solid which results.

(27) The region between $y = \sin 2x$ and the x axis for $x \subset [0, \pi]$ is revolved about the line $x = -1$. Find the volume of the solid which results.

8.6   Lengths Of Curves And Areas Of Surfaces Of Revolution

8.6.1   *Lengths*

The same techniques can be used to compute lengths of the graph of a function $y = f(x)$. Consider the following picture.

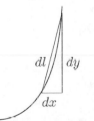

which depicts a small right triangle attached as shown to the graph of a function $y = f(x)$ for $x \in [a, b]$. If the triangle is small enough, this shows the length of the curve joined by the hypotenuse of the right triangle is essentially equal to the length of the hypotenuse. Thus, $(dl)^2 = (dx)^2 + (dy)^2$ and dividing by $(dx)^2$ yields

$$\frac{dl}{dx} = \sqrt{1 + \left(\frac{dy}{dx}\right)^2} = \sqrt{1 + f'(x)^2}, \; l(a) = 0 \tag{8.6}$$

as an initial value problem for the function $l(x)$ which gives the length of this curve on $[a, x]$. Thus the length of the curve is given by the definite integral,

$$\int_a^b \sqrt{1 + f'(x)^2} dx$$

This definition gives the right answer for the length of a straight line. To see this, consider a straight line through the points (a, b) and (c, d) where $a < c$. Then the right answer is given by the Pythagorean theorem or distance formula and is $\sqrt{(a-c)^2 + (d-b)^2}$. What is obtained from the above initial value problem? The equation of the line is $f(x) = b + \left(\frac{d-b}{c-a}\right)(x-a)$ and so $f'(x) = \left(\frac{d-b}{c-a}\right)$. Therefore, by the new procedure, the length is

$$\int_a^c \sqrt{1 + \left(\frac{d-b}{c-a}\right)^2}\, dx = (c-a)\sqrt{1 + \left(\frac{d-b}{c-a}\right)^2} = \sqrt{(a-c)^2 + (d-b)^2}$$

as hoped. Thus the new procedure gives the right answer in the familiar cases but it also can be used to find lengths for more general curves than straight lines. Summarizing,

Procedure 8.6. To find the length of the graph of the function $y = f(x)$ for $x \in [a, b]$, compute

$$\int_a^b \sqrt{1 + f'(x)^2}\, dx.$$

Here is another familiar example.

Example 8.17. Find the length of the part of the circle having radius r which is between the points $(0, r)$ and $\left(\frac{\sqrt{2}}{2}r, \frac{\sqrt{2}}{2}r\right)$.

Here the function is $f(x) = \sqrt{r^2 - x^2}$ and so $f'(x) = -x/\sqrt{r^2 - x^2}$. Therefore, the length is

$$\int_0^{\pi/4} \sqrt{1 + \left(-x/\sqrt{r^2 - x^2}\right)^2}\, dx = \int_0^{\pi/4} r\sqrt{\left(\frac{1}{r^2 - x^2}\right)}\, dx$$

Using a trig substitution $x = r\sin\theta$, it follows $dx = r\cos(\theta)\, d\theta$ and so

$$\int \frac{r}{\sqrt{r^2 - x^2}}\, dx = \int \frac{1}{\sqrt{1 - \sin^2\theta}} r\cos(\theta)\, d\theta = r\int d\theta = r\theta + C$$

Hence changing back to the variable x it follows an antiderivative is $l(x) = r\arcsin\left(\frac{x}{r}\right)$. Then the length is

$$r\arcsin\left(\frac{r}{r}\right) - r\arcsin\left(\frac{1}{r}\frac{\sqrt{2}}{2}r\right) = r\frac{\pi}{2} - r\frac{\pi}{4} = r\frac{\pi}{4}.$$

Note this gives the length of one eighth of the circle and so from this the length of the whole circle should be $2r\pi$. Here is another example

Example 8.18. Find the length of the graph of $y = x^2$ between $x = 0$ and $x = 1$.

Here $f'(x) = 2x$ and so the initial value problem to be solved is

$$\frac{dl}{dx} = \sqrt{1 + 4x^2}, \ l(0) = 0.$$

Thus, in terms of the definite integral, the length of this curve is

$$\int_0^1 \sqrt{1 + 4x^2}\,dx = \frac{1}{2}\sqrt{5} - \frac{1}{4}\ln\left(-2 + \sqrt{5}\right) = \frac{1}{2}\sqrt{5} + \frac{1}{4}\ln\left(\sqrt{5} + 2\right)$$

To find an antiderivative, you use the trig. substitution $2x = \tan u$ so $dx = \frac{1}{2}(\sec^2 u)\,du$. Then you find the antiderivative in terms of u and change back to x by using an appropriate triangle as described earlier.

8.6.2 *Surfaces Of Revolution*

The problem of finding the surface area of a solid of revolution is closely related to that of finding the length of a graph. First consider the following picture of the frustum of a cone in which it is desired to find the lateral surface area. In this picture, the frustum of the cone is the left part which has an l next to it and the lateral surface area is this part of the area of the cone.

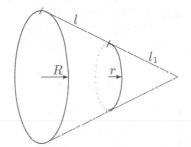

To do this, imagine painting the sides and rolling the shape on the floor for exactly one revolution. The wet paint would make the following shape.

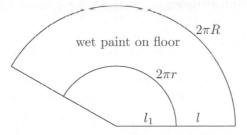

What would be the area of this wet paint? Its area would be the difference between the areas of the two sectors shown, one having radius l_1 and the other having radius $l + l_1$. Both of these have the same central angle equal to

$$\frac{2\pi R}{2\pi(l + l_1)}2\pi = \frac{2\pi R}{l + l_1}.$$

Therefore, by Example 8.7 on Page 200, this area is

$$(l + l_1)^2 \frac{\pi R}{(l + l_1)} - l_1^2 \frac{\pi R}{(l + l_1)} = \pi R l \frac{l + 2l_1}{l + l_1}$$

The view from the side is

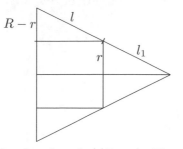

and so by similar triangles, $l_1 = lr/(R - r)$. Therefore, substituting this into the above, the area of this frustum is

$$\pi R l \frac{l + 2 \left(\frac{lr}{R-r} \right)}{l + \left(\frac{lr}{R-r} \right)} = \pi l (R + r) = 2\pi l \left(\frac{R+r}{2} \right).$$

Now consider a function f, defined on an interval $[a, b]$ and suppose it is desired to find the area of the surface which results when the graph of this function is revolved about the x axis. Consider the following picture of a piece of this graph.

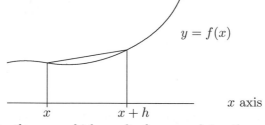

Let $A(x)$ denote the area which results from revolving the graph of the function restricted to $[a, x]$ about the x axis. Then from the above formula for the area of a frustum,

$$\frac{A(x + h) - A(x)}{h} \approx 2\pi \frac{1}{h} \sqrt{h^2 + (f(x + h) - f(x))^2} \left(\frac{f(x + h) + f(x)}{2} \right)$$

where $\approx$ denotes that these are close to being equal and the approximation gets increasingly good as $h \to 0$. Therefore, rewriting this a little yields

$$\frac{A(x + h) - A(x)}{h} \approx 2\pi \sqrt{1 + \left(\frac{f(x + h) - f(x)}{h} \right)^2} \left(\frac{f(x + h) + f(x)}{2} \right)$$

Therefore, taking the limit as $h \to 0$, and using $A(a) = 0$, this yields the following initial value problem for A which can be used to find the area of a surface of revolution.

$$A'(x) = 2\pi f(x) \sqrt{1 + f'(x)^2}, \ A(a) = 0.$$

What would happen if you revolved about the y axis? I will leave it to you to verify this would lead to the initial value problem

$$A'(x) = 2\pi x \sqrt{1 + f'(x)^2}, \ A(a) = 0.$$

As before, this results in the following simple procedure for finding the surface area of a surface of revolution.

Procedure 8.7. To find the surface area of a surface obtained by revolving the graph of $y = f(x)$ for $x \in [a, b]$ about the x axis, compute $\int_a^b 2\pi f(x) \sqrt{1 + f'(x)} dx$ Similarly, to get the area of the graph rotated about the y axis, compute $\int_a^b 2\pi x \sqrt{1 + f'(x)^2} dx$.

Example 8.19. Find the surface area of the surface obtained by revolving the function $y = r$ for $x \subset [a, b]$ about the x axis. Of course this is just the cylinder of radius r and height $b - a$ so this area should equal $2\pi r (b - a)$. (Imagine painting it and rolling it on the floor and then taking the area of the rectangle which results.)

Using the above initial value problem, solve

$$A'(x) = 2\pi r \sqrt{1 + 0^2}, \ A(a) = 0.$$

The solution is $A(x) = 2\pi r(x - a)$. Therefore, $A(b) = 2\pi r(b - a)$ as expected.

Example 8.20. Find the surface area of a sphere of radius r.

Here the function involved is $f(x) = \sqrt{r^2 - x^2}$ for $x \in [-r, r]$ and it is to be revolved about the x axis. In this case

$$f'(x) = \frac{-x}{\sqrt{r^2 - x^2}}$$

and so, by the procedure described above, the surface area is

$$\int_{-r}^{r} 2\pi \sqrt{r^2 - x^2} \sqrt{1 + \frac{x^2}{r^2 - x^2}} dx = 4r^2 \pi$$

8.7 Exercises

(1) Find the length of the graph of $y = \ln(\cos x)$ for $x \in [0, \pi/4]$.
(2) The curve defined by $y = \ln(\cos x)$ for $x \in [0, 1]$ is revolved about the y axis. Find an integral for the area of the surface of revolution.
(3) Find the length of the graph of $y = x^{1/2} - \frac{x^{3/2}}{3}$ for $x \in [1, 3]$.
(4) The graph of the function $y = x^3$ is revolved about the x axis for $x \in [0, 1]$. Find the area of the resulting surface of revolution.
(5) The graph of the function $y = x^3$ is revolved about the y axis for $x \in [0, 1]$. Find the area of the resulting surface of revolution. **Hint:** Formulate this in terms of x and use a change of variables.

(6) The graph of the function $y = \ln x$ is revolved about the y axis for $x \in [1, 2]$. Find the area of the resulting surface of revolution. **Hint:** Consider x as a function of y.

(7) The graph of the function $y = \ln x$ is revolved about the x axis for $x \in [1, 2]$. Find the area of the resulting surface of revolution. If you cannot do the integral, set it up.

(8) Find the length of $y = \cosh(x)$ for $x \in [0, 1]$.

(9) Find the length of $y = 2x^2 - \frac{1}{16} \ln x$ for $x \in [1, 2]$.

(10) The curve defined by $y = 2x^2 - \frac{1}{16} \ln x$ for $x \in [1, 2]$ is revolved about the y axis. Find the area of the resulting surface of revolution.

(11) Find the length of $y = x^2 - \frac{1}{8} \ln x$ for $x \in [1, 2]$.

(12) The curve defined by $y = x^2 - \frac{1}{8} \ln x$ for $x \in [1, 2]$ is revolved about the y axis. Find the area of the resulting surface of revolution.

(13) The curve defined by $y = \cosh(x)$ for $x \in [0, 1]$ is revolved about the x axis. Find the area of the resulting surface of revolution.

(14) The curve defined by $y = \cosh(x)$ for $x \in [0, 1]$ is revolved about the line $y = -3$. Find the area of the resulting surface of revolution.

(15) For a a positive real number, find the length of $y = \frac{ax^2}{2} - \frac{1}{4a} \ln x$ for $x \in [1, 2]$. Of course your answer should depend on a.

(16) The graph of the function $y = x^2$ for $x \in [0, 1]$ is revolved about the x axis. Find the area of the surface of revolution.

(17) The graph of the function $y = \sqrt{x}$ for $x \in [0, 1]$ is revolved about the y axis. Find the area of the surface of revolution. **Hint:** Switch x and y and then use the previous problem.

(18) The graph of the function $y = x^{1/2} - \frac{x^{3/2}}{3}$ is revolved about the y axis. Find the area of the surface of revolution if $x \in [0, 2]$.

(19) The graph of the function $y = \sinh x$ for $x \in [0, 1]$ is revolved about the x axis. Find the area of the surface of revolution.

(20) * The ellipse $\frac{x^2}{a^2} + \frac{y^2}{b^2} = 1$ is revolved about the x axis. Find the area of the surface of revolution.

(21) Find the length of the graph of $y = \frac{2}{3}(x - 1)^{3/2}$ for $x \in [2, 3]$.

(22) The curve defined by $y = \frac{2}{3}(x - 1)^{3/2}$ for $x \in [1, 2]$ is revolved about the y axis. Find the area of the resulting surface of revolution.

(23) Suppose $f'(x) = \sqrt{\sec^2 x - 1}$ and $f(0) = 0$. Find the length of the graph of $y = f(x)$ for $x \in [0, 1]$.

(24) The curve defined by $y = f(x)$ for $x \in [0, \pi]$ is revolved about the y axis where $f'(x) = \sqrt{(2 + \sin x)^2 - 1}$, $f(0) = 1$. Find the area of the resulting surface of revolution.

(25) Revolve $y = 1/x$ for $x \in [1, R]$ about the x axis. Find the area of this surface of revolution. Now show the limit of what you got as $R \to \infty$ does not exist. Next find the volume of this solid of revolution. Show the limit of this are $R \to \infty$ is finite. This infinite solid has infinite area but finite volume.

(26) The surface area of a sphere of radius r was shown to be $4\pi r^2$. Note that if $V(r) = \frac{4}{3}\pi r^3$, then $V'(r)$ equals the area of the sphere. Why is this reasonable based on geometrical considerations?

8.8 Force On A Dam And Work

8.8.1 *Force On A Dam*

Imagine you are a fish swimming in a lake behind a dam and you are interested in the total force acting on the dam. The following picture is what you would see.

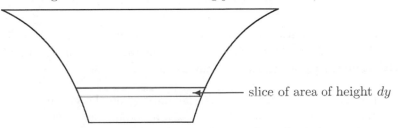

slice of area of height dy

The reason you would be interested in that long thin slice of area having essentially the same depth, say at y feet is because the pressure in the water at that depth is constant and equals $62.5y$ pounds per square foot[1]. Therefore, the total force the water exerts on the long thin slice is

$$dF = 62.5yL(y)\,dy$$

where $L(y)$ denotes the length of the slice. Therefore, the total force on the dam up to depth y is obtained as a solution to the initial value problem

$$\frac{dF}{dy} = 62.5yL(y),\ F(0) = 0.$$

Example 8.21. Suppose the width of a dam at depth y feet equals $L(y) = 1000 - y$ and its depth is 500 feet. Find the total force in pounds exerted on the dam.

From the above, this is obtained as the solution to the initial value problem

$$\frac{dF}{dy} = 62.5y(1000 - y),\ F(0) = 0$$

which is $F(y) = -20.83y^3 + 31250y^2$. The total force on the dam would be

$$F(500) = -20.83(500)^3 + 31250(500)^2 = 5,208,750,000.0$$

pounds. In tons this is $2,604,375$. That is a lot of force.

---

[1]Later on in Volume 2 a nice result on hydrostatic pressure will be presented which will verify this assertion. Here 62.5 is the weight in pounds of a cubic foot of water. If you like, think of a column of water of height y having base area equal to 1 square foot. Then the total force acting on this base area would be $62.5 \times y$ pounds. It turns out that since the water is incompressible, this is roughly the force acting on a square foot situated in any manner at this depth.

8.8.2   *Work*

Now suppose you are pumping water from a tank of depth d to a height of H feet above the top of the water in the tank. Suppose also that at depth y below the surface, the area of a cross section having constant depth is $A(y)$. The total weight of a slice of water having thickness dy at this depth is $62.5A(y)\,dy$ and the pump needs to lift this weight a distance of $y + H$ feet. Therefore, the work done is $dW = (y + H)\,62.5A(y)\,dy$. An initial value problem for the work done to pump the water down to a depth of y feet would be

$$\frac{dW}{dy} = (y + H)\,62.5A(y), \quad W(0) = 0.$$

The reason for the initial condition is that the pump has done no work to pump no water. If the weight of the fluid per cubic foot were different than 62.5 you would do the same things but replace the number.

Example 8.22. A spherical storage tank sitting on the ground having radius 30 feet is half filled with a fluid which weighs 50 pounds per cubic foot. How much work is done to pump this fluid to a height of 100 feet?

Letting r denote the radius of a cross section y feet below the level of the fluid, $r^2 + y^2 = 900$. Therefore,

$$r = \sqrt{900 - y^2}.$$

It follows the area of the cross section at depth y is $\pi\left(900 - y^2\right)$. Here $H = 70$ and so the initial value problem to solve is

$$\frac{dW}{dy} = (y + 70)\,50\pi\left(900 - y^2\right), \quad W(0) = 0.$$

Therefore, $W(y) = 50\pi\left(-\frac{1}{4}y^4 - \frac{70}{3}y^3 + 450y^2 + 63\,000y\right)$ and the total work in foot pounds equals

$$W(30) = 50\pi\left(-\frac{1}{4}(30)^4 - \frac{70}{3}(30)^3 + 450(30)^2 + 63\,000(30)\right) = 73\,,125,\,000\pi$$

In general, the work done by a constant force in a straight line equals the product of the force times the distance over which it acts. This is an over simplification and it will be made more correct later. If the force is varying with respect to position, then you have to use calculus to compute the work. For now, consider the following examples.

Example 8.23. A 500 pound safe is lifted 10 feet. How much work is done?

The work is $500 \times 10 = 5000$ foot pounds.

Example 8.24. The force needed to stretch a spring x feet past its equilibrium position is kx. This is known as Hooke's law and is a good approximation as long as the spring is not stretched too far. If $k = 3$, how much work is needed to stretch the spring a distance of 2 feet beyond its equilibrium position? The constant k is called the spring constant. Different springs would have different spring constants. The units on k are pounds/foot.

This is a case of a variable force. To stretch the spring from x to $x + dx$ requires $3xdx$ foot pounds of work. Therefore, letting W denote the work up till time x, $dW = 3xdx$ and so the initial value problem is

$$\frac{dW}{dx} = 3x, \ W(0) = 0.$$

Thus $W(2) = \frac{3}{2}\left(2^2\right) = 6$ foot pounds because an antiderivative for $3x$ is $\frac{3}{2}x^2$. In terms of the definite integral, this is written as $\int_0^2 3xdx$.

8.9 Exercises

(1) The main span of the Portage Lake lift bridge[2] weighs 4,400,000 pounds. How much work is done in raising this main span to a height of 100 feet?

(2) A cylindrical storage tank having radius 20 feet and length 40 feet is filled with a fluid which weighs 50 pounds per cubic foot. This tank is lying on its side on the ground. Find the total force acting on the ends of the tank by the fluid.

(3) Suppose the tank in Problem 2 is filled to a depth of 8 feet. Find an integral for the work needed to pump the fluid to a height of 50 feet.

(4) A conical hole is filled with water which has weight 62.5 pounds per cubic feet. If the depth of the hole is 20 feet and the radius of the hole is 10 feet, how much work is needed to pump the water to a height of 10 feet above the ground?

(5) Suppose the spring constant is 2 pounds per foot. Find the work needed to stretch the spring 3 feet beyond equilibrium.

(6) A 20 foot chain lies on the ground. It weighs 5 pounds per foot. How much work is done to lift one end of the chain to a height of 20 feet?

(7) A 200 foot chain dangles from the top of a tall building. How much work is needed to haul it to the top of the building if it weighs 1 pound per foot?

---

[2]This is the heaviest lift bridge in the world. It joins the towns of Houghton and Hancock in the upper peninsula of Michigan spanning Portage lake. It provides 250 feet of clear channel for ships and can provide as much as 100 feet of vertical clearance. The lifting machinery is at the top of two massive towers 180 feet above the water. Aided by 1,100 ton counter weights on each tower, sixteen foot gears pull on 42 cables to raise the bridge. This usually creates impressive traffic jams on either side of the lake. The motion up and down of this span is quite slow.

(8) A dam 500 feet high has a width at depth y equal to $4000 - 2y$ feet. What is the total force on the dam if it is filled?

(9) *When the bucket is filled with water it weighs 30 pounds and when empty it weighs 2 pounds and the person on top of a 100 foot building exerts a constant force of 40 pounds. The bucket is full at the bottom but leaks at the rate of .1 cubic feet per second. How much work does the person on the top of the building do in lifting the bucket to the top? Will the bucket be empty when it reaches the top? You can use Newton's law that force equals mass times acceleration. You can neglect the weight of the rope.

(10) In the situation of the above problem, suppose the person on the top maintains a constant velocity of 1 foot per second and the bucket leaks at the rate of .1 pound per second. How much work does he do and is the bucket empty when it reaches the top?

(11) A silo is 10 feet in diameter and at a height of 30 feet there is a hemispherical top. The silage weighs 10 pounds per cubic foot. How much work was done in filling it to the very top?

(12) A cylindrical storage tank having radius 10 feet is filled with water to a depth of 20 feet. If the storage tank stands upright on its circular base, what is the total force the water exerts on the sides of the tank? **Hint:** The pressure in the water at depth y is $62.5y$ pounds per square foot.

(13) A spherical storage tank having radius 10 feet is filled with water. What is the total force the water exerts on the storage tank? **Hint:** The pressure in the water at depth y is $62.5y$ consider the area corresponding to a slice at height y. This is a surface of revolution and you know how to deal with these. The area of this slice times the pressure gives the total force acting on it.

(14) A water barrel which is 11 inches in radius and 34 inches high is filled with water. If it is standing on end, what is the total force acting on the circular sides of the barrel?

(15) Find the total force acting on the circular sides of the cylinder in Problem 2.

(16) A cylindrical tank having radius 10 feet is contains water which weight 62.5 pounds per cubic foot. Find the force on one end of this tank if it is filled to a depth of y feet.

(17) Here is a calculator problem. In the above problem, to what depth may the tank be filled if the total force on an end is not to exceed 40000 pounds?

(18) The force on a satellite of mass m slugs in pounds is
$$\frac{mk}{r^2}$$
where k is approximately $k = 1.427\,374\,08 \times 10^{16}$ and r is the distance from the center of the earth. Assuming the radius of the earth is 4000 miles, find the work in foot pounds needed to place a satellite weighing 500 pounds on the surface of the earth into an orbit 18,000 miles above the surface of the earth. You should use a calculator on this problem.

Chapter 9

Other Differential Equations*

These are equations in which more is involved than simply taking an antiderivative.

9.1 The Equation $y' + a(t) y = b(t)$

The homogeneous first order constant coefficient linear differential equation is a differential equation of the form

$$y' + ay = 0. \tag{9.1}$$

It is arguably the most important differential equation in existence. Generalizations of it include the entire subject of linear differential equations and even many of the most important partial differential equations occurring in applications.

Here is how to find the solutions to this equation. Multiply both sides of the equation by e^{at}. Then use the product and chain rules to verify that

$$e^{at}(y' + ay) = \frac{d}{dt}(e^{at}y) = 0.$$

Therefore, since the derivative of the function $t \to e^{at}y(t)$ equals zero, it follows this function must equal some constant C. Consequently, $ye^{at} = C$ and so $y(t) = Ce^{-at}$. This shows that if there is a solution of the equation $y' + ay = 0$, then it must be of the form Ce^{-at} for some constant C. You should verify that every function of the form $y(t) = Ce^{-at}$ is a solution of the above differential equation, showing this yields all solutions. This proves the following theorem.

Theorem 9.1. *The solutions to the equation $y' + ay = 0$ consist of all functions of the form Ce^{-at} where C is some constant.*

Example 9.1. Radioactive substances decay in the following way. The rate of decay is proportional to the amount present. In other words, letting $A(t)$ denote the amount of the radioactive substance at time t, $A(t)$ satisfies the following initial value problem.

$$A'(t) = -k^2 A(t), \quad A(0) = A_0$$

where A_0 is the initial amount of the substance. What is the solution to the initial value problem?

Write the differential equation as $A'(t) + k^2 A(t) = 0$. From Theorem 9.1 the solution is

$$A(t) = Ce^{-k^2 t}$$

and it only remains to find C. Letting $t = 0$, it follows $A_0 = A(0) = C$. Thus $A(t) = A_0 \exp(-k^2 t)$.

Now consider a slightly harder equation.

$$y' + a(t) y = b(t).$$

In the easier case, you multiplied both sides by e^{at}. In this case, you multiply both sides by $e^{A(t)}$ where $A'(t) = a(t)$. In other words, you find an antiderivative of $a(t)$ and multiply both sides of the equation by e raised to that function. (It will be shown later in Theorem 10.5 on Page 242 that such an A always exists provided a is continuous.) Thus

$$e^{A(t)}(y' + a(t) y) = e^{A(t)} b(t).$$

Now you notice that this becomes

$$\frac{d}{dt}\left(e^{A(t)} y\right) = e^{A(t)} b(t). \tag{9.2}$$

This follows from the chain rule.

$$\frac{d}{dt}\left(e^{A(t)} y\right) = A'(t) e^{A(t)} y + e^{A(t)} y' = e^{A(t)}(y' + a(t) y).$$

Then from (9.2),

$$e^{A(t)} y \in \int e^{A(t)} b(t)\, dt.$$

Therefore, to find the solution, you find a function in $\int e^{A(t)} b(t)\, dt$, say $F(t)$, and

$$e^{A(t)} y = F(t) + C$$

for some constant C, so the solution is given by $y = e^{-A(t)} F(t) + e^{-A(t)} C$. This proves the following theorem.

Theorem 9.2. *The solutions to the equation $y' + a(t) y = b(t)$ consist of all functions of the form*

$$y = e^{-A(t)} F(t) + e^{-A(t)} C$$

where $F(t) \in \int e^{A(t)} b(t)\, dt$ and C is a constant.

Example 9.2. Find the solution to the initial value problem

$$y' + 2ty = \sin(t) e^{-t^2}, \quad y(0) = 3.$$

Multiply both sides by e^{t^2} because $t^2 \in \int t\,dt$. Then $\frac{d}{dt}\left(e^{t^2}y\right) = \sin(t)$ and so $e^{t^2}y = -\cos(t) + C$. Hence the solution is of the form $y(t) = -\cos(t)\,e^{-t^2} + Ce^{-t^2}$. It only remains to choose C in such a way that the initial condition is satisfied. From the initial condition, $3 = y(0) = -1 + C$ and so $C = 4$. Therefore, the solution is $y = -\cos(t)\,e^{-t^2} + 4e^{-t^2}$. Now at this point, you should check and see if it works. It needs to solve both the initial condition and the differential equation.

Finally, here is a uniqueness theorem.

Theorem 9.3. *If $a(t)$ is a continuous function, there is at most one solution to the initial value problem $y' + a(t)y = b(t)$, $y(r) = y_0$.*

Proof: If there were two solutions y_1, and y_2, then letting $w = y_1 - y_2$, it follows $w' + a(t)w = 0$ and $w(r) - 0$. Then multiplying both sides of the differential equation by $e^{A(t)}$ where $A'(t) = a(t)$, (It will be shown later in Theorem 10.5 on Page 242 that such an A always exists provided a is continuous.) it follows

$$\left(e^{A(t)}w\right)' = 0$$

and so $e^{A(t)}w(t) = C$ for some constant C. However, $w(r) = 0$ and so this constant can only be 0. Hence $w = 0$, and so $y_1 = y_2$. ∎

9.2 Separable Differential Equations

Definition 9.1. Separable differential equations are those which can be written in the form

$$\frac{dy}{dx} = \frac{f(x)}{g(y)}.$$

The reason these are called separable is that if you formally cross multiply,

$$g(y)\,dy = f(x)\,dx$$

and the variables are "separated". The x variables are on one side and the y variables are on the other.

Proposition 9.1. *If $G'(y) = g(y)$ and $F'(x) = f(x)$, then if the equation $F(x) - G(y) = c$ specifies y as a differentiable function of x, then $x \to y(x)$ solves the separable differential equation*

$$\frac{dy}{dx} = \frac{f(x)}{g(y)}. \tag{9.3}$$

Proof: Differentiate both sides of $F(x) - G(y) = c$ with respect to x. Using the chain rule,

$$F'(x) - G'(y)\frac{dy}{dx} = 0.$$

Therefore, since $F'(x) = f(x)$ and $G'(y) = g(y)$, $f(x) = g(y)\frac{dy}{dx}$ which is equivalent to (9.3). ■

Example 9.3. Find the solution to the initial value problem

$$y' = \frac{x}{y^2}, \; y(0) = 1.$$

This is a separable equation and in fact, $y^2 dy = x dx$ so the solution to the differential equation is of the form

$$\frac{y^3}{3} - \frac{x^2}{2} = C \tag{9.4}$$

and it only remains to find the constant C. To do this, you use the initial condition. Letting $x = 0$, it follows $\frac{1}{3} = C$ and so

$$\frac{y^3}{3} - \frac{x^2}{2} = \frac{1}{3}$$

Example 9.4. What is the equation of a hanging chain?

Consider the following picture of a portion of this chain.

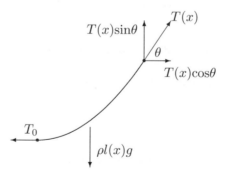

In this picture, ρ denotes the density of the chain which is assumed to be constant and g is the acceleration due to gravity. $T(x)$ and T_0 represent the magnitude of the tension in the chain at x and at 0 respectively, as shown. Let the bottom of the chain be at the origin as shown. If this chain does not move, then all these forces acting on it must balance. In particular,

$$T(x)\sin\theta = l(x)\rho g, \; T(x)\cos\theta = T_0.$$

Therefore, dividing these yields

$$\frac{\sin\theta}{\cos\theta} = l(x) \overbrace{\rho g/T_0}^{\equiv c}.$$

Now letting $y(x)$ denote the y coordinate of the hanging chain corresponding to x,

$$\frac{\sin\theta}{\cos\theta} = \tan\theta = y'(x).$$

Therefore, this yields

$$y'(x) = cl(x).$$

Now differentiating both sides of the differential equation

$$y''(x) = cl'(x) = c\sqrt{1 + y'(x)^2}$$

and so

$$\frac{y''(x)}{\sqrt{1 + y'(x)^2}} = c.$$

Let $z(x) = y'(x)$ so the above differential equation becomes

$$\frac{z'(x)}{\sqrt{1 + z^2}} = c.$$

Therefore, $\int \frac{z'(x)}{\sqrt{1+z^2}}\, dx = cx + d$. Change the variable in the antiderivative letting $u = z(x)$ and this yields

$$\int \frac{z'(x)}{\sqrt{1 \mid z^2}}\, dx = \int \frac{du}{\sqrt{1 + u^2}} = \sinh^{-1}(u) + C$$

$$= \sinh^{-1}(z(x)) + C$$

by (5.10) on Page 134. Therefore, combining the constants of integration,

$$\sinh^{-1}(y'(x)) = cx + d$$

and so

$$y'(x) = \sinh(cx + d).$$

Therefore,

$$y(x) = \frac{1}{c}\cosh(cx + d) + k$$

where d and k are some constants and $c = \rho g / T_0$. Curves of this sort are called catenaries. Note these curves result from an assumption that the only forces acting on the chain are as shown.

9.3 Exercises

(1) For x sufficiently large, let $f(x) = (\ln x)^{\ln x}$. Find $f'(x)$. How big does x need to be in order for this to make sense. **Hint:** You should have $\ln x > 0$.

(2) In the hanging chain problem the picture and the derivation involved an assumption that at its lowest point, the chain was horizontal. Imagine lifting the end higher and higher and you will see this might not be the case in general. Can you modify the above derivation for the hanging chain to show that even in this case the chain will be in the form of a catenary?

(3) Find the solution to the initial value problem

$$y' = 1 + y^2, \ y(0) = 0.$$

(4) Verify that for any constant C, the function $y(t) = Ce^{-at}$ solves the differential equation $y' + ay = 0$.

(5) In Example 9.1 the half life is the time it takes for half of the original amount of the substance to decay. Find a formula for the half life assuming you know k^2.

(6) There are ten grams of a radioactive substance which is allowed to decay for five years. At the end of the five years there are 9.5 grams of the substance left. Find the half life of the substance. **Hint:** Use the given information to find k^2 and then use Problem 5.

(7) The giant arch in St. Louis is in the form of an inverted catenary. Why?

(8) Sometimes banks compound interest continuously. One way to think of this is to let the amount in the account satisfy the initial value problem

$$A'(t) = rA(t), \ A(0) = A_0$$

where here $A(t)$ is the amount at time t measured in years, r is the interest rate per year, and A_0 is the initial amount. Find $A(t)$ explicitly. If $100 is placed in an account which is compounded continuously at 6% per year, how many years will it be before there is $200 in the account? **Hint:** In this case, $r = .06$.

(9) An object falling through the air experiences a force of gravity and air resistance. Later, it will be shown that this implies the velocity satisfies a differential equation of the form

$$v' = \frac{g}{m} - kv$$

where k is a positive constant and $\frac{g}{m}$ is also a positive constant. Find the solutions to this differential equation and determine what happens to v as t gets large. If you do it right you will find the terminal velocity.

(10) Solve the initial value problem

$$A'(t) = rA(t) + 1 + \sin(t), A(0) = 1$$

and describe its behavior as $t \to \infty$. Assume r is a positive constant. what r is a negative constant?

(11) A population is growing at the rate of 11% per year. This means it satisfies the differential equation $A' = .11A$. Find the time it takes for the population to double.

(12) A substance is decaying at the rate of .01 per year. Find the half life of the substance.

(13) The half live of a substance is 40 years. Find the rate of decay of the substance.

(14) A sample of 4 grams of a radioactive substance has decayed to 3 grams in 5 days. Find the half life of the substance. Give your answer in terms of logarithms.

(15) $1000 is deposited in an account that earns interest at the rate of 5% per year compounded continuously. How much will be in the account after 10 years?

(16) A sample of one ounce of water from a water supply is cultured in a Petri dish. After four hours, there are 3000.0 bacteria present and after six hours there are 7000 bacteria present. How many were present in the original sample?

(17) Carbon 14 is a radioactive isotope of carbon and it is produced at a more or less constant rate in the earth's atmosphere by radiation from the sun. It also decays at a rate proportional to the amount present. Show this implies that, assuming this has been going on for billions of years, it is reasonable to assume the amount of Carbon 14 in the atmosphere is essentially constant over time. **Hint:** By assumption, if A is the amount of Carbon 14 in the atmosphere, $\frac{dA}{dt} = -k^2 A + r$ where k^2 is the constant of decay described above and r is the constant rate of production. Now show $A'(t) = Ce^{-k^2 t}$. Conclude $A(t) = D - (C/k^2) e^{-k^2 t}$. What happens to this over a long period of time?

(18) The method of carbon dating is based on the result of Problem 17. When an animal or plant is alive it absorbs carbon from the atmosphere and so when it is living it has a known percentage of carbon 14. When it dies, it quits absorbing carbon and the carbon 14 begins to decay. After some time, t, the amount of carbon 14 can be measured and on this basis, the time since the death of the animal or plant can be estimated. Given the half life of carbon 14 is 5730 years and the amount of carbon 14 in a mummy is .3 what it was at the time of death, how long has it been since the mummy was alive? (To see how to come up with this figure for the half life, see Problem 6. You do experiments and take measurements over a smaller period of time.)

(19) The half life of carbon 14 is known to be 5730 years. A certain tree stump is known to be 4000 years old. What percentage of the original carbon 14 should it contain?

(20) One model for population growth is to assume the rate of growth is proportional to both the population and the difference between some maximum sustainable population and the population. The reason for this is that when the population gets large enough, there begin to be insufficient resources. Thus $\frac{dA}{dt} = kA(M - A)$, where k and M are positive constants. Show this is a separable differential equation and its solutions are of the form

$$A(t) = \frac{M}{1 + CMe^{-kMt}}$$

where C is a constant. Given three measurements of population at three equally spaced times, show how to predict the maximum sustainable population[1].

---

[1] This has been done with the earth's population and the maximum sustainable population has been exceeded. Therefore, the model is far too simplistic for human population growth. However, it would work somewhat better for predicting the growth of things like bacteria.

(21) Homogeneous differential equations are those which can be written in the form $y' = f\left(\frac{y}{x}\right)$. For example, $y' = \frac{x^2}{y^2+x^2}$. There is a trick to solving such equations. You define a new variable $v = y/x$ and then write the differential equation in terms of v rather than y. Show that $xv' + v = y'$ and so the differential equation reduces to $xv' = f(v) - v$, a separable differential equation. Use the technique to solve $y' = 1 + 2\frac{y}{x}$, $y(1) = 1$.

(22) Solve the following initial value problems involving homogeneous differential equations.

(a) $y' = 1 + \left(\frac{y}{x}\right) + \left(\frac{y}{x}\right)^2$, $y(1) = 1$

(b) $y' = \tan\left(\frac{y}{x}\right) + \frac{y}{x}$, $y(1) = 1$

(c) $y' = \frac{x^3 + xy^2 + y^3}{x^2 y + xy^2}$, $y(1) = 1$

(23) * Suppose f is a function defined on $\mathbb{R}$ and it satisfies the functional equation $f(a+b) = f(a) f(b) + 3ab$. Show $f(0)$ must equal either 1 or 0. Show that if f is to be differentiable, then $f(0) = 1$. Assuming $f(0) = 1$ and $f'(0) = 2$, find $f(x)$ if it exists.

9.4 The Equations Of Undamped And Damped Oscillation

Consider a garage door spring. These springs exert a force which resists extension. Attach such a spring to the ceiling and attach a mass m, to the bottom end of the spring as shown in the following picture. Any mass will do. It does not have to be a small elephant.

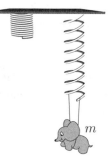

The weight of this mass mg, is a downward force which extends the spring, moving the bottom end of the spring downward a distance l where the upward force exerted by the spring exactly balances the downward force exerted on the mass by gravity. It has been experimentally observed that as long as the extension z, of such a spring is not too great, the restoring force exerted by the spring is of the form kz where k is some constant which depends on the spring. (It would be different for a slinky than for a garage door spring.) This is known as Hooke's law which is the simplest model for elastic materials. Therefore, $mg = kl$. Now let y be the displacement from this equilibrium position of the bottom of the spring with the positive direction being up. Thus the acceleration of the spring is y''. The extension of the spring in terms of y is $(l - y)$. Then Newton's second law[2] along with Hooke's law imply

$$my'' = k(l - y) - mg$$

and since $kl - mg = 0$, this yields

$$my'' + ky = 0.$$

Dividing by m and letting $\omega^2 = k/m$ yields the equation for undamped oscillation,

$$y'' + \omega^2 y = 0.$$

Based on physical reasoning just presented, there should be a solution to this equation. It is the displacement of the bottom end of a spring from the equilibrium position. However, it is not enough to base questions of existence in mathematics on physical intuition, although it is sometimes done. The following theorem gives the necessary existence and uniqueness results. The equation is the equation of undamped oscillations. It occurs in modeling a weight on a spring but it also occurs in many other physical settings.

Theorem 9.4. *The initial value problem*

$$y'' + \omega^2 y = 0, \; y(0) = y_0, y'(0) = y_1 \tag{9.5}$$

has a unique solution and this solution is

$$y(t) = y_0 \cos(\omega t) + \frac{y_1}{\omega} \sin(\omega t). \tag{9.6}$$

Proof: You should verify that (9.6) does indeed provide a solution to the initial value problem. It only remains to verify uniqueness. Suppose then that y_1 and y_2 both solve the initial value problem 9.5. Let $w = y_1 - y_2$. Then you should verify that $w'' + \omega^2 w = 0$, $w(0) = 0 = w'(0)$. Then multiplying both sides of the differential equation by w' it follows

$$w''w' + \omega^2 ww' = 0.$$

However, $w''w' = \frac{1}{2}\frac{d}{dt}(w')^2$ and $w'w = \frac{1}{2}\frac{d}{dt}(w)^2$ so the above equation reduces to

$$\frac{1}{2}\frac{d}{dt}\left((w')^2 + w^2\right) = 0.$$

Therefore, $(w')^2 + w^2$ is equal to some constant. However, when $t = 0$, this shows the constant must be zero. Therefore, $y_1 - y_2 = w = 0$. ∎

Now consider another sort of differential equation

$$y'' - a^2 y = 0, \; a > 0 \tag{9.7}$$

To give the complete solution, let $Dy \equiv y'$. Then the differential equation may be written as

$$(D + a)(D - a)y = 0.$$

---

[2]This important law is discussed more thoroughly in an appendix. I assume you have seen it in a physics class by now.

Let $z = (D - a) y$. Thus $(D + a) z = 0$ and so $z(t) = C_1 e^{-at}$ from Theorem 9.1 on Page 221. Therefore,

$$(D - a) y \equiv y' - ay = C_1 e^{-at}.$$

Multiply both sides of this last equation by e^{-at}. By the product and chain rules,

$$\frac{d}{dt} \left(e^{-at} y \right) = C_1 e^{-2at}.$$

Therefore,

$$e^{-at} y = \frac{C_1}{-2a} e^{-2at} + C_2$$

and so

$$y = \frac{C_1}{-2a} e^{-at} + C_2 e^{at}.$$

Now since C_1 is arbitrary, it follows any solution of (9.7) is of the form $y = C_1 e^{-at} + C_2 e^{at}$. Now you should verify that any expression of this form actually solves the equation (9.7). This proves most of the following theorem.

Theorem 9.5. *Every solution of the differential equation $y'' - a^2 y = 0$ is of the form $C_1 e^{-at} + C_2 e^{at}$ for some constants C_1 and C_2, provided $a > 0$. In the case when $a = 0$, every solution of $y'' = 0$ is of the form $C_1 t + C_2$ for some constants C_1 and C_2.*

All that remains of the proof is to do the part when $a = 0$ which is left as an exercise involving the mean value theorem.

Now consider the differential equation of damped oscillation. In the example of the object bobbing on the end of a spring,

$$my'' = -ky$$

where k was the spring constant and m was the mass of the object. Suppose the object is also attached to a dash pot. This is a device which resists motion like a shock absorber on a car. You know how these work. If the car is just sitting still the shock absorber applies no force to the car. It only gives a force in response to up and down motion of the car and you assume this force is proportional to the velocity and opposite the velocity. Thus in our spring example, you would have

$$my'' = -ky - \delta^2 y'$$

where δ^2 is the constant of proportionality of the resisting force. Dividing by m and adjusting the coefficients, such damped oscillation satisfies an equation of the form

$$y'' + 2by' + ay = 0. \tag{9.8}$$

Actually this is a general homogeneous second order equation, more general than what results from damped oscillation. Concerning the solutions to this equation, the following theorem is given. In this theorem the first case is referred to as the under-damped case. The second case is called the critically damped case and the third is called the over-damped case.

Theorem 9.6. *Suppose $b^2 - a < 0$. Then all solutions of (9.8) are of the form*

$$e^{-bt} \left(C_1 \cos (\omega t) + C_2 \sin (\omega t) \right) \tag{9.9}$$

where $\omega = \sqrt{a - b^2}$ *and* C_1 *and* C_2 *are constants. In the case that* $b^2 - a = 0$ *the solutions of (9.8) are of the form*

$$e^{-bt} \left(C_1 + C_2 t \right). \tag{9.10}$$

In the case that $b^2 - a > 0$ *the solutions are of the form*

$$e^{-bt} \left(C_1 e^{-rt} + C_2 e^{rt} \right), \tag{9.11}$$

where $r = \sqrt{b^2 - a}$.

Proof: Let $z = e^{bt} y$ and write (9.8) in terms of z. Thus, z is a solution to the equation

$$z'' + \left(a - b^2 \right) z = 0. \tag{9.12}$$

If $b^2 - a < 0$, then by Theorem 9.4, $z(t) = C_1 \cos (\omega t) + C_2 \sin (\omega t)$ where $\omega = \sqrt{a - b^2}$. Therefore,

$$y = e^{-bt} \left(C_1 \cos (\omega t) + C_2 \sin (\omega t) \right)$$

as claimed. The other two cases are completely similar. They use Theorem 9.5 rather than Theorem 9.4. ■

Example 9.5. An important example of these equations occurs in an electrical circuit having a capacitor, a resistor, and an inductor in series as shown in the following picture.

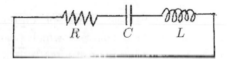

The voltage drop across the inductor is $L\frac{di}{dt}$ where i is the current and L is the inductance. The voltage drop across the resistor is Ri where R is the resistance. This is according to Ohm's law. The voltage drop across the capacitor is $v = \frac{Q}{C}$ where Q is the charge on the capacitor and C is a constant called the capacitance. The current equals the rate of change of the charge on the capacitor. Thus $i = Q' = Cv'$. When these voltages are summed, you must get zero because there is no voltage source in the circuit. Thus $L\frac{di}{dt} + Ri + \frac{Q}{C} = 0$ and written in terms of the voltage drop across the capacitor, this becomes $LCv'' + CRv' + v = 0$, a second order linear differential equation of the sort discussed above.

9.5   Exercises

(1) Verify that $y = C_1 e^{-at} + C_2 e^{at}$ solves the differential equation (9.7).
(2) Verify (9.12).

(3) Verify that all solutions to the differential equation $y'' = 0$ are of the form $y = C_1 t + C_2$.

(4) Show that for all $x \geq 0$,

$$\sin x \leq x - \frac{x^3}{6} + \frac{x^5}{120}. \tag{9.13}$$

Also verify $\sin x \leq x - \frac{x^3}{6}$. **Hint:** Let $f(x) = x - \frac{x^3}{6} + \frac{x^5}{120} - \sin x$. Then $f(0) = 0$. You need to show $f'(x) > 0$ for $x > 0$. You will not be able to do this directly. Consider $f'(x) = 1 - \frac{x^2}{2} + \frac{x^4}{24} - \cos x$. Then $f'(0) = 0$. You need to show $f''(x) > 0$. You will not be able to do this directly. Consider $f''(x) = -x + \frac{x^3}{6} + \sin x$. Then $f''(0) = 0$. You need to show $f'''(x) > 0$. You won't be able to do this directly. Continue this way. You will eventually have $f^{(k)}(0) = 0$ and it will be obvious that $f^{(k+1)}(x) > 0$.

(5) Using Problem 4, along with similar estimates for lower bounds and upper bounds for sine and cosine, estimate $\sin(.1)$ and $\cos(.1)$. Give upper and lower bounds for these numbers.

(6) Using Problem 4 and Theorem 2.7 on Page 60, establish the limit,

$$\lim_{x \to 0} \frac{\sin(x)}{x} = 1.$$

(7) A mass of ten Kilograms is suspended from a spring attached to the ceiling. This mass causes the end of the spring to be displaced a distance of 39.2 cm. The mass end of the spring is then pulled down a distance of one cm. and released. Find the displacement from the equilibrium position of the end of the spring as a function of time. Assume the acceleration of gravity is 9.8 meters/ sec^2.

(8) Keep everything the same in Problem 7 except suppose the suspended end of the spring is also attached to a dash pot which provides a force opposite the direction of the velocity having magnitude $10\sqrt{19}\,|v|$ Newtons for $|v|$ the speed. Give the displacement as before.

(9) In (9.8) consider the equation in which $b = -1$ and $a = 3$. Explain why this equation describes a physical system which has some dubious properties.

(10) Solve the initial value problem $y'' + 5y' - y = 0$, $y(0) = 1$, $y'(0) = 0$.

(11) Solve the initial value problem $y'' + 2y' + 2y = 0$, $y(0) = 0$, $y'(0) = 1$.

(12) Suppose $f(t) = a \cos \omega t + b \sin \omega t$. Show there exists ϕ such that

$$f(t) = \sqrt{a^2 + b^2} \sin(\omega t + \phi).$$

Hint: $f(t) = \sqrt{a^2 + b^2} \left(\frac{a}{\sqrt{a^2+b^2}} \cos \omega t + \frac{b}{\sqrt{a^2+b^2}} \sin \omega t \right)$, and

$$\left(\frac{b}{\sqrt{a^2 + b^2}}, \frac{a}{\sqrt{a^2 + b^2}} \right)$$

is a point on the unit circle, so it is of the form $(\cos \phi, \sin \phi)$. Now recall the formulas for the sine and cosine of sums of angles. Can you also write $f(t) = \sqrt{a^2 + b^2} \cos(\omega t + \phi)$? Explain.

(13) In the case of undamped oscillation show the solution can be written in the form $A \cos(\omega t - \phi)$ where ϕ is some angle called a phase shift and a constant A, called the amplitude.

(14) Using Problem 4 and Theorem 2.7, establish the limit,

$$\lim_{x \to 0} \frac{1 - \cos x}{x} = 0.$$

(15) Is the derivative of a function always continuous? **Hint:** Consider

$$h(x) = \begin{cases} x^2 \sin\left(\frac{1}{x}\right) & \text{if } x \neq 0 \\ 0 & \text{if } x = 0 \end{cases}.$$

Show $h'(0) = 0$. What is $h'(x)$ for $x \neq 0$?

9.6   Review Exercises

(1) Find $\int x^2 e^{-5x} dx$.

(2) Find $\int x^{15} \cos\left(x^8\right) dx$.

(3) Find $\int \sqrt{4 - 9x^2} dx$.

(4) Find $\int \frac{1}{\sqrt{4 - 3x^2}} dx$.

(5) Find $\int \frac{1}{x^2 + 2x + 2} dx$.

(6) Find $\int \frac{x}{\sqrt{(4 + 5x)}} dx$.

(7) Find $\int x \arctan(-x) dx$.

(8) Find the area between e^x and $\cos(2x)$ for $x \in [0, \pi]$.

(9) Find the volume of the solid of revolution which results from revolving $\sin\left(\frac{1}{4}x\right)$ about the x axis for $x \in [0, \pi]$.

(10) Find $\int \sec(-3x) dx$

(11) Find all solutions to the differential equation $y'' - 10y' - 3y = 0$.

(12) Find $\int \sqrt{x^2 - 1} dx$.

(13) Find $\int \sec^2(3x) \tan(3x) dx$.

(14) Find $\int \left(\frac{1}{4x^2 + 8}\right)^2 dx$.

(15) Find $\int \frac{1}{2x^2 + 5} dx$.

(16) Find $\int \sqrt{9x^2 + 9} dx$.

(17) A circle of radius 3 is the base of a solid which has the property that cross sections perpendicular to the x axis are equilateral triangles. Find its volume.

(18) A spherical storage tank having radius 14 feet is filled with water to a depth of 7 feet. Find the work needed to pump the water which weighs 62.5 pounds per cubic foot out of the top of the tank.

(19) Find $\int \left(\frac{7x + 11 + x^2}{(x^2 + 7)(7x + 4)}\right) dx$.

(20) The graph of $y = 4x^2$ for $x \in [0, 2]$ is revolved about the x axis. Find the area of the surface of revolution.

(21) Find $\int \sec^3(5x) dx$.

(22) Find the area between the graphs of $x = y^2$ and $x = 2 - y$.

(23) Find $\int \left(\ln \left(|x| \right) \right) dx$.

(24) Find the length of the graph of $y = \ln \left(\cos \left(x \right) \right)$ for $x \in [0, \pi/4]$.

(25) Find $\int x^{11} \sqrt{(x^3 + 3)} dx$.

(26) Find $\int -4x \left(-3x^2 + 2 \right)^{11} dx$.

(27) Find all solutions to the differential equation $y'' - 2y' + 8y = 0$.

(28) Find $\int \left(\frac{7x+17}{(x+2)(x+5)} \right) dx$.

(29) Solve the initial value problem $y' = x^7 \sin \left(6x^8 \right)$, $y \left(1 \right) = 3$

(30) Find $\int \left(\frac{28x+8+7x^3+23x^2}{(x+2)(7x+2)} \right) dx$.

(31) Find $\int \arcsin \left(3x \right) dx$.

(32) The graph of the function $y = \ln \left(2x \right)$ for $x \in [1, 5]$ is revolved about the y axis. Find the area of the surface of revolution.

(33) The graph of $y = 2x$ for $x \in [0, 4]$ is revolved about the x axis. Find the area of the surface of revolution.

(34) Find all solutions to the differential equation $y'' + 8y' + 16y = 0$.

(35) Find $\int \arctan \left(x \right) dx$.

(36) Find $\int \frac{1}{\sqrt{2+2x^2}} dx$.

(37) Find the solution to the initial value problem

$$y' = ay + by^2, \ y \left(0 \right) = 1$$

(38) The region between $y = \arctan \left(x \right)$ and the x axis for $x \in [0, 1]$ is revolved about the line $x = -5$. Find the volume of the solid of revolution.

(39) Find the area between the graphs of $y = 3 \sin 3x$ and $y = 3 \cos 3x$ for $x \in [0, \pi]$.

(40) Find the length of $y = \frac{1}{2}x^2 - \frac{1}{4} \ln x$ for $x \in [1, 2]$.

Chapter 10

The Integral

The integral originated in attempts to find areas of various shapes and the ideas involved in finding integrals are much older than the ideas related to finding derivatives. In fact, Archimedes[1] was finding areas of various curved shapes about 250 B.C. The integral is needed to remove some of the mathematical loose ends and also to enable the study of more general problems involving differential equations. It will also be useful for formulating other physical models. The technique used for finding the area of a circular segment presented early in the book was essentially that employed by Archimedes and contains the essential ideas for the integral. The main difference is that here the triangles will be replaced with rectangles. You may be wondering what the fuss is about. Areas have already been found as solutions of differential equations. However, there is a profound difference between what is about to be presented and what has just been done. It is related to the fundamental mathematical question of existence. As an illustration, consider the problem of finding the area between $y - e^{x^2}$ and the x axis for $x \in [0, 1]$. As pointed out earlier, the area is obtained as a solution to the initial value problem

$$A'(x) = e^{x^2}, \ A(0) = 0.$$

So what is the solution to this initial value problem? By Theorem 7.1 there is at most one solution, but what is the solution? Does it even exist? More generally, for which functions f does there exist a solution to the initial value problem $y'(x) = f(x), y(0) = y_0$? These questions are typical of mathematics. There are usually two aspects to a mathematical concept. One is the question of existence and the other is how to find that which exists. The two questions are often very different and one can have a good understanding of one without having any idea how to go about considering the other. However, both are absolutely essential. In the preceding chapter the only thing considered was the second question.

---

[1] Archimedes 287-212 B.C. found areas of curved regions by stuffing them with simple shapes which he knew the area of and taking a limit. He also made fundamental contributions to physics. The story is told about how he determined that a gold smith had cheated the king by giving him a crown which was not solid gold as had been claimed. He did this by finding the amount of water displaced by the crown and comparing with the amount of water it should have displaced if it had been solid gold.

10.1   Upper And Lower Sums

The Riemann integral pertains to bounded functions which are defined on a bounded interval. Let $[a, b]$ be a closed interval. A set of points in $[a, b]$, $\{x_0, \cdots, x_n\}$ is a partition if

$$a = x_0 < x_1 < \cdots < x_n = b.$$

Such partitions are denoted by P or Q. For f a bounded function defined on $[a, b]$, let

$$M_i(f) \equiv \sup\{f(x) : x \in [x_{i-1}, x_i]\}, \ m_i(f) \equiv \inf\{f(x) : x \in [x_{i-1}, x_i]\}.$$

Also let $\Delta x_i \equiv x_i - x_{i-1}$. Then define upper and lower sums as

$$U(f, P) \equiv \sum_{i=1}^{n} M_i(f) \Delta x_i \text{ and } L(f, P) \equiv \sum_{i=1}^{n} m_i(f) \Delta x_i$$

respectively. The numbers, $M_i(f)$ and $m_i(f)$, are well defined real numbers because f is assumed to be bounded and $\mathbb{R}$ is complete. Thus the set $S = \{f(x) : x \in [x_{i-1}, x_i]\}$ is bounded above and below. In the following picture, the sum of the areas of the rectangles in the picture on the left is a lower sum for the function in the picture and the sum of the areas of the rectangles in the picture on the right is an upper sum for the same function which uses the same partition.

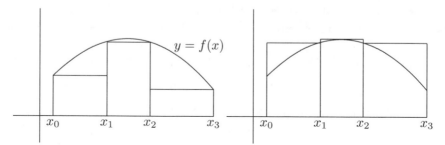

What happens when you add in more points in a partition? The following pictures illustrate in the context of the above example. In this example a single additional point, labeled z has been added in.

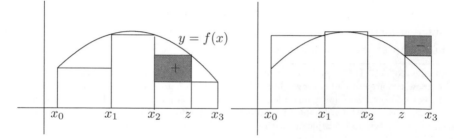

Note how the lower sum got larger by the amount of the area in the shaded rectangle and the upper sum got smaller by the amount in the other shaded rectangle. In general this is the way it works and this is shown in the following lemma.

Lemma 10.1. *If $P \subseteq Q$ then*

$$U(f,Q) \leq U(f,P), \quad \text{and} \quad L(f,P) \leq L(f,Q).$$

Proof: This is verified by adding in one point at a time. Thus let $P = \{x_0, \cdots, x_n\}$ and let $Q = \{x_0, \cdots, x_k, y, x_{k+1}, \cdots, x_n\}$. Thus exactly one point y, is added between x_k and x_{k+1}. Now the term in the upper sum which corresponds to the interval $[x_k, x_{k+1}]$ in $U(f,P)$ is

$$\sup\{f(x) : x \in [x_k, x_{k+1}]\}(x_{k+1} - x_k) \tag{10.1}$$

and the terms which corresponds to the interval $[x_k, x_{k+1}]$ in $U(f,Q)$ are

$$\sup\{f(x) : x \in [x_k, y]\}(y - x_k) + \sup\{f(x) : x \in [y, x_{k+1}]\}(x_{k+1} - y) \tag{10.2}$$
$$\equiv M_1(y - x_k) + M_2(x_{k+1} - y) \tag{10.3}$$

All the other terms in the two sums coincide. Now $\sup\{f(x) : x \in [x_k, x_{k+1}]\} > \max(M_1, M_2)$ and so the expression in (10.2) is no larger than

$$\sup\{f(x) : x \in [x_k, x_{k+1}]\}(x_{k+1} - y) + \sup\{f(x) : x \in [x_k, x_{k+1}]\}(y - x_k)$$

$$= \sup\{f(x) : x \in [x_k, x_{k+1}]\}(x_{k+1} - x_k),$$

the term corresponding to the interval $[x_k, x_{k+1}]$ and $U(f,P)$. This proves the first part of the lemma pertaining to upper sums because if $Q \supseteq P$, one can obtain Q from P by adding in one point at a time and each time a point is added, the corresponding upper sum either gets smaller or stays the same. The second part for lower sums is similar and is left as an exercise. ∎

Lemma 10.2. *If P and Q are two partitions, then*

$$L(f,P) \leq U(f,Q).$$

Proof: By Lemma 10.1,

$$L(f,P) \leq L(f, P \cup Q) \leq U(f, P \cup Q) \leq U(f,Q). \qquad \blacksquare$$

Definition 10.1.

$$\overline{I} \equiv \inf\{U(f,Q) \text{ where } Q \text{ is a partition}\}$$

$$\underline{I} \equiv \sup\{L(f,P) \text{ where } P \text{ is a partition}\}.$$

Note that $\underline{I}$ and $\overline{I}$ are well defined real numbers.

Theorem 10.1. $\underline{I} \leq \overline{I}$.

Proof: From Lemma 10.2,

$$\underline{I} = \sup\{L\,(f,P) \text{ where } P \text{ is a partition}\} \leq U\,(f,Q)$$

because $U\,(f,Q)$ is an upper bound to the set of all lower sums and so it is no smaller than the least upper bound. Therefore, since Q is arbitrary,

$$\underline{I} = \sup\{L\,(f,P) \text{ where } P \text{ is a partition}\}$$
$$\leq \inf\{U\,(f,Q) \text{ where } Q \text{ is a partition}\} \equiv \overline{I}$$

where the inequality holds because it was just shown that $\underline{I}$ is a lower bound to the set of all upper sums and so it is no larger than the greatest lower bound of this set. ∎

Definition 10.2. A bounded function f is Riemann integrable, written as

$$f \in R\,([a,b])$$

if $\underline{I} = \overline{I}$ and in this case,

$$\int_a^b f\,(x)\ dx \equiv \underline{I} = \overline{I}.$$

Thus, in words, the Riemann integral is the unique number which lies between all upper sums and all lower sums if there is such a unique number. Recall the following in Proposition 1.1 which was proved earlier.

Proposition 10.1. Let S be a nonempty set and suppose $\sup(S)$ exists. Then for every $\delta > 0$,

$$S \cap (\sup(S) - \delta, \sup(S)] \neq \emptyset.$$

If $\inf(S)$ exists, then for every $\delta > 0$,

$$S \cap [\inf(S), \inf(S) + \delta) \neq \emptyset.$$

This proposition implies the following theorem which is used to determine the question of Riemann integrability.

Theorem 10.2. *A bounded function f is Riemann integrable if and only if for all $\varepsilon > 0$, there exists a partition P such that*

$$U\,(f,P) - L\,(f,P)\ < \varepsilon. \tag{10.4}$$

Proof: First assume f is Riemann integrable. Then let P and Q be two partitions such that

$$U\,(f,Q) < \overline{I} + \varepsilon/2,\ \ L\,(f,P) > \underline{I} - \varepsilon/2.$$

Then since $\underline{I} = \overline{I}$,

$$U\,(f,Q \cup P) - L\,(f,P \cup Q) \leq U\,(f,Q) - L\,(f,P) < \overline{I} + \varepsilon/2 - (\underline{I} - \varepsilon/2) = \varepsilon.$$

Now suppose that for all $\varepsilon > 0$ there exists a partition such that (10.4) holds. Then for given ε and partition P corresponding to ε

$$\overline{I} - \underline{I} \leq U(f, P) - L(f, P) \leq \varepsilon.$$

Since ε is arbitrary, this shows $\underline{I} = \overline{I}$. ∎

The condition described in the theorem is called the Riemann criterion .

Not all bounded functions are Riemann integrable. For example, let

$$f(x) \equiv \begin{cases} 1 \text{ if } x \in \mathbb{Q} \\ 0 \text{ if } x \in \mathbb{R} \setminus \mathbb{Q} \end{cases} \tag{10.5}$$

Then if $[a, b] = [0, 1]$ all upper sums for f equal 1 while all lower sums for f equal 0. Therefore the Riemann criterion is violated for $\varepsilon = 1/2$.

10.2 Exercises

(1) Prove the second half of Lemma 10.1 about lower sums.

(2) Verify that for f given in (10.5), the lower sums on the interval $[0, 1]$ are all equal to zero while the upper sums are all equal to one.

(3) Let $f(x) - 1 + x^2$ for $x \in [-1, 3]$ and let $P = \{-1, -\frac{1}{3}, 0, \frac{1}{2}, 1, 2\}$. Find $U(f, P)$ and $L(f, P)$.

(4) Show that if $f \in R([a, b])$, there exists a partition $\{x_0, \cdots, x_n\}$ such that for any $z_k \in [x_k, x_{k+1}]$, $\left| \int_a^b f(x)\, dx - \sum_{k=1}^{n} f(z_k)(x_k - x_{k-1}) \right| < \varepsilon$. This sum $\sum_{k=1}^{n} f(z_k)(x_k - x_{k-1})$, is called a Riemann sum and this exercise shows that the integral can always be approximated by a Riemann sum.

(5) Let $P = \{1, 1\frac{1}{4}, 1\frac{1}{2}, 1\frac{3}{4}, 2\}$. Find upper and lower sums for the function $f(x) = \frac{1}{x}$ using this partition What does this tell you about $\ln(2)$?

(6) If $f \in R([a, b])$ and f is changed at finitely many points, show the new function is also in $R([a, b])$ and has the same integral as the unchanged function.

(7) Consider the function $y = x^2$ for $x \in [0, 1]$. Show this function is Riemann integrable and find the integral using the definition and the formula

$$\sum_{k=1}^{n} k^2 = \frac{1}{3}(n+1)^3 - \frac{1}{2}(n+1)^2 + \frac{1}{6}(n+1)$$

which you should verify by using math induction. This is not a practical way to find integrals in general.

(8) Define a "left sum" as $\sum_{k=1}^{n} f(x_{k-1})(x_k - x_{k-1})$ and a "right sum", $\sum_{k=1}^{n} f(x_k)(x_k - x_{k-1})$. Also suppose that all partitions have the property that $x_k - x_{k-1}$ equals a constant $(b - a)/n$ so the points in the partition are equally spaced, and define the integral to be the number these right and left sums get close to as n gets larger and larger. Show that for f given in (10.5), $\int_0^x f(t)\, dt = 1$ if x is rational and $\int_0^x f(t)\, dt = 0$ if x is irrational. It turns out that the correct answer should always equal zero for that function, regardless

of whether x is rational. This is shown in more advanced courses when the Lebesgue integral is studied. This illustrates why the method of defining the integral in terms of left and right sums is nonsense.

(9) *Suppose f is Riemann integrable on $[a, b]$ and suppose also that for every $x \in (a, b)$, $f(x) = F'(x)$ where F is a continuous function on $[a, b]$. Show that

$$\int_a^b f(x)\, dx = F(b) - F(a).$$

Hint: There exists a partition P such that $U(f, P) - L(f, P) < \varepsilon$. Let $P = \{x_0, x_1, x_2, \cdots, x_n\}$. Then

$$F(b) - F(a) = \sum_{k=1}^n (F(x_k) - F(x_{k-1}))$$

Now use the mean value theorem and obtain a Riemann sum between the upper and lower sums just described. Explain why $\int_a^b f(x)\, dx$, and $F(b) - F(a)$ are both contained in an interval of length less than ε. Since ε was arbitrary, this requires the two numbers to be equal. This result is sometimes called the fundamental theorem of calculus. Note that this shows that the integral described in this chapter coincides with the earlier one which was **defined** in terms of antiderivatives.

(10) Using the result of Problem 9 compute the following definite integrals.

(a) $\int_0^1 \cos(x)\, dx$

(b) $\int_0^1 \sec^2(x)\, dx$

(c) $\int_0^2 \sqrt{2x^2 + 1}\, dx$

(d) $\int_{-2}^2 \sqrt{4 - x^2}\, dx$

(e) $\int_0^{\pi/2} x \sin(2x)\, dx$

(f) $\int_1^2 \frac{x^2 + 1 + x}{(x+1)(x^2+1)}\, dx$

(g) $\int_0^1 \frac{3x+2}{(x+1)(2x+1)}\, dx$

(11) Using Problem 9 find the volume of the region between the x axis and the graph of $y = \sin x$ for $x \in [0, \pi]$ revolved around the line $x = -1$.

(12) Using Problem 9 and Problem 4, recognize each of the following limits as the limit of a Riemann sum for a function and using this, find the limit. You can use the fact that every continuous function is Riemann integrable.

(a) $\lim_{n \to \infty} \sum_{k=1}^n \left(\frac{k}{n}\right)^2 \frac{1}{n}$

(b) $\lim_{n \to \infty} \sum_{k=1}^n \cos\left(\frac{k\pi}{2n}\right) \frac{1}{n}$

(c) $\lim_{n \to \infty} \sum_{k=1}^n \frac{1}{n+k}$

10.3   Functions Of Riemann Integrable Functions

It is often necessary to consider functions of Riemann integrable functions, and a natural question is whether these are Riemann integrable. The following theorem gives a partial answer to this question. This is not the most general theorem which will relate to this question but it will be enough for the needs of this book.

Theorem 10.3. *Let f, g be bounded functions and let $f([a, b]) \subseteq [c_1, d_1]$ and $g([a, b]) \subseteq [c_2, d_2]$. Let $H : [c_1, d_1] \times [c_2, d_2] \to \mathbb{R}$ satisfy,*

$$|H(a_1, b_1) - H(a_2, b_2)| \leq K[|a_1 - a_2| + |b_1 - b_2|]$$

for some constant K. Then if $f, g \in R([a, b])$ it follows that $H \circ (f, g) \in R([a, b])$.

Proof: In the following claim, $M_i(h)$ and $m_i(h)$ have the meanings assigned above with respect to some partition of $[a, b]$ for the function h.

Claim: The following inequality holds.

$$|M_i(H \circ (f, g)) - m_i(H \circ (f, g))| \leq$$

$$K[|M_i(f) - m_i(f)| + |M_i(g) - m_i(g)|].$$

Proof of the claim: By the above proposition, there exist $x_1, x_2 \in [x_{i-1}, x_i]$ be such that

$$H(f(x_1), g(x_1)) + \eta > M_i(H \circ (f, g)),$$

and

$$H(f(x_2), g(x_2)) - \eta < m_i(H \circ (f, g)).$$

Then

$$|M_i(H \circ (f, g)) - m_i(H \circ (f, g))|$$

$$< 2\eta + |H(f(x_1), g(x_1)) - H(f(x_2), g(x_2))|$$
$$< 2\eta + K[|f(x_1) - f(x_2)| + |g(x_1) - g(x_2)|]$$
$$\leq 2\eta + K[|M_i(f) - m_i(f)| + |M_i(g) - m_i(g)|].$$

Since $\eta > 0$ is arbitrary, this proves the claim.

Now continuing with the proof of the theorem, let P be such that

$$\sum_{i=1}^{n}(M_i(f) - m_i(f))\Delta x_i < \frac{\varepsilon}{2K}, \quad \sum_{i=1}^{n}(M_i(g) - m_i(g))\Delta x_i < \frac{\varepsilon}{2K}.$$

Then from the claim,

$$\sum_{i=1}^{n}(M_i(H \circ (f, g)) - m_i(H \circ (f, g)))\Delta x_i$$

$$< \sum_{i=1}^{n} K[|M_i(f) - m_i(f)| + |M_i(g) - m_i(g)|]\Delta x_i < \varepsilon.$$

Since $\varepsilon > 0$ is arbitrary, this shows $H \circ (f, g)$ satisfies the Riemann criterion and hence $H \circ (f, g)$ is Riemann integrable as claimed. ∎

This theorem implies that if f, g are Riemann integrable, then so is $af + bg, |f|, f^2$, along with infinitely many other such continuous combinations of Riemann integrable functions. For example, to see that $|f|$ is Riemann integrable, let $H(a, b) = |a|$. Clearly this function satisfies the conditions of the above theorem and so $|f| = H(f, f) \in R([a, b])$ as claimed. The following theorem gives an example of many functions which are Riemann integrable.

Theorem 10.4. *Let* $f : [a, b] \to \mathbb{R}$ *be either increasing or decreasing on* $[a, b]$. *Then* $f \in R([a, b])$.

Proof: Let $\varepsilon > 0$ be given and let

$$x_i = a + i\left(\frac{b-a}{n}\right), \quad i = 0, \cdots, n.$$

Then let **f** be increasing. It follows that

$$U(f, P) - L(f, P) = \sum_{i=1}^{n} (f(x_i) - f(x_{i-1}))\left(\frac{b-a}{n}\right)$$

$$= (f(b) - f(a))\left(\frac{b-a}{n}\right) < \varepsilon$$

whenever n is large enough. Thus the Riemann criterion is satisfied and so the function is Riemann integrable. The proof for decreasing f is similar. ■

Corollary 10.1. *Let* $[a, b]$ *be a bounded closed interval and let* $\phi : [a, b] \to \mathbb{R}$ *be Lipschitz continuous. Then* $\phi \in R([a, b])$. *Recall that a function* ϕ, *is Lipschitz continuous if there is a constant* K, *such that for all* x, y,

$$|\phi(x) - \phi(y)| < K|x - y|.$$

Proof: Let $f(x) = x$. Then by Theorem 10.4, f is Riemann integrable. Let $H(a, b) \equiv \phi(a)$. Then by Theorem 10.3 $H \circ (f, f) = \phi \circ f = \phi$ is also Riemann integrable. ■

In fact, it is enough to assume ϕ is continuous, although this is harder. This is the content of the next theorem which is where the difficult theorems about continuity and uniform continuity are used.

Theorem 10.5. *Suppose* $f : [a, b] \to \mathbb{R}$ *is continuous. Then* $f \in R([a, b])$.

Proof: By Corollary 2.3 on Page 71, f is uniformly continuous on $[a, b]$. Therefore, if $\varepsilon > 0$ is given, there exists a $\delta > 0$ such that if $|x_i - x_{i-1}| < \delta$, then $M_i - m_i < \frac{\varepsilon}{b-a}$. Let

$$P \equiv \{x_0, \cdots, x_n\}$$

be a partition with $|x_i - x_{i-1}| < \delta$. Then

$$U(f, P) - L(f, P) < \sum_{i=1}^{n} (M_i - m_i)(x_i - x_{i-1}) < \frac{\varepsilon}{b-a}(b-a) = \varepsilon.$$

By the Riemann criterion, $f \in R([a, b])$. ■

10.4 Properties Of The Integral

The integral has many important algebraic properties. First here is a simple lemma.

Lemma 10.3. *Let S be a nonempty set which is bounded above and below. Then if $-S \equiv \{-x : x \in S\}$,*

$$\sup(-S) = -\inf(S) \tag{10.6}$$

and

$$\inf(-S) = -\sup(S). \tag{10.7}$$

Proof: Consider 10.6. Let $x \in S$. Then $-x \le \sup(-S)$ and so $x \ge -\sup(-S)$. If follows that $-\sup(-S)$ is a lower bound for S and therefore, $-\sup(-S) \le \inf(S)$. This implies $\sup(-S) \ge -\inf(S)$. Now let $-x \in -S$. Then $x \in S$ and so $x \ge \inf(S)$ which implies $-x \le -\inf(S)$. Therefore, $-\inf(S)$ is an upper bound for $-S$ and so $-\inf(S) \ge \sup(-S)$. This shows (10.6). Formula (10.7) is similar and is left as an exercise. ∎

In particular, the above lemma implies that for $M_i(f)$ and $m_i(f)$ defined above $M_i(-f) = -m_i(f)$, and $m_i(-f) = -M_i(f)$.

Lemma 10.4. *If $f \in R([a,b])$ then $-f \in R([a,b])$ and*

$$-\int_a^b f(x)\, dx = \int_a^b -f(x)\, dx.$$

Proof: The first part of the conclusion of this lemma follows from Theorem 10.4 since the function $\phi(y) = -y$ is Lipschitz continuous. Now choose P such that

$$\int_a^b -f(x)\, dx - L(-f, P) < \varepsilon.$$

Then since $m_i(-f) = -M_i(f)$,

$$\varepsilon > \int_a^b -f(x)\, dx - \sum_{i=1}^n m_i(-f)\, \Delta x_i = \int_a^b -f(x)\, dx + \sum_{i=1}^n M_i(f)\, \Delta x_i$$

which implies

$$\varepsilon > \int_a^b -f(x)\, dx + \sum_{i=1}^n M_i(f)\, \Delta x_i \ge \int_a^b -f(x)\, dx + \int_a^b f(x)\, dx.$$

Thus, since ε is arbitrary,

$$\int_a^b -f(x)\, dx \le -\int_a^b f(x)\, dx$$

whenever $f \in R([a,b])$. It follows

$$\int_a^b -f(x)\, dx \le -\int_a^b f(x)\, dx = -\int_a^b -(-f(x))\, dx \le \int_a^b -f(x)\, dx. \quad ∎$$

Theorem 10.6. *The integral is linear,*

$$\int_a^b (\alpha f + \beta g)(x) \, dx = \alpha \int_a^b f(x) \, dx + \beta \int_a^b g(x) \, dx.$$

whenever $f, g \in R([a, b])$ *and* $\alpha, \beta \in \mathbb{R}$.

Proof: First note that by Theorem 10.3, $\alpha f + \beta g \in R([a, b])$. To begin with, consider the claim that if $f, g \in R([a, b])$ then

$$\int_a^b (f + g)(x) \, dx = \int_a^b f(x) \, dx + \int_a^b g(x) \, dx. \qquad (10.8)$$

Let P_1, Q_1 be such that

$$U(f, Q_1) - L(f, Q_1) < \varepsilon/2, \ U(g, P_1) - L(g, P_1) < \varepsilon/2.$$

Then letting $P \equiv P_1 \cup Q_1$, Lemma 10.1 implies

$$U(f, P) - L(f, P) < \varepsilon/2, \text{ and } U(g, P) - U(g, P) < \varepsilon/2.$$

Next note that

$$m_i(f + g) \geq m_i(f) + m_i(g), \ M_i(f + g) \leq M_i(f) + M_i(g).$$

Therefore,

$$L(g + f, P) \geq L(f, P) + L(g, P), \ U(g + f, P) \leq U(f, P) + U(g, P).$$

For this partition,

$$\int_a^b (f + g)(x) \, dx \in [L(f + g, P), U(f + g, P)]$$

$$\subseteq [L(f, P) + L(g, P), U(f, P) + U(g, P)]$$

and

$$\int_a^b f(x) \, dx + \int_a^b g(x) \, dx \in [L(f, P) + L(g, P), U(f, P) + U(g, P)].$$

Therefore,

$$\left| \int_a^b (f + g)(x) \, dx - \left(\int_a^b f(x) \, dx + \int_a^b g(x) \, dx \right) \right| \leq$$

$$U(f, P) + U(g, P) - (L(f, P) + L(g, P)) < \varepsilon/2 + \varepsilon/2 = \varepsilon.$$

This proves (10.8) since ε is arbitrary.

It remains to show that

$$\alpha \int_a^b f(x) \, dx = \int_a^b \alpha f(x) \, dx.$$

Suppose first that $\alpha \geq 0$. Then

$$\int_a^b \alpha f(x) \, dx \equiv \sup\{L(\alpha f, P) : P \text{ is a partition}\} =$$

$$\alpha \sup\{L(f, P) : P \text{ is a partition}\} \equiv \alpha \int_a^b f(x) \, dx.$$

If $\alpha < 0$, then this and Lemma 10.4 imply

$$\int_a^b \alpha f(x) \, dx = \int_a^b (-\alpha)(-f(x)) \, dx$$

$$= (-\alpha) \int_a^b (-f(x)) \, dx = \alpha \int_a^b f(x) \, dx. \qquad \blacksquare$$

Theorem 10.7. *If $f \in R([a,b])$ and $f \in R([b,c])$, then $f \in R([a,c])$ and*

$$\int_a^c f(x) \, dx = \int_a^b f(x) \, dx + \int_b^c f(x) \, dx. \qquad (10.9)$$

Proof: Let P_1 be a partition of $[a,b]$ and P_2 be a partition of $[b,c]$ such that

$$U(f, P_i) - L(f, P_i) < \varepsilon/2, \quad i = 1, 2.$$

Let $P = P_1 \cup P_2$. Then P is a partition of $[a,c]$ and

$$U(f, P) - L(f, P)$$

$$= U(f, P_1) - L(f, P_1) + U(f, P_2) - L(f, P_2) < \varepsilon/2 + \varepsilon/2 = \varepsilon. \qquad (10.10)$$

Thus, $f \in R([a,c])$ by the Riemann criterion and also for this partition,

$$\int_a^b f(x) \, dx + \int_b^c f(x) \, dx \in [L(f, P_1) + L(f, P_2), U(f, P_1) + U(f, P_2)]$$

$$= [L(f, P), U(f, P)]$$

and

$$\int_a^c f(x) \, dx \in [L(f, P), U(f, P)].$$

Hence by (10.10),

$$\left| \int_a^c f(x) \, dx - \left(\int_a^b f(x) \, dx + \int_b^c f(x) \, dx \right) \right| < U(f, P) - L(f, P) < \varepsilon$$

which shows that since ε is arbitrary, (10.9) holds. $\blacksquare$

Corollary 10.2. *Let $[a,b]$ be a closed and bounded interval and suppose that*

$$a = y_1 < y_2 \cdots < y_l = b$$

and that f is a bounded function defined on $[a,b]$ which has the property that f is either increasing on $[y_j, y_{j+1}]$ or decreasing on $[y_j, y_{j+1}]$ for $j = 1, \cdots, l-1$. Then $f \in R([a,b])$.

Proof: This follows from Theorem 10.7 and Theorem 10.4. ∎

The symbol $\int_a^b f(x)\, dx$ when $a > b$ has not yet been defined.

Definition 10.3. Let $[a, b]$ be an interval and let $f \in R([a, b])$. Then

$$\int_b^a f(x)\, dx \equiv -\int_a^b f(x)\, dx.$$

Note that with this definition,

$$\int_a^a f(x)\, dx = -\int_a^a f(x)\, dx$$

and so

$$\int_a^a f(x)\, dx = 0.$$

Theorem 10.8. *Assuming all the integrals make sense,*

$$\int_a^b f(x)\, dx + \int_b^c f(x)\, dx = \int_a^c f(x)\, dx.$$

Proof: This follows from Theorem 10.7 and Definition 10.3. For example, assume $c \in (a, b)$. Then from Theorem 10.7,

$$\int_a^c f(x)\, dx + \int_c^b f(x)\, dx = \int_a^b f(x)\, dx$$

and so by Definition 10.3,

$$\int_a^c f(x)\, dx = \int_a^b f(x)\, dx - \int_c^b f(x)\, dx = \int_a^b f(x)\, dx + \int_b^c f(x)\, dx.$$

The other cases are similar. ∎

The following properties of the integral have either been established or they follow quickly from what has been shown so far.

$$\text{If } f \in R([a, b]) \text{ then if } c \in [a, b],\ f \in R([a, c]), \tag{10.11}$$

$$\int_a^b \alpha\, dx = \alpha(b - a), \tag{10.12}$$

$$\int_a^b (\alpha f + \beta g)(x)\, dx = \alpha \int_a^b f(x)\, dx + \beta \int_a^b g(x)\, dx, \tag{10.13}$$

$$\int_a^b f(x)\, dx + \int_b^c f(x)\, dx = \int_a^c f(x)\, dx, \tag{10.14}$$

$$\int_a^b f(x)\, dx \geq 0 \text{ if } f(x) \geq 0 \text{ and } a < b, \tag{10.15}$$

$$\left| \int_a^b f(x)\, dx \right| \leq \left| \int_a^b |f(x)|\, dx \right|. \tag{10.16}$$

The only one of these claims which may not be completely obvious is the last one. To show this one, note that

$$|f(x)| - f(x) \geq 0, \ |f(x)| + f(x) \geq 0.$$

Therefore, by (10.15) and (10.13), if $a < b$,

$$\int_a^b |f(x)|\, dx \geq \int_a^b f(x)\, dx$$

and

$$\int_a^b |f(x)|\, dx > - \int_a^b f(x)\, dx.$$

Therefore,

$$\int_a^b |f(x)|\, dx \geq \left| \int_a^b f(x)\, dx \right|.$$

If $b < a$ then the above inequality holds with a and b switched. This implies (10.16).

10.5   Fundamental Theorem Of Calculus

With these properties, it is easy to prove the fundamental theorem of calculus[2]. Let $f \in R([a, b])$. Then by (10.11), $f \subset R([a, x])$ for each $x \in [a, b]$. The first version of the fundamental theorem of calculus is a statement about the derivative of the function

$$x \to \int_a^x f(t)\, dt.$$

Theorem 10.9. *Let* $f \in R([a, b])$ *and let*

$$F(x) \equiv \int_a^x f(t)\, dt.$$

Then if f *is continuous at* $x \in (a, b)$,

$$F'(x) = f(x).$$

---

[2]This theorem is why Newton and Leibniz are credited with inventing calculus. The integral had been around for thousands of years and the derivative was by their time well known. However the connection between these two ideas had not been fully made although Newton's predecessor, Isaac Barrow had made some progress in this direction.

Proof: Let $x \in (a, b)$ be a point of continuity of f and let h be small enough that $x + h \in [a, b]$. Then by using (10.14),

$$h^{-1} \left(F \left(x + h \right) - F \left(x \right) \right) = h^{-1} \int_x^{x+h} f \left(t \right) \, dt.$$

Also, using (10.12),

$$f \left(x \right) = h^{-1} \int_x^{x+h} f \left(x \right) \, dt.$$

Therefore, by (10.16),

$$\left| h^{-1} \left(F \left(x + h \right) - F \left(x \right) \right) - f \left(x \right) \right| = \left| h^{-1} \int_x^{x+h} \left(f \left(t \right) - f \left(x \right) \right) \, dt \right|$$

$$\leq \left| h^{-1} \int_x^{x+h} \left| f \left(t \right) - f \left(x \right) \right| \, dt \right|.$$

Let $\varepsilon > 0$ and let $\delta > 0$ be small enough that if $|t - x| < \delta$, then

$$\left| f \left(t \right) - f \left(x \right) \right| < \varepsilon.$$

Therefore, if $|h| < \delta$, the above inequality and (10.12) shows that

$$\left| h^{-1} \left(F \left(x + h \right) - F \left(x \right) \right) - f \left(x \right) \right| \leq |h|^{-1} \varepsilon |h| = \varepsilon.$$

Since $\varepsilon > 0$ is arbitrary, this shows

$$\lim_{h \to 0} h^{-1} \left(F \left(x + h \right) - F \left(x \right) \right) = f \left(x \right). \qquad \blacksquare$$

Note this gives existence for the initial value problem

$$F' \left(x \right) = f \left(x \right), \ F \left(a \right) = 0$$

whenever f is Riemann integrable and continuous.[3]

The next theorem is also called the fundamental theorem of calculus.

Theorem 10.10. *Let $f \in R \left([a, b] \right)$ and suppose there exists an antiderivative for f, G, such that*

$$G' \left(x \right) = f \left(x \right)$$

for every point of (a, b) and G is continuous on $[a, b]$. Then

$$\int_a^b f \left(x \right) \, dx = G \left(b \right) - G \left(a \right). \tag{10.17}$$

---

[3]Of course it was proved that if f is continuous on a closed interval, $[a, b]$, then $f \in R \left([a, b] \right)$ but this is a hard theorem using the difficult result about uniform continuity.

Proof: Let $P = \{x_0, \cdots, x_n\}$ be a partition satisfying

$$U(f, P) - L(f, P) < \varepsilon.$$

Then

$$G(b) - G(a) = G(x_n) - G(x_0)$$

$$= \sum_{i=1}^{n} G(x_i) - G(x_{i-1}).$$

By the mean value theorem,

$$G(b) - G(a) = \sum_{i=1}^{n} G'(z_i)(x_i - x_{i-1})$$

$$= \sum_{i=1}^{n} f(z_i) \Delta x_i$$

where z_i is some point in $[x_{i-1}, x_i]$. It follows, since the above sum lies between the upper and lower sums, that

$$G(b) - G(a) \in [L(f, P), U(f, P)],$$

and also

$$\int_a^b f(x) \, dx \subset [L(f, P), U(f, P)].$$

Therefore,

$$\left| G(b) - G(a) - \int_a^b f(x) \, dx \right| < U(f, P) - L(f, P) < \varepsilon.$$

Since $\varepsilon > 0$ is arbitrary, (10.17) holds. ∎

The following notation is often used in this context. Suppose F is an antiderivative of f as just described with F continuous on $[a, b]$ and $F' = f$ on (a, b). Then

$$\int_a^b f(x) \, dx = F(b) - F(a) = F(x) \big|_a^b.$$

Recall how many interesting problems can be reduced to initial value problems. This was true of work, area, arc length and many other examples. The examples given were cooked up to work out and you could actually solve the initial value problem using known functions. What if you could not do this? The next theorem is a significant existence theorem which tells you that solutions of the initial value problem exist.

Theorem 10.11. *Suppose f is a continuous function defined on an interval (a, b), $c \in (a, b)$, and $y_0 \in \mathbb{R}$. Then there exists a unique solution to the initial value problem*

$$F'(x) = f(x), \ F(c) = y_0.$$

This solution is given by

$$F(x) = y_0 + \int_c^x f(t) \, dt. \tag{10.18}$$

Proof: From Theorem 10.5, it follows the integral in (10.18) is well defined. Now by the fundamental theorem of calculus, $F'(x) = f(x)$. Therefore, F solves the given differential equation. Also, $F(c) = y_0 + \int_c^c f(t) \, dt = y_0$ so the initial condition is also satisfied. This establishes the existence part of the theorem.

Suppose F and G both solve the initial value problem. Then $F'(x) - G'(x) = f(x) - f(x) = 0$ and so $F(x) - G(x) = C$ for some constant C. However, $F(c) - G(c) = y_0 - y_0 = 0$, and so the constant C can only equal 0. This proves the uniqueness part of the theorem. ∎

Example 10.1. Find the area between $y = x^2$ and $y = x^3$ for $x \in [0, 1]$.

You need to solve the initial value problem $A'(x) = x^2 - x^3$, $A'(0) = 0$. The answer is then $A(1)$. By Theorem 10.11, $A(x) = 0 + \int_0^x (t^2 - t^3) \, dt$ and so the answer is

$$A(1) = \int_0^1 (t^2 - t^3) \, dt = \frac{1}{12}$$

Example 10.2. A cylinder standing on its end is filled with water which weighs 62.5 pounds per cubic foot is 5 feet in radius and is 10 feet tall. Find the total force on the sides of the cylinder.

Remember the force acting on the sides of a horizontal slice of the cylinder y feet below the top of the fluid is $(2\pi \times 5 \times 62.5 \times y) \, dy = dF$. Thus you need to find a solution to the initial value problem

$$\frac{dF}{dy} = (2\pi \times 5 \times 62.5 \times y), \quad F(0) = 0$$

and in this case, the answer equals $F(10)$. By Theorem 10.11 and Theorem 10.10,

$$F(10) = \int_0^{10} (2\pi)(5)(62.5)(y) \, dy = 98175 \text{ pounds.}$$

An informal way of looking at this is $dF = (2\pi \times 5 \times 62.5 \times y) \, dy$ for y between 0 and 10. The dF is an "infinitesimal" piece of the total force. To get the total force you just sum the $dF's$. Thus $F = \int_0^{10} (2\pi)(5)(62.5)(y) \, dy$. The reason the integral sign looks like an S is that it is a sort of a sum. You are summing up "infinitesimal" contributions to obtain the total. This last statement is mathematical gobbledygook. I have told you nothing about infinitesimals. I have only used the term in an evocative manner. However, it turns out to be a useful way of thinking about things.

Example 10.3. Find the area of the surface of revolution formed by revolving $y = x^2 - \frac{1}{8} \ln x$ about the x axis for $x \in [1, 2]$.

An infinitesimal contribution is of the form

$$dA = 2\pi \left(x^2 - \frac{1}{8} \ln x \right) \sqrt{1 + \left(\frac{1}{8} \frac{16x^2 - 1}{x} \right)^2} \, dx = 2\pi \left(x^2 - \frac{1}{8} \ln x \right) \frac{1}{8} \frac{1 + 16x^2}{x} \, dx$$

Summing these with the integral and using Theorem 10.10 yields

$$\int_1^2 2\pi \left(x^2 - \frac{1}{8}\ln x \right) \frac{1}{8}\frac{1+16x^2}{x}dx = \frac{63}{4}\pi - \frac{1}{64}\pi \ln^2 2 - \pi \ln 2$$

Sometimes you will be unable to do the sort of computations done above. For example, consider the following example.

Example 10.4. Find the area of the surface of revolution formed by revolving $y = x^2 - \ln x$ about the x axis for $x \in [1, 2]$.

In this case an infinitesimal contribution to this area is

$$dA = 2\pi \left(x^2 - \ln x \right) \frac{1}{x}\sqrt{4x^4 - 3x^2 + 1}dx$$

and so the total area is given by the integral

$$A = \int_1^2 2\pi \left(x^2 - \ln x \right) \frac{1}{x}\sqrt{4x^4 - 3x^2 + 1}dx.$$

What now? Can you find an antiderivative in terms of functions you have names for and use Theorem 10.10 to evaluate it? If you want to waste lots of time, give it a try. If you can do this one, it is very easy to find important examples that you can't do by finding antiderivatives in terms of known functions. Examples are integrals which have integrands equal to $\sin x/x$, c^{-x^2}, $\sqrt{1 + x^4}$, $\sin\left(x^2\right)$, and many others. Sometimes these occur in important applications. What do you do then when Theorem 10.10 has failed? The answer is, you use a numerical method to compute the integral. The most rudimentary numerical method goes right to the definition of the integral. It is described in the following definition. A much better method is described in the exercises. An extensive study of numerical methods for finding integrals will not be attempted in this book but it is very important you be aware that such methods are available. When you use a calculator or computer algebra system to compute an integral numerically, the machine is using a numerical method, probably one which is much more sophisticated than any presented in beginning calculus.

Definition 10.4. Let f be a bounded function defined on a closed interval $[a, b]$ and let $P = \{x_0, \cdots, x_n\}$ be a partition of the interval. Suppose $z_i \in [x_{i-1}, x_i]$ is chosen. Then the sum

$$\sum_{i=1}^n f(z_i)(x_i - x_{i-1})$$

is known as a Riemann sum. Also,

$$\|P\| \equiv \max\{|x_i - x_{i-1}| : i = 1, \cdots, n\}.$$

Proposition 10.2. Suppose $f \in R([a,b])$. Then there exists a partition $P \equiv \{x_0, \cdots, x_n\}$ with the property that for any choice of $z_k \in [x_{k-1}, x_k]$,

$$\left| \int_a^b f(x)\, dx - \sum_{k=1}^n f(z_k)(x_k - x_{k-1}) \right| < \varepsilon.$$

Proof: Choose P such that $U(f, P) - L(f, P) < \varepsilon$ and then both $\int_a^b f(x)\, dx$ and $\sum_{k=1}^n f(z_k)(x_k - x_{k-1})$ are contained in $[L(f, P), U(f, P)]$ and so the claimed inequality must hold. ∎

This proposition is significant because it gives a way of approximating the integral.

Example 10.5. Use a Riemann sum to approximate $\int_0^1 \sin(x^2)\, dx$.

I will use the partition $\{0, \frac{1}{4}, \frac{1}{2}, \frac{3}{4}, 1\}$ and I will pick the midpoint of each subinterval to evaluate the function. This particular Riemann sum is called the **midpoint sum**. Thus the Riemann sum approximating the integral is

$$\sin\left(\frac{1}{64}\right)\left(\frac{1}{4}\right) + \sin\left(\left(\frac{3}{8}\right)^2\right)\frac{1}{4} + \sin\left(\left(\frac{5}{8}\right)^2\right)\frac{1}{4} + \sin\left(\left(\frac{7}{8}\right)^2\right)\frac{1}{4} = .307\,39$$

Using a computer algebra system to evaluate this integral gives

$$\int_0^1 \sin(x^2)\, dx = .310\,27$$

so you see in this case, the primitive Riemann sum approach yielded a pretty good answer. Of course it is important to know how close you are to the true answer and this involves the concept of error estimates. The following is a rudimentary error estimate which tells how well such a mid point sum approximates an integral.

Theorem 10.12. *Let f be a continuous function defined on the interval $[a, b]$ and suppose also that f' exists on (a, b) and there exists $K \geq |f'(x)|$ for all $x \in (a, b)$. Suppose also that $\{x_0, x_1, \cdots, x_n\}$ is a partition in which the points are equally spaced. Thus $x_k = a + k\frac{b-a}{n}$. Then*

$$\left| midpoint\ sum - \int_a^b f(x)\, dx \right| < K\frac{(b-a)^2}{n}.$$

Proof: You estimate the error on each interval $[x_{k-1}, x_k]$ and then sum up these estimates. The contribution of the midpoint sum on this interval equals $f\left(\frac{x_{k-1}+x_k}{2}\right)\left(\frac{b-a}{n}\right)$. Thus the error corresponding to this interval is $\left| f\left(\frac{x_{k-1}+x_k}{2}\right)\left(\frac{b-a}{n}\right) - \int_{x_{k-1}}^{x_k} f(t)\, dt \right|$. By the mean value theorem for integrals, Problem 11 on Page 256 there exists $z_k \in (x_{k-1}, x_k)$ such that

$$\int_{x_{k-1}}^{x_k} f(t)\, dt = f(z_k)(x_k - x_{k-1}) = f(z_k)\left(\frac{b-a}{n}\right).$$

Therefore, this error equals

$$\left| f\left(\frac{x_{k-1}+x_k}{2}\right) - f(z_k) \right|\left(\frac{b-a}{n}\right)$$

which by the mean value theorem is no more than

$$|f'(w_k)|\left| \frac{x_{k-1}+x_k}{2} - z_k \right|\left(\frac{b-a}{n}\right) \leq K\left(\frac{b-a}{2n}\right)\left(\frac{b-a}{n}\right).$$

Since there are n of these, it follows the total error is no more than $K\frac{(b-a)^2}{n}$ as claimed. ∎

Note that to get an error estimate, you need information on the derivative of the function, something which has nothing to do with the existence of the integral. This is typical of error estimates in numerical methods. You need to know something extra beyond what you need to get existence.

10.6   The Riemann Integral

The definition of Riemann integrability given in this chapter is also called Darboux integrability and the integral defined as the unique number which lies between all upper sums and all lower sums which is given in this chapter is called the Darboux integral . The definition of the Riemann integral in terms of Riemann sums is given next.

Definition 10.5. A bounded function f defined on $[a, b]$ is said to be Riemann integrable if there exists a number I with the property that for every $\varepsilon > 0$, there exists $\delta > 0$ such that if

$$P \equiv \{x_0, x_1, \cdots, x_n\}$$

is any partition having $||P|| < \delta$, and $z_i \in [x_{i-1}, x_i]$,

$$\left| I - \sum_{i=1}^{n} f(z_i)(x_i - x_{i-1}) \right| < \varepsilon.$$

The number $\int_a^b f(x)\, dx$ is defined as I.

Thus, there are two definitions of the integral, this one in terms of Riemann sums, and the earlier one which defined the integral to be the number which is between all the upper sums and lower sums. It turns out they are equivalent which is the following theorem of Darboux.

Theorem 10.13. *A bounded function defined on $[a, b]$ is Riemann integrable in the sense of Definition 10.5 if and only if it is integrable in the sense of Darboux. Furthermore the two integrals coincide.*

The proof of this theorem is left for the exercises in Problems 16 - 18. It is not essential that you understand this theorem, so if it does not interest you, leave it out. Note that it implies that given a Riemann integrable function f in either sense, it can be approximated by Riemann sums whenever $||P||$ is sufficiently small. Both versions of the integral are obsolete but entirely adequate for most applications and as a point of departure for a more up to date and satisfactory integral. The reason for using the Darboux approach to the integral is that all the existence theorems are easier to prove in this context.

10.7   Exercises

(1) Let $F(x) = \int_{x^2}^{x^3} \frac{t^5+7}{t^7+87t^6+1}\, dt$. Find $F'(x)$.

(2) Let $F(x) = \int_2^x \frac{1}{1+t^4}\, dt$. Sketch a graph of F and explain why it looks the way it does.

(3) Use a midpoint sum with 4 subintervals to estimate $\int_1^2 \frac{1}{t}\, dt$ and give an estimate of how close your approximation is to the true answer.

(4) Use a midpoint sum with 4 subintervals to estimate $\int_0^1 \frac{1}{t^2+1}\, dt$ and give an estimate of how close your approximation is to the true answer.

(5) There is a general procedure for estimating the integral of a function f on an interval $[a, b]$. Form a uniform partition $P = \{x_0, x_1, \cdots, x_n\}$ where for each j, $x_j - x_{j-1} = h$. Let $f_i = f(x_i)$ and assuming $f \geq 0$ on the interval $[x_{i-1}, x_i]$, approximate the area above this interval and under the curve with the area of a trapezoid having vertical sides, f_{i-1}, and f_i as shown in the following picture.

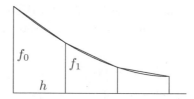

Thus $\frac{1}{2}\left(\frac{f_i+f_{i-1}}{2}\right)$ approximates the area under the curve. Show that adding these up yields

$$\frac{h}{2}\left[f_0 + 2f_1 + \cdots + 2f_{n-1} + f_n\right]$$

as an approximation to $\int_a^b f(x)\, dx$. This is known as the trapezoid rule. Verify that if $f(x) = mx+b$, the trapezoid rule gives the exact answer for the integral. Would this be true of upper and lower sums for such a function? Can you show that in the case of the function $f(t) = 1/t$ the trapezoid rule will always yield an answer which is too large for $\int_1^2 \frac{1}{t}\, dt$?

(6) Let there be three equally spaced points $x_{i-1}, x_{i-1}+h \equiv x_i$, and $x_i+2h \equiv x_{i+1}$. Suppose also a function f, has the value f_{i-1} at x, f_i at $x + h$, and f_{i+1} at $x + 2h$. Then consider

$$g_i(x) \equiv \frac{f_{i-1}}{2h^2}(x - x_i)(x - x_{i+1}) - \frac{f_i}{h^2}(x - x_{i-1})(x - x_{i+1})$$
$$+ \frac{f_{i+1}}{2h^2}(x - x_{i-1})(x - x_i).$$

Check that this is a second degree polynomial which equals the values f_{i-1}, f_i, and f_{i+1} at the points x_{i-1}, x_i, and x_{i+1} respectively. The function g_i is an approximation to the function f on the interval $[x_{i-1}, x_{i+1}]$. Also,

$$\int_{x_{i-1}}^{x_{i+1}} g_i(x)\, dx$$

is an approximation to $\int_{x_{i-1}}^{x_{i+1}} f(x)\, dx$. Show $\int_{x_{i-1}}^{x_{i+1}} g_i(x)\, dx$ equals

$$\frac{hf_{i-1}}{3} + \frac{hf_i 4}{3} + \frac{hf_{i+1}}{3}.$$

Now suppose n is even and $\{x_0, x_1, \cdots, x_n\}$ is a partition of the interval $[a, b]$ and the values of a function f defined on this interval are $f_i = f(x_i)$. Adding these approximations for the integral of f on the succession of intervals,

$$[x_0, x_2], [x_2, x_4], \cdots, [x_{n-2}, x_n],$$

show that an approximation to $\int_a^b f(x)\, dx$ is

$$\frac{h}{3}\left[f_0 + 4f_1 + 2f_2 + 4f_3 + 2f_2 + \cdots + 4f_{n-1} + f_n\right].$$

This is called Simpson's rule. Use Simpson's rule to compute an approximation to $\int_1^2 \frac{1}{t}\, dt$ letting $n = 4$. Compare with the answer from a calculator or computer.

(7) A mine shaft has the shape $y = \cosh(.01x) - 5000$ where the units are in feet. Thus this mine shaft has a depth of a little less than one mile occurring when $x - 0$. The surface corresponds to $y - 0$ which occurs when $x = 921$ feet (approximately). A cable which weighs 5 pounds per foot extends from this point on the surface down to the very bottom of this mine shaft. Set up an integral which will give the work required to haul this cable up to the surface and use a computer algebra system or calculator to compute this integral. Next suppose there is a 5 ton ore skip on the bottom of this cable. How much work would be required to haul up the cable and the 5 ton skip of ore[4]?

(8) Let a and b be positive numbers and consider the function

$$F(x) = \int_0^{ax} \frac{1}{a^2 + t^2}\, dt + \int_b^{a/x} \frac{1}{a^2 + t^2}\, dt.$$

Show that F is a constant.

(9) Solve the following initial value problem from ordinary differential equations which is to find a function y such that

$$y'(x) = \frac{x^7 + 1}{x^6 + 97x^5 + 7}, \quad y(10) = 5.$$

---

[4]In the upper peninsula of Michigan, there are copper mines having inclined shafts which achieve a depth of roughly one mile. These shafts start off very steep and become less so toward the bottom to allow for the sagging in the cable. One such is shaft 2 for the Quincy mine in Hancock Michigan. The problem of hauling up so much heavy wire cable was solved by letting out cable and an empty ore skip on one of the dual skipways while taking up the cable and the full ore skip on the other. The Hoist which did the work of hauling the skip weighs 880 tons and is the largest steam hoist ever built. It was operated between 1920 and 1931. This hoist could bring up a 10 ton skip of ore at over 30 miles per hour. The huge steam hoist is housed in its own building and the cables from this hoist passed over head on large pulleys to the mine shaft. Once the cable broke. Imagine what happened at the top and at the bottom.

(10) If $F, G \in \int f(x) \, dx$ for all $x \in \mathbb{R}$, show $F(x) = G(x) + C$ for some constant C. Use this to give a different proof of the fundamental theorem of calculus which has for its conclusion $\int_a^b f(t) \, dt = G(b) - G(a)$ where $G'(x) = f(x)$.

(11) Suppose f is continuous on $[a, b]$. Show there exists $c \in (a, b)$ such that

$$f(c) = \frac{1}{b-a} \int_a^b f(x) \, dx.$$

Hint: You might consider the function $F(x) \equiv \int_a^x f(t) \, dt$ and use the mean value theorem for derivatives and the fundamental theorem of calculus.

(12) Suppose f and g are continuous functions on $[a, b]$ and that $g(x) \neq 0$ on (a, b). Show there exists $c \in (a, b)$ such that

$$f(c) \int_a^b g(x) \, dx = \int_a^b f(x) g(x) \, dx.$$

Hint: Define $F(x) \equiv \int_a^x f(t) g(t) \, dt$ and let $G(x) \equiv \int_a^x g(t) \, dt$. Then use the Cauchy mean value theorem on these two functions.

(13) Consider the function

$$f(x) \equiv \begin{cases} \sin\left(\frac{1}{x}\right) & \text{if } x \neq 0 \\ 0 & \text{if } x = 0 \end{cases}.$$

Is f Riemann integrable? Explain why or why not.

(14) Prove the second part of Theorem 10.4 about decreasing functions.

(15) Find the following limits by identifying them as a Riemann sum for a certain integral.

(a) $\lim_{n \to \infty} \sum_{k=1}^n \frac{1}{n} \sin\left(\frac{k\pi}{n}\right)$

(b) $\lim_{n \to \infty} \sum_{k=1}^n \frac{1}{n+k}$

(c) $\lim_{n \to \infty} \frac{4}{n^3} \sum_{k=1}^n k^2$

(d) $\lim_{n \to \infty} \sum_{k=1}^{2n} \frac{n}{2n^2 - 2kn + k^2}$

(16) * Suppose f is a bounded function defined on $[a, b]$ and $|f(x)| < M$ for all $x \in [a, b]$. Now let Q be a partition having n points $\{x_0^*, \cdots, x_n^*\}$ and let P be any other partition. Show that

$$|U(f, P) - L(f, P)| \leq 2Mn \, \|P\| + |U(f, Q) - L(f, Q)|.$$

Hint: Write the sum for $U(f, P) - L(f, P)$ and split this sum into two sums, the sum of terms for which $[x_{i-1}, x_i]$ contains at least one point of Q, and terms for which $[x_{i-1}, x_i]$ does not contain any points of Q. In the latter case, $[x_{i-1}, x_i]$ must be contained in some interval $[x_{k-1}^*, x_k^*]$. Therefore, the sum of these terms should be no larger than $|U(f, Q) - L(f, Q)|$.

(17) * ↑ If $\varepsilon > 0$ is given and f is a Darboux integrable function defined on $[a, b]$, show there exists $\delta > 0$ such that whenever $\|P\| < \delta$, then

$$|U(f, P) - L(f, P)| < \varepsilon.$$

(18) * ↑ Prove Theorem 10.13.

10.8 Return Of The Wild Assumption

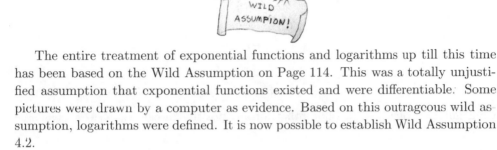

The entire treatment of exponential functions and logarithms up till this time has been based on the Wild Assumption on Page 114. This was a totally unjustified assumption that exponential functions existed and were differentiable. Some pictures were drawn by a computer as evidence. Based on this outrageous wild assumption, logarithms were defined. It is now possible to establish Wild Assumption 4.2.

Define

$$L_1(x) \equiv \int_1^x \frac{1}{t}\, dt. \tag{10.19}$$

There is no problem in writing this integral because the function $f(t) = 1/t$ is decreasing.

Theorem 10.14. *The function $L_1 : (0, \infty) \to \mathbb{R}$ satisfies the following properties.*

$$L_1(xy) = L_1(x) + L_1(y), \quad L_1(1) = 0, \tag{10.20}$$

The function L_1 is one to one and onto, strictly increasing, and its graph is concave downward. In addition to this, whenever $\frac{m}{n} \in \mathbb{Q}$

$$L_1\left(\sqrt[n]{x^m}\right) = \frac{m}{n} L_1(x). \tag{10.21}$$

Proof: Fix $y > 0$ and let

$$f(x) = L_1(xy) - (L_1(x) + L_1(y))$$

Then by Theorem 3.2 on Page 90 and the Fundamental theorem of calculus, Theorem 10.9,

$$f'(x) = y\left(\frac{1}{xy}\right) - \frac{1}{x} = 0.$$

Therefore, by Corollary 3.2 on Page 103, $f(x)$ is a constant. However, $f(1) = 0$ and so this proves (10.20).

From the Fundamental theorem of calculus, Theorem 10.9, $L_1'(x) = \frac{1}{x} > 0$ and so L_1 is a strictly increasing function and is therefore one to one. Also the second derivative equals

$$L_1''(x) = \frac{-1}{x^2} < 0$$

showing that the graph of L_1 is concave down.

Now consider the assertion that L_1 is onto. First note that from the definition, $L_1(2) > 0$. In fact,

$$L_1(2) \geq 1/2$$

as can be seen by looking at a lower sum for $\int_1^2 (1/t)\, dt$. Now if $x > 0$

$$L_1(x \times x) = L_1 x + L_1 x = 2L_1 x.$$

Also,

$$L_1((x)(x)(x)) = L_1((x)(x)) + L_1(x)$$
$$= L_1(x) + L_1(x) + L_1(x)$$
$$= 3L_1(x).$$

Continuing in this way, it follows that for any positive integer n,

$$L_1(x^n) = nL_1(x). \tag{10.22}$$

Therefore, $L_1(x)$ achieves arbitrarily large values as x gets increasingly large because you can take $x = 2$ in (10.22) and use the definition of L_1 to verify that $L_1(2) > 0$. Now if $x > 0$

$$0 = L_1\left(\overbrace{\left(\frac{1}{x}\right)(x)}^{=1} \right) = L_1\left(\frac{1}{x}\right) + L_1(x)$$

showing that

$$L_1(x^{-1}) = L_1\left(\frac{1}{x}\right) = -L_1 x. \tag{10.23}$$

You see that $\left(\frac{1}{2}\right)^n$ can be made as close to zero as desired by taking n sufficiently large. Also, from (10.22),

$$L_1\left(\frac{1}{2^n}\right) = nL_1\left(\frac{1}{2}\right) = -nL_1(2)$$

showing that $L_1(x)$ gets arbitrarily large in the negative direction provided that x is a sufficiently small positive number. Since L_1 is continuous, the intermediate value theorem may be used to fill in all the numbers in between. Thus the picture of the graph of L_1 looks like the following

where the graph approaches the y axis as x gets close to 0.

It only remains to verify the claim about raising x to a rational power. From (10.22) and (10.23), for any integer n, positive or negative,

$$L_1\left(x^n\right) = nL_1\left(x\right).$$

Therefore, letting m, n be integers,

$$mL_1\left(x\right) = L_1\left(x^m\right) = L_1\left(\left(\sqrt[n]{x^m}\right)^n\right) = nL_1\left(\sqrt[n]{x^m}\right)$$

and so

$$\frac{m}{n}L_1\left(x\right) = L_1\left(\sqrt[n]{x^m}\right).$$ ∎

Now Wild Assumption 4.2 on Page 114 can be fully justified.

Definition 10.6. For $b > 0$ define

$$\exp_b\left(x\right) \equiv L_1^{-1}\left(xL_1\left(b\right)\right). \tag{10.24}$$

Proposition 10.3. The function just defined in (10.24) satisfies all conditions of Wild Assumption 4.2 on Page 114 and $L_1\left(b\right) - \ln b$ as defined on Page 114.

Proof: First let $x = \frac{m}{n}$ where m and n are integers. To verify that $\exp_b\left(m/n\right) = \sqrt[n]{b^m}$,

$$\exp_b\left(\frac{m}{n}\right) \equiv L_1^{-1}\left(\frac{m}{n}L_1\left(b\right)\right) = L_1^{-1}\left(L_1\left(\sqrt[n]{b^m}\right)\right) = \sqrt[n]{b^m}$$

by (10.21).

That $\exp_b\left(x\right) > 0$ follows immediately from the definition of the inverse function. (L_1 is defined on positive real numbers and so L_1^{-1} has values in the positive real numbers.)

Consider the claim that if $h \neq 0$ and $b \neq 1$, then $\exp_b\left(h\right) \neq 1$. Suppose then that $\exp_b\left(h\right) = 1$. Then doing L_1 to both sides, $hL_1\left(b\right) = L_1\left(1\right) = 0$. Hence either $h = 0$ or $b = 1$ which are both excluded.

What of the laws of exponents for arbitrary values of x and y? As part of Wild Assumption 4.2, these were assumed to hold.

$$L_1\left(\exp_b\left(x + y\right)\right) = \left(x + y\right)L_1\left(b\right)$$

and

$$\begin{aligned} L_1\left(\exp_b\left(x\right)\exp_b\left(y\right)\right) &= L_1\left(\exp_b\left(x\right)\right) + L_1\left(\exp_b\left(y\right)\right) \\ &= xL_1\left(b\right) + yL_1\left(b\right) = \left(x + y\right)L_1\left(b\right). \end{aligned}$$

Since L_1 is one to one, this shows the first law of exponents holds,

$$\exp_b\left(x + y\right) = \exp_b\left(x\right)\exp_b\left(y\right).$$

From the definition, $\exp_b\left(1\right) = b$ and $\exp_b\left(0\right) = 1$. Therefore,

$$1 = \exp_b\left(1 + \left(-1\right)\right) = \exp_b\left(-1\right)\exp_b\left(1\right) = \exp_b\left(-1\right)b$$

showing that $\exp_b(-1) = b^{-1}$. Now

$$L_1(\exp_{ab}(x)) = xL_1(ab) = xL_1(a) + xL_1(b)$$

while

$$
\begin{aligned}
L_1(\exp_a(x)\exp_b(x)) &= L_1(\exp_a(x)) + L_1(\exp_b(x)) \\
&= xL_1(a) + xL_1(b).
\end{aligned}
$$

Again, since L_1 is one to one, $\exp_{ab}(x) = \exp_a(x)\exp_b(x)$. Finally,

$$L_1\left(\exp_{\exp_b(x)}(y)\right) = yL_1(\exp_b(x)) = yxL_1(b)$$

while

$$L_1(\exp_b(xy)) = xyL_1(b)$$

and so since L_1 is one to one, $\exp_{\exp_b(x)}(y) = \exp_b(xy)$. This establishes all the laws of exponents for arbitrary real values of the exponent.

$\exp'_b$ exists by Theorem 5.2 on Page 123. Therefore, $\exp_b$ defined in (10.24) satisfies all the conditions of Wild Assumption 4.2.

It remains to consider the derivative of $\exp_b$ and verify $L_1(b) = \ln b$. First,

$$L_1\left(L_1^{-1}(x)\right) = x$$

and so

$$L_1'\left(L_1^{-1}(x)\right)\left(L_1^{-1}\right)'(x) = 1.$$

By (10.19),

$$\frac{\left(L_1^{-1}\right)'(x)}{L_1^{-1}(x)} = 1$$

showing that $\left(L_1^{-1}\right)'(x) = \left(L_1^{-1}\right)(x)$. Now by the chain rule,

$$\exp'_b(x) = \left(L_1^{-1}\right)'(xL_1(b))L_1(b) = L_1^{-1}(xL_1(b))L_1(b) = \exp_b(x)L_1(b).$$

From the definition of $\ln b$ on Page 114, it follows $\ln = L_1$. ∎

10.9   Exercises

(1) Show $\ln 2 \in [.5, 1]$.
(2) Apply the trapezoid rule, (see Problem 5 on Page 254) to estimate $\ln 2$ in the case where $h = 1/5$. Now use a calculator or table to find the exact value of $\ln 2$.
(3) Suppose it is desired to find a function $L : (0, \infty) \to \mathbb{R}$ which satisfies

$$L(xy) = Lx + Ly, \quad L(1) = 0. \tag{10.25}$$

Show the only differentiable solutions to this equation are functions of the form $L_k(x) = \int_1^x \frac{k}{t}\, dt$. **Hint:** Fix $x > 0$ and differentiate both sides of the above equation with respect to y. Then let $y = 1$.

(4) Recall that $\ln e = 1$. In fact, this was how e was defined. Show that
$$\lim_{y \to 0+} (1 + yx)^{1/y} = e^x.$$
Hint: Consider $\ln (1 + yx)^{1/y} = \frac{1}{y} \ln (1 + yx) = \frac{1}{y} \int_1^{1+yx} \frac{1}{t} \, dt$, use upper and lower sums and then the squeezing theorem to verify $\ln (1 + yx)^{1/y} \to x$. Recall that $x \to e^x$ is continuous.

(5) Logarithms were invented before calculus, one of the inventors being Napier, a Scottish nobleman. His interest in logarithms was computational. Describe how one could use logarithms to find 7th roots for example. Also describe how one could use logarithms to do computations involving large numbers. Describe how you could construct a table for $\log_{10} x$ for x various numbers. Next, how do you suppose Napier did it? Remember, he did not have calculus. **Hint:** $\log_b \left(35^{1/7} \right) = \frac{1}{7} \log_b (35)$. You can find the logarithm of the number you are after. If you had a table of logarithms to some base, you could then see what number this corresponded to. Napier essentially found a table of logarithms to the base $b = .999999$ or some such number like that. The reason for the strange choice is that successive integer powers of this number are quite close to each other and the graph of the function b^x is close to a straight line between two successive integers. This is not true for a number like 10. Shortly after calculus was invented, it was realized that the best way to consider logarithms was in terms of integrals. This occurred in the 1600's.

10.10 Techniques Of Integration

The techniques for finding antiderivatives may be used to find integrals.

10.10.1 The Method Of Substitution

Recall
$$\int f(g(x)) g'(x) \, dx = F(g(x)) + C, \tag{10.26}$$
where $F'(y) = f(y)$.

How does this relate to finding definite integrals? This is based on the following formula in which all the functions are integrable and $F'(y) = f(y)$.
$$\int_a^b f(g(x)) g'(x) \, dx = \int_{g(a)}^{g(b)} f(y) \, dy. \tag{10.27}$$
This formula follows from the observation that, by the fundamental theorem of calculus, both sides equal $F(g(b)) - F(g(a))$.

How can you remember this? The easiest way is to use the Leibniz notation. In (10.27) let $y = g(x)$. Then
$$\frac{dy}{dx} = g'(x)$$

and so formally $dy = g'(x)\,dx$. Then making the substitution

$$\int_a^b \overbrace{f(g(x))}^{f(y)}\overbrace{g'(x)\,dx}^{dy} = \int_?^? f(y)\,dy.$$

What should go in as the top and bottom limits of the integral? The important thing to remember is that **if you change the variable, you must change the limits!** When $x = a$, it follows that $y = g(a)$ so the bottom limit must equal $g(a)$. Similarly the top limit should be $g(b)$.

Example 10.6. Find $\int_1^2 x\sin\left(x^2\right)\,dx$

Let $u = x^2$ so $du = 2x\,dx$ and so $\frac{du}{2} = x\,dx$. Therefore, changing the variables gives

$$\int_1^2 x\sin\left(x^2\right)\,dx = \frac{1}{2}\int_1^4 \sin(u)\,du = -\frac{1}{2}\cos(4) + \frac{1}{2}\cos(1)$$

Sometimes people prefer not to worry about the limits. This is fine provided you do not write anything which is false. The above problem can be done in the following way.

$$\int x\sin\left(x^2\right)\,dx = -\frac{1}{2}\cos\left(x^2\right) + C$$

and so from an application of the fundamental theorem of calculus

$$\int_1^2 x\sin\left(x^2\right)\,dx = -\frac{1}{2}\cos\left(x^2\right)\Big|_1^2$$

$$= -\frac{1}{2}\cos(4) + \frac{1}{2}\cos(1).$$

Example 10.7. Find the area of the ellipse

$$\frac{(y-\beta)^2}{b^2} + \frac{(x-\alpha)^2}{a^2} = 1.$$

If you sketch the ellipse, you see that it suffices to find the area of the top right quarter for $y \geq \beta$ and $x \geq \alpha$ and multiply by 4 since the bottom half is just a reflection of the top half about the line, $y = \beta$ and the left top quarter is just the reflection of the top right quarter reflected about the line, $x = \alpha$. Thus the area of the ellipse is

$$4\int_\alpha^{\alpha+a} b\sqrt{1 - \frac{(x-\alpha)^2}{a^2}}\,dx$$

Change the variables, letting $u = \frac{x-\alpha}{a}$. Then $du = \frac{1}{a}\,dx$ and so upon changing the limits to correspond to the new variables, this equals

$$4ba\overbrace{\int_0^1 \sqrt{1-u^2}\,du}^{\pi/4} = 4 \times ab \times \frac{1}{4}\pi = \pi ab$$

because the integral in the above is just one quarter of the unit circle and so has area equal to $\pi/4$.

10.10.2   Integration By Parts

Recall the following proposition for finding antiderivative.

Proposition 10.4. Let u and v be differentiable functions for which

$$\int u\left(x\right)v'\left(x\right)\,dx, \quad \int u'\left(x\right)v\left(x\right)\,dx$$

are nonempty. Then

$$uv - \int u'\left(x\right)v\left(x\right)\,dx = \int u\left(x\right)v'\left(x\right)\,dx. \tag{10.28}$$

In terms of integrals, this is stated in the following proposition.

Proposition 10.5. Let u and v be differentiable functions on $[a,b]$ such that uv', $u'v \in R\left(\left[a,b\right]\right)$. Then

$$\int_a^b u\left(x\right)v'\left(x\right)\,dx = uv\left(x\right)\big|_a^b - \int_a^b u'\left(x\right)v\left(x\right)\,dx \tag{10.29}$$

Proof: Use the product rule and properties of integrals to write

$$\int_a^b u\left(x\right)v'\left(x\right)\,dx = \int_a^b \left(uv\right)'\left(x\right)\,dx - \int_a^b u'\left(x\right)v\left(x\right)\,dx$$

$$= uv\left(x\right)\big|_a^b - \int_a^b u'\left(x\right)v\left(x\right)\,dx. \qquad \blacksquare$$

Example 10.8. Find $\int_0^\pi x\sin\left(x\right)\,dx$.

Let $u\left(x\right) = x$ and $v'\left(x\right) = \sin\left(x\right)$. Then applying (10.28),

$$\int_0^\pi x\sin\left(x\right)\,dx = \left(-\cos\left(x\right)\right)x\big|_0^\pi - \int_0^\pi \left(-\cos\left(x\right)\right)\,dx$$

$$= -\pi\cos\left(\pi\right) = \pi.$$

Example 10.9. Find $\int_0^1 xe^{2x}\,dx$.

Let $u\left(x\right) = x$ and $v'\left(x\right) = e^{2x}$. Then from (10.29)

$$\int_0^1 xe^{2x}\,dx = \frac{e^{2x}}{2}x\big|_0^1 - \int_0^1 \frac{e^{2x}}{2}\,dx = \frac{e^{2x}}{2}x\big|_0^1 - \frac{e^{2x}}{4}\big|_0^1$$

$$= \frac{e^2}{2} - \left(\frac{e^2}{4} - \frac{1}{4}\right) = \frac{e^2}{4} + \frac{1}{4}.$$

10.11   Exercises

(1) Find the integrals.

   (a) $\int_0^4 xe^{-3x}\,dx$

   (b) $\int_2^3 \frac{1}{x(\ln(|x|))^2}\,dx$

   (c) $\int_0^1 x\sqrt{2-x}\,dx$

   (d) $\int_2^3 (\ln|x|)^2\,dx$ **Hint:** Let $u(x) = (\ln|x|)^2$ and $v'(x) = 1$.

   (e) $\int_0^\pi x^3 \cos(x^2)\,dx$

(2) Find $\int_1^2 x\ln(x^2)\,dx$.

(3) Find $\int_0^1 e^x \sin(x)\,dx$.

(4) Find $\int_0^1 2^x \cos(x)\,dx$.

(5) Find $\int_0^2 x^3 \cos(x)\,dx$.

(6) Find the integrals.

   (a) $\int_5^6 \frac{x}{\sqrt{2x-3}}\,dx$

   (b) $\int_2^4 x\left(3x^2+6\right)^5\,dx$

   (c) $\int_0^\pi x\sin(x^2)\,dx$

   (d) $\int_0^{\pi/4} \sin^3(2x)\cos(2x)$

   (e) $\int_0^7 \frac{1}{\sqrt{1+4x^2}}\,dx$ **Hint:** Remember the $\sinh^{-1}$ function and its derivative.

(7) Find the integrals.

   (a) $\int_0^{\pi/9} \sec(3x)\,dx$

   (b) $\int_0^{\pi/9} \sec^2(3x)\tan(3x)\,dx$

   (c) $\int_0^5 \frac{1}{3+5x^2}\,dx$

   (d) $\int_0^1 \frac{1}{\sqrt{5-4x^2}}\,dx$

   (e) $\int_2^6 \frac{3}{x\sqrt{4x^2-5}}\,dx$

(8) Find the integrals.

   (a) $\int_0^3 x\cosh(x^2+1)\,dx$

   (b) $\int_0^2 x^3 5^{x^4}\,dx$

   (c) $\int_{-\pi}^\pi \sin(x)\,7^{\cos(x)}\,dx$

   (d) $\int_0^\pi x\sin(x^2)\,dx$

   (e) $\int_1^2 x^5\sqrt{2x^2+1}\,dx$ **Hint:** Let $u = 2x^2+1$.

(9) Find $\int_0^{\pi/4} \sin^2(x)\,dx$. **Hint:** Derive and use $\sin^2(x) = \frac{1-\cos(2x)}{2}$.

(10) Find the area between the graphs of $y = \sin(2x)$ and $y = \cos(2x)$ for $x \in [0, 2\pi]$.

(11) Find the following integrals.

   (a) $\int_1^\pi x^2 \sin(x^3)\,dx$

   (b) $\int_1^6 \frac{x}{1+x^2}\,dx$

(c) $\int_0^{.5} \frac{1}{1+4x^2} \, dx$

(d) $\int_1^4 x^2 \sqrt{1+x} \, dx$

(12) The most important of all differential equations is the first order linear equation $y' + p(t) y = f(t)$. Show the solution to the initial value problem consisting of this equation and the initial condition $y(a) = y_a$ is

$$y(t) = e^{-P(t)} y_a$$

$$+ e^{-P(t)} \int_a^t e^{P(s)} f(s) \, ds,$$

where $P(t) = \int_a^t p(s) \, ds$. Give conditions under which everything is correct. **Hint:** You use the integrating factor approach. Multiply both sides by $e^{P(t)}$, verify the left side equals

$$\frac{d}{dt} \left(e^{P(t)} y(t) \right),$$

and then take the integral $\int_a^t$ of both sides.

(13) Suppose f is a continuous function which is not equal to zero on $[0, b]$. Show that

$$\int_0^b \frac{f(x)}{f(x) + f(b-x)} \, dx = \frac{b}{2}.$$

Hint: First change the variables to obtain the integral equals

$$\int_{-b/2}^{b/2} \frac{f(y+b/2)}{f(y+b/2) + f(b/2 - y)} \, dy$$

Next show by another change of variables that this integral equals

$$\int_{-b/2}^{b/2} \frac{f(b/2 - y)}{f(y+b/2) + f(b/2 - y)} \, dy$$

Thus the sum of these equals b.

(14) Suppose $x_0 \in (a, b)$ and that f is a function which has $n+1$ continuous derivatives on this interval. Consider the following.

$$f(x) = f(x_0) + \int_{x_0}^x f'(t) \, dt = f(x_0) + (t-x) f'(t) \big|_{x_0}^x$$

$$+ \int_{x_0}^x (x-t) f''(t) \, dt = f(x_0) + f'(x_0)(x-x_0) + \int_{x_0}^x (x-t) f''(t) \, dt.$$

Explain the above steps and continue the process to eventually obtain Taylor's formula

$$f(x) = f(x_0) + \sum_{k=1}^{n} \frac{f^{(k)}(x_0)}{k!} (x - x_0)^k + \frac{1}{n!} \int_{x_0}^{x} (x - t)^n f^{(n+1)}(t) \, dt$$

where $n! \equiv n(n-1)\cdots 3 \cdot 2 \cdot 1$ if $n \geq 1$ and $0! \equiv 1$.

(15) In the above Taylor's formula, use Problem 12 on Page 256 to obtain the existence of some z between x_0 and x such that

$$f(x) = f(x_0) + \sum_{k=1}^{n} \frac{f^{(k)}(x_0)}{k!} (x - x_0)^k + \frac{f^{(n+1)}(z)}{(n+1)!} (x - x_0)^{n+1}.$$

Hint: You might consider two cases, the case when $x > x_0$ and the case when $x < x_0$.

(16) * There is a general procedure for constructing these methods of approximate integration like the trapezoid rule and Simpson's rule. Consider $[0, 1]$ and divide this interval into n pieces using a uniform partition $\{x_0, \cdots, x_n\}$ where $x_i - x_{i-1} = 1/n$ for each i. The approximate integration scheme for a function f, will be of the form

$$\left(\frac{1}{n}\right) \sum_{i=0}^{n} c_i f_i \approx \int_0^1 f(x) \, dx$$

where $f_i = f(x_i)$ and the constants, c_i are chosen in such a way that the above sum gives the exact answer for $\int_0^1 f(x) \, dx$ where $f(x) = 1, x, x^2, \cdots, x^n$. When this has been done, change variables to write

$$\int_a^b f(y) \, dy = (b - a) \int_0^1 f(a + (b - a)x) \, dx$$

$$\approx \frac{b - a}{n} \sum_{i=1}^{n} c_i f\left(a + (b - a)\left(\frac{i}{n}\right)\right) = \frac{b - a}{n} \sum_{i=1}^{n} c_i f_i$$

where $f_i = f\left(a + (b - a)\left(\frac{i}{n}\right)\right)$. Consider the case where $n = 1$. It is necessary to find constants c_0 and c_1 such that

$$c_0 + c_1 = 1 = \int_0^1 1 \, dx$$
$$0c_0 + c_1 = 1/2 = \int_0^1 x \, dx.$$

Show that $c_0 = c_1 = 1/2$, and that this yields the trapezoid rule. Next take $n = 2$ and show the above procedure yields Simpson's rule. Show also that if this integration scheme is applied to any polynomial of degree 3 the result will be exact. That is,

$$\frac{1}{2}\left(\frac{1}{3}f_0 + \frac{4}{3}f_1 + \frac{1}{3}f_2\right) = \int_0^1 f(x) \, dx$$

whenever $f(x)$ is a polynomial of degree three. Show that if f_i are the values of f at a, $\frac{a+b}{2}$, and b with $f_1 = f\left(\frac{a+b}{2}\right)$, it follows that the above formula gives $\int_a^b f(x) \, dx$ exactly whenever f is a polynomial of degree three. Obtain an integration scheme for $n = 3$.

(17) *Let f have four continuous derivatives on $[x_{i-1}, x_{i+1}]$ where $x_{i+1} = x_{i-1} + 2h$ and $x_i = x_{i-1} + h$. Show using Problem 15, there exists a polynomial of degree three $p_3(x)$, such that

$$f(x) = p_3(x) + \frac{1}{4!} f^{(4)}(\xi)(x - x_i)^4$$

Now conclude

$$\left| \int_{x_{i-1}}^{x_{i+1}} f(x)\, dx - \left(\frac{hf_{i-1}}{3} + \frac{hf_i 4}{3} + \frac{hf_{i+1}}{3} \right) \right| < \frac{10}{3} \frac{M}{4!} h^5,$$

where M satisfies, $M \geq \max\left\{ \left| f^{(4)}(t) \right| : t \in [x_{i-1}, x_i] \right\}$. Now let $S(a, b, f, 2m)$ denote the approximation to $\int_a^b f(x)\, dx$ obtained from Simpson's rule using $2m$ equally spaced points. Show

$$\left| \int_a^b f(x)\, dx - S(a, b, f, 2m) \right| < C(b-a)^5 \frac{1}{m^4}$$

where C is some constant depending on $M \equiv \max\left\{ \left| f^{(4)}(t) \right| : t \in [a, b] \right\}$. Consider the first inequality. The polynomial $p_3(x)$ is the Taylor polynomial of degree 3 for f expanded about x_1. It follows that

$$|p_3(x_2) - f_2| < \frac{M}{4!} h^4, \quad |p_3(x_0) - f_1| < \frac{M}{4!} h^4$$

Also, since the polynomial is of degree 3,

$$\int_{x_{i-1}}^{x_{i+1}} p_3(x)\, dx = \frac{1}{3} p_3(x_0) + \frac{4}{3} p_3(x_1) + \frac{1}{3} p_3(x_2)$$

Now $p_3(x_1) = f(x_1)$ and so

$$\left| \frac{1}{3} p_3(x_0) + \frac{4}{3} p_3(x_1) + \frac{1}{3} p_3(x_2) - \left(\frac{1}{3} f_0 + \frac{4}{3} f_1 + \frac{1}{3} f_2 \right) \right| \leq \frac{2}{3} \frac{M}{4!} h^4$$

Hence

$$\left| \int_{x_{i-1}}^{x_{i+1}} f(x)\, dx - \left(\frac{hf_{i-1}}{3} + \frac{hf_i 4}{3} + \frac{hf_{i+1}}{3} \right) \right| \leq \int_{x_{i-1}}^{x_{i+1}} \frac{5}{3} \frac{M}{4!} h^4 dx = \frac{10}{3} \frac{M}{4!} h^5$$

Thus

$$\left| \int_a^b f(x)\, dx - S(a, b, f, 2m) \right|$$

$$\leq \sum_{k=1}^{m} \left| \int_{x_{k-1}}^{x_{k+1}} f(x)\, dx - \left(\frac{hf_{i-1}}{3} + \frac{hf_i 4}{3} + \frac{hf_{i+1}}{3} \right) \right|$$

$$\leq \sum_{k=1}^{m} \frac{10}{3} \frac{M}{4!} h^5 = \frac{10}{3} \frac{M}{4!} m \left(\frac{b-a}{m} \right)^5$$

$$= \frac{5}{36} \frac{M}{m^4} (b-a)^5 \leq CM(b-a)^5 \frac{1}{m^4}.$$

(18) * A **regular Sturm Liouville problem** involves the differential equation for an unknown function of x which is denoted here by y,

$$(p(x) y')' + (\lambda q(x) + r(x)) y = 0, \ x \in [a, b]$$

and it is assumed that $p(t), q(t) > 0$ for any t along with boundary conditions,

$$C_1 y(a) + C_2 y'(a) = 0$$
$$C_3 y(b) + C_4 y'(b) = 0$$

where

$$C_1^2 + C_2^2 > 0, \text{ and } C_3^2 + C_4^2 > 0.$$

There is an immense theory connected to these important problems. The constant λ is called an eigenvalue. Show that if y is a solution to the above problem corresponding to $\lambda = \lambda_1$ and if z is a solution corresponding to $\lambda = \lambda_2 \neq \lambda_1$, then

$$\int_a^b q(x) y(x) z(x) \, dx = 0. \tag{10.30}$$

Hint: Do something like this:

$$(p(x) y')' z + (\lambda_1 q(x) + r(x)) yz = 0,$$

$$(p(x) z')' y + (\lambda_2 q(x) + r(x)) zy = 0.$$

Now subtract and either use integration by parts or show

$$(p(x) y')' z - (p(x) z')' y = ((p(x) y') z - (p(x) z') y)'$$

and then integrate. Use the boundary conditions to show that $y'(a) z(a) - z'(a) y(a) = 0$ and $y'(b) z(b) - z'(b) y(b) = 0$. The formula (10.30) is called an orthogonality relation and it makes possible an expansion in terms of certain functions called eigenfunctions.

(19) Letting $[a, b] = [-\pi, \pi]$, consider an example of a regular Sturm Liouville problem which is of the form

$$y'' + \lambda y = 0, \ y(-\pi) = 0, \ y(\pi) = 0.$$

Show that if $\lambda = n^2$ and $y_n(x) = \sin(nx)$ for n a positive integer, then y_n is a solution to this regular Sturm Liouville problem. In this case, $q(x) = 1$ and so from Problem 18, it must be the case that

$$\int_{-\pi}^{\pi} \sin(nx) \sin(mx) \, dx = 0$$

if $n \neq m$. Show directly using integration by parts that the above equation is true.

(20) Find $\int_0^{\sqrt{2}/2} \frac{x}{\sqrt{1-x^4}} \, dx$.

10.12   Improper Integrals

The integral is only defined for certain bounded functions which are defined on closed and bounded intervals. Nevertheless people do consider things like the following: $\int_0^\infty f(t)\, dt$. Whenever things like this occur they require a special definition. They are called **improper integrals**. In this section a few types of improper integrals will be discussed.

Definition 10.7. The symbol $\int_a^\infty f(t)\, dt$ is defined to equal

$$\lim_{R\to\infty} \int_a^R f(t)\, dt$$

whenever this limit exists. If $\lim_{x\to a+} f(t) = \pm\infty$ but f is integrable on $[a+\delta, b]$ for all small δ, then

$$\int_a^b f(t)\, dt \equiv \lim_{\delta\to 0+} \int_{a+\delta}^b f(t)\, dt$$

whenever this limit exists. Similarly, if $\lim_{x\to b-} f(t) = \pm\infty$ but f is integrable on $[a, b-\delta]$ for all small δ, then

$$\int_a^b f(t)\, dt \equiv \lim_{\delta\to 0+} \int_a^{b-\delta} f(t)\, dt$$

whenever the limit exists. Finally, if $\lim_{x\to a+} f(t) = \pm\infty$, then

$$\int_a^\infty f(t)\, dt \equiv \lim_{R\to\infty} \int_a^R f(t)\, dt$$

where the improper integral $\int_a^R f(t)\, dt$ is defined above as $\lim_{\delta\to 0+} \int_{a+\delta}^R f(t)\, dt$.

You can probably construct other examples of improper integrals such as integrals of the form $\int_{-\infty}^a f(t)\, dt$. The definitions are analogous to the above.

Example 10.10. Find $\int_0^\infty e^{-t}\, dt$.

From the definition, this equals $\lim_{R\to\infty} \int_0^R e^{-t}\, dt = \lim_{R\to\infty} \left(1 - e^{-R}\right) = 1$.

Example 10.11. Find $\int_0^1 \frac{1}{\sqrt{x}}\, dx$.

From the definition, this equals $\lim_{\delta\to 0+} \int_\delta^1 x^{-1/2}\, dx = \lim_{\delta\to 0+} \left(2 - 2\sqrt{\delta}\right) = 2$.

Sometimes you can argue the improper integral exists even though you cannot find it. The following theorem is about this question of existence.

Theorem 10.15. *Suppose* $f(t) \geq 0$ *for all* $t \in [a, \infty)$ *and that* $f \in R([a+\delta, R])$ *whenever* $\delta > 0$ *is small enough, for every* $R > a$. *Suppose also there exists a number* M *such that for all* $R > a$, *the integral* $\int_a^R f(t)\, dt$ *exists and*

$$\int_a^R f(t)\, dt \leq M. \tag{10.31}$$

Then $\int_a^\infty f(t)\, dt$ exists.

If $f(t) \geq 0$ for all $t \in (a, b]$ and there exists M such that

$$\int_{a+\delta}^b f(t)\, dt \leq M \tag{10.32}$$

for all $\delta > 0$, then $\int_a^b f(t)\, dt$ exists.

If $f(t) \geq 0$ for all $t \in [a, b)$ and there exists M such that

$$\int_a^{b-\delta} f(t)\, dt \leq M \tag{10.33}$$

for all $\delta > 0$, then $\int_a^b f(t)\, dt$ exists.

Proof: Suppose (10.31). Then $I \equiv \text{lub}\left\{\int_a^R f(t)\, dt : R > a\right\} \leq M$. It follows that if $\varepsilon > 0$ is given, there exists R_0 such that $\int_a^{R_0} f(t)\, dt \in (I - \varepsilon, I]$. Then, since $f(t) \geq 0$, it follows that for $R \geq R_0$, and small $\delta > 0$,

$$\int_{a+\delta}^R f(t)\, dt = \int_{a+\delta}^{R_0} f(t)\, dt + \int_{R_0}^R f(t)\, dt.$$

Letting $\delta \to 0+$,

$$\int_a^R f(t)\, dt = \int_a^{R_0} f(t)\, dt + \int_{R_0}^R f(t)\, dt \geq \int_a^{R_0} f(t)\, dt.$$

Therefore, whenever $R > R_0, \left|\int_a^R f(t)\, dt - I\right| < \varepsilon$. Since ε is arbitrary, the conditions for

$$I = \lim_{R \to \infty} \int_a^R f(t)\, dt$$

are satisfied and so $I = \int_a^\infty f(t)\, dt$.

Now suppose (10.32). Then $I \equiv \sup\left\{\int_{a+\delta}^b f(t)\, dt : \delta > 0\right\} \leq M$. It follows that if $\varepsilon > 0$ is given, there exists $\delta_0 > 0$ such that

$$\int_{a+\delta_0}^b f(t)\, dt \in (I - \varepsilon, I].$$

Therefore, if $\delta < \delta_0$,

$$I - \varepsilon < \int_{a+\delta_0}^b f(t)\, dt \leq \int_{a+\delta}^b f(t)\, dt \leq I$$

showing that for such $\delta, \left|\int_{a+\delta}^b f(t)\, dt - I\right| < \varepsilon$. This is what is meant by the expression

$$\lim_{\delta \to 0+} \int_{a+\delta}^b f(t)\, dt = I$$

and so $I = \int_a^b f(t)\, dt$.

The last case is entirely similar to this one. ∎

There is an easy way to use known convergence of some integrals to verify convergence of others. It is called the limit comparison test.

Theorem 10.16. *Suppose* $f(x), g(x) \geq 0$ *for all* $x \geq a$ *and* f, g *are Riemann integrable on every interval of the form* $[a, b]$. *Suppose also that*

$$\lim_{x \to \infty} \frac{f(x)}{g(x)} = L$$

where L *is a positive number. Then* $\int_a^\infty f(x)\, dx$ *and* $\int_a^\infty g(x)$ *converge or diverge together. In case the functions may be unbounded at* a, *then if*

$$\lim_{x \to a+} \frac{f(x)}{g(x)} = L$$

where L *is a positive number, then* $\int_a^b f(x)\, dx$ *and* $\int_a^b g(x)\, dx$ *converge or diverge together. The same can be said if the functions may be unbounded at* b *except in this case, the condition is*

$$\lim_{x \to b-} \frac{f(x)}{g(x)} = L.$$

Proof: Suppose the limit condition and that $\int_a^\infty f(x)\, dx$ converges. Then from the limit condition, it follows that there exists R_1 such that if $x > R_1$, then

$$\frac{f(x)}{g(x)} > \frac{L}{2}$$

Therefore, if $R > R_1$,

$$\int_a^R g(x)\, dx \leq \int_a^{R_1} g(x)\, dx + \frac{2}{L}\int_{R_1}^R f(x)\, dx$$

$$\leq \int_a^{R_1} g(x)\, dx + \frac{2}{L}\int_a^\infty f(x)\, dx \equiv M$$

Therefore, from Theorem 10.15 $\int_a^\infty g(x)\, dx$ exists. Next suppose that $\int_a^\infty f(x)\, dx$ diverges. Since $f(x) \geq 0$, this can only happen if $\lim_{R \to \infty} \int_a^R f(x)\, dx = \infty$. There exists R_1 such that if $x > R_1$, then

$$\frac{f(x)}{g(x)} < 2L$$

This must occur because of the limit condition. Therefore, for all $R > R_1$,

$$\int_{R_1}^R f(x)\, dx \leq 2L \int_{R_1}^R g(x)\, dx$$

and so, since $\lim_{R \to \infty} \int_{R_1}^R f(x)\, dx = \infty$, the same must be true of $\int_{R_1}^R g(x)\, dx$. The proof of the last assertion is similar. For $\delta > 0$ small enough, then for $\varepsilon < \delta$

$$\frac{L}{2}\int_{a+\varepsilon}^{a+\delta} g(x)\, dx \leq \int_{a+\varepsilon}^{a+\delta} f(x)\, dx < 2L \int_{a+\varepsilon}^{a+\delta} g(x)\, dx$$

and so $\int_a^{a+\delta} f(x)\, dx$ and $\int_a^{a+\delta} g(x)\, dx$ have the same convergence properties. The case where the functions are unbounded at b is similar. ∎

Example 10.12. Does $\int_1^\infty \sin\left(\frac{1}{x^2}\right) dx$ exist?

Yes, it does. To see this note $\lim_{x\to\infty} \sin\left(\frac{1}{x^2}\right)/1/x^2 = 1$. Therefore, from the limit comparison test, this integral exists because $\int_1^\infty \frac{1}{x^2} dx$ exists. Note I had absolutely no way of finding the antiderivative of $\sin\left(\frac{1}{x^2}\right)$.

Example 10.13. Does $\int_0^1 \frac{1}{\sqrt{\sin x}} dx$ exist?

I do not know how to find an antiderivative for this function but the question of existence can still be resolved. Since $\lim_{x\to 0+} \frac{x}{\sin x} = 1$, it follows that for x small enough, $\frac{x}{\sin x} < \frac{3}{2}$, say for $x < \delta_1$. Then for such x, it follows

$$\frac{2}{3} x < \sin x$$

and so if $\delta < \delta_1$,

$$\int_\delta^1 \frac{1}{\sqrt{\sin x}} dx \leq \int_\delta^{\delta_1} \sqrt{\frac{3}{2}} \frac{1}{\sqrt{x}} dx + \int_{\delta_1}^1 \frac{1}{\sqrt{\sin \delta_1}} dx.$$

Now using the argument of Example 10.11, the first integral in the above is bounded above by $\left(\sqrt{\delta_1} - \sqrt{\delta}\right)\sqrt{6}$. The second integral equals $\frac{1-\delta_1}{\sqrt{\sin \delta_1}}$. Therefore, the improper integral exists because the conditions of Theorem 10.15 with

$$M = \frac{1 - \delta_1}{\sqrt{\sin \delta_1}} + \left(\sqrt{\delta_1} - \sqrt{\delta}\right)\sqrt{6}.$$

Example 10.14. The gamma function is defined by $\Gamma(\alpha) \equiv \int_0^\infty e^{-t} t^{\alpha-1} dt$ whenever $\alpha > 0$. Does the improper integral exist?

You should supply the details to the following estimate in which δ is a small positive number less than 1 and R is a large positive number.

$$\int_\delta^R e^{-t} t^{\alpha-1} dt \leq \int_\delta^k e^{-t} t^{\alpha-1} dt + \int_k^R e^{-t} t^{\alpha-1} dt$$

$$\leq \int_0^k t^{\alpha-1} dt + \int_k^\infty e^{-t/2} dt.$$

Here k is chosen such that if $t \geq k$,

$$e^{-t} t^{\alpha-1} < e^{-t/2}.$$

Such a k exists because

$$\lim_{t\to\infty} \frac{e^{-t} t^{\alpha-1}}{e^{-t/2}} = 0.$$

Therefore, let $M \equiv \int_0^k t^{\alpha-1} dt + \int_k^\infty e^{-t/2} dt$ and this shows from the above theorem that

$$\int_0^R e^{-t} t^{\alpha-1} dt \leq M$$

for all large R and so $\int_0^\infty e^{-t} t^{\alpha-1} dt$ exists.

Sometimes the existence of the improper integral is a little more subtle. This is the case when functions are not all the same sign for example.

Example 10.15. Does $\int_0^\infty \frac{\sin x}{x} dx$ exist?

You should verify $\int_0^1 \frac{\sin x}{x} dx$ exists and that

$$\int_0^R \frac{\sin x}{x} dx = \int_0^1 \frac{\sin x}{x} dx + \int_1^R \frac{\sin x}{x} dx$$

$$= \int_0^1 \frac{\sin x}{x} dx + \cos 1 - \frac{\cos R}{R} - \int_1^R \frac{\cos x}{x^2} dx.$$

Thus the improper integral exists if it can be shown that $\int_1^\infty \frac{\cos x}{x^2} dx$ exists. However,

$$\int_1^R \frac{\cos x}{x^2} dx = \int_1^R \frac{\cos x + |\cos x|}{x^2} dx - \int_1^R \frac{(|\cos x| - \cos x)}{x^2} dx \qquad (10.34)$$

and

$$\int_1^R \frac{\cos x + |\cos x|}{x^2} dx \leq \int_1^R \frac{2}{x^2} dx \leq \int_0^\infty \frac{2}{x^2} dx < \infty$$

$$\int_1^R \frac{(|\cos x| - \cos x)}{x^2} dx \leq \int_1^R \frac{2}{x^2} dx \leq \int_0^\infty \frac{2}{x^2} dx < \infty.$$

Since both integrands are positive, Theorem 10.15 applies and the limits

$$\lim_{R \to \infty} \int_1^R \frac{(|\cos x| - \cos x)}{x^2} dx, \quad \lim_{R \to \infty} \int_1^R \frac{\cos x + |\cos x|}{x^2} dx$$

both exist and so from (10.34) $\lim_{R \to \infty} \int_1^R \frac{\cos x}{x^2} dx$ also exists and so $\int_0^\infty \frac{\sin x}{x} dx$ exists.

This is an important example. There are at least two ways to show that $\int_0^\infty \frac{\sin x}{x} dx = \frac{1}{2}\pi$. However, they involve techniques which will not be discussed in this book. It is a standard problem in the subject of complex analysis. The above argument is a special case of the following corollary to Theorem 10.15.

Definition 10.8. Let f be a real valued function. Then $f^+(t) \equiv \frac{|f(t)| + f(t)}{2}$ and $f^-(t) \equiv \frac{|f(t)| - f(t)}{2}$. Thus $|f(t)| = f^+(t) + f^-(t)$ and $f(t) = f^+(t) - f^-(t)$ while both f^+ and f^- are nonnegative functions.

Corollary 10.3. *Suppose f is a real valued function, Riemann integrable on every finite interval, and the conditions of Theorem 10.15 hold for both f^+ and f^-. Then $\int_0^\infty f(t) dt$ exists.*

Corollary 10.4. *Suppose f is a real valued function, Riemann integrable on every finite interval, and $\int_0^\infty |f(x)| dx$ exists. Then $\int_0^\infty f(x) dx$ exists.*

Proof: $0 \leq f^+(x) \leq |f(x)|$, and $0 \leq f^-(x) \leq |f(x)|$ and so for every $R > 0$,

$$\int_0^R f^+(x) dx \leq \int_0^\infty |f(x)| dx$$

and

$$\int_0^R f^-(x)\, dx \leq \int_0^\infty |f(x)|\, dx$$

and so it follows from Theorem 10.15 that $\int_0^\infty f^+(x)\, dx$ and $\int_0^\infty f^-(x)\, dx$ both exist. Therefore,

$$\int_0^\infty f(x)\, dx \equiv \lim_{R\to\infty} \left(\int_0^R f^+(x)\, dx - \int_0^R f^-(x)\, dx \right)$$

also exists. ∎

Example 10.16. Does $\int_0^\infty \cos(x^2)\, dx$ exist?

This is called a **Fresnel integral** and it has also been evaluated exactly using techniques from complex analysis. In fact $\int_0^\infty \cos(x^2)\, dx = \frac{1}{4}\sqrt{2}\sqrt{\pi}$. The verification that this integral exists is left to you. First change the variable letting $x^2 = u$ and then integrate by parts. You will eventually get an integral of the form $\int_0^\infty \frac{\sin u}{u^{3/2}}\, du$. Now consider $\int_\delta^1 \frac{\sin u}{u^{3/2}}\, du$ where δ is a small positive number. On $[\delta, 1]$, $\sin u$ is nonnegative. You also know that $\sin u \leq u$. Therefore, $\frac{\sin u}{u^{3/2}} \leq \frac{1}{u^{1/2}}$. Thus

$$\int_\delta^1 \frac{\sin u}{u^{3/2}}\, du \leq \int_\delta^1 \frac{1}{u^{1/2}}\, du = 2 - 2\delta^{1/2} \leq 2.$$

Therefore, $\int_0^1 \frac{\sin u}{u^{3/2}}\, du$ exists. Similarly, $\int_0^R \frac{\sin u}{u^{3/2}}\, du$ exists. Now

$$\int_0^R \frac{|\sin u|}{u^{3/2}} \leq \int_0^1 \frac{|\sin u|}{u^{3/2}}\, du + \int_1^R \frac{1}{u^{3/2}}\, du \leq 2 + 2$$

and so $\int_0^\infty \frac{|\sin u|}{u^{3/2}}\, du$ exists which shows by Corollary 10.4 that $\int_0^\infty \frac{\sin u}{u^{3/2}}\, du$ exists also. I have been a little sketchy on the details. You finish them. This is an especially exciting example because $\lim_{x\to\infty} \cos(x^2)$ does not exist. This is a contrast to the situation for infinite sums discussed later.

10.13   Exercises

(1) Verify all the details in Example 10.14.
(2) Verify all the details of Example 10.15.
(3) Verify all the details of Example 10.16.
(4) Find the values of p for which $\int_1^\infty \frac{1}{t^p}\, dt$ exists and compute the integral when it does exist.
(5) Find the values of p for which $\int_2^\infty \frac{1}{t(\ln t)^p}\, dt$ exists and compute the integral when it does exist.
(6) Determine whether $\int_1^\infty \frac{\sin t}{\sqrt{t}}\, dt$ exists.
(7) Determine whether $\int_3^\infty \frac{\sin t}{\ln t}\, dt$ exists. **Hint:** You might try integrating by parts.

(8) Determine whether $\int_0^1 \frac{1}{\sqrt{x}+\sin x} dx$ exists.

(9) Determine whether $\int_1^\infty \frac{1}{\sqrt{x+x^5}} dx$ exists.

(10) Determine whether $\int_0^1 \frac{\sin t}{t} dt$ exists.

(11) Determine whether $\int_0^1 \frac{\cot t}{t} dt$ exists.

(12) Determine whether $\int_0^1 \frac{1}{1-t^3} dt$ exists.

(13) Determine whether $\int_0^1 \frac{1}{1-\sqrt{x}} dx$ exists.

(14) Determine whether $\int_0^1 \frac{1}{\sqrt{1-x}} dx$ exists.

(15) Find $\int_0^1 \frac{1}{\sqrt[3]{1-x}} dx$ if it exists.

(16) Find $\int_0^{\pi/2} \frac{\cos x}{\sqrt{1-\sin x}} dx$ if it exists.

(17) Find $\int_0^\infty \frac{1}{4x^2+9} dx$ if it exists.

(18) Define and find $\int_{-\infty}^0 e^x dx$. Note the lower limit of integration is $-\infty$.

(19) Find $\int_0^\infty e^{-x} dx$ and then define and find the volume obtained by revolving the graph of $y = e^{-x}$ about the x axis. Define the surface area of the shape obtained by revolving about the x axis and determine whether it is finite.

(20) When $\int_0^\infty f(x) \, dx$ and $\int_{-\infty}^0 f(x) \, dx$ both exist, it follows $\int_{-\infty}^\infty f(x) \, dx$ also exists and equals the sum of the two first integrals,

$$\int_{-\infty}^\infty f(x) \, dx = \int_{-\infty}^0 f(x) \, dx + \int_0^\infty f(x) \, dx.$$

The normal distribution function is $\frac{1}{\sqrt{2\pi}\sigma} e^{-\frac{(x-\mu)^2}{2\sigma^2}}$. In this formula, μ is the mean and σ is a positive number called the standard deviation. In statistics, there are things called random variables. These are really just a kind of function and one of these random variables is said to be normally distributed if the probability that it has a value between a and b is given by the integral $\int_a^b \frac{1}{\sqrt{2\pi}\sigma} e^{-\frac{(x-\mu)^2}{2\sigma^2}} \, dx$. You may be wondering why $\sqrt{2\pi}$ occurs in this. It is because it is what is needed to have $\int_{-\infty}^\infty \frac{1}{\sqrt{2\pi}\sigma} e^{-\frac{(x-\mu)^2}{2\sigma^2}} \, dx = 1$. Later in the book you will learn how to show this. For now, show $\int_{-\infty}^\infty \frac{1}{\sqrt{2\pi}\sigma} e^{-\frac{(x-\mu)^2}{2\sigma^2}} \, dx$ exists. The importance of this distribution function cannot be over stated.

(21) Show $\Gamma(1) = 1 = \Gamma(2)$. Next show $\Gamma(\alpha+1) = \alpha\Gamma(\alpha)$ and prove that for n a nonnegative integer, $\Gamma(n+1) = n!$.

(22) It can be shown that $\Gamma\left(\frac{1}{2}\right) = \sqrt{\pi}$. Using this and Problem 21, find $\Gamma\left(\frac{5}{2}\right)$. What is the advantage of the gamma function over the notion of factorials? **Hint:** For what values of x is $\Gamma(x)$ defined?

(23) Prove $\int_0^\infty \sin(x^2) \, dx$ exists. This is also called a Fresnel integral.

(24) For $\alpha > 0$ find $\int_0^1 t^\alpha \ln(t) \, dt$ if it exists and if it does not exist, explain why.

(25) Prove that for every $\alpha > 0$, $\int_0^1 t^{\alpha-1} dt$ exists and find the answer.

(26) Prove that for every $\alpha > 0$,

$$\int_0^1 (\tan t)^{\alpha-1}\, dt$$

exists.

(27) Prove that for every $\alpha > 0$,

$$\int_0^1 (\sin t)^{\alpha-1}\, dt$$

exists.

(28) Recall the area of the surface obtained by revolving the graph of $y = f(x)$ about the x axis for $x \in [a, b]$ for f a positive continuous function having continuous derivative is given by

$$2\pi \int_a^b f(x) \sqrt{1 + (f'(x))^2}\, dx.$$

Also the volume of the solid obtained in the same way is given by the integral

$$\pi \int_a^b (f(x))^2\, dx.$$

It seems reasonable to define the surface area and volume for $x \in [a, \infty)$ in terms of an improper integral. Try it on the function $f(x) = 1/x$ for $x \geq 1$. Show the resulting solid has finite volume but infinite surface area. Thus you could fill it but you could not paint it. Sometimes people call this Gabriel's horn.

(29) Show $\int_0^\infty \frac{2x}{1+x^2}\, dx$ does not exist but that $\lim_{R\to\infty} \int_{-R}^R \frac{2x}{1+x^2}\, dx = 0$. This last limit is called the Cauchy principle value integral. It is not a very respectable thing. Later you might study a subject called complex analysis in which techniques for finding hard integrals are developed. These methods often give the Cauchy principle value. Try to show the following: For every number A, there exist sequences $a_n, b_n \to \infty$ such that

$$\lim_{n\to\infty} \int_{-a_n}^{b_n} \frac{2x}{1+x^2}\, dx = A.$$

This is true and shows why such principle value integrals are somewhat disreputable.

(30) Suppose f is a continuous function which is bounded and defined on $\mathbb{R}$. Show

$$\int_{-\infty}^{\infty} \frac{\varepsilon}{\pi \left(\varepsilon^2 + (x - x_1)^2\right)}\, dx = 1.$$

Next show that

$$\lim_{\varepsilon\to 0+} \int_{-\infty}^{\infty} \frac{\varepsilon f(x)}{\pi \left(\varepsilon^2 + (x - x_1)^2\right)}\, dx$$

$$= f(x_1).$$

(31) The ideas in these next two problems come from Wooley, who is an engineer I met. He was interested in nonlinear springs and he devised this model for describing collisions. It is a very interesting application of improper integrals. Recall the equation of undamped oscillation,

$$y'' + \beta y = 0$$

where $\beta > 0$ is related to a spring constant. This describes a linear spring. Of course there is no such thing. It is a mathematical idealization which is convenient to use because the solutions involve known functions sin and cos. Wooley proposed the following equation for a nonlinear spring,

$$y'' + \beta y^\alpha = 0$$

where $\alpha \in (0, 1]$ and $\beta > 0$. As part of the analysis, it was interesting to find β such that the solution to

$$y'' + \beta y^\alpha \ = \ 0 \tag{10.35}$$
$$y(0) \ = \ 0, y'(0) = 1 \tag{10.36}$$

also has the property that $y'(1) = 0$ and $y'(t) > 0$ on $(0, 1)$. Multiply the differential equation by y' and then show that

$$\left(\frac{(y')^2}{2} + \frac{\beta y^{\alpha+1}}{\alpha + 1} \right)' = 0.$$

Then use the conditions given to conclude

$$(y')^2 + \frac{2\beta y^{\alpha+1}}{\alpha + 1} = 1 \tag{10.37}$$

(32) ↑ Now argue that from this it follows

$$\frac{dy}{dt} = \sqrt{1 - \frac{2\beta y^{\alpha+1}}{\alpha + 1}}$$

and this is a separable differential equation. Now explain why

$$\int_0^y \frac{dz}{\sqrt{1 - \frac{2\beta z^{\alpha+1}}{\alpha+1}}} = t \tag{10.38}$$

for $t \subset (0, 1)$. Now change the variables, letting

$$w = \frac{2\beta z^{\alpha+1}}{\alpha + 1}$$

Show (10.38) reduces to

$$\int_0^{\frac{2\beta y^{\alpha+1}}{\gamma}} \frac{dw}{2\beta \left(\left(\frac{\gamma w}{2\beta} \right)^{1/\gamma} \right)^\alpha \sqrt{1 - w}} = t$$

where $\gamma = \alpha + 1$, which simplifies to

$$\frac{1}{2^{\frac{1}{\alpha+1}} \beta^{\frac{1}{\alpha+1}} (\alpha+1)^{\frac{\alpha}{\alpha+1}}} \int_0^{\frac{2\beta y^{\alpha+1}}{\alpha+1}} \frac{dw}{w^{\frac{\alpha}{\alpha+1}} \sqrt{1-w}} = t$$

Now explain how (10.37) and the condition that $y'(1) = 0$ implies

$$\frac{1}{2^{\frac{1}{\alpha+1}} \beta^{\frac{1}{\alpha+1}} (\alpha+1)^{\frac{\alpha}{\alpha+1}}} \int_0^1 \frac{dw}{w^{\frac{\alpha}{\alpha+1}} \sqrt{1-w}} = 1$$

Explain why this improper integral converges and solve for β to find

$$\left(\frac{1}{(\alpha+1)^{\frac{\alpha}{\alpha+1}} 2^{\frac{1}{\alpha+1}}} \int_0^1 \frac{dw}{w^{\frac{\alpha}{\alpha+1}} \sqrt{1-w}} \right)^{\alpha+1} = \beta.$$

What happens in the case that $\alpha = 1$? What should be the correct value of β in this case? If you know β, then there are well known numerical methods which will allow you to find the solution to the initial value problem. Note that the correct value of β involves the computation of an improper integral.

(33) For $x > 0$, let $g(x) = \int_0^1 \frac{e^{-x^2(1+t^2)}}{1+t^2} dt$. Show that $g'(x) = \int_0^1 -2xe^{-x^2(1+t^2)} dt$.

(34) ↑For $x > 0$, let $f(x) = \left(\int_0^x e^{-t^2} dt \right)^2$. Show that

$$f'(x) = 2e^{-x^2} \int_0^x e^{-t^2} dt = 2e^{-x^2} \int_0^1 e^{-t^2 x^2} x dt.$$

(35) ↑Show that $\int_0^\infty e^{-t^2} dt = \frac{\sqrt{\pi}}{2}$. The method for showing this, I found in the advanced calculus book by Apostol. **Hint:** Let $h(x) = f(x) + g(x)$ where $f(x)$ and $g(x)$ are given above. Then show $h'(x) = 2e^{-x^2} \int_0^1 e^{-t^2 x^2} x dt + \left(\int_0^1 -2xe^{-x^2(1+t^2)} dt \right) = 0$ so $h(x)$ is a constant. What is this constant? We can find it by letting $x = 0$. Do so, and then evaluate the resulting integral. Next take a limit as $x \to \infty$.

(36) ↑Show $\Gamma\left(\frac{1}{2}\right) = \sqrt{\pi}$. Recall $\Gamma(\alpha) \equiv \int_0^\infty e^{-t} t^{\alpha-1} dt$.

Chapter 11

Infinite Series

11.1 Approximation By Taylor Polynomials

By now, you have noticed there are two sorts of functions, those which come from a formula like $f(x) = x^2 + 2$ which are easy to evaluate by following a simple procedure, and those which come as short words; things like $\ln(x)$ or $\sin(x)$. This latter type of function is not so easy to evaluate. For example, what is $\sin 2$? Can you get it by doing a simple sequence of operations like you can with $f(x) = x^2 + 2$? How can you find $\sin 2$? It turns out there are many ways to do so. In this section, the method of Taylor polynomials is discussed. The following theorem is called Taylor's theorem. Before presenting it, recall the meaning of $n!$ for n a positive integer. Define $0! \equiv 1 = 1!$ and $(n+1)! \equiv (n+1) n!$ so that $n! = n(n-1)\cdots 1$. In particular, $2! - 2, 3! - 3 \times 2! = 6, 4! = 4 \times 3! = 24$, etc. A version of the following theorem is due to Lagrange, about 1790.

Theorem 11.1. *Suppose f has $n+1$ derivatives on an interval (a, b) and let $c \in (a, b)$. Then if $x \in (a, b)$, there exists ξ between c and x such that*

$$f(x) = f(c) + \sum_{k=1}^{n} \frac{f^{(k)}(c)}{k!}(x-c)^k + \frac{f^{(n+1)}(\xi)}{(n+1)!}(x-c)^{n+1}.$$

(In this formula, the symbol $\sum_{k=1}^{0} a_k$ will denote the number 0.)

Proof: If $n = 0$ then the theorem is true because it is just the mean value theorem. Suppose the theorem is true for $n-1, n \geq 1$. It can be assumed $x \neq c$ because if $x = c$ there is nothing to show. Then there exists K such that

$$f(x) - \left(f(c) + \sum_{k=1}^{n} \frac{f^{(k)}(c)}{k!}(x-c)^k + K(x-c)^{n+1} \right) = 0 \qquad (11.1)$$

In fact,

$$K = \frac{-f(x) + \left(f(c) + \sum_{k=1}^{n} \frac{f^{(k)}(c)}{k!}(x-c)^k \right)}{(x-c)^{n+1}}.$$

It remains to find K. Define $F(t)$ for t in the closed interval determined by x and c by

$$F(t) \equiv f(x) - \left(f(t) + \sum_{k=1}^{n} \frac{f^{(k)}(c)}{k!} (x-t)^k + K(x-t)^{n+1} \right).$$

The c in (11.1) got replaced by t.

Therefore, $F(c) = 0$ by the way K was chosen and also $F(x) = 0$. By the mean value theorem or Rolle's theorem, there exists t_1 between x and c such that $F'(t_1) = 0$. Therefore,

$$0 = f'(t_1) - \sum_{k=1}^{n} \frac{f^{(k)}(c)}{k!} k (x-t_1)^{k-1} - K(n+1)(x-t_1)^n$$

$$= f'(t_1) - \left(f'(c) + \sum_{k=1}^{n-1} \frac{f^{(k+1)}(c)}{k!} (x-t_1)^k \right) - K(n+1)(x-t_1)^n$$

$$= f'(t_1) - \left(f'(c) + \sum_{k=1}^{n-1} \frac{f'^{(k)}(c)}{k!} (x-t_1)^k \right) - K(n+1)(x-t_1)^n$$

By induction applied to f', there exists ξ between x and t_1 such that the above simplifies to

$$0 = \frac{f'^{(n)}(\xi)(x-t_1)^n}{n!} - K(n+1)(x-t_1)^n$$

$$= \frac{f^{(n+1)}(\xi)(x-t_1)^n}{n!} - K(n+1)(x-t_1)^n$$

therefore,

$$K = \frac{f^{(n+1)}(\xi)}{(n+1)\,n!} = \frac{f^{(n+1)}(\xi)}{(n+1)!}$$

and the formula is true for n. ∎

The term $\frac{f^{(n+1)}(\xi)}{(n+1)!} (x-c)^{n+1}$, is called the remainder, and this particular form of the remainder is called the Lagrange form of the remainder.

Example 11.1. Approximate $\sin x$ for x in some open interval containing 0.

Use Taylor's formula just presented and let $c = 0$. Then for $f(x) = \sin x$,

$$f'(x) = \cos x, \quad f''(x) = -\sin x, \quad f'''(x) = -\cos x,$$

etc. Therefore, $f(0) = 0, f'(0) = 1, f''(0) = 0, f'''(0) = -1$, etc. Thus the Taylor polynomial for $\sin x$ is of the form

$$x - \frac{x^3}{3!} + \cdots \pm \frac{x^{2n+1}}{(2n+1)!} = \sum_{k=1}^{n+1} (-1)^{k-1} \frac{x^{2k-1}}{(2k-1)!}$$

while the remainder is of the form

$$\frac{f^{(2n+2)}(\xi) x^{2n+2}}{(2n+2)!}$$

for some ξ between 0 and x. For $n = 2$ in the above, the resulting polynomial is

$$x - \frac{x^3}{3!} + \frac{x^5}{5!}$$

and the error between this polynomial and $\sin x$ must be measured by the remainder term. Therefore,

$$\left| \sin x - \left(x - \frac{x^3}{3!} + \frac{x^5}{5!} \right) \right| \leq \left| \frac{f^{(6)}(\xi) x^6}{6!} \right| \leq \frac{x^6}{6!}.$$

For small x, this error is very small, but if x is large, no such conclusion can be drawn. This is illustrated in the following picture drawn by a computer algebra system.

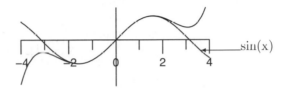

You see from the picture that the polynomial is a very good approximation for the function $\sin x$ as long as $|x|$ is small but that if $|x|$ gets very large, the approximation is lousy. The above estimate indicates the good approximation holds as long as $|x|$ is small and it quantifies how good the approximation is. Suppose for example, you wanted to find $\sin (.5)$. Then from the above error estimate,

$$\left| \sin (.5) - \left((.5) - \frac{(.5)^3}{3!} + \frac{(.5)^5}{5!} \right) \right| \leq \frac{(.5)^6}{6!} = \frac{1}{46\,080}$$

so difference between the approximation and $\sin (.5)$ is less than 10^{-4}. If this is used to find $\sin (.1)$ the polynomial approximation would be even closer.

11.2   Exercises

(1) Let $p_n(x) = a_0 + \sum_{k=1}^{n} a_k (x - c)^k$. Show that if you require that $p_n(c) = f(c), p'_n(c) = f'(c), \cdots, p_n^{(n)}(c) = f^{(n)}(c)$, then this requirement is achieved if and only if $a_0 = f(c), a_1 = f'(c), \cdots, a_n = \frac{f^{(n)}(c)}{n!}$. Thus the Taylor polynomial of degree n and its first n derivatives agree with the function and its first n derivatives when $x = c$.

(2) Find the Taylor polynomials for $\cos x$ for x near 0 along with a formula for the remainder. Use your approximate polynomial to compute $\cos (.5)$ to 3 decimal places and prove your approximation is this good.

(3) Find the Taylor polynomials for $x^4 + 2x^3 + x - 7$ for x near 1. Prove that the Taylor polynomial of degree 4 equals the function.

(4) Find the Taylor polynomials for $3x^4 + 2x^3 + x^2 - 7$ for x near -1. Prove that the Taylor polynomial of degree 4 equals the function.

(5) Find the Taylor polynomials for $\cosh x$ for x near 0.

(6) Find the Taylor polynomials for $\sinh x$ for x near 0.

(7) Find the Taylor polynomials for $\ln(1+x)$ for x near 0.

(8) Find the Taylor polynomials for $\ln(1-x)$ for x near 0.

(9) Find a Taylor polynomial for $\ln\left(\frac{1+x}{1-x}\right)$ and use it to compute $\ln 5$ to three decimal places.

(10) Verify that $\lim_{n\to\infty} \frac{M^n}{n!} = 0$ whenever M is a positive real number. **Hint:** Prove by induction that $M^n/n! \le (2M)^n / (\sqrt{n})^n$. Now consider what happens when $\sqrt{n}$ is much larger than $2M$.

(11) Show that for every $x \in \mathbb{R}$, $\sin(x) = \lim_{n\to\infty} \sum_{k=1}^{n} (-1)^{n-1} \frac{x^{2n-1}}{(2n-1)!}$. **Hint:** Use the formula for the error to conclude that

$$\left| \sin x - \sum_{k=1}^{n} (-1)^{k-1} \frac{x^{2k-1}}{(2k-1)!} \right| \le \frac{|x|^{2n}}{(2n)!}$$

and then use the result of Problem 10.

(12) Show $\cos(x) = \lim_{n\to\infty} \sum_{k=1}^{n} (-1)^{n-1} \frac{x^{2n-2}}{(2n-2)!}$ by finding a suitable formula for the remainder and then using an argument similar to that done in Problem 11.

(13) Using Problem 10, show $e^x = \lim_{n\to\infty} \sum_{k=0}^{n} \frac{x^k}{k!}$ for all $x \in \mathbb{R}$.

(14) Find a_n such that $\arctan(x) = \lim_{n\to\infty} \sum_{k=0}^{n} a_k x^k$ for some values of x. Find the values of x for which the limit is true and prove your result. **Hint:** It is a good idea to use $\arctan(x) = \int_0^x \frac{1}{1+t^2}\, dt$, show $\frac{1}{1+t^2} = \sum_{k=0}^{n} (-1)^k t^{2k} \pm \frac{t^{2n+2}}{1+t^2}$, and then integrate this finite sum from 0 to x. Thus the error would be no larger than $\left| \int_0^x \frac{t^{2n+2}}{1+t^2}\, dt \right| \le \left| \int_0^x t^{2n+2}\, dt \right|$.

(15) If you did Problem 14 correctly, you found

$$\arctan x = \lim_{n\to\infty} \sum_{k=1}^{n} (-1)^{k-1} \frac{x^{2k-1}}{2k-1}$$

and that this limit will hold for $x \in [-1,1]$. Use this to verify that $\frac{\pi}{4} = \lim_{n\to\infty} \sum_{k=1}^{n} (-1)^{k-1} \frac{1}{2k-1}$.

(16) Do for $\ln(1+x)$ what was done for $\arctan(x)$ and find a formula of this sort for $\ln 2$. Use $\ln(1+x) = \int_0^x \frac{1}{1+t}\, dt$.

(17) Repeat 16 for the function $\ln(1-x)$.

(18) Suppose a function $y(x)$ satisfies the initial value problem $y' = y$, $y(0) = 1$. Find Taylor polynomials for this function. Do you know this function which satisfies the given initial value problem?

(19) Suppose a function $y(x)$ satisfies the initial value problem $y'' + y = 0$, $y(0) = 0$, $y'(0) = 1$. Find Taylor polynomials for this function using only the differential equation and initial conditions. Do you know this function which satisfies the given initial value problem?

11.3 Infinite Series Of Numbers

11.3.1 *Basic Considerations*

Earlier in Definition 2.13 on Page 63 the notion of limit of a sequence was discussed. There is a very closely related concept called an infinite series which is dealt with in this section.

Definition 11.1. Define

$$\sum_{k=m}^{\infty} a_k \equiv \lim_{n \to \infty} \sum_{k=m}^{n} a_k$$

whenever the limit exists and is finite. In this case the series is said to converge. If it does not converge, it is said to diverge. The sequence $\{\sum_{k=m}^{n} a_k\}_{n=m}^{\infty}$ in the above is called the sequence of partial sums.

From this definition, it should be clear that infinite sums do not always make sense. Sometimes they do and sometimes they do not, depending on the behavior of the partial sums. As an example, consider $\sum_{k=1}^{\infty} (-1)^k$. The partial sums corresponding to this symbol alternate between -1 and 0. Therefore, there is no limit for the sequence of partial sums. It follows the symbol just written is meaningless and the infinite sum diverges.

Example 11.2. Find the infinite sum $\sum_{n=1}^{\infty} \frac{1}{n(n+1)}$.

Note $\frac{1}{n(n+1)} = \frac{1}{n} - \frac{1}{n+1}$ and so $\sum_{n=1}^{N} \frac{1}{n(n+1)} = \sum_{n=1}^{N} \left(\frac{1}{n} - \frac{1}{n+1} \right) = -\frac{1}{N+1} + 1$.
Therefore,

$$\lim_{N \to \infty} \sum_{n=1}^{N} \frac{1}{n(n+1)} = \lim_{N \to \infty} \left(-\frac{1}{N+1} + 1 \right) = 1.$$

Proposition 11.1. Let $a_k \geq 0$. Then $\{\sum_{k=m}^{n} a_k\}_{n=m}^{\infty}$ is an increasing sequence. If this sequence is bounded above, then $\sum_{k=m}^{\infty} a_k$ converges and its value equals

$$\sup \left\{ \sum_{k=m}^{n} a_k : n = m, m+1, \cdots \right\}.$$

When the sequence is not bounded above, $\sum_{k=m}^{\infty} a_k$ diverges.

Proof: It follows $\{\sum_{k=m}^{n} a_k\}_{n=m}^{\infty}$ is an increasing sequence because

$$\sum_{k=m}^{n+1} a_k - \sum_{k=m}^{n} a_k = a_{n+1} \geq 0.$$

If it is bounded above, let

$$l \equiv \sup \left\{ \sum_{k=m}^{n} a_k : n = m, m+1, \cdots \right\}$$

be the least upper bound. Then for every $\varepsilon > 0$ there exists p such that $l - \varepsilon < \sum_{k=m}^{p} a_k \leq l$ since otherwise l would not be the least upper bound of the partial sums. Then since each $a_k \geq 0$, it follows for all $n \geq p$, $l - \varepsilon < \sum_{k=m}^{p} a_k \leq \sum_{k=m}^{n} a_k \leq l$, so by definition, the partial sums converge to l. It follows the sequence of partial sums converges to $\sup \{\sum_{k=m}^{n} a_k : n = m, m+1, \cdots\}$. If the sequence of partial sums is not bounded, then it is not a Cauchy sequence and so it does not converge. See Theorem 2.15 on Page 66. ∎

In the case where $a_k \geq 0$, the above proposition shows there are only two alternatives available. Either the sequence of partial sums is bounded above or it is not bounded above. In the first case convergence occurs and in the second case, the infinite series diverges. For this reason, people will sometimes write $\sum_{k=m}^{\infty} a_k < \infty$ to denote the case where convergence occurs and $\sum_{k=m}^{\infty} a_k = \infty$ for the case where divergence occurs. Be very careful you never think this way in the case where it is not true that all $a_k \geq 0$. For example, the partial sums of $\sum_{k=1}^{\infty} (-1)^k$ are bounded because they are all either -1 or 0 but the series does not converge.

One of the most important examples of a convergent series is the geometric series. This series is $\sum_{n=0}^{\infty} r^n$. The study of this series depends on simple high school algebra and Theorem 2.12 on Page 65. Let $S_n \equiv \sum_{k=0}^{n} r^k$. Then

$$S_n = \sum_{k=0}^{n} r^k, \quad rS_n = \sum_{k=0}^{n} r^{k+1} = \sum_{k=1}^{n+1} r^k.$$

Therefore, subtracting the second equation from the first yields $(1 - r) S_n = 1 - r^{n+1}$, and so a formula for S_n is available. In fact, if $r \neq 1$,

$$S_n = \frac{1 - r^{n+1}}{1 - r}.$$

By Theorem 2.12, $\lim_{n \to \infty} S_n = \frac{1}{1-r}$ in the case when $|r| < 1$. Now if $|r| \geq 1$, the limit clearly does not exist because S_n fails to be a Cauchy sequence (Why?). This shows the following.

Theorem 11.2. *The geometric series $\sum_{n=0}^{\infty} r^n$ converges and equals $\frac{1}{1-r}$ if $|r| < 1$ and diverges if $|r| \geq 1$.*

If the series do converge, the following holds about combinations of infinite series.

Theorem 11.3. *If $\sum_{k=m}^{\infty} a_k$ and $\sum_{k=m}^{\infty} b_k$ both converge and x, y are numbers, then*

$$\sum_{k=m}^{\infty} a_k = \sum_{k=m+j}^{\infty} a_{k-j} \tag{11.2}$$

$$\sum_{k=m}^{\infty} x a_k + y b_k = x \sum_{k=m}^{\infty} a_k + y \sum_{k=m}^{\infty} b_k \tag{11.3}$$

$$\left| \sum_{k=m}^{\infty} a_k \right| \le \sum_{k=m}^{\infty} |a_k| \tag{11.4}$$

where in the last inequality, the last sum equals $+\infty$ if the partial sums are not bounded above.

Proof: The above theorem is really only a restatement of Theorem 2.10 on Page 64 and the above definitions of infinite series. Thus

$$\sum_{k=m}^{\infty} a_k = \lim_{n \to \infty} \sum_{k=m}^{n} a_k = \lim_{n \to \infty} \sum_{k=m+j}^{n+j} a_{k-j} = \sum_{k=m+j}^{\infty} a_{k-j}.$$

To establish (11.3), use Theorem 2.10 on Page 64 to write

$$\sum_{k=m}^{\infty} xa_k + yb_k = \lim_{n \to \infty} \sum_{k=m}^{n} xa_k + yb_k = \lim_{n \to \infty} \left(x \sum_{k=m}^{n} a_k + y \sum_{k=m}^{n} b_k \right)$$

$$= x \sum_{k=m}^{\infty} a_k + y \sum_{k=m}^{\infty} b_k,$$

Formula (11.4) follows from the observation that, from the triangle inequality,

$$\left| \sum_{k-m}^{n} a_k \right| \le \sum_{k-m}^{\infty} |a_k|,$$

and so

$$\left| \sum_{k=m}^{\infty} a_k \right| = \lim_{n \to \infty} \left| \sum_{k=m}^{n} a_k \right| \le \sum_{k-m}^{\infty} |a_k|. \qquad \blacksquare$$

Example 11.3. Find $\sum_{n=0}^{\infty} \left(\frac{5}{2^n} + \frac{6}{3^n} \right)$.

From the above theorem and Theorem 11.2,

$$\sum_{n=0}^{\infty} \left(\frac{5}{2^n} + \frac{6}{3^n} \right) = 5 \sum_{n=0}^{\infty} \frac{1}{2^n} + 6 \sum_{n=0}^{\infty} \frac{1}{3^n} = 5 \frac{1}{1-(1/2)} + 6 \frac{1}{1-(1/3)} = 19.$$

The following criterion is useful in checking convergence.

Theorem 11.4. *The sum $\sum_{k=m}^{\infty} a_k$ converges if and only if for all $\varepsilon > 0$, there exists n_ε such that if $q \ge p \ge n_\varepsilon$, then*

$$\left| \sum_{k=p}^{q} a_k \right| < \varepsilon. \tag{11.5}$$

Proof: Suppose first that the series converges. Then $\{\sum_{k=m}^{n} a_k\}_{n=m}^{\infty}$ is a Cauchy sequence by Theorem 2.15 on Page 66. Therefore, there exists $n_\varepsilon > m$ such that if $q \geq p - 1 \geq n_\varepsilon > m$,

$$\left| \sum_{k=m}^{q} a_k - \sum_{k=m}^{p-1} a_k \right| = \left| \sum_{k=p}^{q} a_k \right| < \varepsilon. \tag{11.6}$$

Next suppose (11.5) holds. Then from (11.6), it follows upon letting p be replaced with $p + 1$ that $\{\sum_{k=m}^{n} a_k\}_{n=m}^{\infty}$ is a Cauchy sequence and so, by the completeness axiom, it converges. By the definition of infinite series, this shows the infinite sum converges as claimed. ∎

Definition 11.2. A series $\sum_{k=m}^{\infty} a_k$ is said to converge absolutely if $\sum_{k=m}^{\infty} |a_k|$ converges. If the series does converge but does not converge absolutely, then it is said to converge conditionally.

Theorem 11.5. *If $\sum_{k=m}^{\infty} a_k$ converges absolutely, then it converges. Also if each $a_k \geq 0$ and the partial sums are bounded above, then $\sum_{k=m}^{\infty} a_k$ converges.*

Proof: Let $\varepsilon > 0$ be given. Then by assumption and Theorem 11.4, there exists n_ε such that whenever $q \geq p \geq n_\varepsilon$, $\sum_{k=p}^{q} |a_k| < \varepsilon$. Therefore, from the triangle inequality, $\varepsilon > \sum_{k=p}^{q} |a_k| \geq \left| \sum_{k=p}^{q} a_k \right|$. By Theorem 11.4, $\sum_{k=m}^{\infty} a_k$ converges and this proves the first part. The second claim follows from Proposition 11.1. ∎

In fact, the above theorem is really another version of the completeness axiom. Thus its validity implies completeness. You might try to show this.

Theorem 11.6. *(comparison test) Suppose $\{a_n\}$ and $\{b_n\}$ are sequences of nonnegative real numbers and suppose for all n sufficiently large, $a_n \leq b_n$. Then*

(1) If $\sum_{n=k}^{\infty} b_n$ converges, then $\sum_{n=m}^{\infty} a_n$ converges.
(2) If $\sum_{n=k}^{\infty} a_n$ diverges, then $\sum_{n=m}^{\infty} b_n$ diverges.

Proof: Consider the first claim. From the assumption there exists n^* such that $n^* > \max(k, m)$ and for all $n \geq n^*$ $b_n \geq a_n$. Then if $p \geq n^*$,

$$\sum_{n=m}^{p} a_n \leq \sum_{n=m}^{n^*} a_n + \sum_{n=n^*+1}^{k} b_n \leq \sum_{n=m}^{n^*} a_n + \sum_{n=k}^{\infty} b_n.$$

Thus the sequence $\{\sum_{n=m}^{p} a_n\}_{p=m}^{\infty}$ is bounded above and increasing. Therefore, it converges by Theorem 11.5. The second claim is left as an exercise. ∎

Example 11.4. Determine the convergence of $\sum_{n=1}^{\infty} \frac{1}{n^2}$.

For $n > 1$,

$$\frac{1}{n^2} \leq \frac{1}{n(n-1)}.$$

Now

$$\sum_{n=2}^{p} \frac{1}{n(n-1)} = \sum_{n=2}^{p} \left[\frac{1}{n-1} - \frac{1}{n} \right] = 1 - \frac{1}{p}$$

which converges to 1 as $p \to \infty$. Therefore, letting $a_n = \frac{1}{n^2}$ and $b_n = \frac{1}{n(n-1)}$, it follows by the comparison test, the given series converges.

A convenient way to implement the comparison test is to use the limit comparison test. This is considered next.

Theorem 11.7. *Let $a_n, b_n > 0$ and suppose for all n large enough,*

$$0 < a < \frac{a_n}{b_n} < b < \infty.$$

Then $\sum a_n$ and $\sum b_n$ converge or diverge together.

Proof: Let n^* be such that $n \geq n^*$, then

$$\frac{a_n}{b_n} > a \text{ and } \frac{a_n}{b_n} < b$$

and so for all such n,

$$ab_n < a_n < bb_n$$

and so the conclusion follows from the comparison test. ∎

Corollary 11.1. *If $a_n, b_n > 0$ and $\lim_{n \to \infty} \frac{a_n}{b_n} = r \in (0, \infty)$, then $\sum a_n, \sum b_n$ converge or diverge together.*

Proof: The assumption implies that for all n large enough,

$$\frac{r}{2} < \frac{a_n}{b_n} < 2r$$

The conclusion now follows from Theorem 11.7. ∎

Example 11.5. Determine the convergence of $\sum_{k=1}^{\infty} \frac{1}{\sqrt{n^4 + 2n + 7}}$.

This series converges by the limit comparison test. Compare with the series of Example 11.4.

$$\lim_{n \to \infty} \frac{\left(\frac{1}{n^2} \right)}{\left(\frac{1}{\sqrt{n^4 + 2n + 7}} \right)} = \lim_{n \to \infty} \frac{\sqrt{n^4 + 2n + 7}}{n^2} = \lim_{n \to \infty} \sqrt{1 + \frac{2}{n^3} + \frac{7}{n^4}} = 1.$$

Therefore, the series converges with the series of Example 11.4. How did I know what to compare with? I noticed that $\sqrt{n^4 + 2n + 7}$ is essentially like $\sqrt{n^4} = n^2$ for large enough n. You see, the higher order term n^4 dominates the other terms in $n^4 + 2n + 7$. Therefore, reasoning that $1/\sqrt{n^4 + 2n + 7}$ is a lot like $1/n^2$ for large n, it was easy to see what to compare with. Of course this is not always easy and there is room for acquiring skill through practice.

To really exploit this limit comparison test, it is desirable to get lots of examples of series, some which converge and some which do not. The tool for obtaining these examples here will be the following wonderful theorem known as the Cauchy condensation test. It is an extremely simple idea. You group the terms in blocks and use the assumption that the terms of the series are decreasing.

Theorem 11.8. *Let $a_n \geq 0$ and suppose the terms of the sequence $\{a_n\}$ are decreasing. Thus $a_n \geq a_{n+1}$ for all n. Then the two series*

$$\sum_{n=1}^{\infty} a_n, \quad \sum_{n=0}^{\infty} 2^n a_{2^n}$$

converge or diverge together.

Proof: This follows from the inequality of the following claim.
Claim:

$$\sum_{k=0}^{n} 2^k a_{2^k} \geq \sum_{k=1}^{2^n} a_k \geq \frac{1}{2} \sum_{k=0}^{n} 2^k a_{2^k}.$$

Proof of the Claim: Consider the following two ways of grouping the terms of the sum.

$$\overbrace{a_1 + a_2}^{\geq \frac{1}{2} a_1 + a_2} + \overbrace{a_3 + a_4}^{\geq 2a_4} + \overbrace{a_5 + a_6 + a_7 + a_8}^{\geq 2^2 a_{2^3}} + \cdots$$

Second way:

$$a_1 + \overbrace{a_2 + a_3}^{\leq 2^1 a_{2^1}} + \overbrace{a_4 + a_5 + a_6 + a_7}^{\leq 2^2 a_{2^2}} +$$

$$\overbrace{a_8 + a_9 + a_{10} + a_{11} + a_{12} + a_{13} + a_{14} + a_{15}}^{\leq 2^3 a_{2^3}} + a_{16} + \cdots$$

Therefore, writing the above in terms of sums, it would seem that

$$\sum_{k=0}^{n} 2^k a_{2^k} \geq \sum_{k=1}^{2^n} a_k \geq \frac{1}{2} \sum_{k=0}^{n} 2^k a_{2^k}.$$

In fact, this is true for all positive n. The proof is by induction.

It is clear that the above inequality is true if $n = 1$ because it says

$$a_1 + 2a_2 \geq a_1 + a_2 \geq \frac{1}{2} a_1 + a_2,$$

which is obvious. Suppose the inequality is true for n. Then by induction,

$$\sum_{k=0}^{n+1} 2^k a_{2^k} = 2^{n+1} a_{2^{n+1}} + \sum_{k=0}^{n} 2^k a_{2^k} \geq 2^{n+1} a_{2^{n+1}} + \sum_{k=1}^{2^n} a_k$$

$$\geq 2^{n+1} a_{2^{n+1}} + \frac{1}{2} \sum_{k=0}^{n} 2^k a_{2^k} \geq \frac{1}{2} \sum_{k=0}^{n+1} 2^k a_{2^k},$$

the last inequality because $2^{n+1} a_{2^{n+1}} \geq \frac{1}{2} 2^{n+1} a_{2^{n+1}}$. ∎

Example 11.6. Determine the convergence of $\sum_{k=1}^{\infty} \frac{1}{k^p}$ where p is a positive number. These are called the p series.

Let $a_n = \frac{1}{n^p}$. Then $a_{2^n} = \left(\frac{1}{2^p}\right)^n$. From the Cauchy condensation test the two series

$$\sum_{n=1}^{\infty} \frac{1}{n^p} \text{ and } \sum_{n=0}^{\infty} 2^n \left(\frac{1}{2^p}\right)^n = \sum_{n=0}^{\infty} \left(2^{(1-p)}\right)^n$$

converge or diverge together. If $p > 1$, the last series above is a geometric series having common ratio less than 1, and so it converges. If $p \le 1$, it is still a geometric series but in this case the common ratio is either 1 or greater than 1, so the series diverges. It follows that the p series converges if $p > 1$ and diverges if $p \le 1$. In particular, $\sum_{n=1}^{\infty} n^{-1}$ diverges while $\sum_{n=1}^{\infty} n^{-2}$ converges. The following table summarizes the above results.

The p series table

	$p \le 1$	$p > 1$
$\sum \frac{1}{n^p}$	diverges	converges

Example 11.7. Determine the convergence of $\sum_{k=1}^{\infty} \frac{1}{\sqrt{n^2+100n}}$.

Use the limit comparison test.

$$\lim_{n \to \infty} \frac{\left(\frac{1}{n}\right)}{\left(\frac{1}{\sqrt{n^2+100n}}\right)} = 1,$$

and so this series diverges with $\sum_{k=1}^{\infty} \frac{1}{k}$.

Example 11.8. Determine the convergence of $\sum_{k=2}^{\infty} \frac{1}{k \ln k}$.

Use the Cauchy condensation test. The above series does the same thing in terms of convergence as the series

$$\sum_{n=1}^{\infty} 2^n \frac{1}{2^n \ln(2^n)} = \sum_{n=1}^{\infty} \frac{1}{n \ln 2}$$

and this series diverges by limit comparison with the series $\sum \frac{1}{n}$.

Sometimes it is good to be able to say a series does not converge. The nth term test gives such a condition which is sufficient for this. Sometimes this is called the divergence test. It is really a corollary of Theorem 11.4.

Theorem 11.9. *If $\sum_{n=m}^{\infty} a_n$ converges, then $\lim_{n\to\infty} a_n = 0$.*

Proof: By Theorem 11.4, $\lim_{n\to\infty} a_n = \lim_{n\to\infty} \sum_{k=n}^{n} a_k = 0.$ ∎

It is very important to observe that this theorem goes only in one direction. That is, you **cannot conclude** the series converges if $\lim_{n\to\infty} a_n = 0$. If this happens, you do not know anything from this information. Recall $\lim_{n\to\infty} n^{-1} = 0$ but $\sum_{n=1}^{\infty} n^{-1}$ diverges. The following picture is descriptive of the situation.

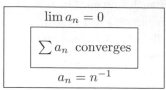

$$\lim a_n = 0$$
$$\sum a_n \text{ converges}$$
$$a_n = n^{-1}$$

11.4 Exercises

(1) Determine whether the following series converge and give reasons for your answers.

    (a) $\sum_{n=1}^{\infty} \frac{1}{\sqrt{n^2+n+1}}$

    (b) $\sum_{n=1}^{\infty} \left(\sqrt{n+1} - \sqrt{n} \right)$

    (c) $\sum_{n=1}^{\infty} \frac{(2n)!}{(n!)^2}$

    (d) $\sum_{n=1}^{\infty} \frac{1}{2n+2}$

    (e) $\sum_{n=1}^{\infty} \left(\frac{n}{n+1} \right)^n$

(2) Determine whether the following series converge and give reasons for your answers.

    (a) $\sum_{n=1}^{\infty} \frac{\ln(k^5)}{k}$

    (b) $\sum_{n=1}^{\infty} \frac{\ln(k^5)}{k^{1.01}}$

    (c) $\sum_{n=1}^{\infty} \sin\left(\frac{1}{n} \right)$

    (d) $\sum_{n=1}^{\infty} \tan\left(\frac{1}{n^2} \right)$

    (e) $\sum_{n=1}^{\infty} \cos\left(\frac{1}{n^2} \right)$

    (f) $\sum_{n=1}^{\infty} \sin\left(\frac{\sqrt{n}}{n^2+1} \right)$

(3) Determine whether the following series converge and give reasons for your answers.

    (a) $\sum_{n=1}^{\infty} \frac{2^n+n}{n2^n}$

    (b) $\sum_{n=1}^{\infty} \frac{2^n+n}{n^2 2^n}$

    (c) $\sum_{n=1}^{\infty} \frac{n}{2n+1}$

    (d) $\sum_{n=1}^{\infty} \frac{\ln n}{n^2}$

(4) Suppose a_n and b_n are just numbers, either positive or negative and that for all n large enough, $0 < a < a_n/b_n < b < \infty$. Show that either one of $\sum_{n=1}^{\infty} a_n$ and $\sum_{n=1}^{\infty} b_n$ converges absolutely if and only if the other converges absolutely. This little fact, generalizing Theorem 11.7 was pointed out to me by Rodney Forcade.

(5) Suppose f is a nonnegative continuous decreasing function defined on $[1, \infty)$. Show the improper integral $\int_1^{\infty} f(t) \, dt$ and the sum $\sum_{k=1}^{\infty} f(k)$ converge or diverge together. This is called the integral test. Use this test to verify convergence of $\sum_{k=1}^{\infty} \frac{1}{k^{\alpha}}$ whenever $\alpha > 1$ and divergence whenever $\alpha \leq 1$. In showing this integral test, it might be helpful to consider the following picture.

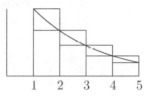

$$\begin{array}{ccccc} 1 & 2 & 3 & 4 & 5 \end{array}$$

In this picture, the graph of the continuous decreasing function is represented. There are rectangles below this curve and rectangles above it. $f(1)$ is the area of the first rectangle on the left which is above the curve while $f(2)$ is the area of the first rectangle on the left which lies below the curve. From the picture, $\int_1^5 f(t) \, dt$ lies between $f(2) + f(3) + f(4) + f(5)$ and $f(1) + f(2) + f(3) + f(4)$. Generalize to conclude that for any $n \in \mathbb{N}$, it follows that $\sum_{k=2}^{n} f(k) \leq \int_1^n f(t) \, dt \leq \sum_{k=1}^{n-1} f(k)$. Now explain why this inequality implies the integral and sum have the same convergence properties.

(6) Show, using either the Cauchy condensation test or the integral test, that

$$\sum_{k=4}^{\infty} \frac{1}{\ln(\ln(k)) \ln(k) \, k}$$

diverges. Estimate how big N must be in order that $\sum_{k=4}^{N} \frac{1}{\ln(\ln(k)) \ln(k)k} > 10$. What does this tell you about the wisdom of attempting to determine questions of convergence through experimentation?

(7) For p a positive number, determine the convergence of $\sum_{n=2}^{\infty} \frac{1}{n(\ln(n))^p}$ for various values of p.

(8) For p a positive number, determine the convergence of

$$\sum_{n=2}^{\infty} \frac{\ln n}{n^p}$$

for various values of p.

(9) Determine the convergence of the series $\sum_{n=1}^{\infty} \left(\sum_{k=1}^{n} \frac{1}{k} \right)^{-n/2}$.

(10) Is it possible that there could exist a decreasing sequence of positive numbers, $\{a_n\}$ such that $\lim_{n \to \infty} a_n = 0$ but $\sum_{n=1}^{\infty} \left(1 - \frac{a_{n+1}}{a_n} \right)$ converges? **Hint:** You might do something like this. Show

$$\lim_{x \to 1} \frac{1-x}{-\ln(x)} = \frac{1-x}{\ln(1/x)} = 1$$

Next use a limit comparison test with

$$\sum_{n=1}^{\infty} \ln\left(\frac{a_n}{a_{n+1}}\right)$$

(11) Recall that a number larger than 1 is prime if the only numbers which divide it are itself and 1. Also recall that every positive integer can be factored as a product of primes. The following little fact is due to Euler. Show that if $\{p_i\}_{i=1}^{\infty}$ are the prime numbers, then $\sum_{i=1}^{\infty} \frac{1}{p_i} = \infty$. That is, not only are there infinitely many primes, there are also enough primes that the sum of their reciprocals diverges. In what follows, the symbol Π means to multiply all the indicated terms from $k = 1$ to $k = \phi(n)$ **Hint:** Let $\phi(n)$ denote the number of primes less than equal to n. Then explain why

$$\sum_{k=1}^{n} \frac{1}{k} \leq \prod_{k=1}^{\phi(n)} \left(1 + \frac{1}{p_k}\right) \leq \prod_{k=1}^{\phi(n)} e^{1/p_k} = e^{\sum_{k=1}^{\phi(n)} 1/p_k}$$

and consequently why $\lim_{n\to\infty} \phi(n) = \infty$ and $\sum_{i=1}^{\infty} \frac{1}{p_i} = \infty$.

(12) *Stirling's formula is a very significant formula for understanding $n!$. It says that $\lim_{n\to\infty} \frac{n!}{\sqrt{2\pi} n^{n+1/2} e^{-n}} = 1$. In this problem, you will get the most important part of this formula following the presentation in the old calculus book by Courant. First show that

$$\int_{1}^{n} \ln(x) \, dx = n \ln n - n + 1 \equiv A_n.$$

Next use the trapezoid rule to obtain the corresponding trapezoid rule approximation which is

$$T_n \equiv \ln(n!) - \frac{1}{2} \ln n.$$

This approximation is too small. Let $a_n \equiv A_n - T_n$. You want to show that a_n converges to some positive number β. To do this, make use of the following picture which can be used to estimate the error between the integral and the trapezoid approximating it.

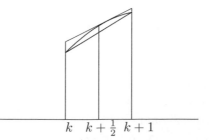

$$k \quad\quad k + \tfrac{1}{2} \quad k + 1$$

Using this picture, argue that

$$a_n \leq \sum_{k=1}^{n} \left(\ln\left(k + \frac{1}{2}\right) - \frac{\ln(k+1) + \ln(k)}{2} \right).$$

Now explain why

$$\ln\left(k+\frac{1}{2}\right) - \frac{\ln(k+1)+\ln(k)}{2}$$

$$= \frac{1}{2}\left(\ln\left(k+\frac{1}{2}\right) - \ln(k)\right) - \frac{1}{2}\left(\ln(k+1) - \ln\left(k+\frac{1}{2}\right)\right)$$

$$= \frac{1}{2}\left(\ln\left(1+\frac{1}{2k}\right) - \ln\left(1+\frac{1}{2k+1}\right)\right)$$

$$\leq \frac{1}{2}\left(\ln\left(1+\frac{1}{2k}\right) - \ln\left(1+\frac{1}{2(k+1)}\right)\right)$$

Now it follows that $a_n \leq \frac{1}{2}\ln\left(\frac{3}{2}\right)$. Tell why this implies the limit of a_n exists and equals β say. Next explain why $\frac{e}{e^{a_n}} = \frac{n!}{n^{n+1/2}e^{-n}}$, and so letting $\alpha = \lim_{n\to\infty}\frac{e}{e^{a_n}}$, it follows that

$$\lim_{n\to\infty}\frac{n!}{\alpha n^{n+1/2}e^{-n}} = 1.$$

It will be shown below that $\alpha = \sqrt{2\pi}$. (Answers to these problems can be found in the supplementary exercises in the key to the tenth homework set.)

(13) Consider the following series. $\sum_{n=1}^{\infty}\frac{n!e^n}{n^n}$ Find whether this series converges using Stirling's formula.

(14) Consider the series $\sum_{n=1}^{\infty}\frac{n^n}{n!e^n}$. Find whether this series converges using Stirling's formula.

(15) Find $\lim_{n\to\infty}\frac{(n!)^{1/n}}{n}$.

(16) Now we will figure out α in Stirling's formula.

(a) Show $\int_0^{\pi/2}\sin^n x\,dx = \frac{n-1}{n}\int_0^{\pi/2}\sin^{n-2}(x)\,dx$, $n>1$

(b) Using induction, show $\int_0^{\pi/2}\sin^{2m}(x)\,dx = \frac{(2m-1)(2m-3)\cdots 1}{(2m)(2m-2)\cdots 2}\frac{\pi}{2}$ and

$\int_0^{\pi/2}\sin^{2m+1}(x)\,dx = \frac{(2m)(2m-2)\cdots 2}{(2m+1)(2m-1)\cdots 3}$.

(c) Dividing the two above, explain why

$\frac{\pi}{2} = \frac{2m}{2m+1}2m\frac{(2m-2)^2\cdots 4}{(2m-1)^2(2m-3)^2\cdots 3^2}\frac{\int_0^{\pi/2}\sin^{2m}(x)dx}{\int_0^{\pi/2}\sin^{2m+1}(x)dx}$

(d) Use the above reduction identity and the fact that $\sin(x)<1$ to write

$1 \leq \frac{\int_0^{\pi/2}\sin^{2m}(x)dx}{\int_0^{\pi/2}\sin^{2m+1}(x)dx} \leq \frac{\int_0^{\pi/2}\sin^{2m}(x)dx}{\frac{2m}{2m+1}\int_0^{\pi/2}\sin^{2m-1}(x)dx} \leq \frac{2m+1}{2m}$ and conclude that

$\lim_{m\to\infty}\frac{\int_0^{\pi/2}\sin^{2m}(x)dx}{\int_0^{\pi/2}\sin^{2m+1}(x)dx} = 1.$

(e) Explain why $\frac{\pi}{2} = \lim_{m\to\infty}\frac{2m}{2m+1}2m\frac{(2m-2)^2\cdots 4}{(2m-1)^2(2m-3)^2\cdots 3^2}$

$= \lim_{m\to\infty}2m\frac{(2m-2)^2\cdots 2^2}{(2m-1)^2(2m-3)^2\cdots 3^2}$. This is Wallis's formula.

(f) Next show that $\sqrt{\frac{\pi}{2}} = \lim_{m\to\infty}\sqrt{2m}\frac{(2m-2)\cdots 2}{(2m-1)(2m-3)\cdots 3}$

$= \lim_{m\to\infty}\sqrt{2m}\frac{(2m-2)^2\cdots 2^2}{(2m-1)(2m-2)(2m-3)(2m-4)\cdots 3\cdot 2}$

$$= \lim_{m\to\infty} \sqrt{2m} \frac{1}{(2m)^2} \frac{(2m)^2(2m-2)^2\cdots 2^2}{(2m-1)(2m-2)(2m-3)(2m-4)\cdots 3\cdot 2}$$

$$= \lim_{m\to\infty} \frac{1}{(2m)^{1/2}} \frac{2^{2m}(m!)^2}{(2m)!}$$

(g) Recall that $\dfrac{m!}{\alpha m^{m+1/2}e^{-m}} \to 1$ and so also $\dfrac{(2m)!}{\alpha(2m)^{2m+1/2}e^{-2m}} \to 1$. Explain why this implies

$$\sqrt{\tfrac{\pi}{2}} = \lim_{m\to\infty} \frac{1}{(2m)^{1/2}} \frac{2^{2m}\left(\alpha m^{m+1/2}e^{-m}\right)^2}{\alpha(2m)^{2m+1/2}e^{-2m}} = \tfrac{1}{2}\alpha \text{ and so } \alpha = \sqrt{2\pi}.$$

11.5 More Tests For Convergence

11.5.1 *Convergence Because Of Cancelation*

So far, the tests for convergence have been applied to nonnegative terms only. Sometimes, a series converges, not because the terms of the series get small fast enough, but because of cancelation taking place between positive and negative terms. A discussion of this involves some simple algebra.

Let $\{a_n\}$ and $\{b_n\}$ be sequences and let

$$A_n \equiv \sum_{k=1}^{n} a_k, \ A_{-1} \equiv A_0 \equiv 0.$$

Then if $p < q$

$$\sum_{n=p}^{q} a_n b_n = \sum_{n=p}^{q} b_n (A_n - A_{n-1}) = \sum_{n=p}^{q} b_n A_n - \sum_{n=p}^{q} b_n A_{n-1}$$

$$= \sum_{n=p}^{q} b_n A_n - \sum_{n=p-1}^{q-1} b_{n+1} A_n = b_q A_q - b_p A_{p-1} + \sum_{n=p}^{q-1} A_n (b_n - b_{n+1})$$

This formula is called the partial summation formula. It is just like integration by parts.

Theorem 11.10. *(Dirichlet's test) Suppose A_n is bounded and $\lim_{n\to\infty} b_n = 0$, with $b_n \geq b_{n+1}$. Then*

$$\sum_{n=1}^{\infty} a_n b_n$$

converges.

Proof: This follows quickly from Theorem 11.4. Indeed, letting $|A_n| \leq C$, and using the partial summation formula above along with the assumption that the b_n are decreasing,

$$\left| \sum_{n=p}^{q} a_n b_n \right| = \left| b_q A_q - b_p A_{p-1} + \sum_{n=p}^{q-1} A_n (b_n - b_{n+1}) \right|$$

$$\leq C\left(|b_q| + |b_p|\right) + C \sum_{n=p}^{q-1} \left(b_n - b_{n+1}\right)$$

$$= C\left(|b_q| + |b_p|\right) + C\left(b_p - b_q\right)$$

and by assumption, this last expression is small whenever p and q are sufficiently large. ∎

Definition 11.3. If $b_n > 0$ for all n, a series of the form $\sum_k (-1)^k b_k$ or $\sum_k (-1)^{k-1} b_k$ is known as an **alternating series**.

The following corollary is known as the alternating series test.

Corollary 11.2. *(alternating series test) If* $\lim_{n \to \infty} b_n = 0$, *with* $b_n \geq b_{n+1}$, *then* $\sum_{n=1}^{\infty} (-1)^n b_n$ *converges.*

Proof: Let $a_n = (-1)^n$. Then the partial sums of $\sum_n a_n$ are bounded, and so Theorem 11.10 applies. ∎

In the situation of Corollary 11.2 there is a convenient error estimate available.

Theorem 11.11. *Let* $b_n > 0$ *for all* n *such that* $b_n \geq b_{n+1}$ *for all* n *and* $\lim_{n \to \infty} b_n = 0$ *and consider either* $\sum_{n=1}^{\infty} (-1)^n b_n$ *or* $\sum_{n=1}^{\infty} (-1)^{n-1} b_n$. *Then*

$$\left| \sum_{n=1}^{\infty} (-1)^n b_n - \sum_{n=1}^{N} (-1)^n b_n \right| \leq |b_{N+1}|, \quad \left| \sum_{n=1}^{\infty} (-1)^{n-1} b_n - \sum_{n=1}^{N} (-1)^{n-1} b_n \right| \leq |b_{N+1}|$$

Example 11.9. How many terms must I take in the sum $\sum_{n=1}^{\infty} (-1)^n \frac{1}{n^2+1}$ to be closer than $\frac{1}{10}$ to $\sum_{n=1}^{\infty} (-1)^n \frac{1}{n^2+1}$?

From Theorem 11.11, I need to find n such that $\frac{1}{n^2+1} \leq \frac{1}{10}$ and then $n-1$ is the desired value. Thus $n = 3$, and so

$$\left| \sum_{n=1}^{\infty} (-1)^n \frac{1}{n^2+1} - \sum_{n=1}^{2} (-1)^n \frac{1}{n^2+1} \right| \leq \frac{1}{10}$$

11.5.2 Ratio And Root Tests

A favorite test for convergence is the ratio test. This is discussed next. It is the at the other extreme. This test is completely oblivious to any sort of cancelation. It only gives absolute convergence or spectacular divergence.

Theorem 11.12. *Suppose* $|a_n| > 0$ *for all* n *and suppose*

$$\lim_{n \to \infty} \frac{|a_{n+1}|}{|a_n|} = r.$$

Then

$$\sum_{n=1}^{\infty} a_n \begin{cases} \text{diverges if } r > 1 \\ \text{converges absolutely if } r < 1 \\ \text{test fails if } r = 1 \end{cases}.$$

Proof: Suppose $r < 1$. Then there exists n_1 such that if $n \geq n_1$, then

$$0 < \left| \frac{a_{n+1}}{a_n} \right| < R$$

where $r < R < 1$. Then

$$|a_{n+1}| < R|a_n|$$

for all such n. Therefore,

$$|a_{n_1+p}| < R|a_{n_1+p-1}| < R^2|a_{n_1+p-2}| < \cdots < R^p|a_{n_1}| \qquad (11.7)$$

and so if $m > n$, then $|a_m| < R^{m-n_1}|a_{n_1}|$. By the comparison test and the theorem on geometric series, $\sum|a_n|$ converges. This proves the convergence part of the theorem.

To verify the divergence part, note that if $r > 1$, then (11.7) can be turned around for some $R > 1$. Showing $\lim_{n\to\infty}|a_n| = \infty$. Since the nth term fails to converge to 0, it follows the series diverges.

To see the test fails if $r = 1$, consider $\sum n^{-1}$ and $\sum n^{-2}$. The first series diverges while the second one converges but in both cases, $r = 1$. (Be sure to check this last claim.) ∎

The ratio test is very useful for many different examples but it is somewhat unsatisfactory mathematically. One reason for this is the assumption that $a_n > 0$, necessitated by the need to divide by a_n, and the other reason is the possibility that the limit might not exist. The next test, called the root test removes both of these objections.

Theorem 11.13. *Suppose* $|a_n|^{1/n} < R < 1$ *for all n sufficiently large. Then*

$$\sum_{n=1}^{\infty} a_n \text{ converges absolutely.}$$

If there are infinitely many values of n such that $|a_n|^{1/n} \geq 1$, then

$$\sum_{n=1}^{\infty} a_n \text{ diverges.}$$

Proof: Suppose first that $|a_n|^{1/n} < R < 1$ for all n sufficiently large. Say this holds for all $n \geq n_R$. Then for such n,

$$\sqrt[n]{|a_n|} < R.$$

Therefore, for such n,

$$|a_n| \leq R^n,$$

and so the comparison test with a geometric series applies and gives absolute convergence as claimed.

Next suppose $|a_n|^{1/n} \geq 1$ for infinitely many values of n. Then for those values of n, $|a_n| \geq 1$, and so the series fails to converge by the nth term test. ∎

Corollary 11.3. *Suppose* $\lim_{n \to \infty} |a_n|^{1/n}$ *exists and equals* r. *Then*

$$\sum_{k=m}^{\infty} a_k \begin{cases} \text{converges absolutely if } r < 1 \\ \text{test fails if } r = 1 \\ \text{diverges if } r > 1 \end{cases}$$

Proof: The first and last alternatives follow from Theorem 11.13. To see the test fails if $r = 1$, consider the two series $\sum_{n=1}^{\infty} \frac{1}{n}$ and $\sum_{n=1}^{\infty} \frac{1}{n^2}$ both of which have $r = 1$ but having different convergence properties. ∎

Example 11.10. Show that for all $x \in \mathbb{R}$,

$$e^x = \sum_{k=0}^{\infty} \frac{x^k}{k!}. \tag{11.8}$$

By Taylor's theorem

$$\begin{aligned} e^x &= 1 + x + \frac{x^2}{2!} + \cdots + \frac{x^n}{n!} + e^{\xi_n} \frac{x^{n+1}}{(n+1)!} \\ &= \sum_{k=0}^{n} \frac{x^k}{k!} + e^{\xi_n} \frac{x^{n+1}}{(n+1)!} \end{aligned} \tag{11.9}$$

where $|\xi_n| \leq |x|$. Now for any $x \in \mathbb{R}$

$$\left| e^{\xi_n} \frac{x^{n+1}}{(n+1)!} \right| \leq e^{|x|} \frac{|x|^{n+1}}{(n+1)!}$$

and an application of the ratio test shows

$$\sum_{n=0}^{\infty} e^{|x|} \frac{|x|^{n+1}}{(n+1)!} < \infty.$$

Therefore, the nth term converges to zero by the nth term test, and so for each $x \in \mathbb{R}$,

$$\lim_{n \to \infty} e^{\xi_n} \frac{x^{n+1}}{(n+1)!} = 0.$$

Therefore, taking the limit in (11.9) it follows (11.8) holds.

11.6 Double Series*

Sometimes it is required to consider double series which are of the form

$$\sum_{k=m}^{\infty} \sum_{j=m}^{\infty} a_{jk} \equiv \sum_{k=m}^{\infty} \left(\sum_{j=m}^{\infty} a_{jk} \right).$$

In other words, first sum on j yielding something which depends on k and then sum these. The major consideration for these double series is the question of when

$$\sum_{k=m}^{\infty}\sum_{j=m}^{\infty} a_{jk} = \sum_{j=m}^{\infty}\sum_{k=m}^{\infty} a_{jk}.$$

In other words, when does it make no difference which subscript is summed over first? In the case of finite sums there is no issue here. You can always write

$$\sum_{k=m}^{M}\sum_{j=m}^{N} a_{jk} = \sum_{j=m}^{N}\sum_{k=m}^{M} a_{jk}$$

because addition is commutative. However, there are limits involved with infinite sums and the interchange in order of summation involves taking limits in a different order. Therefore, it is not always true that it is permissible to interchange the two sums. A general rule of thumb is this: If something involves changing the order in which two limits are taken, you may not do it without agonizing over the question. In general, limits foul up algebra and also introduce things which are counter intuitive. Here is an example. This example is a little technical. It is placed here just to prove conclusively there is a question which needs to be considered.

Example 11.11. Consider the following picture which depicts some of the ordered pairs (m, n) where m, n are positive integers.

```
0.  0.  0.  0.  0.  c.  0.  -c.
0.  0.  0.  0.  c.  0.  -c.  0.
0.  0.  0.  c.  0.  -c.  0.  0.
0.  0.  c.  0.  -c.  0.  0.  0.
0.  c.  0.  -c.  0.  0.  0.  0.
b.  0.  -c.  0.  0.  0.  0.  0.
0.  a.  0.  0.  0.  0.  0.  0.
```

The numbers next to the point are the values of a_{mn}. You see $a_{nn} = 0$ for all n, $a_{21} = a$, $a_{12} = b$, $a_{mn} = c$ for (m, n) on the line $y = 1 + x$ whenever $m > 1$, and $a_{mn} = -c$ for all (m, n) on the line $y = x - 1$ whenever $m > 2$.

Then $\sum_{m=1}^{\infty} a_{mn} = a$ if $n = 1$, $\sum_{m=1}^{\infty} a_{mn} = b - c$ if $n = 2$ and if $n > 2$, $\sum_{m=1}^{\infty} a_{mn} = 0$. Therefore,

$$\sum_{n=1}^{\infty}\sum_{m=1}^{\infty} a_{mn} = a + b - c.$$

Next observe that $\sum_{n=1}^{\infty} a_{mn} = b$ if $m = 1$, $\sum_{n=1}^{\infty} a_{mn} = a + c$ if $m = 2$, and $\sum_{n=1}^{\infty} a_{mn} = 0$ if $m > 2$. Therefore,

$$\sum_{m=1}^{\infty}\sum_{n=1}^{\infty} a_{mn} = b + a + c,$$

and so the two sums are different. Moreover, you can see that by assigning different values of a, b, and c, you can get an example for any two different numbers desired.

Do not become upset by this. It happens because, as indicated above, limits are taken in two different orders. An infinite sum always involves a limit and this illustrates why you must always remember this. This example in no way violates the commutative law of addition which has nothing to do with limits. However, it turns out that if $a_{ij} \geq 0$ for all i, j, then you can always interchange the order of summation. This is shown next and is based on the following lemma. First, some notation should be discussed.

Definition 11.4. Let $f(a, b) \in [-\infty, \infty]$ for $a \in A$ and $b \in B$ where A, B are sets which means that $f(a, b)$ is either a number, ∞, or $-\infty$. The symbol $+\infty$ is interpreted as a point out at the end of the number line which is larger than every real number and read as plus infinity. Of course there is no such number. That is why it is called ∞. The symbol $-\infty$ is interpreted similarly. Then $\sup_{a \in A} f(a, b)$ means $\sup(S_b)$ where $S_b \equiv \{f(a, b) : a \in A\}$.

Unlike limits, you can take the sup in different orders.

Lemma 11.1. *Let $f(a, b) \in [-\infty, \infty]$ for $a \in A$ and $b \in B$ where A, B are sets. Then*

$$\sup_{a \in A} \sup_{b \in B} f(a, b) = \sup_{b \in B} \sup_{a \in A} f(a, b).$$

Proof: Note that for all a, b, $f(a, b) \leq \sup_{b \in B} \sup_{a \in A} f(a, b)$ and therefore, for all a, $\sup_{b \in B} f(a, b) \leq \sup_{b \in B} \sup_{a \in A} f(a, b)$. Therefore,

$$\sup_{a \in A} \sup_{b \in B} f(a, b) \leq \sup_{b \in B} \sup_{a \in A} f(a, b).$$

Repeat the same argument interchanging a and b, to get the conclusion of the lemma. ∎

Lemma 11.2. *If $\{A_n\}$ is an increasing sequence in $[-\infty, \infty]$, then*

$$\sup\{A_n\} = \lim_{n \to \infty} A_n.$$

Proof: Let $\sup(\{A_n : n \in \mathbb{N}\}) = r$. In the first case, suppose $r < \infty$. Then letting $\varepsilon > 0$ be given, there exists n such that $A_n \in (r - \varepsilon, r]$. Since $\{A_n\}$ is increasing, it follows if $m > n$, then $r - \varepsilon < A_n \leq A_m \leq r$ and so $\lim_{n \to \infty} A_n = r$ as claimed. In the case where $r = \infty$, then if a is a real number, there exists n such that $A_n > a$. Since $\{A_k\}$ is increasing, it follows that if $m > n$, $A_m > a$. But this is what is meant by $\lim_{n \to \infty} A_n = \infty$. The other case is that $r = -\infty$. But in this case, $A_n = -\infty$ for all n and so $\lim_{n \to \infty} A_n = -\infty$. ∎

Theorem 11.14. *Let $a_{ij} \geq 0$. Then $\sum_{i=1}^{\infty} \sum_{j=1}^{\infty} a_{ij} = \sum_{j=1}^{\infty} \sum_{i=1}^{\infty} a_{ij}$.*

Proof: First note there is no trouble in defining these sums because the a_{ij} are all nonnegative. If a sum diverges, it only diverges to ∞ and so ∞ is the value of the sum. Next note that

$$\sum_{j=r}^{\infty}\sum_{i=r}^{\infty} a_{ij} \geq \sup_{n} \sum_{j=r}^{\infty}\sum_{i=r}^{n} a_{ij}$$

because for all j,

$$\sum_{i=r}^{\infty} a_{ij} \geq \sum_{i=r}^{n} a_{ij}.$$

Therefore,

$$\sum_{j=r}^{\infty}\sum_{i=r}^{\infty} a_{ij} \geq \sup_{n} \sum_{j=r}^{\infty}\sum_{i=r}^{n} a_{ij} = \sup_{n} \lim_{m\to\infty} \sum_{j=r}^{m}\sum_{i=r}^{n} a_{ij}$$

$$= \sup_{n} \lim_{m\to\infty} \sum_{i=r}^{n}\sum_{j=r}^{m} a_{ij} = \sup_{n} \sum_{i=r}^{n} \lim_{m\to\infty} \sum_{j=r}^{m} a_{ij}$$

$$= \sup_{n} \sum_{i=r}^{n}\sum_{j=r}^{\infty} a_{ij} = \lim_{n\to\infty} \sum_{i=r}^{n}\sum_{j=r}^{\infty} a_{ij} = \sum_{i=r}^{\infty}\sum_{j=r}^{\infty} a_{ij}$$

Interchanging the i and j in the above argument proves the theorem. ∎

The following is the fundamental result on double sums.

Theorem 11.15. *Let a_{ij} be a number and suppose*

$$\sum_{i=r}^{\infty}\sum_{j=r}^{\infty} |a_{ij}| < \infty .$$

Then

$$\sum_{i=r}^{\infty}\sum_{j=r}^{\infty} a_{ij} = \sum_{j=r}^{\infty}\sum_{i=r}^{\infty} a_{ij}$$

and every infinite sum encountered in the above equation converges.

Proof: By Theorem 11.14

$$\sum_{j=r}^{\infty}\sum_{i=r}^{\infty} |a_{ij}| = \sum_{i=r}^{\infty}\sum_{j=r}^{\infty} |a_{ij}| < \infty$$

Therefore, for each j, $\sum_{i=r}^{\infty} |a_{ij}| < \infty$ and for each i, $\sum_{j=r}^{\infty} |a_{ij}| < \infty$. By Theorem 11.5 on Page 286, $\sum_{i=r}^{\infty} a_{ij}$, $\sum_{j=r}^{\infty} a_{ij}$ both converge, the first one for every j and the second for every i. Also,

$$\sum_{j=r}^{\infty} \left|\sum_{i=r}^{\infty} a_{ij}\right| \leq \sum_{j=r}^{\infty}\sum_{i=r}^{\infty} |a_{ij}| < \infty$$

and

$$\sum_{i=r}^{\infty}\left|\sum_{j=r}^{\infty}a_{ij}\right| \le \sum_{i=r}^{\infty}\sum_{j=r}^{\infty}|a_{ij}| < \infty$$

so by Theorem 11.5 again,

$$\sum_{j=r}^{\infty}\sum_{i=r}^{\infty}a_{ij}, \quad \sum_{i=r}^{\infty}\sum_{j=r}^{\infty}a_{ij}$$

both exist. It only remains to verify they are equal. Note $0 \le (|a_{ij}| + a_{ij}) \le |a_{ij}|$. Therefore, by Theorem 11.14 and Theorem 11.3 on Page 284

$$\sum_{j=r}^{\infty}\sum_{i=r}^{\infty}|a_{ij}| + \sum_{j=r}^{\infty}\sum_{i=r}^{\infty}a_{ij} = \sum_{j=r}^{\infty}\sum_{i=r}^{\infty}(|a_{ij}| + a_{ij}) = \sum_{i=r}^{\infty}\sum_{j=r}^{\infty}(|a_{ij}| + a_{ij})$$

$$= \sum_{i=r}^{\infty}\sum_{j=r}^{\infty}|a_{ij}| + \sum_{i=r}^{\infty}\sum_{j=r}^{\infty}a_{ij} - \sum_{j=r}^{\infty}\sum_{i=r}^{\infty}|a_{ij}| + \sum_{i=r}^{\infty}\sum_{j=r}^{\infty}a_{ij}$$

and so $\sum_{j=r}^{\infty}\sum_{i=r}^{\infty}a_{ij} = \sum_{i=r}^{\infty}\sum_{j=r}^{\infty}a_{ij}$ as claimed. ∎

One of the most important applications of this theorem is to the problem of multiplication of series.

Definition 11.5. Let $\sum_{i=r}^{\infty}a_i$ and $\sum_{i=r}^{\infty}b_i$ be two series. For $n \ge r$, define

$$c_n \equiv \sum_{k=r}^{n}a_k b_{n-k+r}.$$

The series $\sum_{n=r}^{\infty}c_n$ is called the Cauchy product of the two series.

It is not hard to see where this comes from. Formally write the following in the case $r = 0$:

$$(a_0 + a_1 + a_2 + a_3 \cdots)(b_0 + b_1 + b_2 + b_3 \cdots)$$

and start multiplying in the usual way. This yields

$$a_0 b_0 + (a_0 b_1 + b_0 a_1) + (a_0 b_2 + a_1 b_1 + a_2 b_0) + \cdots$$

and you see the expressions in parentheses above are just the c_n for $n = 0, 1, 2, \cdots$. Therefore, it is reasonable to conjecture that

$$\sum_{i=r}^{\infty}a_i\sum_{j=r}^{\infty}b_j = \sum_{n=r}^{\infty}c_n$$

and of course there would be no problem with this in the case of finite sums but in the case of infinite sums, it is necessary to prove a theorem. The following is a special case of Merten's theorem.

Theorem 11.16. *Suppose $\sum_{i=r}^{\infty}a_i$ and $\sum_{j=r}^{\infty}b_j$ both converge absolutely[1]. Then*

$$\left(\sum_{i=r}^{\infty}a_i\right)\left(\sum_{j=r}^{\infty}b_j\right) = \sum_{n=r}^{\infty}c_n$$

---

[1]Actually, it is only necessary to assume one of the series converges and the other converges absolutely. This is known as Merten's theorem and may be read in the 1974 book by Apostol listed in the bibliography.

where

$$c_n = \sum_{k=r}^{n} a_k b_{n-k+r}.$$

Proof: Let $p_{nk} = 1$ if $r \le k \le n$ and $p_{nk} = 0$ if $k > n$. Then

$$c_n = \sum_{k=r}^{\infty} p_{nk} a_k b_{n-k+r}.$$

Also,

$$\sum_{k=r}^{\infty} \sum_{n=r}^{\infty} p_{nk} |a_k| |b_{n-k+r}| = \sum_{k=r}^{\infty} |a_k| \sum_{n=r}^{\infty} p_{nk} |b_{n-k+r}| = \sum_{k=r}^{\infty} |a_k| \sum_{n=k}^{\infty} |b_{n-k+r}|$$

$$= \sum_{k=r}^{\infty} |a_k| \sum_{n=k}^{\infty} |b_{n-(k-r)}| = \sum_{k=r}^{\infty} |a_k| \sum_{m=r}^{\infty} |b_m| < \infty.$$

Therefore, by Theorem 11.15

$$\sum_{n=r}^{\infty} c_n = \sum_{n=r}^{\infty} \sum_{k=r}^{n} a_k b_{n-k+r} = \sum_{n=r}^{\infty} \sum_{k=r}^{\infty} p_{nk} a_k b_{n-k+r}$$

$$= \sum_{k=r}^{\infty} a_k \sum_{n=r}^{\infty} p_{nk} b_{n-k+r} = \sum_{k=r}^{\infty} a_k \sum_{n=k}^{\infty} b_{n-k+r} = \sum_{k=r}^{\infty} a_k \sum_{m=r}^{\infty} b_m \qquad \blacksquare$$

11.7   Exercises

(1) Determine whether the following series converge absolutely, conditionally, or not at all and give reasons for your answers.

   (a) $\sum_{n=1}^{\infty} (-1)^n \frac{1}{\sqrt{n^2+n+1}}$

   (b) $\sum_{n=1}^{\infty} (-1)^n \left(\sqrt{n+1} - \sqrt{n}\right)$

   (c) $\sum_{n=1}^{\infty} (-1)^n \frac{(n!)^2}{(2n)!}$

   (d) $\sum_{n=1}^{\infty} (-1)^n \frac{(2n)!}{(n!)^2}$

   (e) $\sum_{n=1}^{\infty} \frac{(-1)^n}{2n+2}$

   (f) $\sum_{n=1}^{\infty} (-1)^n \left(\frac{n}{n+1}\right)^n$

   (g) $\sum_{n=1}^{\infty} (-1)^n \left(\frac{n}{n+1}\right)^{n^2}$

(2) Determine whether the following series converge absolutely, conditionally, or not at all and give reasons for your answers.

   (a) $\sum_{k=1}^{\infty} (-1)^n \frac{\ln(k^5)}{k}$

   (b) $\sum_{k=1}^{\infty} (-1)^n \frac{\ln(k^5)}{k^{1.01}}$

   (c) $\sum_{n=1}^{\infty} (-1)^n \frac{10^n}{(1.01)^n}$

(d) $\sum_{n=1}^{\infty} (-1)^n \sin\left(\frac{1}{n}\right)$

(e) $\sum_{n=1}^{\infty} (-1)^n \tan\left(\frac{1}{n^2}\right)$

(f) $\sum_{n=1}^{\infty} (-1)^n \cos\left(\frac{1}{n^2}\right)$

(g) $\sum_{n=1}^{\infty} (-1)^n \sin\left(\frac{\sqrt{n}}{n^2+1}\right)$

(3) Determine whether the following series converge absolutely, conditionally, or not at all and give reasons for your answers.

(a) $\sum_{n=1}^{\infty} (-1)^n \frac{2^n+n}{n2^n}$

(b) $\sum_{n=1}^{\infty} (-1)^n \frac{2^n+n}{n^2 2^n}$

(c) $\sum_{n=1}^{\infty} (-1)^n \frac{n}{2n+1}$

(d) $\sum_{n=1}^{\infty} (-1)^n \frac{10^n}{n!}$

(e) $\sum_{n=1}^{\infty} (-1)^n \frac{n^{100}}{1.01^n}$

(f) $\sum_{n=1}^{\infty} (-1)^n \frac{\ln n}{n^2}$

(g) $\sum_{n=1}^{\infty} (-1)^n \frac{3^n}{n^3}$

(h) $\sum_{n=1}^{\infty} (-1)^n \frac{n^3}{3^n}$

(i) $\sum_{n=1}^{\infty} (-1)^n \frac{n^3}{n!}$

(j) $\sum_{n=1}^{\infty} (-1)^n \frac{n!}{n^{100}}$

(4) Find the exact values of the following infinite series if they converge.

(a) $\sum_{k=3}^{\infty} \frac{1}{k(k-2)}$

(b) $\sum_{k=1}^{\infty} \frac{1}{k(k+1)}$

(c) $\sum_{k=3}^{\infty} \frac{1}{(k+1)(k-2)}$

(d) $\sum_{k=1}^{\infty} \left(\frac{1}{\sqrt{k}} - \frac{1}{\sqrt{k+1}}\right)$

(e) $\sum_{n=1}^{\infty} \ln\left(\frac{(n+1)^2}{n(n+2)}\right)$

(5) Suppose $\sum_{n=1}^{\infty} a_n$ converges absolutely. Can the same thing be said about $\sum_{n=1}^{\infty} a_n^2$? Explain.

(6) A person says a series converges conditionally by the ratio test. Explain why his statement is total nonsense.

(7) A person says a series diverges by the alternating series test. Explain why his statement is total nonsense.

(8) Find a series which diverges using one test but converges using another if possible. If this is not possible, tell why.

(9) If $\sum_{n=1}^{\infty} a_n$ and $\sum_{n=1}^{\infty} b_n$ both converge, can you conclude the sum $\sum_{n=1}^{\infty} a_n b_n$ converges?

(10) If $\sum_{n=1}^{\infty} a_n$ converges absolutely, and b_n is bounded, can you conclude $\sum_{n=1}^{\infty} a_n b_n$ converges? What if it is only the case that $\sum_{n=1}^{\infty} a_n$ converges?

(11) The logarithm test states the following. Suppose $a_k \neq 0$ for large k and that

$$p = \lim_{k \to \infty} \frac{\ln\left(\frac{1}{|a_k|}\right)}{\ln k}$$

exists. If $p > 1$, then $\sum_{k=1}^{\infty} a_k$ converges absolutely. If $p < 1$, then the series $\sum_{k=1}^{\infty} a_k$ does not converge absolutely. Prove this theorem.

(12) Prove Theorem 11.11. **Hint:** For $\sum_{k=1}^{\infty} (-1)^k b_k$, show the odd partial sums are all at least as small as $\sum_{k=1}^{\infty} (-1)^k b_k$ and are increasing while the even partial sums are at least as large as $\sum_{k=1}^{\infty} (-1)^k b_k$ and are decreasing.

(13) Use Theorem 11.11 in the following alternating series to tell how large n must be so that

$$\left| \sum_{k=1}^{\infty} (-1)^k a_k - \sum_{k=1}^{n} (-1)^k a_k \right|$$

is no larger than the given number.

(a) $\sum_{k=1}^{\infty} (-1)^k \frac{1}{k}$, .001

(b) $\sum_{k=1}^{\infty} (-1)^k \frac{1}{k^2}$, .001

(c) $\sum_{k=1}^{\infty} (-1)^k \sin\left(\frac{1}{k}\right)$, .001

(d) $\sum_{k=1}^{\infty} (-1)^{k-1} \frac{1}{\sqrt{k}}$, .001

(e) $\sum_{k=1}^{\infty} (-1)^k \frac{\ln k}{k}$, .001 Note that this one satisfies b_k is decreasing if $k > 3$ but not for all k. Does this matter?

(14) For $1 \geq x \geq 0$, and $p \geq 1$, show that $(1 - x)^p \geq 1 - px$. **Hint:** This can be done using the mean value theorem from calculus. Define $f(x) \equiv (1 - x)^p - 1 + px$ and show that $f(0) = 0$ while $f'(x) \geq 0$ for all $x \in (0, 1)$.

(15) Using the result of Problem 14 establish Raabe's Test, an interesting variation on the ratio test. This test says the following. Suppose there exists a constant C and a number p such that

$$\left| \frac{a_{k+1}}{a_k} \right| \leq 1 - \frac{p}{k + C}$$

for all k large enough. Then if $p > 1$, it follows that $\sum_{k=1}^{\infty} a_k$ converges absolutely. **Hint:** Let $b_k \equiv k - 1 + C$ and note that for all k large enough, $b_k > 1$. Now conclude that there exists an integer k_0 such that $b_{k_0} > 1$ and for all $k \geq k_0$ the given inequality above holds. Use Problem 14 to conclude that

$$\left| \frac{a_{k+1}}{a_k} \right| \leq 1 - \frac{p}{k + C}$$

$$\leq \left(1 - \frac{1}{k + C} \right)^p = \left(\frac{b_k}{b_{k+1}} \right)^p$$

showing that $|a_k| b_k^p$ is decreasing for $k \geq k_0$. Thus for some constant K

$$|a_k| \leq K / b_k^p$$
$$= K / (k - 1 + C)^p.$$

Now use comparison theorems and the p series to obtain the conclusion of the theorem.

(16) Consider the series $\sum_{k=0}^{\infty} (-1)^n \frac{1}{\sqrt{n+1}}$. Show this series converges and so it makes sense to write $\left(\sum_{k=0}^{\infty} (-1)^n \frac{1}{\sqrt{n+1}}\right)^2$. What about the Cauchy product of this series? Does it even converge? What does this mean about using algebra on infinite sums as though they were finite sums?

(17) Verify Theorem 11.16 on the two series $\sum_{k=0}^{\infty} 2^{-k}$ and $\sum_{k=0}^{\infty} 3^{-k}$.

(18) You can define infinite series of complex numbers in exactly the same way as infinite series of real numbers. That is $w = \sum_{k=1}^{\infty} z_k$ means: For every $\varepsilon > 0$ there exists N such that if $n \geq N$, then $\left| w - \sum_{k=1}^{N} z_k \right| < \varepsilon$. Here the absolute value is the one which applies to complex numbers. That is, $|a + ib| = \sqrt{a^2 + b^2}$. Show that if $\{b_n\}$ is a decreasing sequence of nonnegative numbers with the property that $\lim_{n \to \infty} b_n = 0$ and if w is any complex number which is not equal to 1 but which satisfies $|w| = 1$, then $\sum_{k=1}^{\infty} w^n b_n$ must converge. Note a sequence of complex numbers, $\{a_n + ib_n\}$ converges to $a + ib$ if and only if $a_n \to a$ and $b_n \to b$. See Problem 12 on Page 69. There are quite a few things in this problem you should think about.

(19) Consider the following series. $\sum_{n=1}^{\infty} (-1)^n \frac{n^n}{n! e^n}$. Find whether this series converges using Stirling's formula, $\lim_{n \to \infty} \frac{n!}{\alpha n^{n+1/2} e^{-n}} = 1$ for some positive number α.

11.8   Power Series

11.8.1   *Functions Defined In Terms Of Series*

Earlier Taylor polynomials were used to approximate known functions such as $\sin x$ and $\ln (1 + x)$. A much more exciting idea is to use infinite series as definitions of possibly new functions.

Definition 11.6. Let $\{a_k\}_{k=0}^{\infty}$ be a sequence of numbers. The expression

$$\sum_{k=0}^{\infty} a_k (x - a)^k \tag{11.10}$$

is called a Taylor series centered at a. This is also called a power series centered at a.

In the above definition, x is a variable. Thus you can put in various values of x and ask whether the resulting series of numbers converges. Defining, D to be the set of all values of x such that the resulting series does converge, define a new function f defined on D as

$$f(x) = \sum_{k=0}^{\infty} a_k (x - a)^k .$$

This might be a totally new function, one which has no name. Nevertheless, much can be said about such functions. The following lemma is fundamental in considering the form of D which always turns out to be an interval centered at a which may or may not contain either end point.

Lemma 11.3. *Suppose* $z \in D$. *Then if* $|x - a| < |z - a|$, *then* $x \in D$ *also and furthermore, the series* $\sum_{k=0}^{\infty} |a_k| |x - a|^k$ *converges.*

Proof: Let $1 > r = |x - a| / |z - a|$. The nth term test implies

$$\lim_{n \to \infty} |a_n| |z - a|^n = 0$$

and so for all n large enough,

$$|a_n| |z - a|^n < 1$$

so for such n,

$$|a_n| |x - a|^n = |a_n| |z - a|^n \frac{|x - a|^n}{|z - a|^n} \le \frac{|x - a|^n}{|z - a|^n} < r^n$$

Therefore, $\sum_{k=0}^{\infty} |a_k| |x - a|^k$ converges by comparison with the geometric series $\sum r^n$. ∎

With this lemma, the following fundamental theorem is obtained.

Theorem 11.17. *Let* $\sum_{k=0}^{\infty} a_k (x - a)^k$ *be a Taylor series. Then there exists* $r \le \infty$ *such that the Taylor series converges absolutely if* $|x - a| < r$. *Furthermore, if* $|x - a| > r$, *the Taylor series diverges.*

Proof: Let

$$r \equiv \sup \{|y - a| : y \in D\}.$$

Then if $|x - a| < r$, it follows there exists $z \in D$ such that $|z - a| > |x - a|$ since otherwise, r wouldn't be as defined. In fact $|x - a|$ would then be an upper bound to $\{|y - a| : y \in D\}$. Therefore, by the above lemma $\sum_{k=0}^{\infty} |a_k| |x - a|^k$ converges and this proves the first part of this theorem.

Now suppose $|x - a| > r$. If $\sum_{k=0}^{\infty} a_k (x - a)^k$ converges then by the above lemma, r fails to be an upper bound to $\{|y - a| : y \in D\}$ and so the Taylor series must diverge as claimed. ∎

From now on D will be referred to as the interval of convergence and r of the above theorem as the radius of convergence. Determining which points of $\{x : |x - a| = r\}$ are in D requires the use of specific convergence tests and can be quite hard. However, the determination of r tends to be pretty easy.

Example 11.12. Find the interval of convergence of the Taylor series $\sum_{n=1}^{\infty} \frac{x^n}{n}$.

Use Corollary 11.3.

$$\lim_{n \to \infty} \left(\frac{|x|^n}{n} \right)^{1/n} = \lim_{n \to \infty} \frac{|x|}{\sqrt[n]{n}} = |x|$$

because $\lim_{n \to \infty} \sqrt[n]{n} = 1$ and so if $|x| < 1$ the series converges. The endpoints require special attention. When $x = 1$ the series diverges because it reduces to $\sum_{n=1}^{\infty} \frac{1}{n}$. At the other endpoint however, the series converges, because it reduces to $\sum_{n=1}^{\infty} \frac{(-1)^n}{n}$ and the alternating series test applies and gives convergence.

Example 11.13. Find the radius of convergence of $\sum_{n=1}^{\infty} \frac{n^n}{n!} x^n$.

Apply the ratio test. Taking the ratio of the absolute values of the $(n+1)^{th}$ and the nth terms

$$\frac{\frac{(n+1)^{(n+1)}}{(n+1)n!} |x|^{n+1}}{\frac{n^n}{n!} |x|^n} = (n+1)^n |x| n^{-n} = |x| \left(1 + \frac{1}{n} \right)^n \to |x| e$$

Therefore the series converges absolutely if $|x| e < 1$ and diverges if $|x| e > 1$. Consequently, $r = 1/e$.

11.8.2   *Operations On Power Series*

It is desirable to be able to differentiate, integrate, and multiply power series. The following theorem says one can differentiate power series in the most natural way on the interval of convergence, just as you would differentiate a polynomial. This theorem may seem obvious, but it is a serious mistake to think this. You usually cannot differentiate an infinite series whose terms are functions even if the functions are themselves polynomials. The following is special and pertains to power series. It is another example of the interchange of two limits, in this case, the limit involved in taking the derivative and the limit of the sequence of finite sums.

Theorem 11.18. *Let $\sum_{n=0}^{\infty} a_n (x - a)^n$ be a Taylor series having radius of convergence $r > 0$ and let*

$$f(x) \equiv \sum_{n=0}^{\infty} a_n (x - a)^n \tag{11.11}$$

for $|x - a| < r$. Then

$$f'(x) = \sum_{n=0}^{\infty} a_n n (x - a)^{n-1} = \sum_{n=1}^{\infty} a_n n (x - a)^{n-1} \tag{11.12}$$

and this new differentiated power series, the derived series, has radius of convergence equal to r.

Proof: First it will be shown that the series on the right in (11.12) has the same radius of convergence as the original series. Thus let $|x - a| < r$ and pick y such that

$$|x - a| < |y - a| < r.$$

Then

$$\lim_{n \to \infty} |a_n| |y - a|^{n-1} = \lim_{n \to \infty} |a_n| |y - a|^n = 0$$

because

$$\sum_{n=0}^{\infty} |a_n| |y - a|^n < \infty$$

and so, for n large enough,

$$|a_n| |y - a|^{n-1} < 1.$$

Therefore, for large enough n,

$$|a_n| n |x - a|^{n-1} = |a_n| |y - a|^{n-1} n \left| \frac{x-a}{y-a} \right|^{n-1} \leq n \left| \frac{x-a}{y-a} \right|^{n-1}$$

and so $\sum_{n=1}^{\infty} n \left| \frac{x-a}{y-a} \right|^{n-1}$ converges by the ratio test. By the comparison test, it follows $\sum_{n=1}^{\infty} a_n n (x - a)^{n-1}$ converges absolutely for any x satisfying $|x - a| < r$. Therefore, the radius of convergence of the derived series is at least as large as that of the original series. On the other hand, if $\sum_{n=1}^{\infty} |a_n| n |x - a|^{n-1}$ converges then by the comparison test, $\sum_{n=1}^{\infty} |a_n| |x - a|^{n-1}$ and therefore $\sum_{n=1}^{\infty} |a_n| |x - a|^n$ also converges which shows the radius of convergence of the derived series is no larger than that of the original series. It remains to verify the assertion about the derivative.

Let $|x - a| < r$ and let $r_1 < r$ be close enough to r that

$$x \in (a - r_1, a + r_1) \subseteq [a - r_1, a + r_1] \subseteq (a - r, a + r).$$

Thus, letting $r_2 \in (r_1, r)$,

$$\sum_{n=0}^{\infty} |a_n| r_1^n, \ \sum_{n=0}^{\infty} |a_n| r_2^n < \infty \tag{11.13}$$

Letting y be close enough to x, it follows both x and y are in $[a - r_1, a + r_1]$. Then considering the difference quotient,

$$\frac{f(y) - f(x)}{y - x} = \sum_{n=0}^{\infty} a_n (y - x)^{-1} [(y - a)^n - (x - a)^n] = \sum_{n=1}^{\infty} a_n n z_n^{n-1} \tag{11.14}$$

where the last equation follows from the mean value theorem and z_n is some point between $x - a$ and $y - a$. Therefore,

$$\frac{f(y) - f(x)}{y - x} = \sum_{n=1}^{\infty} a_n n z_n^{n-1}$$

$$= \sum_{n=1}^{\infty} a_n n \left(z_n^{n-1} - (x-a)^{n-1} \right) + \sum_{n=1}^{\infty} a_n n \, (x-a)^{n-1}$$

$$= \sum_{n=2}^{\infty} a_n n \, (n-1) \, w_n^{n-2} \, (z_n - (x-a)) + \sum_{n=1}^{\infty} a_n n \, (x-a)^{n-1} \qquad (11.15)$$

where w_n is between z_n and $x-a$. Thus w_n is between $x-a$ and $y-a$ and so

$$w_n + a \in [a - r_1, a + r_1]$$

which implies $|w_n| \le r_1$. The first sum on the right in (11.15) therefore satisfies

$$\left| \sum_{n=2}^{\infty} a_n n \, (n-1) \, w_n^{n-2} \, (z_n - (x-a)) \right| \le |y-x| \sum_{n=2}^{\infty} |a_n| \, n \, (n-1) \, |w_n|^{n-2}$$

$$\le |y-x| \sum_{n=2}^{\infty} |a_n| \, n \, (n-1) \, r_1^{n-2}$$

$$= |y-x| \sum_{n=2}^{\infty} |a_n| \, r_2^{n-2} \, n \, (n-1) \left(\frac{r_1}{r_2} \right)^{n-2}$$

Now from (11.13), $|a_n| \, r_2^{n-2} < 1$ for all n large enough. Therefore, for such n,

$$|a_n| \, r_2^{n-2} n \, (n-1) \left(\frac{r_1}{r_2} \right)^{n-2} \le n \, (n-1) \left(\frac{r_1}{r_2} \right)^{n-2}$$

and the series $\sum n \, (n-1) \left(\frac{r_1}{r_2} \right)^{n-2}$ converges by the ratio test. Therefore, there exists a constant C independent of y such that

$$\sum_{n=2}^{\infty} |a_n| \, n \, (n-1) \, r_1^{n-2} = C < \infty$$

Consequently, from (11.15)

$$\left| \frac{f(y) - f(x)}{y - x} - \sum_{n=1}^{\infty} a_n n \, (x-a)^{n-1} \right| \le C \, |y-x| .$$

Taking the limit as $y \to x$ (11.12) follows. ∎

As an immediate corollary, it is possible to characterize the coefficients of a Taylor series.

Corollary 11.4. *Let $\sum_{n=0}^{\infty} a_n \, (x-a)^n$ be a Taylor series with radius of convergence $r > 0$ and let*

$$f(x) \equiv \sum_{n=0}^{\infty} a_n \, (x-a)^n . \qquad (11.16)$$

Then

$$a_n = \frac{f^{(n)}(a)}{n!} . \qquad (11.17)$$

Proof: From (11.16), $f(a) = a_0 \equiv f^{(0)}(a)/0!$. From Theorem 11.18,

$$f'(x) = \sum_{n=1}^{\infty} a_n n (x-a)^{n-1} = a_1 + \sum_{n=2}^{\infty} a_n n (x-a)^{n-1}.$$

Now let $x = a$ and obtain that $f'(a) = a_1 = f'(a)/1!$. Next use Theorem 11.18 again to take the second derivative and obtain

$$f''(x) = 2a_2 + \sum_{n=3}^{\infty} a_n n (n-1)(x-a)^{n-2}$$

let $x = a$ in this equation and obtain $a_2 = f''(a)/2 = f''(a)/2!$. Continuing this way proves the corollary.∎

This also shows the coefficients of a Taylor series are unique. That is, if

$$\sum_{k=0}^{\infty} a_k (x-a)^k = \sum_{k=0}^{\infty} b_k (x-a)^k$$

for all x in some interval, then $a_k = b_k$ for all k.

Example 11.14. Find the power series for $\sin(x)$, and $\cos(x)$ centered at 0 and give the interval of convergence.

First consider $f(x) = \sin(x)$. Then $f'(x) = \cos(x)$, $f''(x) = -\sin(x)$, $f'''(x) = -\cos(x)$ etc. Therefore, from Taylor's formula, Theorem 11.1 on Page 279,

$$f(x) = 0 + x + 0 - \frac{x^3}{3!} + 0 + \frac{x^5}{5!} + \cdots + \frac{x^{2n+1}}{(2n+1)!} + \frac{f^{(2n+2)}(\xi_n)}{(2n+2)!}$$

where ξ_n is some number between 0 and x. Furthermore, this equals either $\pm\sin(\xi_n)$ or $\pm\cos(\xi_n)$ and so its absolute value is no larger than 1. Thus

$$\left| \frac{f^{(2n+2)}(\xi_n)}{(2n+2)!} \right| \leq \frac{1}{(2n+2)!}.$$

By the ratio test, it follows that

$$\sum_{n=0}^{\infty} \frac{1}{(2n+2)!} < \infty$$

and so by the comparison test,

$$\sum_{n=0}^{\infty} \left| \frac{f^{(2n+2)}(\xi_n)}{(2n+2)!} \right| < \infty$$

also. Therefore, by the nth term test $\lim_{n\to\infty} \frac{f^{(2n+2)}(\xi_n)}{(2n+2)!} = 0$. This implies

$$\sin(x) = \sum_{k=0}^{n} (-1)^k \frac{x^{2k+1}}{(2k+1)!} + \frac{f^{(2n+2)}(\xi_n)}{(2n+2)!}$$

and the last term converges to zero as $n \to \infty$ for any value of x and therefore,

$$\sin(x) = \sum_{k=0}^{\infty} (-1)^k \frac{x^{2k+1}}{(2k+1)!}$$

for all $x \in \mathbb{R}$. By Theorem 11.18, you can differentiate both sides, doing the series term by term and obtain

$$\cos(x) = \sum_{k=0}^{\infty} (-1)^k \frac{x^{2k}}{(2k)!}$$

for all $x \in \mathbb{R}$.

Example 11.15. Find the sum $\sum_{k=1}^{\infty} k2^{-k}$.

It may not be obvious what this sum equals but with the above theorem it is easy to find. From the formula for the sum of a geometric series, $\frac{1}{1-t} = \sum_{k=0}^{\infty} t^k$ if $|t| < 1$. Differentiate both sides to obtain

$$(1-t)^{-2} = \sum_{k=1}^{\infty} kt^{k-1}$$

whenever $|t| < 1$. Let $t - 1/2$. Then

$$4 - \frac{1}{(1-(1/2))^2} = \sum_{k=1}^{\infty} k2^{-(k-1)}$$

and so if you multiply both sides by 2^{-1},

$$2 = \sum_{k=1}^{\infty} k2^{-k}.$$

The following is a very important example known as the binomial series.

Example 11.16. Find a Taylor series for the function $(1+x)^{\alpha}$ centered at 0 valid for $|x| < 1$.

Use Theorem 11.18 to do this. First note that if $y(x) \equiv (1+x)^{\alpha}$, then y is a solution of the following initial value problem.

$$y' - \frac{\alpha}{(1+x)} y = 0, \ y(0) = 1. \tag{11.18}$$

Next it is necessary to observe there is only one solution to this initial value problem. To see this, multiply both sides of the differential equation in (11.18) by $(1+x)^{-\alpha}$. When this is done one obtains

$$\frac{d}{dx}\left((1+x)^{-\alpha} y\right) = (1+x)^{-\alpha}\left(y' - \frac{\alpha}{(1+x)}y\right) = 0. \tag{11.19}$$

Therefore, from (11.19), there must exist a constant C, such that

$$(1+x)^{-\alpha} y = C.$$

However, $y(0) = 1$ and so it must be that $C = 1$. Therefore, there is exactly one solution to the initial value problem in (11.18) and it is $y(x) = (1+x)^\alpha$. The strategy for finding the Taylor series of this function consists of finding a series which solves the initial value problem above. Let

$$y(x) \equiv \sum_{n=0}^{\infty} a_n x^n \tag{11.20}$$

be a solution to (11.18). Of course it is not known at this time whether such a series exists. However, the process of finding it will demonstrate its existence. From Theorem 11.18 and the initial value problem

$$(1+x)\sum_{n=0}^{\infty} a_n n x^{n-1} - \sum_{n=0}^{\infty} \alpha a_n x^n = 0$$

and so

$$\sum_{n=1}^{\infty} a_n n x^{n-1} + \sum_{n=0}^{\infty} a_n (n-\alpha) x^n = 0$$

Changing the order variable of summation in the first sum

$$\sum_{n=0}^{\infty} a_{n+1}(n+1) x^n + \sum_{n=0}^{\infty} a_n (n-\alpha) x^n = 0$$

and from Corollary 11.4 and the initial condition for (11.18) this requires

$$a_{n+1} = \frac{a_n (\alpha - n)}{n+1}, a_0 = 1. \tag{11.21}$$

Therefore, from (11.21) and letting $n = 0$, $a_1 = \alpha$. Then using (11.21) again along with this information, $a_2 = \frac{\alpha(\alpha-1)}{2}$. Using the same process, $a_3 = \frac{\left(\frac{\alpha(\alpha-1)}{2}\right)(\alpha-2)}{3} = \frac{\alpha(\alpha-1)(\alpha-2)}{3!}$. By now you can spot the pattern. In general,

$$a_n = \frac{\overbrace{\alpha(\alpha-1)\cdots(\alpha-n+1)}^{n \text{ of these factors}}}{n!}.$$

Therefore, the candidate for the Taylor series is

$$y(x) = \sum_{n=0}^{\infty} \frac{\alpha(\alpha-1)\cdots(\alpha-n+1)}{n!} x^n.$$

Furthermore, the above discussion shows this series solves the initial value problem on its interval of convergence. It only remains to show the radius of convergence of this series equals 1. It will then follow that this series equals $(1+x)^\alpha$ because of uniqueness of the initial value problem. To find the radius of convergence, use the ratio test. Thus the ratio of the absolute values of the $(n+1)^{st}$ term to the absolute value of the nth term is

$$\frac{\left|\frac{\alpha(\alpha-1)\cdots(\alpha-n+1)(\alpha-n)}{(n+1)n!}\right| |x|^{n+1}}{\left|\frac{\alpha(\alpha-1)\cdots(\alpha-n+1)}{n!}\right| |x|^n} = |x| \frac{|\alpha-n|}{n+1} \to |x|$$

showing that the radius of convergence is 1 since the series converges if $|x| < 1$ and diverges if $|x| > 1$.

The expression $\frac{\alpha(\alpha-1)\cdots(\alpha-n+1)}{n!}$ is often denoted as $\binom{\alpha}{n}$. With this notation, the following theorem has been established.

Theorem 11.19. *Let α be a real number and let $|x| < 1$. Then*

$$(1+x)^\alpha = \sum_{n=0}^{\infty} \binom{\alpha}{n} x^n.$$

There is a very interesting issue related to the above theorem which illustrates the limitation of power series. The function $f(x) = (1+x)^\alpha$ makes sense for all $x > -1$ but one is only able to describe it with a power series on the interval $(-1, 1)$. Think about this. The above technique is a standard one for obtaining solutions of differential equations and this example illustrates a deficiency in the method. To completely understand power series, it is necessary to take a course in complex analysis. You may have noticed the prominent role played by geometric series. This is no accident. It turns out that the right way to consider Taylor series is through the use of geometric series and something called the Cauchy integral formula of complex analysis. However, these are topics for another course.

You can also integrate power series on their interval of convergence.

Theorem 11.20. *Let $f(x) = \sum_{n=0}^{\infty} a_n (x-a)^n$ and suppose the radius of convergence is $r > 0$. Then if $|y - a| < r$,*

$$\int_a^y f(x)\, dx = \sum_{n=0}^{\infty} \int_a^y a_n (x-a)^n\, dx = \sum_{n=0}^{\infty} \frac{a_n (y-a)^{n+1}}{n+1}.$$

Proof: Define $F(y) = \int_a^y f(x)\, dx$ and $G(y) \equiv \sum_{n=0}^{\infty} \frac{a_n(y-a)^{n+1}}{n+1}$. By Theorem 11.18 and the Fundamental theorem of calculus,

$$G'(y) = \sum_{n=0}^{\infty} a_n (y-a)^n = f(y) = F'(y).$$

Therefore, $G(y) - F(y) = C$ for some constant. But $C = 0$ because $F(a) - G(a) = 0$. ∎

Next consider the problem of multiplying two power series.

Theorem 11.21. *Let $\sum_{n=0}^{\infty} a_n (x-a)^n$ and $\sum_{n=0}^{\infty} b_n (x-a)^n$ be two power series having radii of convergence r_1 and r_2, both positive. Then*

$$\left(\sum_{n=0}^{\infty} a_n (x-a)^n \right) \left(\sum_{n=0}^{\infty} b_n (x-a)^n \right) = \sum_{n=0}^{\infty} \left(\sum_{k=0}^{n} a_k b_{n-k} \right) (x-a)^n$$

whenever $|x - a| < r \equiv \min(r_1, r_2)$.

Proof: By Theorem 11.17 both series converge absolutely if $|x - a| < r$. Therefore, by Theorem 11.16

$$\left(\sum_{n=0}^{\infty} a_n (x - a)^n \right) \left(\sum_{n=0}^{\infty} b_n (x - a)^n \right) =$$

$$\sum_{n=0}^{\infty} \sum_{k=0}^{n} a_k (x - a)^k b_{n-k} (x - a)^{n-k} = \sum_{n=0}^{\infty} \left(\sum_{k=0}^{n} a_k b_{n-k} \right) (x - a)^n. \quad \blacksquare$$

The significance of this theorem in terms of applications is that it states you can multiply power series just as you would multiply polynomials and everything will be all right on the common interval of convergence.

This theorem can be used to find Taylor series which would perhaps be hard to find without it. Here is an example.

Example 11.17. Find the Taylor series for $e^x \sin x$ centered at $x = 0$.

Using Problems 11 - 13 on Page 282 or Example 11.14 on Page 310, and Example 11.10 on Page 297, all that is required is to multiply

$$\left(\overbrace{1 + x + \frac{x^2}{2!} + \frac{x^3}{3!} \cdots}^{e^x} \right) \left(\overbrace{x - \frac{x^3}{3!} + \frac{x^5}{5!} + \cdots}^{\sin x} \right)$$

From the above theorem the result should be

$$x + x^2 + \left(-\frac{1}{3!} + \frac{1}{2!} \right) x^3 + \cdots = x + x^2 + \frac{1}{3} x^3 + \cdots$$

You can continue this way and get the following to a few more terms.

$$x + x^2 + \frac{1}{3} x^3 - \frac{1}{30} x^5 - \frac{1}{90} x^6 - \frac{1}{630} x^7 + \cdots$$

I do not see a pattern in these coefficients but I can go on generating them as long as I want. (In practice this tends to not be very long.) I also know the resulting power series will converge for all x because both the series for e^x and the one for $\sin x$ converge for all x.

Example 11.18. Find the Taylor series for $\tan x$ centered at $x = 0$.

Lets suppose it has a Taylor series $a_0 + a_1 x + a_2 x^2 + \cdots$. Then

$$\left(a_0 + a_1 x + a_2 x^2 + \cdots \right) \left(\overbrace{1 - \frac{x^2}{2} + \frac{x^4}{4!} + \cdots}^{\cos x} \right) = \left(x - \frac{x^3}{3!} + \frac{x^5}{5!} + \cdots \right).$$

Using the above, $a_0 = 0$, $a_1 x = x$ so $a_1 = 1$, $\left(0 \left(\frac{-1}{2} \right) + a_2 \right) x^2 = 0$ so $a_2 = 0$.

$$\left(a_3 - \frac{a_1}{2} \right) x^3 = \frac{-1}{3!} x^3$$

so $a_3 - \frac{1}{2} = -\frac{1}{6}$ so $a_3 = \frac{1}{3}$. Clearly one can continue in this manner. Thus the first several terms of the power series for tan are

$$\tan x = x + \frac{1}{3}x^3 + \cdots .$$

You can go on calculating these terms and find the next two yielding

$$\tan x = x + \frac{1}{3}x^3 + \frac{2}{15}x^5 + \frac{17}{315}x^7 + \cdots$$

This is a very significant technique because, as you see, there does not appear to be a very simple pattern for the coefficients of the power series for $\tan x$. Of course there are some issues here about whether $\tan x$ even has a power series, but if it does, the above must be it. In fact, $\tan(x)$ will have a power series valid on some interval centered at 0 and this becomes completely obvious when one uses methods from complex analysis but it is not too obvious at this point. If you are interested in this issue, read the last section of the chapter. Note also that what has been accomplished is to divide the power series for $\sin x$ by the power series for $\cos x$ just like they were polynomials.

11.9 Exercises

(1) Find the radius of convergence of the following.

(a) $\sum_{k=1}^{\infty} \left(\frac{x}{2}\right)^n$

(b) $\sum_{k=1}^{\infty} \sin\left(\frac{1}{n}\right) 3^n x^n$

(c) $\sum_{k=0}^{\infty} k! x^k$

(d) $\sum_{n=0}^{\infty} \frac{(3n)^n}{(3n)!} x^n$

(e) $\sum_{n=0}^{\infty} \frac{(2n)^n}{(2n)!} x^n$

(2) Find $\sum_{k=1}^{\infty} k 2^{-k}$.

(3) Find $\sum_{k=1}^{\infty} k^2 3^{-k}$.

(4) Find $\sum_{k=1}^{\infty} \frac{2^{-k}}{k}$.

(5) Find $\sum_{k=1}^{\infty} \frac{3^{-k}}{k}$.

(6) Find where the series $\sum_{k=1}^{\infty} \frac{1-e^{-kx}}{k}$ converges.

(7) Find the power series centered at 0 for the function $1/(1+x^2)$ and give the radius of convergence. Where does the function make sense? Where does the power series equal the function?

(8) Find a power series for the function $f(x) \equiv \frac{\sin(\sqrt{x})}{\sqrt{x}}$ for $x > 0$. Where does $f(x)$ make sense? Where does the power series you found converge?

(9) Use the power series technique which was applied in Example 11.16 to consider the initial value problem $y' = y, y(0) = 1$. This yields another way to obtain the power series for e^x.

(10) Use the power series technique on the initial value problem $y' + y = 0$, $y(0) = 1$. What is the solution to this initial value problem?

(11) Use the power series technique to find the first several nonzero terms in the power series solution to the initial value problem

$$y'' + xy = 0, \; y(0) = 0, y'(0) = 1.$$

Tell where your solution gives a valid description of a solution for the initial value problem. **Hint:** This is a little different but you proceed the same way as in Example 11.16. The main difference is you have to do two differentiations of the power series instead of one.

(12) Suppose the function e^x is defined in terms of a power series, $e^x \equiv \sum_{k=0}^{\infty} \frac{x^k}{k!}$. Use Theorem 11.16 on Page 301 to show directly the usual law of exponents,

$$e^{x+y} = e^x e^y.$$

Be sure to check all the hypotheses.

(13) Define the following function[2]:

$$f(x) \equiv \begin{cases} e^{-(1/x^2)} & \text{if } x \neq 0 \\ 0 \text{ if } x = 0 \end{cases}.$$

Show that $f^{(k)}(x)$ exists for all k and for all x. Show also that $f^{(k)}(0) = 0$ for all $k \in \mathbb{N}$. Therefore, the power series for $f(x)$ is of the form $\sum_{k=0}^{\infty} 0 x^k$ and it converges for all values of x. However, it fails to converge to $f(x)$ except at the single point $x = 0$.

(14) Let $f_n(x) \equiv \left(\frac{1}{n} + x^2\right)^{1/2}$. Show that for all x,

$$||x| - f_n(x)| \leq \frac{1}{\sqrt{n}}.$$

Now show $f_n'(0) = 0$ for all n and so $f_n'(0) \to 0$. However, the function $f(x) \equiv |x|$ has no derivative at $x = 0$. Thus even though $f_n(x) \to f(x)$ for all x, you cannot say that $f_n'(0) \to f'(0)$.

(15) Let the functions $f_n(x)$ be given in Problem 14 and consider

$$g_1(x) = f_1(x), \; g_n(x) = f_n(x) - f_{n-1}(x) \text{ if } n > 1.$$

Show that for all x,

$$\sum_{k=1}^{\infty} g_k(x) = |x|$$

[2]Surprisingly, this function is very important to those who use modern techniques to study differential equations. One needs to consider test functions which have the property they have infinitely many derivatives but vanish outside of some interval. The theory of complex variables can be used to show there are no examples of such functions if they have a valid power series expansion. It even becomes a little questionable whether such strange functions even exist at all. Nevertheless, they do, there are enough of them, and it is this very example which is used to show this.

and that $g'_k(0) = 0$ for all k. Therefore, you cannot differentiate the series term by term and get the right answer[3].

(16) Use the theorem about the binomial series to give a proof of the binomial theorem

$$(a+b)^n = \sum_{k=0}^{n} \binom{n}{k} a^{n-k} b^k$$

whenever n is a positive integer.

(17) You know $\int_0^x \frac{1}{t+1} dt = \ln|1+x|$. Use this and Theorem 11.20 to find the power series for $\ln|1+x|$ centered at 0. Where does this power series converge? Where does it converge to the function $\ln|1+x|$?

(18) You know $\int_0^x \frac{1}{t^2+1} dt = \arctan x$. Use this and Theorem 11.20 to find the power series for $\arctan x$ centered at 0. Where does this power series converge? Where does it converge to the function $\arctan x$?

(19) Find the power series for $\sin(x^2)$ by plugging in x^2 where ever there is an x in the power series for $\sin x$. How do you know this is the power series for $\sin(x^2)$?

(20) Find the first several terms of the power series for $\sin^2(x)$ by multiplying the power series for $\sin(x)$. Next use the trig. identity $\sin^2(x) = \frac{1-\cos(2x)}{2}$ and the power series for $\cos(2x)$ to find the power series.

(21) Find the power series for $f(x) = \frac{1}{\sqrt{1-x^2}}$.

(22) It is hard to find $\int_0^1 e^{x^2} dx$ because you do not have a convenient antiderivative for the integrand. Replace e^{x^2} with an appropriate power series and estimate this integral.

(23) Do the same as the previous problem for $\int_0^1 \sin(x^2) dx$.

(24) Find $\lim_{x\to 0} \frac{\tan(\sin x) - \sin(\tan x)}{x^7}$.[4]

(25) Consider the function $S(x) \equiv \sum_{n=1}^{\infty} (-1)^{n+1} \frac{x^{2n-1}}{(2n-1)!}$. This is the power series for $\sin(x)$ but pretend you don't know this. Show that the series for $S(x)$ converges for all $x \in \mathbb{R}$. Also show that S satisfies the initial value problem $y'' + y = 0$, $y(0) = 0$, $y'(0) = 1$.

(26) Consider the function $C(x) \equiv \sum_{n=0}^{\infty} (-1)^n \frac{x^{2n}}{(2n)!}$. This is the power series for $\cos(x)$ but pretend you do not know this. Show that the series for $C(x)$ converges for all $x \in \mathbb{R}$. Also show that S satisfies the initial value problem $y'' + y = 0$, $y(0) = 1$, $y'(0) = 0$.

(27) Show there is at most one solution to the initial value problem $y'' + y = 0$, $y(0) = a$, $y'(0) = b$ and find the solution to this problem in terms of $C(x)$

---

[3] How bad can this get? It can be much worse than this. In fact, there are functions which are continuous everywhere and differentiable nowhere. We typically do not have names for them but they are there just the same. Every such function can be written as an infinite sum of polynomials which of course have derivatives at every point. Thus it is nonsense to differentiate an infinite sum term by term without a theorem of some sort.

[4] This is a wonderful example. You should plug in small values of x using a calculator and see what you get using modern technology.

and $S(x)$. Also show directly from the series descriptions for $C(x)$ and $S(x)$ that $S'(x) = C(x)$ and $C'(x) = -S(x)$.

(28) *Using problem 27 about uniqueness of the initial value problem, show that

$$C(x + y) = C(x)C(y) - S(x)S(y)$$

and

$$S(x + y) = S(x)C(y) + S(y)C(x).$$

Do this in the following way: Fix y and consider the function $f(x) \equiv C(x + y)$ and

$$g(x) = C(x)C(y) - S(x)S(y).$$

Then show both f and g satisfy the same initial value problem and so they must be equal. Do the other identity the same way. Also show $S(-x) = -S(x)$ and $C(-x) = C(x)$ and $S(x)^2 + C(x)^2 = 1$. This last claim is really easy. Just take the derivative and see $S^2 + C^2$ must be constant.

(29) *You know $S(0) = 0$ and $C(0) = 1$. Show there exists $T > 0$ such that on $(0, T)$ both $S(x)$ and $C(x)$ are positive but $C(T) = 0$ while $S(T) = 1$. (We usually refer to T as $\frac{\pi}{2}$.) To do this, note that $S'(0) > 0$ and so S is an increasing function on some interval. Therefore, C is a decreasing function on that interval because of $S^2 + C^2 = 1$. If C is bounded below by some positive number, then S must be unbounded because $S' = C$. However this would contradict $S^2 + C^2 = 1$. Therefore, $C(T) = 0$ for some T. Let T be the first time this occurs. You fill in the mathematical details of this argument. Next show that on $(T, 2T)$, $S(x) > 0$ and $C(x) < 0$ and on $(2T, 3T)$, both $C(x)$ and $S(x)$ are negative. Finally, show that on $(3T, 4T)$, $C(x) > 0$ and $S(x) < 0$. Also show $C(x + 2T) = C(x)$ and $S(x + 2T) = S(x)$. Do all this without resorting to identifying $S(x)$ with $\sin x$ and $C(x)$ with $\cos x$. Finally explain why $\sin x = S(x)$ for all x and $C(x) = \cos x$ for all x.

(30) Bessel's equation of order n is the differential equation

$$x^2 y'' + xy' + (x^2 - n^2) y = 0.$$

Show that a solution of the form $\sum_{k=0}^{\infty} a_k x^k$ exists in the case where $n = 0$. Show that this function $J_0(x)$ is defined as

$$J_0(x) = \sum_{k=0}^{\infty} (-1)^k \frac{x^{2k}}{(k!)^2 2^{2k}}$$

and verify that the series converges for all real x.

(31) Explain why the function $y(x) = \tan(x)$ solves the initial value problem

$$y' = 1 + y^2, \ y(0) = 0.$$

Thus $y(0) = 0$. Then from the equation, $y'(0) = 1$, $y'' = 2yy'$ and so $y''(0) = 0$, etc. Explain these assertions and then tell how to use this differential equation to find a power series for $\tan(x)$.

Note: Problems 25 - 29 outline a way to define the circular functions with no reference to plane geometry. The job is not finished because these circular functions were defined as the x and y coordinates of a point on the unit circle where the angle was measured in terms of arc length and I have not yet tied it in to arc length. This is very easy to do later. Taking the Pythagorean theorem as the **definition** of length, a precise description of what is meant by arc length in terms of integrals can be presented. When this is done, it is possible to **define** $\sin x$ and $\cos x$ in terms of these power series described above and totally eliminate all references to plane geometry. This approach is vastly superior to the traditional approach presented earlier in this book. Calculus is different than geometry and so it is desirable to obtain descriptions of the important functions which are free of geometry.

11.10   Some Other Theorems

First recall Theorem 11.16 on Page 301. For convenience, the version of this theorem which is of interest here is listed below.

Theorem 11.22. *Suppose $\sum_{i=0}^{\infty} a_i$ and $\sum_{j=0}^{\infty} b_j$ both converge absolutely. Then*

$$\left(\sum_{i=0}^{\infty} a_i\right)\left(\sum_{j=0}^{\infty} b_j\right) = \sum_{n=0}^{\infty} c_n \ \text{where} \ c_n = \sum_{k=0}^{n} a_k b_{n-k}.$$

Furthermore, $\sum_{n=0}^{\infty} c_n$ converges absolutely.

Proof: It only remains to verify the last series converges absolutely. By Theorem 11.14 on Page 299 and letting p_{nk} be as defined there,

$$
\begin{aligned}
\sum_{n=0}^{\infty} |c_n| &= \sum_{n=0}^{\infty} \left| \sum_{k=0}^{n} a_k b_{n-k} \right| \\
&\leq \sum_{n=0}^{\infty} \sum_{k=0}^{n} |a_k| \, |b_{n-k}| = \sum_{n=0}^{\infty} \sum_{k=0}^{\infty} p_{nk} \, |a_k| \, |b_{n-k}| \\
&= \sum_{k=0}^{\infty} \sum_{n=0}^{\infty} p_{nk} \, |a_k| \, |b_{n-k}| = \sum_{k=0}^{\infty} \sum_{n=k}^{\infty} |a_k| \, |b_{n-k}| \\
&= \sum_{k=0}^{\infty} |a_k| \sum_{n=0}^{\infty} |b_n| < \infty. \qquad \blacksquare
\end{aligned}
$$

The theorem is about multiplying two series. What if you wanted to consider $\left(\sum_{n=0}^{\infty} a_n\right)^p$ where p is a positive integer maybe larger than 2? Is there a similar theorem to the above?

Definition 11.7. Define

$$\sum_{k_1 + \cdots + k_p = m} a_{k_1} a_{k_2} \cdots a_{k_p}$$

as follows. Consider all ordered lists of nonnegative integers $k_1, \cdots, k_p$ which have the property that $\sum_{i=1}^{p} k_i = m$. For each such list of integers, form the product $a_{k_1} a_{k_2} \cdots a_{k_p}$ and then add all these products.

Note that

$$\sum_{k=0}^{n} a_k a_{n-k} = \sum_{k_1+k_2=n} a_{k_1} a_{k_2}$$

Therefore, from the above theorem, if $\sum a_i$ converges absolutely, it follows

$$\left(\sum_{i=0}^{\infty} a_i \right)^2 = \sum_{n=0}^{\infty} \left(\sum_{k_1+k_2=n} a_{k_1} a_{k_2} \right).$$

It turns out a similar theorem holds replacing 2 with p.

Theorem 11.23. *Suppose $\sum_{n=0}^{\infty} a_n$ converges absolutely. Then*

$$\left(\sum_{n=0}^{\infty} a_n \right)^p = \sum_{m=0}^{\infty} c_{mp} \text{ where } c_{mp} \equiv \sum_{k_1+\cdots+k_p=m} a_{k_1} \cdots a_{k_p}.$$

Proof: First note this is obviously true if $p = 1$ and is also true if $p = 2$ from the above theorem. Now suppose this is true for p and consider $\left(\sum_{n=0}^{\infty} a_n \right)^{p+1}$. By the induction hypothesis and the above theorem on the Cauchy product,

$$\left(\sum_{n=0}^{\infty} a_n \right)^{p+1} = \left(\sum_{n=0}^{\infty} a_n \right)^p \left(\sum_{n=0}^{\infty} a_n \right) = \left(\sum_{m=0}^{\infty} c_{mp} \right) \left(\sum_{n=0}^{\infty} a_n \right)$$

$$= \sum_{n=0}^{\infty} \left(\sum_{k=0}^{n} c_{kp} a_{n-k} \right) = \sum_{n=0}^{\infty} \sum_{k=0}^{n} \sum_{k_1+\cdots+k_p=k} a_{k_1} \cdots a_{k_p} a_{n-k}$$

$$= \sum_{n=0}^{\infty} \sum_{k_1+\cdots+k_{p+1}=n} a_{k_1} \cdots a_{k_{p+1}} \qquad \blacksquare$$

This theorem implies the following corollary for power series.

Corollary 11.5. *Let*

$$\sum_{n=0}^{\infty} a_n (x-a)^n$$

be a power series having radius of convergence $r > 0$. Then if $|x - a| < r$,

$$\left(\sum_{n=0}^{\infty} a_n (x-a)^n \right)^p = \sum_{n=0}^{\infty} b_{np} (x-a)^n \text{ where } b_{np} \equiv \sum_{k_1+\cdots+k_p=n} a_{k_1} \cdots a_{k_p}.$$

Proof: Since $|x - a| < r$, the series $\sum_{n=0}^{\infty} a_n (x - a)^n$, converges absolutely. Therefore, the above theorem applies and

$$\left(\sum_{n=0}^{\infty} a_n (x - a)^n \right)^p = \sum_{n=0}^{\infty} \left(\sum_{k_1 + \cdots + k_p = n} a_{k_1} (x - a)^{k_1} \cdots a_{k_p} (x - a)^{k_p} \right)$$

$$= \sum_{n=0}^{\infty} \left(\sum_{k_1 + \cdots + k_p = n} a_{k_1} \cdots a_{k_p} \right) (x - a)^n . \qquad \blacksquare$$

With this theorem it is possible to consider the question raised in Example 11.18 on Page 314 about the existence of the power series for $\tan x$. This question is clearly included in the more general question of when $\left(\sum_{n=0}^{\infty} a_n (x - a)^n \right)^{-1}$ has a power series.

Lemma 11.4. *Let* $f(x) = \sum_{n=0}^{\infty} a_n (x - a)^n$, *a power series having radius of convergence* $r > 0$. *Suppose also that* $f(a) = 1$. *Then there exists* $r_1 > 0$ *and* $\{b_n\}$ *such that for all* $|x - a| < r_1$,

$$\frac{1}{f(x)} = \sum_{n=0}^{\infty} b_n (x - a)^n .$$

Proof: By continuity, there exists $r_1 > 0$ such that if $|x - a| < r_1$, then

$$\sum_{n=1}^{\infty} |a_n| |x - a|^n < 1.$$

Now pick such an x. Then

$$\frac{1}{f(x)} = \frac{1}{1 + \sum_{n=1}^{\infty} a_n (x - a)^n} = \frac{1}{1 + \sum_{n=0}^{\infty} c_n (x - a)^n}$$

where $c_n = a_n$ if $n > 0$ and $c_0 = 0$. Then

$$\left| \sum_{n=1}^{\infty} a_n (x - a)^n \right| \leq \sum_{n=1}^{\infty} |a_n| |x - a|^n < 1 \qquad (11.22)$$

and so from the formula for the sum of a geometric series,

$$\frac{1}{f(x)} = \sum_{p=0}^{\infty} (-1)^p \left(\sum_{n=0}^{\infty} c_n (x - a)^n \right)^p .$$

By Corollary 11.5, this equals

$$\sum_{p=0}^{\infty} \sum_{n=0}^{\infty} b_{np} (x - a)^n \text{ where } b_{np} = \sum_{k_1 + \cdots + k_p = n} c_{k_1} \cdots c_{k_p}. \qquad (11.23)$$

Thus $|b_{np}| \leq \sum_{k_1 + \cdots + k_p = n} |c_{k_1}| \cdots |c_{k_p}| \equiv B_{np}$ and so by Theorem 11.23,

$$\sum_{p=0}^{\infty} \sum_{n=0}^{\infty} |b_{np}| |x - a|^n \leq \sum_{p=0}^{\infty} \sum_{n=0}^{\infty} B_{np} |x - a|^n = \sum_{p=0}^{\infty} \left(\sum_{n=0}^{\infty} |c_n| |x - a|^n \right)^p < \infty$$

by (11.22) and the formula for the sum of a geometric series. Since the series of (11.23) converges absolutely, Theorem 11.14 on Page 299 implies the series in (11.23) equals

$$\sum_{n=0}^{\infty} \left(\sum_{p=0}^{\infty} b_{np} \right) (x-a)^n$$

and so, let $\sum_{p=0}^{\infty} b_{np} \equiv b_n$. ∎

With this lemma, the following theorem is easy to obtain.

Theorem 11.24. *Let $f(x) = \sum_{n=0}^{\infty} a_n (x-a)^n$, a power series having radius of convergence $r > 0$. Suppose also that $f(a) \neq 0$. Then there exists $r_1 > 0$ and $\{b_n\}$ such that for all $|x-a| < r_1$,*

$$\frac{1}{f(x)} = \sum_{n=0}^{\infty} b_n (x-a)^n.$$

Proof: Let $g(x) \equiv f(x)/f(a)$ so that $g(x)$ satisfies the conditions of the above lemma. Then by that lemma, there exists $r_1 > 0$ and a sequence $\{b_n\}$ such that

$$\frac{f(a)}{f(x)} = \sum_{n=0}^{\infty} b_n (x-a)^n$$

for all $|x-a| < r_1$. Then

$$\frac{1}{f(x)} = \sum_{n=0}^{\infty} \widetilde{b}_n (x-a)^n$$

where $\widetilde{b}_n = b_n/f(a)$. ∎

There is a very interesting question related to r_1 in this theorem. One might think that if $|x-a| < r$, the radius of convergence of $f(x)$ and if $f(x) \neq 0$ it should be possible to write $1/f(x)$ as a power series centered at a. Unfortunately this is not true. Consider $f(x) = 1 + x^2$. In this case $r = \infty$ but the power series for $1/f(x)$ converges only if $|x| < 1$. What happens is this, $1/f(x)$ will have a power series that will converge for $|x-a| < r_1$ where r_1 is the distance between a and the nearest singularity or zero of $f(x)$ in the complex plane. In the case of $f(x) = 1 + x^2$ this function has a zero at $x = \pm i$. This is just another instance of why the natural setting for the study of power series is the complex plane. To read more on power series, you should see the book by Apostol [3] or any text on complex variable.

Chapter 12

Fundamentals

12.1 $\mathbb{R}^n$

The notation $\mathbb{R}^n$ refers to the collection of ordered lists of n real numbers. More precisely, consider the following definition.

Definition 12.1. Define

$$\mathbb{R}^n \equiv \{(x_1, \cdots, x_n) : x_j \in \mathbb{R} \text{ for } j - 1, \cdots, n\}.$$

$(x_1, \cdots, x_n) = (y_1, \cdots, y_n)$ if and only if for all $j = 1, \cdots, n$, $x_j = y_j$. When

$$(x_1, \cdots, x_n) \in \mathbb{R}^n,$$

it is conventional to denote $(x_1, \cdots, x_n)$ by the single bold face letter, $\mathbf{x}$. The numbers, x_j are called the **coordinates**. The set

$$\{(0, \cdots, 0, t, 0, \cdots, 0) : t \in \mathbb{R}\}$$

for t in the ith slot is called the ith coordinate axis **coordinate axis**, the x_i axis for short. The point $\mathbf{0} \equiv (0, \cdots, 0)$ is called the **origin**. Points in $\mathbb{R}^n$ are also called vectors.

Thus $(1, 2, 4) \in \mathbb{R}^3$ and $(2, 1, 4) \in \mathbb{R}^3$ but $(1, 2, 4) \neq (2, 1, 4)$ because, even though the same numbers are involved, they do not match up. In particular, the first entries are not equal.

Why would anyone be interested in such a thing? First consider the case when $n = 1$. Then from the definition, $\mathbb{R}^1 = \mathbb{R}$. Recall that $\mathbb{R}$ is identified with the points of a line. Look at the number line again. Observe that this amounts to identifying a point on this line with a real number. In other words a real number determines where you are on this line. Now suppose $n = 2$ and consider two lines which intersect each other at right angles as shown in the following picture.

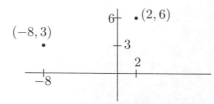

Notice how you can identify a point shown in the plane with the ordered pair $(2,6)$. You go to the right a distance of 2 and then up a distance of 6. Similarly, you can identify another point in the plane with the ordered pair $(-8,3)$. Go to the left a distance of 8 and then up a distance of 3. The reason you go to the left is that there is a $-$ sign on the eight. From this reasoning, every ordered pair determines a unique point in the plane. Conversely, taking a point in the plane, you could draw two lines through the point, one vertical and the other horizontal and determine unique points x_1 on the horizontal line in the above picture and x_2 on the vertical line in the above picture, such that the point of interest is identified with the ordered pair (x_1, x_2). In short, points in the plane can be identified with ordered pairs similar to the way that points on the real line are identified with real numbers. Now suppose $n = 3$. As just explained, the first two coordinates determine a point in a plane. Letting the third component determine how far up or down you go, depending on whether this number is positive or negative, this determines a point in space. Thus, $(1, 4, -5)$ would mean to determine the point in the plane that goes with $(1, 4)$ and then to go below this plane a distance of 5 to obtain a unique point in space. You see that the ordered triples correspond to points in space just as the ordered pairs correspond to points in a plane and single real numbers correspond to points on a line.

You cannot stop here and say that you are only interested in $n \leq 3$. What if you were interested in the motion of two objects? You would need three coordinates to describe where the first object is and you would need another three coordinates to describe where the other object is located. Therefore, you would need to be considering $\mathbb{R}^6$. If the two objects moved around, you would need a time coordinate as well. As another example, consider a hot object which is cooling and suppose you want the temperature of this object. How many coordinates would be needed? You would need one for the temperature, three for the position of the point in the object and one more for the time. Thus you would need to be considering $\mathbb{R}^5$. Many other examples can be given. Sometimes n is very large. This is often the case in applications to business when they are trying to maximize profit subject to constraints. It also occurs in numerical analysis when people try to solve hard problems on a computer.

There are other ways to identify points in space with three numbers but the one presented is the most basic. In this case, the coordinates are known as **Cartesian**

coordinates after Descartes[1] who invented this idea in the first half of the seventeenth century. I will often not bother to draw a distinction between the point in n dimensional space and its Cartesian coordinates.

12.2 Algebra in $\mathbb{R}^n$

There are two algebraic operations done with points of $\mathbb{R}^n$. One is addition and the other is multiplication by numbers, called scalars.

Definition 12.2. If $\mathbf{x} \in \mathbb{R}^n$ and a is a number, also called a **scalar**, then $a\mathbf{x} \in \mathbb{R}^n$ is defined by
$$a\mathbf{x} = a\,(x_1, \cdots, x_n) \equiv (ax_1, \cdots, ax_n). \tag{12.1}$$
This is known as **scalar multiplication**. If $\mathbf{x}, \mathbf{y} \in \mathbb{R}^n$ then $\mathbf{x} + \mathbf{y} \in \mathbb{R}^n$ and is defined by
$$\mathbf{x} + \mathbf{y} = (x_1, \cdots, x_n) + (y_1, \cdots, y_n)$$
$$\equiv (x_1 + y_1, \cdots, x_n + y_n) \tag{12.2}$$
An element of $\mathbb{R}^n$ $\mathbf{x} \equiv (x_1, \cdots, x_n)$ is often called a **vector.** The above definition is known as **vector addition.**

With this definition, the algebraic properties satisfy the conclusions of the following theorem.

Theorem 12.1. *For $\mathbf{v}, \mathbf{w}$ vectors in $\mathbb{R}^n$ and α, β scalars, (real numbers), the following hold.*
$$\mathbf{v} + \mathbf{w} = \mathbf{w} + \mathbf{v}, \tag{12.3}$$
the commutative law of addition,
$$(\mathbf{v} + \mathbf{w}) + \mathbf{z} = \mathbf{v} + (\mathbf{w} + \mathbf{z}), \tag{12.4}$$
the associative law for addition,
$$\mathbf{v} + \mathbf{0} = \mathbf{v}, \tag{12.5}$$
the existence of an additive identity
$$\mathbf{v} + (-\mathbf{v}) = \mathbf{0}, \tag{12.6}$$
the existence of an additive inverse, Also
$$\alpha\,(\mathbf{v} + \mathbf{w}) = \alpha\mathbf{v} + \alpha\mathbf{w}, \tag{12.7}$$
$$(\alpha + \beta)\,\mathbf{v} = \alpha\mathbf{v} + \beta\mathbf{v}, \tag{12.8}$$
$$\alpha\,(\beta\mathbf{v}) = \alpha\beta\,(\mathbf{v}), \tag{12.9}$$
$$1\mathbf{v} = \mathbf{v}. \tag{12.10}$$
In the above $\mathbf{0} = (0, \cdots, 0)$.

---

[1]René Descartes 1596-1650 is often credited with inventing analytic geometry although it seems the ideas were actually known much earlier. He was interested in many different subjects, physiology, chemistry, and physics being some of them. He also wrote a large book in which he tried to explain the book of Genesis scientifically. Descartes ended up dying in Sweden.

You should verify these properties all hold. For example, consider (12.7).

$$\alpha\left(\mathbf{v}+\mathbf{w}\right)=\alpha\left(v_1+w_1,\cdots,v_n+w_n\right)=\left(\alpha\left(v_1+w_1\right),\cdots,\alpha\left(v_n+w_n\right)\right)$$
$$=\left(\alpha v_1+\alpha w_1,\cdots,\alpha v_n+\alpha w_n\right)=\left(\alpha v_1,\cdots,\alpha v_n\right)+\left(\alpha w_1,\cdots,\alpha w_n\right)=\alpha\mathbf{v}+\alpha\mathbf{w}.$$

As usual, subtraction is defined as $\mathbf{x}-\mathbf{y}\equiv\mathbf{x}+\left(-\mathbf{y}\right)$. More generally, the above conclusions are called the vector space axioms. A set V on which is defined addition and multiplication by scalars which satisfy the above axioms is called a vector space.

12.3   Geometric Meaning Of Vector Addition In $\mathbb{R}^3$

It was explained earlier that an element of $\mathbb{R}^n$ is an n tuple of numbers and it was also shown that this can be used to determine a point in three dimensional space in the case where $n=3$ and in two dimensional space, in the case where $n=2$. This point was specified relative to some coordinate axes.

Consider the case where $n=3$ for now. If you draw an arrow from the point in three dimensional space determined by $(0,0,0)$ to the point (a,b,c) with its tail sitting at the point $(0,0,0)$ and its point at the point (a,b,c), this arrow is called the **position vector** of the point determined by $\mathbf{u}\equiv(a,b,c)$. One way to get to this point is to start at $(0,0,0)$ and move in the direction of the x_1 axis to $(a,0,0)$ and then in the direction of the x_2 axis to $(a,b,0)$ and finally in the direction of the x_3 axis to (a,b,c). It is evident that the same arrow (vector) would result if you began at the point $\mathbf{v}\equiv(d,e,f)$, moved in the direction of the x_1 axis to $(d+a,e,f)$, then in the direction of the x_2 axis to $(d+a,e+b,f)$, and finally in the x_3 direction to $(d+a,e+b,f+c)$ only this time, the arrow would have its tail sitting at the point determined by $\mathbf{v}\equiv(d,e,f)$ and its point at $(d+a,e+b,f+c)$. It is said to be the same arrow (vector) because it will point in the same direction and have the same length. It is like you took an actual arrow, the sort of thing you shoot with a bow, and moved it from one location to another keeping it pointing the same direction. This is illustrated in the following picture in which $\mathbf{v}+\mathbf{u}$ is illustrated. Note the parallelogram determined in the picture by the vectors $\mathbf{u}$ and $\mathbf{v}$.

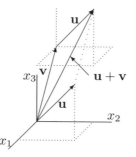

Thus the geometric significance of $(d,e,f)+(a,b,c)=(d+a,e+b,f+c)$ is this. You start with the position vector of the point (d, e, f) and at its point, you place the vector determined by (a, b, c) with its tail at (d, e, f). Then the point of this last vector will be $(d + a, e + b, f + c)$. This is the geometric significance of vector addition. Also, as shown in the picture, $\mathbf{u} + \mathbf{v}$ is the directed diagonal of the parallelogram determined by the two vectors $\mathbf{u}$ and $\mathbf{v}$.

The following example is art.

Example 12.1. Here is a picture of two vectors $\mathbf{u}$ and $\mathbf{v}$.

Sketch a picture of $\mathbf{u} + \mathbf{v}, \mathbf{u} - \mathbf{v}$, and $\mathbf{u}+2\mathbf{v}$.

First here is a picture of $\mathbf{u} + \mathbf{v}$. You first draw $\mathbf{u}$ and then at the point of $\mathbf{u}$ you place the tail of $\mathbf{v}$ as shown. Then $\mathbf{u} + \mathbf{v}$ is the vector which results which is drawn in the following pretty picture.

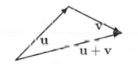

Next consider $\mathbf{u} - \mathbf{v}$. This means $\mathbf{u}+(-\mathbf{v})$. From the above geometric description of vector addition, $-\mathbf{v}$ is the vector which has the same length but which points in the opposite direction to $\mathbf{v}$. Here is a picture.

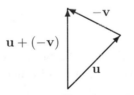

Finally consider the vector $\mathbf{u}+2\mathbf{v}$. Here is a picture of this one also.

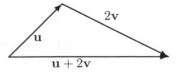

12.4   Lines

To begin with consider the case $n = 1, 2$. In the case where $n = 1$, the only line is just $\mathbb{R}^1 = \mathbb{R}$. Therefore, if x_1 and x_2 are two different points in $\mathbb{R}$, consider

$$x = x_1 + t\left(x_2 - x_1\right)$$

where $t \in \mathbb{R}$ and the totality of all such points will give $\mathbb{R}$. You see that you can always solve the above equation for t, showing that every point on $\mathbb{R}$ is of this form. Now consider the plane. Does a similar formula hold? Let (x_1, y_1) and (x_2, y_2) be two different points in $\mathbb{R}^2$ which are contained in a line, l. Suppose that $x_1 \neq x_2$. Then if (x, y) is an arbitrary point on l,

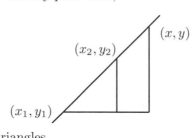

Now by similar triangles,

$$m \equiv \frac{y_2 - y_1}{x_2 - x_1} = \frac{y - y_1}{x - x_1}$$

and so the point slope form of the line, l, is given as

$$y - y_1 = m\left(x - x_1\right).$$

If t is defined by

$$x = x_1 + t\left(x_2 - x_1\right),$$

you obtain this equation along with

$$y = y_1 + mt\left(x_2 - x_1\right)$$
$$= y_1 + t\left(y_2 - y_1\right).$$

Therefore,

$$(x, y) = (x_1, y_1) + t\left(x_2 - x_1, y_2 - y_1\right).$$

If $x_1 = x_2$, then in place of the point slope form above, $x = x_1$. Since the two given points are different, $y_1 \neq y_2$ and so you still obtain the above formula for the line. Because of this, the following is the definition of a line in $\mathbb{R}^n$.

Definition 12.3. A line in $\mathbb{R}^n$ containing the two different points $\mathbf{x}^1$ and $\mathbf{x}^2$ is the collection of points of the form

$$\mathbf{x} = \mathbf{x}^1 + t\left(\mathbf{x}^2 - \mathbf{x}^1\right)$$

where $t \in \mathbb{R}$. This is known as a **parametric equation** and the variable t is called the **parameter**.

Often t denotes time in applications to Physics. Note this definition agrees with the usual notion of a line in two dimensions and so this is consistent with earlier concepts.

Lemma 12.1. *Let* $\mathbf{a}, \mathbf{b} \in \mathbb{R}^n$ *with* $\mathbf{a} \neq \mathbf{0}$. *Then* $\mathbf{x} = t\mathbf{a} + \mathbf{b}$, $t \in \mathbb{R}$, *is a line.*

Proof: Let $\mathbf{x}^1 = \mathbf{b}$ and let $\mathbf{x}^2 - \mathbf{x}^1 = \mathbf{a}$ so that $\mathbf{x}^2 \neq \mathbf{x}^1$. Then $t\mathbf{a} + \mathbf{b} = \mathbf{x}^1 + t\left(\mathbf{x}^2 - \mathbf{x}^1\right)$ and so $\mathbf{x} = t\mathbf{a} + \mathbf{b}$ is a line containing the two different points $\mathbf{x}^1$ and $\mathbf{x}^2$. ■

Definition 12.4. The vector $\mathbf{a}$ in the above lemma is called a **direction vector** for the line.

Definition 12.5. Let $\mathbf{p}$ and $\mathbf{q}$ be two points in $\mathbb{R}^n$, $\mathbf{p} \neq \mathbf{q}$. The **directed line segment** from $\mathbf{p}$ to $\mathbf{q}$, denoted by $\overrightarrow{\mathbf{pq}}$, is defined to be the collection of points
$$\mathbf{x} = \mathbf{p} + t(\mathbf{q} - \mathbf{p}), \ t \in [0, 1]$$
with the direction corresponding to increasing t. In the definition, when $t = 0$, the point $\mathbf{p}$ is obtained and as t increases, other points on this line segment are obtained until when $t = 1$, you get the point $\mathbf{q}$. This is what is meant by saying the direction corresponds to increasing t.

Think of $\overrightarrow{\mathbf{pq}}$ as an arrow whose point is on $\mathbf{q}$ and whose base is at $\mathbf{p}$ as shown in the following picture.

This line segment is a part of a line, from the above Definition.

Example 12.2. Find a parametric equation for the line through the points $(1, 2, 0)$ and $(2, -4, 6)$.

Use the definition of a line given above to write
$$(x, y, z) = (1, 2, 0) + t(1, -6, 6), \ t \in \mathbb{R}.$$
The vector $(1, -6, 6)$ is obtained by $(2, -4, 6) - (1, 2, 0)$ as indicated above.

The reason for the word "a", rather than the word "the" is that there are infinitely many different parametric equations for the same line. To see this, replace t with $3s$. Then you obtain a parametric equation for the same line because the same set of points is obtained. The difference is, they are obtained from different values of the parameter. What happens is this: The line is a set of points but the parametric description gives more information than that. It tells how the points are obtained. Obviously, there are many ways to trace out a given set of points and each of these ways corresponds to a different parametric equation for the line.

Example 12.3. Find a parametric equation for the line which contains the point $(1, 2, 0)$ and has direction vector $(1, 2, 1)$.

From the above this is just

$$(x, y, z) = (1, 2, 0) + t\,(1, 2, 1)\,,\ t \in \mathbb{R}. \tag{12.11}$$

Sometimes people elect to write a line like the above in the form

$$x = 1 + t,\ y = 2 + 2t,\ z = t,\ t \in \mathbb{R}. \tag{12.12}$$

This is a set of scalar parametric equations which amounts to the same thing as (12.11).

There is one other form for a line which is sometimes considered useful. It is the so called symmetric form. Consider the line of (12.12). You can solve for the parameter t to write

$$t = x - 1, t = \frac{y - 2}{2},\ t = z.$$

Therefore,

$$x - 1 = \frac{y - 2}{2} = z.$$

This is the symmetric form of the line.

Example 12.4. Suppose the **symmetric form of a line** is

$$\frac{x - 2}{3} = \frac{y - 1}{2} = z + 3.$$

Find the line in parametric form.

Let $t = \frac{x-2}{3}, t = \frac{y-1}{2}$ and $t = z + 3$. Then solving for x, y, z, you get

$$x = 3t + 2,\ y = 2t + 1,\ z = t - 3,\ t \in \mathbb{R}.$$

Written in terms of vectors, this is

$$(2, 1, -3) + t\,(3, 2, 1) = (x, y, z)\,,\ t \in \mathbb{R}.$$

12.5 Distance in $\mathbb{R}^n$

How is distance between two points in $\mathbb{R}^n$ defined?

Definition 12.6. Let $\mathbf{x} = (x_1, \cdots, x_n)$ and $\mathbf{y} = (y_1, \cdots, y_n)$ be two points in $\mathbb{R}^n$. Then $|\mathbf{x} - \mathbf{y}|$ to indicates the distance between these points and is defined as

$$\text{distance between } \mathbf{x} \text{ and } \mathbf{y} \equiv |\mathbf{x} - \mathbf{y}| \equiv \left(\sum_{k=1}^{n} |x_k - y_k|^2 \right)^{1/2}.$$

This is called the **distance formula**. Thus $|\mathbf{x}| \equiv |\mathbf{x} - \mathbf{0}|$. The symbol $B\,(\mathbf{a}, r)$ is defined by

$$B\,(\mathbf{a}, r) \equiv \{\mathbf{x} \in \mathbb{R}^n : |\mathbf{x} - \mathbf{a}| < r\}\,.$$

This is called an **open ball** of radius r centered at $\mathbf{a}$. It gives all the points in $\mathbb{R}^n$ which are closer to $\mathbf{a}$ than r.

First of all note this is a generalization of the notion of distance in $\mathbb{R}$. There the distance between two points x and y was given by the absolute value of their difference. Thus $|x - y|$ is equal to the distance between these two points on $\mathbb{R}$. Now $|x - y| = \left((x - y)^2\right)^{1/2}$ where the square root is always the positive square root. Thus it is the same formula as the above definition except there is only one term in the sum. Geometrically, this is the right way to define distance which is seen from the Pythagorean theorem. Consider the following picture in the case that $n = 2$.

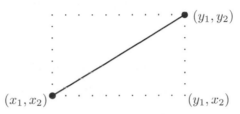

There are two points in the plane whose Cartesian coordinates are (x_1, x_2) and (y_1, y_2) respectively. Then the solid line joining these two points is the hypotenuse of a right triangle which is half of the rectangle shown in dotted lines. What is its length? Note the lengths of the sides of this triangle are $|y_1 - x_1|$ and $|y_2 - x_2|$. Therefore, the Pythagorean theorem implies the length of the hypotenuse equals

$$\left(|y_1 - x_1|^2 + |y_2 - x_2|^2\right)^{1/2} = \left((y_1 - x_1)^2 + (y_2 - x_2)^2\right)^{1/2}$$

which is just the formula for the distance given above.

Now suppose $n = 3$ and let (x_1, x_2, x_3) and (y_1, y_2, y_3) be two points in $\mathbb{R}^3$. Consider the following picture in which one of the solid lines joins the two points and a dotted line joins the points (x_1, x_2, x_3) and (y_1, y_2, x_3).

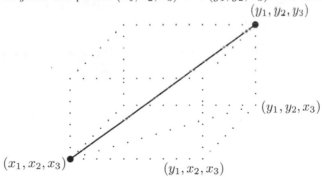

By the Pythagorean theorem, the length of the dotted line joining (x_1, x_2, x_3) and (y_1, y_2, x_3) equals

$$\left((y_1 - x_1)^2 + (y_2 - x_2)^2\right)^{1/2}$$

while the length of the line joining (y_1, y_2, x_3) to (y_1, y_2, y_3) is just $|y_3 - x_3|$. Therefore, by the Pythagorean theorem again, the length of the line joining the points

(x_1, x_2, x_3) and (y_1, y_2, y_3) equals

$$\left\{ \left[\left((y_1 - x_1)^2 + (y_2 - x_2)^2 \right)^{1/2} \right]^2 + (y_3 - x_3)^2 \right\}^{1/2}$$

$$= \left((y_1 - x_1)^2 + (y_2 - x_2)^2 + (y_3 - x_3)^2 \right)^{1/2},$$

which is again just the distance formula above.

This completes the argument that the above definition is reasonable. Of course you cannot continue drawing pictures in ever higher dimensions, but there is no problem with the formula for distance in any number of dimensions. Here is an example.

Example 12.5. Find the distance between the points in $\mathbb{R}^4$,

$$\mathbf{a} = (1, 2, -4, 6), \ \mathbf{b} = (2, 3, -1, 0)$$

Use the distance formula and write

$$|\mathbf{a} - \mathbf{b}|^2 = (1 - 2)^2 + (2 - 3)^2 + (-4 - (-1))^2 + (6 - 0)^2 = 47$$

Therefore, $|\mathbf{a} - \mathbf{b}| = \sqrt{47}$.

All this amounts to defining the distance between two points as the length of a straight line joining these two points. However, there is nothing sacred about using straight lines. One could define the distance to be the length of some other sort of line joining these points. It will not be done in this book but sometimes this sort of thing is done.

Another convention which is usually followed, especially in $\mathbb{R}^2$ and $\mathbb{R}^3$ is to denote the first component of a point in $\mathbb{R}^2$ by x and the second component by y. In $\mathbb{R}^3$ it is customary to denote the first and second components as just described while the third component is called z.

Example 12.6. Describe the points which are at the same distance between $(1, 2, 3)$ and $(0, 1, 2)$.

Let (x, y, z) be such a point. Then

$$\sqrt{(x - 1)^2 + (y - 2)^2 + (z - 3)^2} = \sqrt{x^2 + (y - 1)^2 + (z - 2)^2}.$$

Squaring both sides

$$(x - 1)^2 + (y - 2)^2 + (z - 3)^2 = x^2 + (y - 1)^2 + (z - 2)^2$$

and so

$$x^2 - 2x + 14 + y^2 - 4y + z^2 - 6z = x^2 + y^2 - 2y + 5 + z^2 - 4z$$

which implies

$$-2x + 14 - 4y - 6z = -2y + 5 - 4z$$

hence

$$2x + 2y + 2z = -9. \tag{12.13}$$

Since these steps are reversible, the set of points which is at the same distance from the two given points consists of the points (x, y, z) such that (12.13) holds.

The following lemma is fundamental. It is a form of the Cauchy Schwarz inequality.

Lemma 12.2. *Let* $\mathbf{x} = (x_1, \cdots, x_n)$ *and* $\mathbf{y} = (y_1, \cdots, y_n)$ *be two points in* $\mathbb{R}^n$. *Then*

$$\left| \sum_{i=1}^{n} x_i y_i \right| \leq |\mathbf{x}| \, |\mathbf{y}| \, . \tag{12.14}$$

Proof: Let θ be either 1 or -1 such that

$$\theta \sum_{i=1}^{n} x_i y_i = \sum_{i=1}^{n} x_i \left(\theta y_i \right) = \left| \sum_{i=1}^{n} x_i y_i \right|$$

and consider $p(t) \equiv \sum_{i=1}^{n} \left(x_i + t \theta y_i \right)^2$. Then for all $t \in \mathbb{R}$,

$$0 \leq p(t) = \sum_{i=1}^{n} x_i^2 + 2t \sum_{i=1}^{n} x_i \theta y_i + t^2 \sum_{i=1}^{n} y_i^2$$

$$= |\mathbf{x}|^2 + 2t \sum_{i=1}^{n} x_i \theta y_i + t^2 |\mathbf{y}|^2$$

If $|\mathbf{y}| = 0$ then (12.14) is obviously true because both sides equal zero. Therefore, assume $|\mathbf{y}| \neq 0$. Then $p(t)$ is a polynomial of degree two whose graph opens up. Therefore, it either has no zeroes, two zeros or one repeated zero. If it has two zeros, the above inequality must be violated because in this case the graph must dip below the x axis. Therefore, it either has no zeros or exactly one. From the quadratic formula this happens exactly when

$$4 \left(\sum_{i=1}^{n} x_i \theta y_i \right)^2 - 4 |\mathbf{x}|^2 |\mathbf{y}|^2 \leq 0$$

and so

$$\sum_{i=1}^{n} x_i \theta y_i = \left| \sum_{i=1}^{n} x_i y_i \right| \leq |\mathbf{x}| \, |\mathbf{y}|$$

as claimed. This proves the inequality. ∎

There are certain properties of the distance which are obvious. Two of them which follow directly from the definition are

$$|\mathbf{x} - \mathbf{y}| = |\mathbf{y} - \mathbf{x}| \, ,$$

$$|\mathbf{x} - \mathbf{y}| \geq 0 \text{ and equals } 0 \text{ only if } \mathbf{y} = \mathbf{x}.$$

The third fundamental property of distance is known as the triangle inequality. Recall that in any triangle the sum of the lengths of two sides is always at least as large as the third side. The following corollary is equivalent to this simple statement.

Corollary 12.1. *Let* $\mathbf{x}, \mathbf{y}$ *be points of* $\mathbb{R}^n$. *Then*

$$|\mathbf{x} + \mathbf{y}| \leq |\mathbf{x}| + |\mathbf{y}|.$$

Proof: Using the Cauchy Schwarz inequality, Lemma 12.2,

$$|\mathbf{x} + \mathbf{y}|^2 \equiv \sum_{i=1}^n (x_i + y_i)^2.$$

$$= \sum_{i=1}^n x_i^2 + 2 \sum_{i=1}^n x_i y_i + \sum_{i=1}^n y_i^2$$

$$\leq |\mathbf{x}|^2 + 2|\mathbf{x}||\mathbf{y}| + |\mathbf{y}|^2$$

$$= (|\mathbf{x}| + |\mathbf{y}|)^2$$

and so upon taking square roots of both sides,

$$|\mathbf{x} + \mathbf{y}| \leq |\mathbf{x}| + |\mathbf{y}|$$ ∎

12.6   Geometric Meaning Of Scalar Multiplication In $\mathbb{R}^3$

As discussed earlier, $\mathbf{x} = (x_1, x_2, x_3)$ determines a vector. You draw the line from $\mathbf{0}$ to $\mathbf{x}$ placing the point of the vector on $\mathbf{x}$. What is the length of this vector? The length of this vector is defined to equal $|\mathbf{x}|$ as in Definition 12.6. Thus the length of $\mathbf{x}$ equals $\sqrt{x_1^2 + x_2^2 + x_3^2}$. When you multiply $\mathbf{x}$ by a scalar α, you get $(\alpha x_1, \alpha x_2, \alpha x_3)$ and the length of this vector is defined as $\sqrt{\left((\alpha x_1)^2 + (\alpha x_2)^2 + (\alpha x_3)^2\right)} = |\alpha| \sqrt{x_1^2 + x_2^2 + x_3^2}$. Thus the following holds.

$$|\alpha \mathbf{x}| = |\alpha| |\mathbf{x}|.$$

In other words, multiplication by a scalar magnifies the length of the vector. What about the direction? You should convince yourself by drawing a picture that if α is negative, it causes the resulting vector to point in the opposite direction while if $\alpha > 0$ it preserves the direction the vector points. One way to see this is to first observe that if $\alpha \neq 1$, then $\mathbf{x}$ and $\alpha \mathbf{x}$ are both points on the same line. Note that there is no change in this when you replace $\mathbb{R}^3$ with $\mathbb{R}^n$.

12.7   Exercises

(1) Verify all the properties (12.3)-(12.10).

(2) Compute the following
  (a) $5\,(1,2,3,-2)+6\,(2,1,-2,7)$
  (b) $5\,(1,2,-2)-6\,(2,1,-2)$
  (c) $-3\,(1,0,3,-2)+(2,0,-2,1)$
  (d) $-3\,(1,-2,-3,-2)-2\,(2,-1,-2,7)$
  (e) $-(2,2,-3,-2)+2\,(2,4,-2,7)$

(3) Find symmetric equations for the line through the points $(2,2,4)$ and $(-2,3,1)$.

(4) Find symmetric equations for the line through the points $(1,2,4)$ and $(-2,1,1)$.

(5) Symmetric equations for a line are given. Find parametric equations of the line.
  (a) $\frac{x+1}{3}=\frac{2y+3}{2}=z+7$
  (b) $\frac{2x-1}{3}=\frac{2y+3}{6}=z-7$
  (c) $\frac{x+1}{3}=2y+3=2z-1$
  (d) $\frac{1-2x}{3}-\frac{3-2y}{2}=z+1$
  (e) $\frac{x-1}{3}=\frac{2y-3}{5}=z+2$
  (f) $\frac{x+1}{3}=\frac{3-y}{5}=z+1$

(6) Parametric equations for a line are given. Find symmetric equations for the line if possible. If it is not possible to do it explain why.
  (a) $x=1+2t,\,y=3-t,\,z=5+3l$
  (b) $x=1+l,\,y=3-l,\,z=5-3t$
  (c) $x=1+2t,\,y=3+t,\,z=5+3t$
  (d) $x=1-2l,\,y=1,\,z=1+l$
  (e) $x=1-l,\,y=3+2l,\,z=5-3t$
  (f) $x=t,\,y=3-t,\,z=1+t$

(7) The first point given is a point contained in the line. The second point given is a direction vector for the line. Find parametric equations for the line, determined by this information.
  (a) $(1,2,1),(2,0,3)$
  (b) $(1,0,1),(1,1,3)$
  (c) $(1,2,0),(1,1,0)$
  (d) $(1,0,-6),(-2,-1,3)$
  (e) $(-1,-2,-1),(2,1,-1)$
  (f) $(0,0,0),(2,-3,1)$

(8) Parametric equations for a line are given. Determine a direction vector for this line.
  (a) $x=1+2t,\,y=3-t,\,z=5+3t$
  (b) $x=1+t,\,y=3+3t,\,z=5-l$

(c) $x = 7 + t, y = 3 + 4t, z = 5 - 3t$

(d) $x = 2t, y = -3t, z = 3t$

(e) $x = 2t, y = 3 + 2t, z = 5 + t$

(f) $x = t, y = 3 + 3t, z = 5 + t$

(9) A line contains the given two points. Find parametric equations for this line. Identify the direction vector.

(a) $(0, 1, 0), (2, 1, 2)$

(b) $(0, 1, 1), (2, 5, 0)$

(c) $(1, 1, 0), (0, 1, 2)$

(d) $(0, 1, 3), (0, 3, 0)$

(e) $(0, 1, 0), (0, 6, 2)$

(f) $(0, 1, 2), (2, 0, 2)$

(10) Draw a picture of the points in $\mathbb{R}^2$ which are determined by the following ordered pairs.

(a) $(1, 2)$

(b) $(-2, -2)$

(c) $(-2, 3)$

(d) $(2, -5)$

(11) Does it make sense to write $(1, 2) + (2, 3, 1)$? Explain.

(12) Draw a picture of the points in $\mathbb{R}^3$ which are determined by the following ordered triples.

(a) $(1, 2, 0)$

(b) $(-2, -2, 1)$

(c) $(-2, 3, -2)$

(13) You are given two points in $\mathbb{R}^3$, $(4, 5, -4)$ and $(2, 3, 0)$. Show the distance from the point $(3, 4, -2)$ to the first of these points is the same as the distance from this point to the second of the original pair of points. Note that $3 = \frac{4+2}{2}, 4 = \frac{5+3}{2}$. Obtain a theorem which will be valid for general pairs of points (x, y, z) and (x_1, y_1, z_1) and prove your theorem using the distance formula.

(14) A sphere is the set of all points which are at a given distance from a single given point. Find an equation for the sphere which is the set of all points that are at a distance of 4 from the point $(1, 2, 3)$ in $\mathbb{R}^3$.

(15) A parabola is the set of all points (x, y) in the plane such that the distance from the point (x, y) to a given point (x_0, y_0) equals the distance from (x, y) to a given line. The point (x_0, y_0) is called the **focus** and the line is called the **directrix**. Find the equation of the parabola which results from the line $y = l$ and (x_0, y_0) a given focus with $y_0 < l$. Repeat for $y_0 > l$.

(16) A sphere centered at the point $(x_0, y_0, z_0) \in \mathbb{R}^3$ having radius r consists of all points (x, y, z) whose distance to (x_0, y_0, z_0) equals r. Write an equation for this sphere in $\mathbb{R}^3$.

(17) Suppose the distance between (x, y) and (x', y') were defined to equal the larger of the two numbers $|x - x'|$ and $|y - y'|$. Draw a picture of the sphere centered at the point $(0, 0)$ if this notion of distance is used.

(18) Repeat the same problem except this time let the distance between the two points be $|x - x'| + |y - y'|$.

(19) If (x_1, y_1, z_1) and (x_2, y_2, z_2) are two points such that $|(x_i, y_i, z_i)| = 1$ for $i = 1, 2$, show that in terms of the usual distance, $\left| \left(\frac{x_1 + x_2}{2}, \frac{y_1 + y_2}{2}, \frac{z_1 + z_2}{2} \right) \right| < 1$. What would happen if you used the way of measuring distance given in Problem 17 ($|(x, y, z)| = \text{maximum of } |z|, |x|, |y|$.)?

(20) Give a simple description using the distance formula of the set of points which are at an equal distance between the two points (x_1, y_1, z_1) and (x_2, y_2, z_2).

(21) Suppose you are given two points $(-a, 0)$ and $(a, 0)$ in $\mathbb{R}^2$ and a number $r > 2a$. The set of points described by

$$\{(x, y) \in \mathbb{R}^2 : |(x, y) - (-a, 0)| \\ + |(x, y) - (a, 0)| = r\}$$

is known as an ellipse. The two given points are known as the **focus points** of the ellipse. Find α and β such that this is in the form $\left(\frac{x}{\alpha} \right)^2 + \left(\frac{y}{\beta} \right)^2 = 1$. This is a nice exercise in messy algebra.

(22) Suppose you are given two points $(-a, 0)$ and $(a, 0)$ in $\mathbb{R}^2$ and a number $r < 2a$. The set of points described by

$$\{(x, y) \in \mathbb{R}^2 : |(x, y) - (-a, 0)| \\ - |(x, y) - (a, 0)| = r\}$$

is known as **hyperbola**. The two given points are known as the **focus points** of the hyperbola. Simplify this to the form $\left(\frac{x}{\alpha} \right)^2 - \left(\frac{y}{\beta} \right)^2 = 1$. This is a nice exercise in messy algebra.

(23) Let (x_1, y_1) and (x_2, y_2) be two points in $\mathbb{R}^2$. Give a simple description using the distance formula of the perpendicular bisector of the line segment joining these two points. Thus you want all points (x, y) such that $|(x, y) - (x_1, y_1)| = |(x, y) - (x_2, y_2)|$.

(24) Show that $|\alpha \mathbf{x}| = |\alpha| |\mathbf{x}|$ whenever $\mathbf{x} \in \mathbb{R}^n$ for any positive integer n.

12.8 Physical Vectors

Suppose you push on something. What is important? There are really two things which are important, how hard you push and the direction you push.

Definition 12.7. Force is a vector. The magnitude of this vector is a measure of how hard it is pushing. It is measured in units such as Newtons or pounds or tons. Its direction is the direction in which the push is taking place.

Of course this is a little vague and will be left a little vague until the presentation of Newton's second law later. See the appendix on this or any physics book.

Vectors are used to model force and other physical vectors like velocity. What was just described would be called a force vector. It has two essential ingredients, its magnitude and its direction. Think of vectors as directed line segments or arrows as shown in the following picture in which all the directed line segments are considered to be the same vector because they have the same direction, the direction in which the arrows point, and the same magnitude (length).

Because of this fact that only direction and magnitude are important, it is always possible to put a vector in a certain particularly simple form. Let $\overrightarrow{pq}$ be a directed line segment or vector. Then from Definition 12.5 it follows that $\overrightarrow{pq}$ consists of the points of the form

$$\mathbf{p} + t\,(\mathbf{q} - \mathbf{p})$$

where $t \in [0, 1]$. Subtract $\mathbf{p}$ from all these points to obtain the directed line segment consisting of the points

$$\mathbf{0} + t\,(\mathbf{q} - \mathbf{p}),\ t \in [0, 1].$$

The point in $\mathbb{R}^n$, $\mathbf{q} - \mathbf{p}$, will represent the vector.

Geometrically, the arrow $\overrightarrow{pq}$, was slid so it points in the same direction and the base is at the origin $\mathbf{0}$. For example, see the following picture.

In this way vectors can be identified with points of $\mathbb{R}^n$.

Definition 12.8. Let $\mathbf{x} = (x_1, \cdots, x_n) \in \mathbb{R}^n$. The **position vector** of this point is the vector whose point is at $\mathbf{x}$ and whose tail is at the origin $(0, \cdots, 0)$. If $\mathbf{x} = (x_1, \cdots, x_n)$ is called a vector, the vector which is meant, is this position vector just described. Another term associated with this is **standard position.** A vector is in standard position if the tail is placed at the origin.

It is customary to identify the point in $\mathbb{R}^n$ with its position vector.

The magnitude of a vector determined by a directed line segment $\overrightarrow{pq}$ is just the distance between the point $\mathbf{p}$ and the point $\mathbf{q}$. By the distance formula this equals

$$\left(\sum_{k=1}^{n} (q_k - p_k)^2 \right)^{1/2} = |\mathbf{p} - \mathbf{q}|$$

and for $\mathbf{v}$ any vector in $\mathbb{R}^n$ the magnitude of $\mathbf{v}$ equals $\left(\sum_{k=1}^{n} v_k^2 \right)^{1/2} = |\mathbf{v}|$.

Example 12.7. Consider the vector $\mathbf{v} \equiv (1, 2, 3)$ in $\mathbb{R}^n$. Find $|\mathbf{v}|$.

First, the vector is the directed line segment (arrow) which has its base at $\mathbf{0} \equiv (0, 0, 0)$ and its point at $(1, 2, 3)$. Therefore,

$$|\mathbf{v}| = \sqrt{1^2 + 2^2 + 3^2} = \sqrt{14}.$$

What is the geometric significance of scalar multiplication? If $\mathbf{a}$ represents the vector $\mathbf{v}$ in the sense that when it is slid to place its tail at the origin, the element of $\mathbb{R}^n$ at its point is $\mathbf{a}$, what is $r\mathbf{v}$?

$$|r\mathbf{v}| = \left(\sum_{k=1}^{n} (ra_i)^2 \right)^{1/2} = \left(\sum_{k=1}^{n} r^2 (a_i)^2 \right)^{1/2} = (r^2)^{1/2} \left(\sum_{k=1}^{n} a_i^2 \right)^{1/2} = |r| \, |\mathbf{v}|.$$

Thus the magnitude of $r\mathbf{v}$ equals $|r|$ times the magnitude of $\mathbf{v}$. If r is positive, then the vector represented by $r\mathbf{v}$ has the same direction as the vector $\mathbf{v}$ because multiplying by the scalar r, only has the effect of scaling all the distances. Thus the unit distance along any coordinate axis now has length r and in this re-scaled system the vector is represented by $\mathbf{a}$. If $r < 0$ similar considerations apply except in this case all the a_i also change sign. From now on, $\mathbf{a}$ will be referred to as a vector instead of an element of $\mathbb{R}^n$ representing a vector as just described. The following picture illustrates the effect of scalar multiplication.

Note there are n special vectors which point along the coordinate axes. These are

$$e_i \equiv (0, \cdots, 0, 1, 0, \cdots, 0)$$

where the 1 is in the ith slot and there are zeros in all the other spaces. See the picture in the case of $\mathbb{R}^3$.

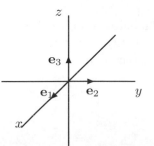

The direction of $\mathbf{e}_i$ is referred to as the ith direction. Given a vector $\mathbf{v} = (a_1, \cdots, a_n)$, $a_i \mathbf{e}_i$ is the ith component of the vector. Thus $a_i \mathbf{e}_i = (0, \cdots, 0, a_i, 0, \cdots, 0)$ and so this vector gives something possibly nonzero only in the ith direction. Also, knowledge of the ith component of the vector is equivalent to knowledge of the vector because it gives the entry in the ith slot and for $\mathbf{v} = (a_1, \cdots, a_n)$, $\mathbf{v} = \sum_{k=1}^{n} a_i \mathbf{e}_i$.

What does addition of vectors mean physically? Suppose two forces are applied to some object. Each of these would be represented by a force vector and the two forces acting together would yield an overall force acting on the object which would also be a force vector known as the resultant. Suppose the two vectors are $\mathbf{a} = \sum_{k=1}^{n} a_i \mathbf{e}_i$ and $\mathbf{b} = \sum_{k=1}^{n} b_i \mathbf{e}_i$. Then the vector $\mathbf{a}$ involves a component in the ith direction, $a_i \mathbf{e}_i$ while the component in the ith direction of $\mathbf{b}$ is $b_i \mathbf{e}_i$. Then it seems physically reasonable that the resultant vector should have a component in the ith direction equal to $(a_i + b_i)\mathbf{e}_i$. This is exactly what is obtained when the vectors $\mathbf{a}$ and $\mathbf{b}$ are added.

$$\mathbf{a} + \mathbf{b} = (a_1 + b_1, \cdots, a_n + b_n) = \sum_{i=1}^{n} (a_i + b_i)\,\mathbf{e}_i$$

Thus the addition of vectors according to the rules of addition in $\mathbb{R}^n$ which were presented earlier, yields the appropriate vector which duplicates the cumulative effect of all the vectors in the sum.

What is the geometric significance of vector addition? Suppose $\mathbf{u}, \mathbf{v}$ are vectors

$$\mathbf{u} = (u_1, \cdots, u_n), \mathbf{v} = (v_1, \cdots, v_n)$$

Then $\mathbf{u} + \mathbf{v} = (u_1 + v_1, \cdots, u_n + v_n)$. How can one obtain this geometrically? Consider the directed line segment, $\overrightarrow{0\mathbf{u}}$ and then, starting at the end of this directed line segment, follow the directed line segment $\overrightarrow{\mathbf{u}(\mathbf{u} + \mathbf{v})}$ to its end $\mathbf{u} + \mathbf{v}$. In other words, place the vector $\mathbf{u}$ in standard position with its base at the origin and then slide the vector $\mathbf{v}$ till its base coincides with the point of $\mathbf{u}$. The point of this slid vector, determines $\mathbf{u} + \mathbf{v}$. To illustrate, see the following picture

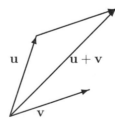

Note the vector $\mathbf{u} + \mathbf{v}$ is the diagonal of a parallelogram determined from the two vectors $\mathbf{u}$ and $\mathbf{v}$ and that identifying $\mathbf{u} + \mathbf{v}$ with the directed diagonal of the parallelogram determined by the vectors $\mathbf{u}$ and $\mathbf{v}$ amounts to the same thing as the above procedure.

An item of notation should be mentioned here. In the case of $\mathbb{R}^n$ where $n \leq 3$, it is standard notation to use $\mathbf{i}$ for $\mathbf{e}_1, \mathbf{j}$ for $\mathbf{e}_2$, and $\mathbf{k}$ for $\mathbf{e}_3$. Now here are some applications of vector addition to some problems.

Example 12.8. There are three ropes attached to a car and three people pull on these ropes. The first exerts a force of $2\mathbf{i}+3\mathbf{j}-2\mathbf{k}$ Newtons, the second exerts a force of $3\mathbf{i}+5\mathbf{j}+\mathbf{k}$ Newtons and the third exerts a force of $5\mathbf{i}-\mathbf{j}+2\mathbf{k}$. Newtons. Find the total force in the direction of $\mathbf{i}$.

To find the total force add the vectors as described above. This gives $10\mathbf{i}+7\mathbf{j}+\mathbf{k}$ Newtons. Therefore, the force in the $\mathbf{i}$ direction is 10 Newtons.

As mentioned earlier, the Newton is a unit of force like pounds.

Example 12.9. An airplane flies North East at 100 miles per hour. Write this as a vector.

A picture of this situation follows.

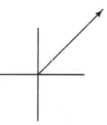

The vector has length 100. Now using that vector as the hypotenuse of a right triangle having equal sides, the sides should be each of length $100/\sqrt{2}$. Therefore, the vector would be $100/\sqrt{2}\mathbf{i} + 100/\sqrt{2}\mathbf{j}$.

This example also motivates the concept of **velocity**.

Definition 12.9. The **speed** of an object is a measure of how fast it is going. It is measured in units of length per unit time. For example, miles per hour, kilometers per minute, feet per second. The **velocity** is a vector having the speed as the magnitude but also specifying the direction.

Thus the velocity vector in the above example is $100/\sqrt{2}\mathbf{i} + 100/\sqrt{2}\mathbf{j}$.

Example 12.10. The velocity of an airplane is $100\mathbf{i}+\mathbf{j}+\mathbf{k}$ measured in kilometers per hour and at a certain instant of time its position is $(1, 2, 1)$. Here imagine a Cartesian coordinate system in which the third component is altitude and the first and second components are measured on a line from West to East and a line from South to North. Find the position of this airplane one minute later.

Consider the vector $(1, 2, 1)$, is the initial position vector of the airplane. As it moves, the position vector changes. After one minute the airplane has moved in the $\mathbf{i}$ direction a distance of $100 \times \frac{1}{60} = \frac{5}{3}$ kilometer. In the $\mathbf{j}$ direction it has moved

$\frac{1}{60}$ kilometer during this same time, while it moves $\frac{1}{60}$ kilometer in the **k** direction. Therefore, the new displacement vector for the airplane is

$$(1,2,1) + \left(\frac{5}{3}, \frac{1}{60}, \frac{1}{60}\right) = \left(\frac{8}{3}, \frac{121}{60}, \frac{121}{60}\right)$$

Example 12.11. A certain river is one half mile wide with a current flowing at 4 miles per hour from East to West. A man swims directly toward the opposite shore from the South bank of the river at a speed of 3 miles per hour. How far down the river does he find himself when he has swam across? How far does he end up swimming?

Consider the following picture.

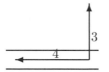

You should write these vectors in terms of components. The velocity of the swimmer in still water would be $3\mathbf{j}$ while the velocity of the river would be $-4\mathbf{i}$. Therefore, the velocity of the swimmer is $-4\mathbf{i}+3\mathbf{j}$. Since the component of velocity in the direction across the river is 3, it follows the trip takes $1/6$ hour or 10 minutes. The speed at which he travels is $\sqrt{4^2 + 3^2} = 5$ miles per hour and so he travels $5 \times \frac{1}{6} = \frac{5}{6}$ miles. Now to find the distance downstream he finds himself, note that if x is this distance, x and $1/2$ are two legs of a right triangle whose hypotenuse equals $5/6$ miles. Therefore, by the Pythagorean theorem the distance downstream is

$$\sqrt{(5/6)^2 - (1/2)^2} = \frac{2}{3} \text{ miles.}$$

12.9   Exercises

(1) The wind blows from the South at 40 kilometers per hour and an airplane which travels at 400 kilometers per hour in still air is heading East. Find the actual velocity of the airplane.

(2) ↑In the above problem, find the position of the airplane after two hours.

(3) ↑In the above problem, if the airplane is to travel due east, in what direction should it head in order to achieve this?

(4) The wind blows from West to East at a speed of 50 miles per hour and an airplane which travels at 300 miles per hour in still air is heading North West. What is the velocity of the airplane relative to the ground? What is the component of this velocity in the direction North?

(5) In the situation of Problem 4 how many degrees to the West of North should the airplane head in order to fly exactly North. What will be the speed of the airplane relative to the ground?

(6) In the situation of 5 suppose the airplane uses 34 gallons of fuel every hour at that air speed and that it needs to fly North a distance of 600 miles. Will the airplane have enough fuel to arrive at its destination given that it has 63 gallons of fuel?

(7) An airplane is flying due north at 150 miles per hour. A wind is pushing the airplane due east at 40 miles per hour. After 1 hour, the plane starts flying 30° East of North. Assuming the plane starts at $(0,0)$, where is it after 2 hours? Let North be the direction of the positive y axis and let East be the direction of the positive x axis.

(8) City A is located at the origin while city B is located at $(300, 500)$ where distances are in miles. An airplane flies at 250 miles per hour in still air. This airplane wants to fly from city A to city B but the wind is blowing in the direction of the positive y axis at a speed of 50 miles per hour. Find a unit vector such that if the plane heads in this direction, it will end up at city B having flown the shortest possible distance. How long will it take to get there?

(9) A certain river is one half mile wide with a current flowing at 2 miles per hour from East to West. A man swims directly toward the opposite shore from the South bank of the river at a speed of 3 miles per hour. How far down the river does he find himself when he has swam across? How far does he end up swimming?

(10) A certain river is one half mile wide with a current flowing at 2 miles per hour from East to West. A man can swim at 3 miles per hour in still water. In what direction should he swim in order to travel directly across the river? What would the answer to this problem be if the river flowed at 3 miles per hour and the man could swim only at the rate of 2 miles per hour?

(11) Three forces are applied to a point which does not move. Two of the forces are $2\mathbf{i} + \mathbf{j} + 3\mathbf{k}$ Newtons and $\mathbf{i} - 3\mathbf{j} + 2\mathbf{k}$ Newtons. Find the third force.

(12) Three forces are applied to a point which does not move. Two of the forces are $\mathbf{i} + \mathbf{j} + 3\mathbf{k}$ Newtons and $\mathbf{i} - 3\mathbf{j} - 2\mathbf{k}$ Newtons. Find the third force.

(13) The total force acting on an object is to be $2\mathbf{i} + \mathbf{j} + \mathbf{k}$ Newtons. A force of $-\mathbf{i} + \mathbf{j} + \mathbf{k}$ Newtons is being applied. What other force should be applied to achieve the desired total force?

(14) The total force acting on an object is to be $\mathbf{i} + \mathbf{j} + 3\mathbf{k}$ Newtons. A force of $-\mathbf{i} - \mathbf{j} + \mathbf{k}$ Newtons is being applied. What other force should be applied to achieve the desired total force?

(15) A bird flies from its nest 5 km. in the direction 60° north of east where it stops to rest on a tree. It then flies 10 km. in the direction due southeast and lands atop a telephone pole. Place an xy coordinate system so that the origin is the bird's nest, and the positive x axis points east and the positive y axis points

north. Find the displacement vector from the nest to the telephone pole.

(16) A car is stuck in the mud. There is a cable stretched tightly from this car to a tree which is 20 feet long. A person grasps the cable in the middle and pulls with a force of 100 pounds perpendicular to the stretched cable. The center of the cable moves two feet and remains still. What is the tension in the cable? The tension in the cable is the force exerted on this point by the part of the cable nearer the car as well as the force exerted on this point by the part of the cable nearer the tree.

Chapter 13

Vector Products

13.1 The Dot Product

There are two ways of multiplying vectors which are of great importance in applications. The first of these is called the **dot product**, also called the **scalar product**
and sometimes the **inner product**.

Definition 13.1. Let $\mathbf{a}, \mathbf{b}$ be two vectors in $\mathbb{R}^n$ define $\mathbf{a} \cdot \mathbf{b}$ as

$$\mathbf{a} \cdot \mathbf{b} = \sum_{k=1}^{n} a_k b_k.$$

With this definition, there are several important properties satisfied by the dot
product. In the statement of these properties, α and β will denote scalars and $\mathbf{a}, \mathbf{b}, \mathbf{c}$
will denote vectors.

Proposition 13.1. The dot product satisfies the following properties.

$$\mathbf{a} \cdot \mathbf{b} = \mathbf{b} \cdot \mathbf{a} \tag{13.1}$$

$$\mathbf{a} \cdot \mathbf{a} \geq 0 \text{ and equals zero if and only if } \mathbf{a} = \mathbf{0} \tag{13.2}$$

$$(\alpha \mathbf{a} + \beta \mathbf{b}) \cdot \mathbf{c} = \alpha \, (\mathbf{a} \cdot \mathbf{c}) + \beta \, (\mathbf{b} \cdot \mathbf{c}) \tag{13.3}$$

$$\mathbf{c} \cdot (\alpha \mathbf{a} + \beta \mathbf{b}) = \alpha \, (\mathbf{c} \cdot \mathbf{a}) + \beta \, (\mathbf{c} \cdot \mathbf{b}) \tag{13.4}$$

$$|\mathbf{a}|^2 = \mathbf{a} \cdot \mathbf{a} \tag{13.5}$$

You should verify these properties. Also be sure you understand that (13.4)
follows from the first three and is therefore redundant. It is listed here for the sake
of convenience.

Example 13.1. Find $(1, 2, 0, -1) \cdot (0, 1, 2, 3)$.

This equals $0 + 2 + 0 + -3 = -1$.

Example 13.2. Find the magnitude of $\mathbf{a} = (2, 1, 4, 2)$. That is, find $|\mathbf{a}|$.

This is $\sqrt{(2,1,4,2) \cdot (2,1,4,2)} = 5$.

The dot product satisfies a fundamental inequality known as the **Cauchy Schwarz inequality.** It has already been proved but here is another proof.

Theorem 13.1. *The dot product satisfies the inequality*

$$|\mathbf{a} \cdot \mathbf{b}| \leq |\mathbf{a}|\,|\mathbf{b}|. \tag{13.6}$$

Furthermore equality is obtained if and only if one of $\mathbf{a}$ or $\mathbf{b}$ is a scalar multiple of the other.

Proof: First note that if $\mathbf{b} = \mathbf{0}$, both sides of (13.6) equal zero and so the inequality holds in this case. Therefore, it will be assumed in what follows that $\mathbf{b} \neq \mathbf{0}$.

Define a function of $t \in \mathbb{R}$

$$f(t) = (\mathbf{a} + t\mathbf{b}) \cdot (\mathbf{a} + t\mathbf{b}).$$

Then by (13.2), $f(t) \geq 0$ for all $t \in \mathbb{R}$. Also from (13.3),(13.4),(13.1), and (13.5)

$$
\begin{aligned}
f(t) &= \mathbf{a} \cdot (\mathbf{a} + t\mathbf{b}) + t\mathbf{b} \cdot (\mathbf{a} + t\mathbf{b}) \\
&= \mathbf{a} \cdot \mathbf{a} + t(\mathbf{a} \cdot \mathbf{b}) + t\mathbf{b} \cdot \mathbf{a} + t^2 \mathbf{b} \cdot \mathbf{b} \\
&= |\mathbf{a}|^2 + 2t(\mathbf{a} \cdot \mathbf{b}) + |\mathbf{b}|^2 t^2.
\end{aligned}
$$

Now

$$
\begin{aligned}
f(t) &= |\mathbf{b}|^2 \left(t^2 + 2t\frac{\mathbf{a} \cdot \mathbf{b}}{|\mathbf{b}|^2} + \frac{|\mathbf{a}|^2}{|\mathbf{b}|^2} \right) \\
&= |\mathbf{b}|^2 \left(t^2 + 2t\frac{\mathbf{a} \cdot \mathbf{b}}{|\mathbf{b}|^2} + \left(\frac{\mathbf{a} \cdot \mathbf{b}}{|\mathbf{b}|^2}\right)^2 - \left(\frac{\mathbf{a} \cdot \mathbf{b}}{|\mathbf{b}|^2}\right)^2 + \frac{|\mathbf{a}|^2}{|\mathbf{b}|^2} \right) \\
&= |\mathbf{b}|^2 \left(\left(t + \frac{\mathbf{a} \cdot \mathbf{b}}{|\mathbf{b}|^2}\right)^2 + \left(\frac{|\mathbf{a}|^2}{|\mathbf{b}|^2} - \left(\frac{\mathbf{a} \cdot \mathbf{b}}{|\mathbf{b}|^2}\right)^2\right) \right) \geq 0
\end{aligned}
$$

for all $t \in \mathbb{R}$. In particular $f(t) \geq 0$ when $t = -\left(\mathbf{a} \cdot \mathbf{b}/|\mathbf{b}|^2\right)$ which implies

$$\frac{|\mathbf{a}|^2}{|\mathbf{b}|^2} - \left(\frac{\mathbf{a} \cdot \mathbf{b}}{|\mathbf{b}|^2}\right)^2 \geq 0. \tag{13.7}$$

Multiplying both sides by $|\mathbf{b}|^4$,

$$|\mathbf{a}|^2 |\mathbf{b}|^2 \geq (\mathbf{a} \cdot \mathbf{b})^2$$

which yields (13.6).

From Theorem 13.1, equality holds in (13.6) whenever one of the vectors is a scalar multiple of the other. It only remains to verify this is the only way equality can occur. If either vector equals zero, then equality is obtained in (13.6) so it can be assumed both vectors are nonzero and that equality is obtained in (13.7). This implies that $f(t) = 0$ when $t = -\left(\mathbf{a} \cdot \mathbf{b}/|\mathbf{b}|^2\right)$ and so from (13.2), it follows that for this value of t, $\mathbf{a} + t\mathbf{b} = \mathbf{0}$ showing $\mathbf{a} = -t\mathbf{b}$. ∎

You should note that the entire argument was based only on the properties of the dot product listed in (13.1) - (13.5). This means that whenever something satisfies these properties, the Cauchy Schwartz inequality holds. There are many other instances of these properties besides vectors in $\mathbb{R}^n$.

The Cauchy Schwartz inequality allows a proof of the **triangle inequality** for distances in $\mathbb{R}^n$ in much the same way as the triangle inequality for the absolute value.

Theorem 13.2. *(Triangle inequality) For* $\mathbf{a}, \mathbf{b} \in \mathbb{R}^n$

$$|\mathbf{a} + \mathbf{b}| \le |\mathbf{a}| + |\mathbf{b}| \tag{13.8}$$

and equality holds if and only if one of the vectors is a nonnegative scalar multiple of the other. Also

$$||\mathbf{a}| - |\mathbf{b}|| \le |\mathbf{a} - \mathbf{b}| \tag{13.9}$$

Proof: By properties of the dot product and the Cauchy Schwarz inequality,

$$|\mathbf{a} + \mathbf{b}|^2 - (\mathbf{a} \mid \mathbf{b}) \, (\mathbf{a} \mid \mathbf{b}) - (\mathbf{a} \cdot \mathbf{a}) + (\mathbf{a} \cdot \mathbf{b}) + (\mathbf{b} \cdot \mathbf{a}) + (\mathbf{h} \cdot \mathbf{h})$$
$$- |\mathbf{a}|^2 + 2(\mathbf{a} \cdot \mathbf{b}) + |\mathbf{h}|^2 \le |\mathbf{a}|^2 + 2|\mathbf{a} \cdot \mathbf{b}| + |\mathbf{b}|^2$$
$$\le |\mathbf{a}|^2 + 2|\mathbf{a}||\mathbf{b}| + |\mathbf{b}|^2 = (|\mathbf{a}| + |\mathbf{b}|)^2.$$

Taking square roots of both sides you obtain (13.8).

It remains to consider when equality occurs. If either vector equals zero, then that vector equals zero times the other vector and the claim about when equality occurs is verified. Therefore, it can be assumed both vectors are nonzero. To get equality in the second inequality above, Theorem 13.1 implies one of the vectors must be a multiple of the other. Say $\mathbf{b} - \alpha\mathbf{a}$. If $\alpha < 0$ then equality cannot occur in the first inequality because in this case

$$(\mathbf{a} \cdot \mathbf{b}) = \alpha|\mathbf{a}|^2 < 0 < |\alpha||\mathbf{a}|^2 = |\mathbf{a} \cdot \mathbf{b}|$$

Therefore, $\alpha \ge 0$.

To get the other form of the triangle inequality, $\mathbf{a} = \mathbf{a} - \mathbf{b} + \mathbf{b}$ so

$$|\mathbf{a}| = |\mathbf{a} - \mathbf{b} + \mathbf{b}| \le |\mathbf{a} - \mathbf{b}| + |\mathbf{b}|.$$

Therefore,

$$|\mathbf{a}| - |\mathbf{b}| \le |\mathbf{a} - \mathbf{b}| \tag{13.10}$$

Similarly,

$$|\mathbf{b}| - |\mathbf{a}| \le |\mathbf{b} - \mathbf{a}| = |\mathbf{a} - \mathbf{b}|. \tag{13.11}$$

It follows from (13.10) and (13.11) that (13.9) holds. This is because $||\mathbf{a}| - |\mathbf{b}||$ equals the left side of either (13.10) or (13.11) and either way, $||\mathbf{a}| - |\mathbf{b}|| \le |\mathbf{a} - \mathbf{b}|$. ∎

13.2   The Geometric Significance Of The Dot Product

13.2.1   *The Angle Between Two Vectors*

Given two vectors **a** and **b**, the included angle is the angle between these two vectors which is less than or equal to 180 degrees. The dot product can be used to determine the included angle between two vectors. To see how to do this, consider the following picture.

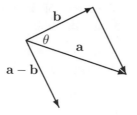

By the law of cosines,

$$|\mathbf{a} - \mathbf{b}|^2 = |\mathbf{a}|^2 + |\mathbf{b}|^2 - 2\,|\mathbf{a}|\,|\mathbf{b}|\cos\theta.$$

Also from the properties of the dot product,

$$|\mathbf{a} - \mathbf{b}|^2 = (\mathbf{a} - \mathbf{b}) \cdot (\mathbf{a} - \mathbf{b}) = |\mathbf{a}|^2 + |\mathbf{b}|^2 - 2\mathbf{a} \cdot \mathbf{b}$$

and so, comparing the above two formulas,

$$\mathbf{a} \cdot \mathbf{b} = |\mathbf{a}|\,|\mathbf{b}|\cos\theta. \tag{13.12}$$

In words, the dot product of two vectors equals the product of the magnitude of the two vectors multiplied by the cosine of the included angle. Note this gives a geometric description of the dot product which does not depend explicitly on the coordinates of the vectors.

Example 13.3. Find the angle between the vectors $2\mathbf{i} + \mathbf{j} - \mathbf{k}$ and $3\mathbf{i} + 4\mathbf{j} + \mathbf{k}$.

The dot product of these two vectors equals $6 + 4 - 1 = 9$ and the norms are

$$\sqrt{4 + 1 + 1} = \sqrt{6}$$

and $\sqrt{9 + 16 + 1} = \sqrt{26}$. Therefore, from (13.12) the cosine of the included angle equals

$$\cos\theta = \frac{9}{\sqrt{26}\sqrt{6}} = .720\,58$$

Now the cosine is known, the angle can be determines by solving the equation $\cos\theta = .720\,58$. This will involve using a calculator or a table of trigonometric functions. The answer is $\theta = .766\,16$ radians or in terms of degrees, $\theta = .766\,16 \times \frac{360}{2\pi} = 43.$ $898°$. Recall how this last computation is done. Set up a proportion $\frac{x}{.76616} = \frac{360}{2\pi}$ because $360°$ corresponds to 2π radians. However, in calculus, you should get used to thinking in terms of radians and not degrees. This is because all the important calculus formulas are defined in terms of radians.

Example 13.4. Let $\mathbf{u}, \mathbf{v}$ be two vectors whose magnitudes are equal to 3 and 4 respectively and such that if they are placed in standard position with their tails at the origin, the angle between $\mathbf{u}$ and the positive x axis equals 30° and the angle between $\mathbf{v}$ and the positive x axis is $-30°$. Find $\mathbf{u} \cdot \mathbf{v}$.

From the geometric description of the dot product in (13.12)

$$\mathbf{u} \cdot \mathbf{v} = 3 \times 4 \times \cos(60°) = 3 \times 4 \times 1/2 = 6.$$

Observation 13.1. Two vectors are said to be **perpendicular** if the included angle is $\pi/2$ radians (90°). You can tell if two nonzero vectors are perpendicular by simply taking their dot product. If the answer is zero, this means they are perpendicular because $\cos\theta = 0$.

Example 13.5. Determine whether the two vectors $2\mathbf{i} + \mathbf{j} - \mathbf{k}$ and $1\mathbf{i} + 3\mathbf{j} + 5\mathbf{k}$ are perpendicular.

When you take this dot product you get $2 + 3 - 5 = 0$ and so these two are indeed perpendicular.

Definition 13.2. When two lines intersect, the angle between the two lines is the smaller of the two angles determined.

Example 13.6. Find the angle between the two lines, $(1, 2, 0) + t(1, 2, 3)$ and $(0, 4, -3) + t(-1, 2, -3)$.

These two lines intersect, when $t = 0$ in the first and $t = -1$ in the second. It is only a matter of finding the angle between the direction vectors. One angle determined is given by

$$\cos\theta = \frac{-6}{14} = \frac{-3}{7}. \tag{13.13}$$

We don't want this angle because it is obtuse. The angle desired is the acute angle given by

$$\cos\theta = \frac{3}{7}.$$

It is obtained by reversing the direction of one of the direction vectors.

13.2.2 *Work And Projections*

Our first application will be to the concept of work. The physical concept of work does not in any way correspond to the notion of work employed in ordinary conversation. For example, if you were to slide a 150 pound weight off a table which is three feet high and shuffle along the floor for 50 yards, sweating profusely and exerting all your strength to keep the weight from falling on your feet, keeping the height always three feet and then deposit this weight on another three foot high table, the physical concept of work would indicate that the force exerted by your arms did no work during this project, even though the muscles in your hands and arms would likely be very tired. The reason for such an unusual definition is that even though your arms exerted considerable force on the weight, enough to keep it from falling, the direction of motion was at right angles to the force they exerted. The only part of a force which does work in the sense of physics is the component of the force in the direction of motion (This is made more precise below.). The work is defined to be the magnitude of the component of this force times the distance over which it acts in the case where this component of force points in the direction of motion and (-1) times the magnitude of this component times the distance in case the force tends to impede the motion. Thus the work done by a force on an object as the object moves from one point to another is a measure of the extent to which the force contributes to the motion. This is illustrated in the following picture in the case where the given force contributes to the motion.

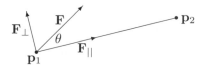

In this picture the force $\mathbf{F}$ is applied to an object which moves on the straight line from $\mathbf{p}_1$ to $\mathbf{p}_2$. There are two vectors shown, $\mathbf{F}_{||}$ and $\mathbf{F}_{\perp}$ and the picture is intended to indicate that when you add these two vectors you get $\mathbf{F}$ while $\mathbf{F}_{||}$ acts in the direction of motion and $\mathbf{F}_{\perp}$ acts perpendicular to the direction of motion. Only $\mathbf{F}_{||}$ contributes to the work done by $\mathbf{F}$ on the object as it moves from $\mathbf{p}_1$ to $\mathbf{p}_2$. $\mathbf{F}_{||}$ is called the **component of the force** in the direction of motion. From trigonometry, you see the magnitude of $\mathbf{F}_{||}$ should equal $|\mathbf{F}|\,|\cos\theta|$. Thus, since $\mathbf{F}_{||}$ points in the direction of the vector from $\mathbf{p}_1$ to $\mathbf{p}_2$, the total work done should equal

$$|\mathbf{F}|\,\left|\overrightarrow{\mathbf{p}_1\mathbf{p}_2}\right|\cos\theta = |\mathbf{F}|\,|\mathbf{p}_2 - \mathbf{p}_1|\cos\theta$$

If the included angle had been obtuse, then the work done by the force $\mathbf{F}$ on the object would have been negative because in this case, the force tends to impede the motion from $\mathbf{p}_1$ to $\mathbf{p}_2$ but in this case, $\cos\theta$ would also be negative and so it is still the case that the work done would be given by the above formula. Thus from the geometric description of the dot product given above, the work equals

$$|\mathbf{F}|\,|\mathbf{p}_2 - \mathbf{p}_1|\cos\theta = \mathbf{F}\cdot(\mathbf{p}_2 - \mathbf{p}_1).$$

This explains the following definition.

Definition 13.3. Let $\mathbf{F}$ be a force acting on an object which moves from the point $\mathbf{p}_1$ to the point $\mathbf{p}_2$. Then the **work** done on the object by the given force equals $\mathbf{F}\cdot(\mathbf{p}_2 - \mathbf{p}_1)$.

Example 13.7. Let $\mathbf{F} = 2\mathbf{i}+7\mathbf{j} - 3\mathbf{k}$ Newtons. Find the work done by this force in moving from the point $(1, 2, 3)$ to the point $(-9, -3, 4)$ along the straight line segment joining these points where distances are measured in meters.

According to the definition, this work is

$$(2\mathbf{i}+7\mathbf{j} - 3\mathbf{k}) \cdot (-10\mathbf{i} - 5\mathbf{j} + \mathbf{k}) = -20 + (-35) + (-3) = -58 \text{ Newton meters.}$$

Note that if the force had been given in pounds and the distance had been given in feet, the units on the work would have been foot pounds. In general, work has units equal to units of a force times units of a length. Instead of writing Newton meter, people write joule because a joule is by definition a Newton meter. That word is pronounced "jewel" and it is the unit of work in the metric system of units. Also be sure you observe that the work done by the force can be negative as in the above example. In fact, work can be either positive, negative, or zero. You just have to do the computations to find out.

The concept of writing a given vector $\mathbf{F}$ in terms of two vectors, one which is parallel to a given vector $\mathbf{D}$ and the other which is perpendicular can also be explained with no reliance on trigonometry, completely in terms of the algebraic properties of the dot product. As before, this is mathematically more significant than any approach involving geometry or trigonometry because it extends to more interesting situations. This is done next.

Theorem 13.3. *Let $\mathbf{F}$ and $\mathbf{D}$ be nonzero vectors. Then there exist unique vectors $\mathbf{F}_{||}$ and $\mathbf{F}_\perp$ such that*

$$\mathbf{F} = \mathbf{F}_{||} + \mathbf{F}_\perp \tag{13.14}$$

where $\mathbf{F}_{||}$ is a scalar multiple of $\mathbf{D}$, also referred to as

$$proj_{\mathbf{D}}(\mathbf{F}) \equiv \frac{\mathbf{F} \cdot \mathbf{D}}{|\mathbf{D}|^2}\mathbf{D},$$

*and $\mathbf{F}_\perp \cdot \mathbf{D} = 0$. The vector $proj_{\mathbf{D}}(\mathbf{F})$ is called the **projection** of $\mathbf{F}$ onto $\mathbf{D}$.*

Proof: Suppose (13.14) and $\mathbf{F}_{||} = \alpha\mathbf{D}$. Taking the dot product of both sides with $\mathbf{D}$ and using $\mathbf{F}_\perp \cdot \mathbf{D} = 0$, this yields

$$\mathbf{F} \cdot \mathbf{D} = \alpha |\mathbf{D}|^2$$

which requires $\alpha = \mathbf{F} \cdot \mathbf{D}/ |\mathbf{D}|^2$. Thus there can be no more than one vector $\mathbf{F}_{||}$. It follows $\mathbf{F}_\perp$ must equal $\mathbf{F} - \mathbf{F}_{||}$. This verifies there can be no more than one choice for both $\mathbf{F}_{||}$ and $\mathbf{F}_\perp$.

Now let

$$\mathbf{F}_{||} \equiv \frac{\mathbf{F} \cdot \mathbf{D}}{|\mathbf{D}|^2}\mathbf{D}$$

and let

$$\mathbf{F}_\perp = \mathbf{F} - \mathbf{F}_{||} = \mathbf{F} - \frac{\mathbf{F} \cdot \mathbf{D}}{|\mathbf{D}|^2}\mathbf{D}$$

Then $\mathbf{F}_{\parallel} = \alpha \, \mathbf{D}$ where $\alpha = \frac{\mathbf{F}\cdot\mathbf{D}}{|\mathbf{D}|^2}$. It only remains to verify $\mathbf{F}_{\perp} \cdot \mathbf{D} = 0$. But

$$\mathbf{F}_{\perp}\cdot\mathbf{D} = \mathbf{F}\cdot\mathbf{D} - \frac{\mathbf{F}\cdot\mathbf{D}}{|\mathbf{D}|^2}\mathbf{D}\cdot\mathbf{D} = \mathbf{F}\cdot\mathbf{D} - \mathbf{F}\cdot\mathbf{D} = 0.\qquad\blacksquare$$

Example 13.8. Find $\text{proj}_{\mathbf{u}}(\mathbf{v})$ if $\mathbf{u} = 2\mathbf{i} + 3\mathbf{j} - 4\mathbf{k}$ and $\mathbf{v} = \mathbf{i} - 2\mathbf{j} + \mathbf{k}$.

From the above discussion in Theorem 13.3, this is just

$$\frac{1}{4+9+16}\,(\mathbf{i} - 2\mathbf{j} + \mathbf{k})\cdot(2\mathbf{i} + 3\mathbf{j} - 4\mathbf{k})\,(2\mathbf{i} + 3\mathbf{j} - 4\mathbf{k})$$

$$= \frac{-8}{29}\,(2\mathbf{i} + 3\mathbf{j} - 4\mathbf{k}) = -\frac{16}{29}\mathbf{i} - \frac{24}{29}\mathbf{j} + \frac{32}{29}\mathbf{k}.$$

Example 13.9. Suppose $\mathbf{a}$, and $\mathbf{b}$ are vectors and $\mathbf{b}_{\perp} = \mathbf{b} - \text{proj}_{\mathbf{a}}(\mathbf{b})$. What is the magnitude of $\mathbf{b}_{\perp}$ in terms of the included angle?

$$|\mathbf{b}_{\perp}|^2 = (\mathbf{b} - \text{proj}_{\mathbf{a}}(\mathbf{b}))\cdot(\mathbf{b} - \text{proj}_{\mathbf{a}}(\mathbf{b})) = \left(\mathbf{b} - \frac{\mathbf{b}\cdot\mathbf{a}}{|\mathbf{a}|^2}\mathbf{a}\right)\cdot\left(\mathbf{b} - \frac{\mathbf{b}\cdot\mathbf{a}}{|\mathbf{a}|^2}\mathbf{a}\right)$$

$$= |\mathbf{b}|^2 - 2\frac{(\mathbf{b}\cdot\mathbf{a})^2}{|\mathbf{a}|^2} + \left(\frac{\mathbf{b}\cdot\mathbf{a}}{|\mathbf{a}|^2}\right)^2|\mathbf{a}|^2 = |\mathbf{b}|^2\left(1 - \frac{(\mathbf{b}\cdot\mathbf{a})^2}{|\mathbf{a}|^2\,|\mathbf{b}|^2}\right)$$

$$= |\mathbf{b}|^2\,(1 - \cos^2\theta) = |\mathbf{b}|^2\sin^2(\theta)$$

where θ is the included angle between $\mathbf{a}$ and $\mathbf{b}$ which is less than π radians. Therefore, taking square roots, $|\mathbf{b}_{\perp}| = |\mathbf{b}|\sin\theta$.

13.2.3 *The Parabolic Mirror, An Application*

When light is reflected the angle of incidence is always equal to the angle of reflection. This is illustrated in the following picture in which a ray of light reflects off something like a mirror.

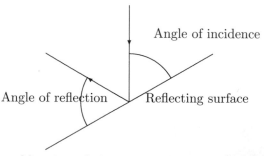

An interesting problem is to design a curved mirror which has the property that it will direct all rays of light coming from a long distance away (essentially parallel rays of light) to a single point. You might be interested in a reflecting telescope for example or some sort of scheme for achieving high temperatures by reflecting the rays of the sun to a small area. Turning things around, you could place a source of light at the single point and desire to have the mirror reflect this in a beam of light consisting of parallel rays. How can you design such a mirror?

It turns out this is pretty easy given the above techniques for finding the angle between vectors. Consider the following picture.

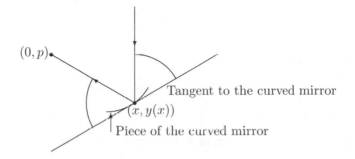

It suffices to consider this in a plane for $x > 0$ and then let the mirror be obtained as a surface of revolution. In the above picture, let $(0, p)$ be the special point at which all the parallel rays of light will be directed. This is set up so the rays of light are parallel to the y axis. The two indicated angles will be equal and the equation of the indicated curve will be $y = y(x)$ while the reflection is taking place at the point $(x, y(x))$ as shown. To say the two angles are equal is to say their cosines are equal. Thus from the above,

$$\frac{(0, 1) \cdot (1, y'(x))}{\sqrt{1 + y'(x)^2}} = \frac{(-x, p - y) \cdot (-1, -y'(x))}{\sqrt{x^2 + (y - p)^2}\sqrt{1 + y'(x)^2}}.$$

This follows because the vectors forming the sides of one of the angles are $(0, 1)$ and $(1, y'(x))$ while the vectors forming the other angle are $(-x, p - y)$ and $(-1, -y'(x))$. Therefore, this yields the differential equation

$$y'(x) = \frac{-y'(x)(p - y) + x}{\sqrt{x^2 + (y - p)^2}}$$

which is written more simply as

$$\left(\sqrt{x^2 + (y - p)^2} + (p - y)\right) y' = x.$$

Now let $y - p = xv$ so that $y' = xv' + v$. Then in terms of v the differential equation is

$$xv' = \frac{1}{\sqrt{1 + v^2} - v} - v.$$

This reduces to

$$\left(\frac{1}{\sqrt{1 + v^2} - v} - v\right) \frac{dv}{dx} = \frac{1}{x}.$$

If $G \in \int \left(\frac{1}{\sqrt{1+v^2} - v} - v \right) dv$, then a solution to the differential equation is of the form

$$G(v) - \ln x = C$$

where C is a constant. This is because if you differentiate both sides with respect to x,

$$G'(v) \frac{dv}{dx} - \frac{1}{x} = \left(\frac{1}{\sqrt{1 + v^2} - v} - v \right) \frac{dv}{dx} - \frac{1}{x} = 0.$$

To find $G \in \int \left(\frac{1}{\sqrt{1+v^2} - v} - v \right) dv$, use a trig. substitution $v = \tan \theta$. Then in terms of θ, the antiderivative becomes

$$\int \left(\frac{1}{\sec \theta - \tan \theta} - \tan \theta \right) \sec^2 \theta \, d\theta = \int \sec \theta \, d\theta = \ln |\sec \theta + \tan \theta| + C.$$

Now in terms of v, this is

$$\ln \left(v + \sqrt{1 + v^2} \right) = \ln x + c.$$

There is no loss of generality in letting $c = \ln C$ because $\ln$ maps onto $\mathbb{R}$. Therefore, from laws of logarithms,

$$\ln \left| v + \sqrt{1 + v^2} \right| = \ln x + c = \ln x + \ln C = \ln Cx.$$

Therefore, $v + \sqrt{1 + v^2} = Cx$ and so $\sqrt{1 + v^2} = Cx - v$. Now square both sides to get

$$1 + v^2 = C^2 x^2 + v^2 - 2Cxv$$

which shows

$$1 = C^2 x^2 - 2Cx \frac{y - p}{x} = C^2 x^2 - 2C (y - p).$$

Solving this for y yields

$$y = \frac{C}{2} x^2 + \left(p - \frac{1}{2C} \right)$$

and for this to correspond to reflection as described above, it must be that $C > 0$. As described in an earlier section, this is just the equation of a parabola. Note it is possible to choose C as desired, adjusting the shape of the mirror.

13.2.4   The Dot Product And Distance In $\mathbb{C}^n$

It is necessary to give a generalization of the dot product for vectors in $\mathbb{C}^n$. This definition reduces to the usual one in the case the components of the vector are real.

Definition 13.4. Let $\mathbf{x}, \mathbf{y} \in \mathbb{C}^n$. Thus $\mathbf{x} = (x_1, \cdots, x_n)$ where each $x_k \in \mathbb{C}$ and a similar formula holding for $\mathbf{y}$. Then the dot product of these two vectors is defined to be

$$\mathbf{x} \cdot \mathbf{y} \equiv \sum_j x_j \overline{y_j} \equiv x_1 \overline{y_1} + \cdots + x_n \overline{y_n}.$$

Notice how you put the conjugate on the entries of the vector $\mathbf{y}$. It makes no difference if the vectors happen to be real vectors but with complex vectors you must do it this way. The reason for this is that when you take the dot product of a vector with itself, you want to get the square of the length of the vector, a positive number. Placing the conjugate on the components of $\mathbf{y}$ in the above definition assures this will take place. Thus

$$\mathbf{x} \cdot \mathbf{x} = \sum_j x_j \overline{x_j} = \sum_j |x_j|^2 \geq 0.$$

If you did not place a conjugate as in the above definition, things would not work out correctly. For example,

$$(1+i)^2 + 2^2 = 4 + 2i$$

and this is not a positive number.

The following properties of the dot product follow immediately from the definition and you should verify each of them.

Properties of the dot product:

(1) $\mathbf{u} \cdot \mathbf{v} = \overline{\mathbf{v} \cdot \mathbf{u}}$.
(2) If a, b are numbers and $\mathbf{u}, \mathbf{v}, \mathbf{z}$ are vectors then $(a\mathbf{u} + b\mathbf{v}) \cdot \mathbf{z} = a(\mathbf{u} \cdot \mathbf{z}) + b(\mathbf{v} \cdot \mathbf{z})$.
(3) $\mathbf{u} \cdot \mathbf{u} \geq 0$ and it equals 0 if and only if $\mathbf{u} = \mathbf{0}$.

The norm is defined in the usual way.

Definition 13.5. For $\mathbf{x} \in \mathbb{C}^n$,

$$|\mathbf{x}| \equiv \left(\sum_{k=1}^n |x_k|^2 \right)^{1/2} = (\mathbf{x} \cdot \mathbf{x})^{1/2}$$

As in the case of $\mathbb{R}^n$, the **Cauchy Schwarz inequality** is of fundamental importance. First here is a simple lemma.

Lemma 13.1. *If $z \in \mathbb{C}$ there exists $\theta \in \mathbb{C}$ such that $\theta z = |z|$ and $|\theta| = 1$.*

Proof: Let $\theta = 1$ if $z = 0$ and otherwise, let $\theta = \dfrac{\overline{z}}{|z|}$. Recall that for $z = x + iy, \overline{z} = x - iy$ and $\overline{z}z = |z|^2$. ∎

Theorem 13.4. *(Cauchy Schwarz) The following inequality holds for x_i and $y_i \in \mathbb{C}$.*

$$|(\mathbf{x} \cdot \mathbf{y})| = \left| \sum_{i=1}^n x_i \overline{y}_i \right| \leq \left(\sum_{i=1}^n |x_i|^2 \right)^{1/2} \left(\sum_{i=1}^n |y_i|^2 \right)^{1/2} = |\mathbf{x}| \, |\mathbf{y}| \qquad (13.15)$$

Proof: Let $\theta \in \mathbb{C}$ such that $|\theta| = 1$ and

$$\theta \sum_{i=1}^{n} x_i \bar{y}_i = \left| \sum_{i=1}^{n} x_i \bar{y}_i \right|$$

Thus

$$\left| \sum_{i=1}^{n} x_i \bar{y}_i \right| = \theta \sum_{i=1}^{n} x_i \bar{y}_i = \sum_{i=1}^{n} x_i \overline{(\bar{\theta} y_i)}.$$

Consider $p(t) \equiv \sum_{i=1}^{n} (x_i + t\bar{\theta} y_i) \overline{(x_i + t\bar{\theta} y_i)}$ where $t \in \mathbb{R}$.

$$\begin{aligned} 0 \leq p(t) &= \sum_{i=1}^{n} |x_i|^2 + 2t\mathrm{Re}\left(\theta \sum_{i=1}^{n} x_i \bar{y}_i \right) + t^2 \sum_{i=1}^{n} |y_i|^2 \\ &= |\mathbf{x}|^2 + 2t \left| \sum_{i=1}^{n} x_i \bar{y}_i \right| + t^2 |\mathbf{y}|^2 \end{aligned}$$

If $|\mathbf{y}| = 0$ then (13.15) is obviously true because both sides equal zero. Therefore, assume $|\mathbf{y}| \neq 0$ and then $p(t)$ is a polynomial of degree two whose graph opens up. Therefore, it either has no zeroes, two zeros or one repeated zero. If it has two zeros, the above inequality must be violated because in this case the graph must dip below the x axis. Therefore, it either has no zeros or exactly one. From the quadratic formula this happens exactly when

$$4 \left| \sum_{i=1}^{n} x_i \bar{y}_i \right|^2 - 4 |\mathbf{x}|^2 |\mathbf{y}|^2 \leq 0$$

and so $\left| \sum_{i=1}^{n} x_i \bar{y}_i \right| \leq |\mathbf{x}| |\mathbf{y}|$ as claimed. ∎

By analogy to the case of $\mathbb{R}^n$, length or magnitude of vectors in $\mathbb{C}^n$ can be defined.

Definition 13.6. Let $\mathbf{z} \in \mathbb{C}^n$. Then $|\mathbf{z}| \equiv (\mathbf{z} \cdot \mathbf{z})^{1/2}$.

Theorem 13.5. *For length defined in Definition 13.6, the following hold.*

$$|\mathbf{z}| \geq 0 \text{ and } |\mathbf{z}| = 0 \text{ if and only if } \mathbf{z} = \mathbf{0} \tag{13.16}$$

$$\text{If } \alpha \text{ is a scalar, } |\alpha \mathbf{z}| = |\alpha| |\mathbf{z}| \tag{13.17}$$

$$|\mathbf{z} + \mathbf{w}| \leq |\mathbf{z}| + |\mathbf{w}|. \tag{13.18}$$

Proof: The first two claims are left as exercises. To establish the third, you use the same argument which was used in $\mathbb{R}^n$.

$$\begin{aligned} |\mathbf{z} + \mathbf{w}|^2 &= (\mathbf{z} + \mathbf{w}, \mathbf{z} + \mathbf{w}) = \mathbf{z} \cdot \mathbf{z} + \mathbf{w} \cdot \mathbf{w} + \mathbf{w} \cdot \mathbf{z} + \mathbf{z} \cdot \mathbf{w} \\ &= |\mathbf{z}|^2 + |\mathbf{w}|^2 + 2\mathrm{Re}\,\mathbf{w} \cdot \mathbf{z} \leq |\mathbf{z}|^2 + |\mathbf{w}|^2 + 2|\mathbf{w} \cdot \mathbf{z}| \\ &\leq |\mathbf{z}|^2 + |\mathbf{w}|^2 + 2|\mathbf{w}||\mathbf{z}| = (|\mathbf{z}| + |\mathbf{w}|)^2. \end{aligned}$$

All other considerations such as open and closed sets and the like are identical in this more general context with the corresponding definition in $\mathbb{R}^n$. The main difference is that here the scalars are complex numbers. ∎

Definition 13.7. Suppose you have a vector space, V and for $z, w \in V$ and α a scalar a norm is a way of measuring distance or magnitude which satisfies the properties (13.16) - (13.18). Thus a norm is something which does the following.

$$||z|| \geq 0 \text{ and } ||z|| = 0 \text{ if and only if } z = 0 \tag{13.19}$$

$$\text{If } \alpha \text{ is a scalar, } ||\alpha z|| = |\alpha| \, ||z|| \tag{13.20}$$

$$||z + w|| \leq ||z|| + ||w||. \tag{13.21}$$

Here is understood that for all $z \in V, ||z|| \in [0, \infty)$.

13.3 Exercises

(1) Find $(1, 2, 3, 4) \cdot (2, 0, 1, 3)$.
(2) Use formula (13.12) to verify the Cauchy Schwartz inequality and to show that equality occurs if and only if one of the vectors is a scalar multiple of the other.
(3) For u, v vectors in $\mathbb{R}^3$, define the product $u * v \equiv u_1 v_1 + 2u_2 v_2 + 3u_3 v_3$. Show the axioms for a dot product all hold for this funny product. Prove $|u * v| \leq (u * u)^{1/2} (v * v)^{1/2}$. **Hint:** Do not try to do this with methods from trigonometry.
(4) Find the angle between the vectors $3i - j - k$ and $i + 4j + 2k$.
(5) Find the angle between the vectors $i - 2j + k$ and $i + 2j - 7k$.
(6) Find $\text{proj}_u (v)$ where $v = (1, 0, -2)$ and $u = (1, 2, 3)$.
(7) Find $\text{proj}_u (v)$ where $v = (1, 2, -2)$ and $u = (1, 0, 3)$.
(8) Find $\text{proj}_u (v)$ where $v = (1, 2, -2, 1)$ and $u = (1, 2, 3, 0)$.
(9) Does it make sense to speak of $\text{proj}_0 (v)$?
(10) If F is a force and D is a vector, show $\text{proj}_D (F) = (|F| \cos \theta) u$ where u is the unit vector in the direction of D, $u = D/|D|$ and θ is the included angle between the two vectors F and D. $|F| \cos \theta$ is sometimes called the component of the force F in the direction, D.
(11) A boy drags a sled for 100 feet along the ground by pulling on a rope which is 20 degrees from the horizontal with a force of 40 pounds. How much work does this force do?

(12) A girl drags a sled for 200 feet along the ground by pulling on a rope which is 30 degrees from the horizontal with a force of 20 pounds. How much work does this force do?

(13) A large dog drags a sled for 300 feet along the ground by pulling on a rope which is 45 degrees from the horizontal with a force of 20 pounds. How much work does this force do?

(14) How much work in Newton meters does it take to slide a crate 20 meters along a loading dock by pulling on it with a 200 Newton force at an angle of $30°$ from the horizontal?

(15) An object moves 10 meters in the direction of $\mathbf{j}$. There are two forces acting on this object $\mathbf{F}_1 = \mathbf{i} + \mathbf{j} + 2\mathbf{k}$, and $\mathbf{F}_2 = -5\mathbf{i} + 2\mathbf{j} - 6\mathbf{k}$. Find the total work done on the object by the two forces. **Hint:** You can take the work done by the resultant of the two forces or you can add the work done by each force. Why?

(16) An object moves 10 meters in the direction of $\mathbf{j} + \mathbf{i}$. There are two forces acting on this object $\mathbf{F}_1 = \mathbf{i} + 2\mathbf{j} + 2\mathbf{k}$, and $\mathbf{F}_2 = 5\mathbf{i} + 2\mathbf{j} - 6\mathbf{k}$. Find the total work done on the object by the two forces. **Hint:** You can take the work done by the resultant of the two forces or you can add the work done by each force. Why?

(17) An object moves 20 meters in the direction of $\mathbf{k} + \mathbf{j}$. There are two forces acting on this object $\mathbf{F}_1 = \mathbf{i} + \mathbf{j} + 2\mathbf{k}$, and $\mathbf{F}_2 = \mathbf{i} + 2\mathbf{j} - 6\mathbf{k}$. Find the total work done on the object by the two forces. **Hint:** You can take the work done by the resultant of the two forces or you can add the work done by each force.

(18) If $\mathbf{a}, \mathbf{b}$, and $\mathbf{c}$ are vectors. Show that $(\mathbf{b} + \mathbf{c})_{\perp} = \mathbf{b}_{\perp} + \mathbf{c}_{\perp}$ where $\mathbf{b}_{\perp} = \mathbf{b} - \text{proj}_{\mathbf{a}}(\mathbf{b})$.

(19) In the discussion of the reflecting mirror which directs all rays to a particular point $(0, p)$. Show that for any choice of positive C this point is the focus of the parabola and the directrix is $y = p - \frac{1}{C}$.

(20) Suppose you wanted to make a solar powered oven to cook food. Are there reasons for using a mirror which is not parabolic? Also describe how you would design a good flash light with a beam which does not spread out too quickly.

(21) Show that $(\mathbf{a} \cdot \mathbf{b}) = \frac{1}{4} \left[|\mathbf{a} + \mathbf{b}|^2 - |\mathbf{a} - \mathbf{b}|^2 \right]$.

(22) Prove from the axioms of the dot product the parallelogram identity $|\mathbf{a} + \mathbf{b}|^2 + |\mathbf{a} - \mathbf{b}|^2 = 2|\mathbf{a}|^2 + 2|\mathbf{b}|^2$.

(23) Suppose f, g are two continuous functions defined on $[0, 1]$. Define $(f \cdot g) = \int_0^1 f(x) g(x)\, dx$. Show this dot product satisfies conditions (13.1) - (13.5). Explain why the Cauchy Schwarz inequality continues to hold in this context and state the Cauchy Schwarz inequality in terms of integrals.

13.4   The Cross Product

The cross product is the other way of multiplying two vectors in $\mathbb{R}^3$. It is very different from the dot product in many ways. First the geometric meaning is discussed and then a description in terms of coordinates is given. Both descriptions of the cross product are important. The geometric description is essential in order to understand the applications to physics and geometry while the coordinate description is the only way to practically compute the cross product.

Definition 13.8. Three vectors $\mathbf{a}, \mathbf{b}, \mathbf{c}$ form a right handed system if when you extend the fingers of your right hand along the vector $\mathbf{a}$ and close them in the direction of $\mathbf{b}$, the thumb points roughly in the direction of $\mathbf{c}$.

For an example of a right handed system of vectors, see the following picture.

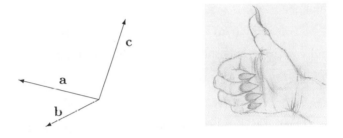

In this picture the vector $\mathbf{c}$ points upwards from the plane determined by the other two vectors. You should consider how a right hand system would differ from a left hand system. Try using your left hand and you will see that the vector $\mathbf{c}$ would need to point in the opposite direction as it would for a right hand system.

From now on, the vectors $\mathbf{i}, \mathbf{j}, \mathbf{k}$ will always form a right handed system. To repeat, if you extend the fingers of your right hand along $\mathbf{i}$ and close them in the direction $\mathbf{j}$, the thumb points in the direction of $\mathbf{k}$.

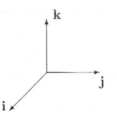

The following is the geometric description of the cross product. It gives both the direction and the magnitude and therefore specifies the vector.

Definition 13.9. Let $\mathbf{a}$ and $\mathbf{b}$ be two vectors in $\mathbb{R}^3$. Then $\mathbf{a} \times \mathbf{b}$ is defined by the following two rules.

(1) $|\mathbf{a} \times \mathbf{b}| = |\mathbf{a}| |\mathbf{b}| \sin \theta$ where θ is the included angle.
(2) $\mathbf{a} \times \mathbf{b} \cdot \mathbf{a} = 0$, $\mathbf{a} \times \mathbf{b} \cdot \mathbf{b} = 0$, and $\mathbf{a}, \mathbf{b}, \mathbf{a} \times \mathbf{b}$ forms a right hand system.

Note that $|\mathbf{a} \times \mathbf{b}|$ is the area of the parallelogram spanned by $\mathbf{a}$ and $\mathbf{b}$.

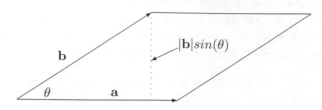

The cross product satisfies the following properties.

$$\mathbf{a} \times \mathbf{b} = -(\mathbf{b} \times \mathbf{a}) \ , \ \mathbf{a} \times \mathbf{a} = \mathbf{0}, \tag{13.22}$$

For α a scalar,

$$(\alpha \mathbf{a}) \times \mathbf{b} = \alpha (\mathbf{a} \times \mathbf{b}) = \mathbf{a} \times (\alpha \mathbf{b}), \tag{13.23}$$

For $\mathbf{a}, \mathbf{b}$, and $\mathbf{c}$ vectors, one obtains the distributive laws,

$$\mathbf{a} \times (\mathbf{b} + \mathbf{c}) = \mathbf{a} \times \mathbf{b} + \mathbf{a} \times \mathbf{c}, \tag{13.24}$$

$$(\mathbf{b} + \mathbf{c}) \times \mathbf{a} = \mathbf{b} \times \mathbf{a} + \mathbf{c} \times \mathbf{a}. \tag{13.25}$$

Formula (13.22) follows immediately from the definition. The vectors $\mathbf{a} \times \mathbf{b}$ and $\mathbf{b} \times \mathbf{a}$ have the same magnitude, $|\mathbf{a}| |\mathbf{b}| \sin \theta$, and an application of the right hand rule shows they have opposite direction. Formula (13.23) is also fairly clear. If α is a nonnegative scalar, the direction of $(\alpha \mathbf{a}) \times \mathbf{b}$ is the same as the direction of $\mathbf{a} \times \mathbf{b}, \alpha (\mathbf{a} \times \mathbf{b})$ and $\mathbf{a} \times (\alpha \mathbf{b})$ while the magnitude is just α times the magnitude of $\mathbf{a} \times \mathbf{b}$ which is the same as the magnitude of $\alpha (\mathbf{a} \times \mathbf{b})$ and $\mathbf{a} \times (\alpha \mathbf{b})$. Using this yields equality in (13.23). In the case where $\alpha < 0$, everything works the same way except the vectors are all pointing in the opposite direction and you must multiply by $|\alpha|$ when comparing their magnitudes. The distributive laws are much harder to establish but the second follows from the first quite easily. Thus, assuming the first, and using (13.22),

$$(\mathbf{b} + \mathbf{c}) \times \mathbf{a} = -\mathbf{a} \times (\mathbf{b} + \mathbf{c}) = -(\mathbf{a} \times \mathbf{b} + \mathbf{a} \times \mathbf{c}) = \mathbf{b} \times \mathbf{a} + \mathbf{c} \times \mathbf{a}.$$

A proof of the distributive law is given in a later section for those who are interested.

Now from the definition of the cross product,

$$\mathbf{i} \times \mathbf{j} = \mathbf{k}, \ \ \mathbf{j} \times \mathbf{i} = -\mathbf{k}$$
$$\mathbf{k} \times \mathbf{i} = \mathbf{j}, \ \ \mathbf{i} \times \mathbf{k} = -\mathbf{j}$$
$$\mathbf{j} \times \mathbf{k} = \mathbf{i}, \ \ \mathbf{k} \times \mathbf{j} = -\mathbf{i}$$

With this information, the following gives the coordinate description of the cross product.

Proposition 13.2. Let $\mathbf{a} = a_1 \mathbf{i} + a_2 \mathbf{j} + a_3 \mathbf{k}$ and $\mathbf{b} = b_1 \mathbf{i} + b_2 \mathbf{j} + b_3 \mathbf{k}$ be two vectors. Then

$$\mathbf{a} \times \mathbf{b} = (a_2 b_3 - a_3 b_2) \mathbf{i} + (a_3 b_1 - a_1 b_3) \mathbf{j} + (a_1 b_2 - a_2 b_1) \mathbf{k}. \tag{13.26}$$

Proof: From the above table and the properties of the cross product listed,
$$(a_1\mathbf{i} + a_2\mathbf{j} + a_3\mathbf{k}) \times (b_1\mathbf{i} + b_2\mathbf{j} + b_3\mathbf{k}) =$$
$$a_1b_2\mathbf{i} \times \mathbf{j} + a_1b_3\mathbf{i} \times \mathbf{k} + a_2b_1\mathbf{j} \times \mathbf{i} + a_2b_3\mathbf{j} \times \mathbf{k} +$$
$$+a_3b_1\mathbf{k} \times \mathbf{i} + a_3b_2\mathbf{k} \times \mathbf{j}$$
$$= a_1b_2\mathbf{k} - a_1b_3\mathbf{j} - a_2b_1\mathbf{k} + a_2b_3\mathbf{i} + a_3b_1\mathbf{j} - a_3b_2\mathbf{i}$$
$$= (a_2b_3 - a_3b_2)\mathbf{i} + (a_3b_1 - a_1b_3)\mathbf{j} + (a_1b_2 - a_2b_1)\mathbf{k} \qquad (13.27)$$
∎

It is probably impossible for most people to remember (13.26). Fortunately, there is a somewhat easier way to remember it.
$$\mathbf{a} \times \mathbf{b} = \begin{vmatrix} \mathbf{i} & \mathbf{j} & \mathbf{k} \\ a_1 & a_2 & a_3 \\ b_1 & b_2 & b_3 \end{vmatrix} \qquad (13.28)$$
where you expand the determinant along the top row. This yields
$$(a_2b_3 - a_3b_2)\mathbf{i} - (a_1b_3 - a_3b_1)\mathbf{j} + (a_1b_2 - a_2b_1)\mathbf{k} \qquad (13.29)$$
which is the same as (13.27). Determinants are discussed in Volume 2 but some people may have seen them. All you need here is how to evaluate 2×2 and 3×3 determinants.
$$\begin{vmatrix} x & y \\ z & w \end{vmatrix} = xw - yz$$
and
$$\begin{vmatrix} a & b & c \\ x & y & z \\ u & v & w \end{vmatrix} = a\begin{vmatrix} y & z \\ v & w \end{vmatrix} - b\begin{vmatrix} x & z \\ u & w \end{vmatrix} + c\begin{vmatrix} x & y \\ u & v \end{vmatrix}.$$
Here is the rule: You look at an entry in the top row and cross out the row and column which contain that entry. If the entry is in the ith column, you multiply $(-1)^{1+i}$ times the determinant of the 2×2 which remains. This is the cofactor. You take the entry in the top row times this cofactor and add all such terms. The rectangular array enclosed by the vertical lines is called a **matrix** and will be discussed more in Volume 2.

Example 13.10. Find $(\mathbf{i} - \mathbf{j} + 2\mathbf{k}) \times (3\mathbf{i} - 2\mathbf{j} + \mathbf{k})$.

Use (13.28) to compute this.
$$\begin{vmatrix} \mathbf{i} & \mathbf{j} & \mathbf{k} \\ 1 & -1 & 2 \\ 3 & -2 & 1 \end{vmatrix} = \begin{vmatrix} -1 & 2 \\ -2 & 1 \end{vmatrix}\mathbf{i} - \begin{vmatrix} 1 & 2 \\ 3 & 1 \end{vmatrix}\mathbf{j} + \begin{vmatrix} 1 & -1 \\ 3 & -2 \end{vmatrix}\mathbf{k}$$
$$= 3\mathbf{i} + 5\mathbf{j} + \mathbf{k}.$$

Example 13.11. Find the area of the parallelogram determined by the vectors
$$(\mathbf{i} - \mathbf{j} + 2\mathbf{k}), \ (3\mathbf{i} - 2\mathbf{j} + \mathbf{k}).$$
These are the same two vectors in Example 13.10.

From Example 13.10 and the geometric description of the cross product, the area is just the norm of the vector obtained in Example 13.10. Thus the area is $\sqrt{9 + 25 + 1} = \sqrt{35}$.

Example 13.12. Find the area of the triangle determined by $(1, 2, 3)$, $(0, 2, 5)$, and $(5, 1, 2)$.

This triangle is obtained by connecting the three points with lines. Picking $(1, 2, 3)$ as a starting point, there are two displacement vectors $(-1, 0, 2)$ and $(4, -1, -1)$ such that the given vector added to these displacement vectors gives the other two vectors. The area of the triangle is half the area of the parallelogram determined by $(-1, 0, 2)$ and $(4, -1, -1)$. Thus $(-1, 0, 2) \times (4, -1, -1) = (2, 7, 1)$ and so the area of the triangle is $\frac{1}{2}\sqrt{4 + 49 + 1} = \frac{3}{2}\sqrt{6}$.

Observation 13.2. In general, if you have three points (vectors) in $\mathbb{R}^3, \mathbf{P}, \mathbf{Q}, \mathbf{R}$ the area of the triangle is given by

$$\frac{1}{2} |(\mathbf{Q} - \mathbf{P}) \times (\mathbf{R} - \mathbf{P})|.$$

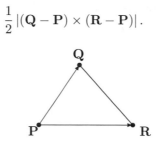

13.4.1 The Distributive Law For The Cross Product

This section gives a proof for (13.24), a fairly difficult topic. It is included here for the interested student. If you are satisfied with taking the distributive law on faith, it is not necessary to read this section. The proof given here is quite clever and follows the one given in [7]. Another approach, based on volumes of parallelepipeds is found in [25] and is discussed a little later.

Lemma 13.2. Let $\mathbf{b}$ and $\mathbf{c}$ be two vectors. Then $\mathbf{b} \times \mathbf{c} = \mathbf{b} \times \mathbf{c}_\perp$ where $\mathbf{c}_{||} + \mathbf{c}_\perp = \mathbf{c}$ and $\mathbf{c}_\perp \cdot \mathbf{b} = 0$.

Proof: Consider the following picture.

Now $\mathbf{c}_\perp = \mathbf{c} - \mathbf{c} \cdot \frac{\mathbf{b}}{|\mathbf{b}|} \frac{\mathbf{b}}{|\mathbf{b}|}$ and so $\mathbf{c}_\perp$ is in the plane determined by $\mathbf{c}$ and $\mathbf{b}$. Therefore, from the geometric definition of the cross product $\mathbf{b} \times \mathbf{c}$ and $\mathbf{b} \times \mathbf{c}_\perp$ have the same direction. Now, referring to the picture,

$$|\mathbf{b} \times \mathbf{c}_\perp| = |\mathbf{b}| \, |\mathbf{c}_\perp| = |\mathbf{b}| \, |\mathbf{c}| \sin\theta = |\mathbf{b} \times \mathbf{c}|.$$

Therefore, $\mathbf{b} \times \mathbf{c}$ and $\mathbf{b} \times \mathbf{c}_\perp$ also have the same magnitude and so they are the same vector. ∎

With this, the proof of the distributive law is in the following theorem.

Theorem 13.6. *Let* $\mathbf{a}, \mathbf{b}$, *and* $\mathbf{c}$ *be vectors in* $\mathbb{R}^3$. *Then*

$$\mathbf{a} \times (\mathbf{b} + \mathbf{c}) = \mathbf{a} \times \mathbf{b} + \mathbf{a} \times \mathbf{c} \tag{13.30}$$

Proof: Suppose first that $\mathbf{a} \cdot \mathbf{b} = \mathbf{a} \cdot \mathbf{c} = 0$. Now imagine $\mathbf{a}$ is a vector coming out of the page and let $\mathbf{b}, \mathbf{c}$ and $\mathbf{b} + \mathbf{c}$ be as shown in the following picture.

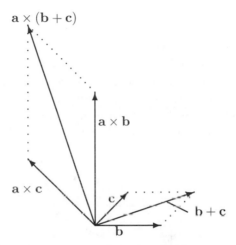

Then $\mathbf{a} \times \mathbf{b}, \mathbf{a} \times (\mathbf{b} + \mathbf{c})$, and $\mathbf{a} \times \mathbf{c}$ are each vectors in the same plane, perpendicular to $\mathbf{a}$ as shown. Thus $\mathbf{a} \times \mathbf{c} \cdot \mathbf{c} = 0, \mathbf{a} \times (\mathbf{b} + \mathbf{c}) \cdot (\mathbf{b} + \mathbf{c}) = 0$, and $\mathbf{a} \times \mathbf{b} \cdot \mathbf{b} = 0$. This implies that to get $\mathbf{a} \times \mathbf{b}$ you move counterclockwise through an angle of $\pi/2$ radians from the vector $\mathbf{b}$. Similar relationships exist between the vectors $\mathbf{a} \times (\mathbf{b} + \mathbf{c})$ and $\mathbf{b} + \mathbf{c}$ and the vectors $\mathbf{a} \times \mathbf{c}$ and $\mathbf{c}$. Thus the angle between $\mathbf{a} \times \mathbf{b}$ and $\mathbf{a} \times (\mathbf{b} + \mathbf{c})$ is the same as the angle between $\mathbf{b} + \mathbf{c}$ and $\mathbf{b}$ and the angle between $\mathbf{a} \times \mathbf{c}$ and $\mathbf{a} \times (\mathbf{b} + \mathbf{c})$ is the same as the angle between $\mathbf{c}$ and $\mathbf{b} + \mathbf{c}$. In addition to this, since $\mathbf{a}$ is perpendicular to these vectors,

$$|\mathbf{a} \times \mathbf{b}| = |\mathbf{a}|\,|\mathbf{b}|\,, |\mathbf{a} \times (\mathbf{b} + \mathbf{c})| = |\mathbf{a}|\,|\mathbf{b} + \mathbf{c}|\,, \text{ and } |\mathbf{a} \times \mathbf{c}| = |\mathbf{a}|\,|\mathbf{c}|\,.$$

Therefore,

$$\frac{|\mathbf{a} \times (\mathbf{b} + \mathbf{c})|}{|\mathbf{b} + \mathbf{c}|} = \frac{|\mathbf{a} \times \mathbf{c}|}{|\mathbf{c}|} = \frac{|\mathbf{a} \times \mathbf{b}|}{|\mathbf{b}|} = |\mathbf{a}|$$

and so

$$\frac{|\mathbf{a} \times (\mathbf{b} + \mathbf{c})|}{|\mathbf{a} \times \mathbf{c}|} = \frac{|\mathbf{b} + \mathbf{c}|}{|\mathbf{c}|}, \frac{|\mathbf{a} \times (\mathbf{b} + \mathbf{c})|}{|\mathbf{a} \times \mathbf{b}|} = \frac{|\mathbf{b} + \mathbf{c}|}{|\mathbf{b}|}$$

showing the triangles making up the parallelogram on the right and the four sided figure on the left in the above picture are similar. It follows the four sided figure on the left is in fact a parallelogram and this implies the diagonal is the vector sum of the vectors on the sides, yielding (13.30).

Now suppose it is not necessarily the case that $\mathbf{a} \cdot \mathbf{b} = \mathbf{a} \cdot \mathbf{c} = 0$. Then write $\mathbf{b} = \mathbf{b}_{||} + \mathbf{b}_{\perp}$ where $\mathbf{b}_{\perp} \cdot \mathbf{a} = 0$. Similarly $\mathbf{c} = \mathbf{c}_{||} + \mathbf{c}_{\perp}$. By the above lemma and what was just shown,

$$\mathbf{a} \times (\mathbf{b} + \mathbf{c}) = \mathbf{a} \times (\mathbf{b} + \mathbf{c})_{\perp} = \mathbf{a} \times (\mathbf{b}_{\perp} + \mathbf{c}_{\perp})$$
$$= \mathbf{a} \times \mathbf{b}_{\perp} + \mathbf{a} \times \mathbf{c}_{\perp} = \mathbf{a} \times \mathbf{b} + \mathbf{a} \times \mathbf{c}. \qquad \blacksquare$$

The result of Problem 18 on Page 358 is used to go from the first to the second line.

13.4.2   *Torque*

Imagine you are using a wrench to loosen a nut. The idea is to turn the nut by applying a force to the end of the wrench. If you push or pull the wrench directly toward or away from the nut, it should be obvious from experience that no progress will be made in turning the nut. The important thing is the component of force perpendicular to the wrench. It is this component of force which will cause the nut to turn. For example see the following picture.

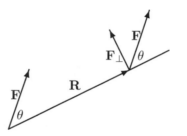

In the picture a force $\mathbf{F}$ is applied at the end of a wrench represented by the position vector $\mathbf{R}$ and the angle between these two is θ. Then the tendency to turn will be $|\mathbf{R}| |\mathbf{F}_{\perp}| = |\mathbf{R}| |\mathbf{F}| \sin \theta$, which you recognize as the magnitude of the cross product of $\mathbf{R}$ and $\mathbf{F}$. If there were just one force acting at one point whose position vector is $\mathbf{R}$, perhaps this would be sufficient, but what if there are numerous forces acting at many different points with neither the position vectors nor the force vectors in the same plane; what then? To keep track of this sort of thing, define for each $\mathbf{R}$ and $\mathbf{F}$, the torque vector

$$\tau \equiv \mathbf{R} \times \mathbf{F}.$$

This is also called the moment of the force $\mathbf{F}$. That way, if there are several forces acting at several points the total torque can be obtained by simply adding up the torques associated with the different forces and positions.

Example 13.13. Suppose $\mathbf{R}_1 = 2\mathbf{i} - \mathbf{j} + 3\mathbf{k}$, $\mathbf{R}_2 = \mathbf{i} + 2\mathbf{j} - 6\mathbf{k}$ meters and at the points determined by these vectors there are forces, $\mathbf{F}_1 = \mathbf{i} - \mathbf{j} + 2\mathbf{k}$ and $\mathbf{F}_2 = \mathbf{i} - 5\mathbf{j} + \mathbf{k}$ Newtons respectively. Find the total torque about the origin produced by these forces acting at the given points.

It is necessary to take $\mathbf{R}_1 \times \mathbf{F}_1 + \mathbf{R}_2 \times \mathbf{F}_2$. Thus the total torque equals

$$\begin{vmatrix} \mathbf{i} & \mathbf{j} & \mathbf{k} \\ 2 & -1 & 3 \\ 1 & -1 & 2 \end{vmatrix} + \begin{vmatrix} \mathbf{i} & \mathbf{j} & \mathbf{k} \\ 1 & 2 & -6 \\ 1 & -5 & 1 \end{vmatrix} = -27\mathbf{i} - 8\mathbf{j} - 8\mathbf{k} \text{ Newton meters}$$

Example 13.14. Find if possible a single force vector $\mathbf{F}$ which if applied at the point $\mathbf{i} + \mathbf{j} + \mathbf{k}$ will produce the same torque as the above two forces acting at the given points.

This is fairly routine. The problem is to find $\mathbf{F} = F_1\mathbf{i} + F_2\mathbf{j} + F_3\mathbf{k}$ which produces the above torque vector. Therefore,

$$\begin{vmatrix} \mathbf{i} & \mathbf{j} & \mathbf{k} \\ 1 & 1 & 1 \\ F_1 & F_2 & F_3 \end{vmatrix} = -27\mathbf{i} - 8\mathbf{j} - 8\mathbf{k}$$

which reduces to $(F_3 - F_2)\mathbf{i} + (F_1 - F_3)\mathbf{j} + (F_2 - F_1)\mathbf{k} = -27\mathbf{i} - 8\mathbf{j} - 8\mathbf{k}$. This amounts to solving the system of three equations in three unknowns, $F_1, F_2,$ and F_3,

$$F_3 - F_2 = -27, \ F_1 - F_3 = -8, \ F_2 - F_1 = -8$$

However, there is no solution to these three equations. (Why?) Therefore no single force acting at the point $\mathbf{i} + \mathbf{j} + \mathbf{k}$ will produce the given torque.

13.4.3 *Center Of Mass*

The mass of an object is a measure of how much stuff there is in the object. An object has mass equal to one kilogram, a unit of mass in the metric system, if it would exactly balance a known one kilogram object when placed on a balance. The known object is one kilogram by definition. The mass of an object does not depend on where the balance is used. It would be one kilogram on the moon as well as on the earth. The weight of an object is something else. It is the force exerted on the object by gravity and has magnitude gm where g is a constant called the acceleration of gravity. Thus the weight of a one kilogram object would be different on the moon which has much less gravity, smaller g, than on the earth. An important idea is that of the center of mass. This is the point at which an object will balance no matter how it is turned.

Definition 13.10. Let an object consist of p point masses $m_1, \cdots, m_p$ with the position of the kth of these at $\mathbf{R}_k$. The center of mass of this object $\mathbf{R}_0$ is the point satisfying

$$\sum_{k=1}^{p} (\mathbf{R}_k - \mathbf{R}_0) \times gm_k\mathbf{u} = \mathbf{0}$$

for all unit vectors $\mathbf{u}$.

The above definition indicates that no matter how the object is suspended, the total torque on it due to gravity is such that no rotation occurs. Using the properties of the cross product

$$\left(\sum_{k=1}^{p} \mathbf{R}_k g m_k - \mathbf{R}_0 \sum_{k=1}^{p} g m_k \right) \times \mathbf{u} = \mathbf{0} \tag{13.31}$$

for any choice of unit vector $\mathbf{u}$. You should verify that if $\mathbf{a} \times \mathbf{u} = \mathbf{0}$ for all $\mathbf{u}$, then it must be the case that $\mathbf{a} = \mathbf{0}$. Then the above formula requires that

$$\sum_{k=1}^{p} \mathbf{R}_k g m_k - \mathbf{R}_0 \sum_{k=1}^{p} g m_k = \mathbf{0}.$$

dividing by g, and then by $\sum_{k=1}^{p} m_k$,

$$\mathbf{R}_0 = \frac{\sum_{k=1}^{p} \mathbf{R}_k m_k}{\sum_{k=1}^{p} m_k}. \tag{13.32}$$

This is the formula for the center of mass of a collection of point masses. To consider the center of mass of a solid consisting of continuously distributed masses, you need the methods of calculus. This is done in Volume 2.

Example 13.15. Let $m_1 = 5, m_2 = 6$, and $m_3 = 3$ where the masses are in kilograms. Suppose m_1 is located at $2\mathbf{i} + 3\mathbf{j} + \mathbf{k}$, m_2 is located at $\mathbf{i} - 3\mathbf{j} + 2\mathbf{k}$ and m_3 is located at $2\mathbf{i} - \mathbf{j} + 3\mathbf{k}$. Find the center of mass of these three masses.

Using (13.32)

$$\mathbf{R}_0 = \frac{5\left(2\mathbf{i} + 3\mathbf{j} + \mathbf{k}\right) + 6\left(\mathbf{i} - 3\mathbf{j} + 2\mathbf{k}\right) + 3\left(2\mathbf{i} - \mathbf{j} + 3\mathbf{k}\right)}{5 + 6 + 3} = \frac{11}{7}\mathbf{i} - \frac{3}{7}\mathbf{j} + \frac{13}{7}\mathbf{k}$$

13.4.4 *Angular Velocity*

Definition 13.11. In a rotating body, a vector $\mathbf{\Omega}$ is called an **angular velocity vector** if the velocity of a point having position vector $\mathbf{u}$ relative to the body is given by $\mathbf{\Omega} \times \mathbf{u}$.

The existence of an angular velocity vector is the key to understanding motion in a moving system of coordinates. It is used to explain the motion on the surface of the rotating earth. For example, have you ever wondered why low pressure areas rotate counter clockwise in the upper hemisphere but clockwise in the lower hemisphere? To quantify these things, you will need the concept of an angular velocity vector. Details are presented later for interesting examples. Here is a simple example.

Example 13.16. A wheel rotates counter clockwise about the vector $\mathbf{i} + \mathbf{j} + \mathbf{k}$ at 60 revolutions per minute. This means that if the thumb of your right hand were to point in the direction of $\mathbf{i} + \mathbf{j} + \mathbf{k}$ your fingers of this hand would wrap in the direction of rotation. Find the angular velocity vector for this wheel. Assume the unit of distance is meters and the unit of time is minutes.

Let $\omega = 60 \times 2\pi = 120\pi$. This is the number of radians per minute corresponding to 60 revolutions per minute. Then the angular velocity vector is $\frac{120\pi}{\sqrt{3}} (\mathbf{i} + \mathbf{j} + \mathbf{k})$. It is a vector of magnitude 120π which points in the direction of $\mathbf{i} + \mathbf{j} + \mathbf{k}$. Note the cross product of this vector with a position vector perpendicular to the angular velocity vector gives what you would expect. Say the position vector has magnitude r. Then by the geometric description of the cross product, the magnitude of the velocity vector is $r120\pi$ meters per minute and corresponds to the speed. An exercise with the right hand shows the direction is correct also. However, if this body is rigid, this will work for every other point in it, even those for which the position vector is not perpendicular to the angular velocity vector.

Example 13.17. A wheel rotates counter clockwise about the vector $\mathbf{i} + \mathbf{j} + \mathbf{k}$ at 60 revolutions per minute exactly as in Example 13.16. Let $\{\mathbf{u}_1, \mathbf{u}_2, \mathbf{u}_3\}$ denote an orthogonal right handed system attached to the rotating wheel in which $\mathbf{u}_3 = \frac{1}{\sqrt{3}} (\mathbf{i} + \mathbf{j} + \mathbf{k})$. Thus $\mathbf{u}_1$ and $\mathbf{u}_2$ depend on time. Find the velocity of the point of the wheel located at the point $2\mathbf{u}_1 + 3\mathbf{u}_2 - \mathbf{u}_3$. Note this point is not fixed in space. It is moving.

Since $\{\mathbf{u}_1, \mathbf{u}_2, \mathbf{u}_3\}$ is a right handed system like $\mathbf{i}, \mathbf{j}, \mathbf{k}$, everything applies to this system in the same way as with $\mathbf{i}, \mathbf{j}, \mathbf{k}$. Thus the cross product is given by

$$(a\mathbf{u}_1 + b\mathbf{u}_2 + c\mathbf{u}_3) \times (d\mathbf{u}_1 + e\mathbf{u}_2 + f\mathbf{u}_3) = \begin{vmatrix} \mathbf{u}_1 & \mathbf{u}_2 & \mathbf{u}_3 \\ a & b & c \\ d & e & f \end{vmatrix}$$

Therefore, in terms of the given vectors $\mathbf{u}_i$, the angular velocity vector is $120\pi\mathbf{u}_3$. The velocity of the given point is

$$\begin{vmatrix} \mathbf{u}_1 & \mathbf{u}_2 & \mathbf{u}_3 \\ 0 & 0 & 120\pi \\ 2 & 3 & 1 \end{vmatrix} = -360\pi\mathbf{u}_1 + 240\pi\mathbf{u}_2$$

in meters per minute. Note how this gives the answer in terms of these vectors which are fixed in the body, not in space. Since $\mathbf{u}_i$ depends on t, this shows the answer in this case does also. Of course this is right. Just think of what is going on with the wheel rotating. Those vectors which are fixed in the wheel are moving in space. The velocity of a point in the wheel should be constantly changing. However, its speed will not change. The speed will be the magnitude of the velocity and this is

$$\sqrt{(-360\pi\mathbf{u}_1 + 240\pi\mathbf{u}_2) \cdot (-360\pi\mathbf{u}_1 + 240\pi\mathbf{u}_2)}$$

which from the properties of the dot product equals

$$\sqrt{(-360\pi)^2 + (240\pi)^2} = 120\sqrt{13}\pi$$

because the $\mathbf{u}_i$ are given to be orthogonal.

13.4.5 The Box Product

Definition 13.12. A parallelepiped determined by the three vectors **a**, **b**, and **c** consists of

$$\{r\mathbf{a}+s\mathbf{b}+t\mathbf{c} : r, s, t \in [0, 1]\}.$$

That is, if you pick three numbers, r, s, and t each in $[0, 1]$ and form $r\mathbf{a}+s\mathbf{b}+t\mathbf{c}$, then the collection of all such points is what is meant by the parallelepiped determined by these three vectors.

The following is a picture of such a thing.

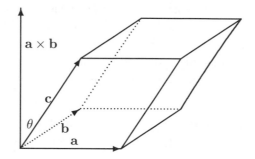

You notice the area of the base of the parallelepiped, the parallelogram determined by the vectors **a** and **b** has area equal to $|\mathbf{a} \times \mathbf{b}|$ while the altitude of the parallelepiped is $|\mathbf{c}| \cos \theta$ where θ is the angle shown in the picture between **c** and $\mathbf{a} \times \mathbf{b}$. Therefore, the volume of this parallelepiped is the area of the base times the altitude which is just

$$|\mathbf{a} \times \mathbf{b}| |\mathbf{c}| \cos \theta = \mathbf{a} \times \mathbf{b} \cdot \mathbf{c}.$$

This expression is known as the box product and is sometimes written as $[\mathbf{a}, \mathbf{b}, \mathbf{c}]$. You should consider what happens if you interchange the **b** with the **c** or the **a** with the **c**. You can see geometrically from drawing pictures that this merely introduces a minus sign. In any case the box product of three vectors always equals either the volume of the parallelepiped determined by the three vectors or else minus this volume.

Example 13.18. Find the volume of the parallelepiped determined by the vectors $\mathbf{i} + 2\mathbf{j} - 5\mathbf{k}, \mathbf{i} + 3\mathbf{j} - 6\mathbf{k}, 3\mathbf{i} + 2\mathbf{j} + 3\mathbf{k}$.

According to the above discussion, pick any two of these, take the cross product and then take the dot product of this with the third of these vectors. The result will be either the desired volume or minus the desired volume.

$$(\mathbf{i} + 2\mathbf{j} - 5\mathbf{k}) \times (\mathbf{i} + 3\mathbf{j} - 6\mathbf{k}) = \begin{vmatrix} \mathbf{i} & \mathbf{j} & \mathbf{k} \\ 1 & 2 & -5 \\ 1 & 3 & -6 \end{vmatrix} = 3\mathbf{i} + \mathbf{j} + \mathbf{k}$$

Now take the dot product of this vector with the third which yields

$$(3\mathbf{i} + \mathbf{j} + \mathbf{k}) \cdot (3\mathbf{i} + 2\mathbf{j} + 3\mathbf{k}) = 9 + 2 + 3 = 14.$$

This shows the volume of this parallelepiped is 14 cubic units.

There is a fundamental observation which comes directly from the geometric definitions of the cross product and the dot product.

Lemma 13.3. *Let* $\mathbf{a}, \mathbf{b},$ *and* $\mathbf{c}$ *be vectors. Then* $(\mathbf{a} \times \mathbf{b}) \cdot \mathbf{c} = \mathbf{a} \cdot (\mathbf{b} \times \mathbf{c}).$

Proof: This follows from observing that either $(\mathbf{a} \times \mathbf{b}) \cdot \mathbf{c}$ and $\mathbf{a} \cdot (\mathbf{b} \times \mathbf{c})$ both give the volume of the parallelepiped or they both give -1 times the volume. ∎

13.4.5.1 *An Alternate Proof Of The Distributive Law*

Here is another proof of the distributive law for the cross product. Let $\mathbf{x}$ be a vector. From the above observation,

$$\mathbf{x} \cdot \mathbf{a} \times (\mathbf{b} + \mathbf{c}) = (\mathbf{x} \times \mathbf{a}) \cdot (\mathbf{b} + \mathbf{c}) - (\mathbf{x} \times \mathbf{a}) \cdot \mathbf{b} + (\mathbf{x} \times \mathbf{a}) \cdot \mathbf{c}$$
$$= \mathbf{x} \cdot \mathbf{a} \times \mathbf{b} + \mathbf{x} \cdot \mathbf{a} \times \mathbf{c} = \mathbf{x} \cdot (\mathbf{a} \times \mathbf{b} + \mathbf{a} \times \mathbf{c}).$$

Therefore,

$$\mathbf{x} \cdot [\mathbf{a} \times (\mathbf{b} + \mathbf{c}) - (\mathbf{a} \times \mathbf{b} + \mathbf{a} \times \mathbf{c})] = 0$$

for all $\mathbf{x}$. In particular, this holds for $\mathbf{x} = \mathbf{a} \times (\mathbf{b} + \mathbf{c}) - (\mathbf{a} \times \mathbf{b} + \mathbf{a} \times \mathbf{c})$ showing that $\mathbf{a} \times (\mathbf{b} + \mathbf{c}) = \mathbf{a} \times \mathbf{b} + \mathbf{a} \times \mathbf{c}$, and this proves the distributive law for the cross product another way.

Observation 13.3. Suppose you have three vectors, $\mathbf{u} = (a, b, c)$, $\mathbf{v} = (d, e, f)$, and $\mathbf{w} = (g, h, i)$. Then $\mathbf{u} \cdot \mathbf{v} \times \mathbf{w}$ is given by the following.

$$\mathbf{u} \cdot \mathbf{v} \times \mathbf{w} = (a, b, c) \cdot \begin{vmatrix} \mathbf{i} & \mathbf{j} & \mathbf{k} \\ d & e & f \\ g & h & i \end{vmatrix}$$

$$= a \begin{vmatrix} e & f \\ h & i \end{vmatrix} - b \begin{vmatrix} d & f \\ g & i \end{vmatrix} + c \begin{vmatrix} d & e \\ g & h \end{vmatrix}$$

$$= \det \begin{pmatrix} a & b & c \\ d & e & f \\ g & h & i \end{pmatrix}.$$

The message is that to take the box product, you can simply take the determinant of the matrix which results by letting the rows be the rectangular components of the given vectors in the order in which they occur in the box product.

13.5 Vector Identities And Notation

To begin with consider $\mathbf{u} \times (\mathbf{v} \times \mathbf{w})$ and it is desired to simplify this quantity. It turns out this is an important quantity which comes up in many different contexts. Let $\mathbf{u} = (u_1, u_2, u_3)$ and let $\mathbf{v}$ and $\mathbf{w}$ be defined similarly.

$$\mathbf{v} \times \mathbf{w} = \begin{vmatrix} \mathbf{i} & \mathbf{j} & \mathbf{k} \\ v_1 & v_2 & v_3 \\ w_1 & w_2 & w_3 \end{vmatrix} = (v_2 w_3 - v_3 w_2)\,\mathbf{i} + (w_1 v_3 - v_1 w_3)\,\mathbf{j} + (v_1 w_2 - v_2 w_1)\,\mathbf{k}$$

Next consider $\mathbf{u} \times (\mathbf{v} \times \mathbf{w})$ which is given by

$$\mathbf{u} \times (\mathbf{v} \times \mathbf{w}) = \begin{vmatrix} \mathbf{i} & \mathbf{j} & \mathbf{k} \\ u_1 & u_2 & u_3 \\ (v_2 w_3 - v_3 w_2) & (w_1 v_3 - v_1 w_3) & (v_1 w_2 - v_2 w_1) \end{vmatrix}.$$

When you multiply this out, you get

$$\mathbf{i}\,(v_1 u_2 w_2 + u_3 v_1 w_3 - w_1 u_2 v_2 - u_3 w_1 v_3) + \mathbf{j}\,(v_2 u_1 w_1 + v_2 w_3 u_3 - w_2 u_1 v_1 - u_3 w_2 v_3)$$

$$+ \mathbf{k}\,(u_1 w_1 v_3 + v_3 w_2 u_2 - u_1 v_1 w_3 - v_2 w_3 u_2)$$

and if you are clever, you see right away that

$$(\mathbf{i} v_1 + \mathbf{j} v_2 + \mathbf{k} v_3)\,(u_1 w_1 + u_2 w_2 + u_3 w_3) - (\mathbf{i} w_1 + \mathbf{j} w_2 + \mathbf{k} w_3)\,(u_1 v_1 + u_2 v_2 + u_3 v_3).$$

Thus

$$\mathbf{u} \times (\mathbf{v} \times \mathbf{w}) = \mathbf{v}\,(\mathbf{u} \cdot \mathbf{w}) - \mathbf{w}\,(\mathbf{u} \cdot \mathbf{v}). \tag{13.33}$$

A related formula is

$$\begin{aligned} (\mathbf{u} \times \mathbf{v}) \times \mathbf{w} &= -[\mathbf{w} \times (\mathbf{u} \times \mathbf{v})] \\ &= -[\mathbf{u}\,(\mathbf{w} \cdot \mathbf{v}) - \mathbf{v}\,(\mathbf{w} \cdot \mathbf{u})] \\ &= \mathbf{v}\,(\mathbf{w} \cdot \mathbf{u}) - \mathbf{u}\,(\mathbf{w} \cdot \mathbf{v}). \end{aligned} \tag{13.34}$$

This derivation is simply wretched and it does nothing for other identities which may arise in applications. Actually, the above two formulas, (13.33) and (13.34) are sufficient for most applications if you are creative in using them, but there is another way. This other way allows you to discover such vector identities as the above without any creativity or any cleverness. Therefore, it is far superior to the above nasty computation. It is a vector identity discovering machine and it is this which is the main topic in what follows.

There are two special symbols, δ_{ij} and ε_{ijk} which are very useful in dealing with vector identities. To begin with, here is the definition of these symbols.

Definition 13.13. The symbol δ_{ij}, called the Kroneker delta symbol is defined as follows.

$$\delta_{ij} \equiv \begin{cases} 1 \text{ if } i = j \\ 0 \text{ if } i \neq j \end{cases}.$$

With the Kroneker symbol i and j can equal any integer in $\{1, 2, \cdots, n\}$ for any $n \in \mathbb{N}$.

Definition 13.14. For i, j, and k integers in the set, $\{1, 2, 3\}$, ε_{ijk} is defined as follows.

$$\varepsilon_{ijk} \equiv \begin{cases} 1 \text{ if } (i, j, k) = (1, 2, 3), (2, 3, 1), \text{ or } (3, 1, 2) \\ -1 \text{ if } (i, j, k) = (2, 1, 3), (1, 3, 2), \text{ or } (3, 2, 1) \\ 0 \text{ if there are any repeated integers} \end{cases}.$$

The subscripts ijk and ij in the above are called indices. A single one is called an index. This symbol ε_{ijk} is also called the permutation symbol.

The way to think of ε_{ijk} is that $\varepsilon_{123} = 1$ and if you switch any two of the numbers in the list i, j, k, it changes the sign. Thus $\varepsilon_{ijk} = -\varepsilon_{jik}$ and $\varepsilon_{ijk} = -\varepsilon_{kji}$ etc. You should check that this rule reduces to the above definition. For example, it immediately implies that if there is a repeated index, the answer is zero. This follows because $\varepsilon_{iij} = -\varepsilon_{iij}$ and so $\varepsilon_{iij} = 0$.

It is useful to use the Einstein summation convention when dealing with these symbols. Simply stated, the convention is that you sum over the repeated index. Thus $a_i b_i$ means $\sum_i a_i b_i$. Also, $\delta_{ij} x_j$ means $\sum_j \delta_{ij} x_j = x_i$. When you use this convention, there is one very important thing to never forget. It is this: Never have an index be repeated more than once. Thus $a_i b_i$ is all right but $a_{ii} b_i$ is not. The reason for this is that you end up getting confused about what is meant. If you want to write $\sum_i a_i b_i c_i$ it is best to simply use the summation notation. There is a very important reduction identity connecting these two symbols.

Lemma 13.4. *The following holds.*

$$\varepsilon_{ijk}\varepsilon_{irs} = (\delta_{jr}\delta_{ks} - \delta_{kr}\delta_{js}).$$

Proof: If $\{j, k\} \neq \{r, s\}$ then every term in the sum on the left must have either ε_{ijk} or ε_{irs} contains a repeated index. Therefore, the left side equals zero. The right side also equals zero in this case. To see this, note that if the two sets are not equal, then there is one of the indices in one of the sets which is not in the other set. For example, it could be that j is not equal to either r or s. Then the right side equals zero.

Therefore, it can be assumed $\{j, k\} = \{r, s\}$. If $i = r$ and $j = s$ for $s \neq r$, then there is exactly one term in the sum on the left and it equals 1. The right also reduces to 1 in this case. If $i = s$ and $j = r$, there is exactly one term in the sum on the left which is nonzero and it must equal -1. The right side also reduces to -1 in this case. If there is a repeated index in $\{j, k\}$, then every term in the sum on the left equals zero. The right also reduces to zero in this case because then $j = k = r = s$ and so the right side becomes $(1)(1) - (-1)(-1) = 0$. $\blacksquare$

Proposition 13.3. Let $\mathbf{u}, \mathbf{v}$ be vectors in $\mathbb{R}^n$ where the Cartesian coordinates of $\mathbf{u}$ are $(u_1, \cdots, u_n)$ and the Cartesian coordinates of $\mathbf{v}$ are $(v_1, \cdots, v_n)$. Then $\mathbf{u} \cdot \mathbf{v} = u_i v_i$. If $\mathbf{u}, \mathbf{v}$ are vectors in $\mathbb{R}^3$, then

$$(\mathbf{u} \times \mathbf{v})_i = \varepsilon_{ijk} u_j v_k.$$

Also, $\delta_{ik} a_k = a_i$.

Proof: The first claim is obvious from the definition of the dot product. The second is verified by simply checking that it works. For example,

$$\mathbf{u} \times \mathbf{v} \equiv \begin{vmatrix} \mathbf{i} & \mathbf{j} & \mathbf{k} \\ u_1 & u_2 & u_3 \\ v_1 & v_2 & v_3 \end{vmatrix}$$

and so

$$(\mathbf{u} \times \mathbf{v})_1 = (u_2 v_3 - u_3 v_2).$$

From the above formula in the proposition,

$$\varepsilon_{1jk} u_j v_k \equiv u_2 v_3 - u_3 v_2,$$

the same thing. The cases for $(\mathbf{u} \times \mathbf{v})_2$ and $(\mathbf{u} \times \mathbf{v})_3$ are verified similarly. The last claim follows directly from the definition. ∎

With this notation, you can easily discover vector identities and simplify expressions which involve the cross product.

Example 13.19. Discover a formula which simplifies $(\mathbf{u} \times \mathbf{v}) \times \mathbf{w}$.

From the above reduction formula,

$$
\begin{aligned}
((\mathbf{u} \times \mathbf{v}) \times \mathbf{w})_i &= \varepsilon_{ijk} (\mathbf{u} \times \mathbf{v})_j w_k = \varepsilon_{ijk} \varepsilon_{jrs} u_r v_s w_k \\
&= -\varepsilon_{jik} \varepsilon_{jrs} u_r v_s w_k = -\left(\delta_{ir} \delta_{ks} - \delta_{is} \delta_{kr} \right) u_r v_s w_k \\
&= -\left(u_i v_k w_k - u_k v_i w_k \right) = \mathbf{u} \cdot \mathbf{w} v_i - \mathbf{v} \cdot \mathbf{w} u_i \\
&= \left((\mathbf{u} \cdot \mathbf{w}) \mathbf{v} - (\mathbf{v} \cdot \mathbf{w}) \mathbf{u} \right)_i .
\end{aligned}
$$

Since this holds for all i, it follows that

$$(\mathbf{u} \times \mathbf{v}) \times \mathbf{w} = (\mathbf{u} \cdot \mathbf{w}) \mathbf{v} - (\mathbf{v} \cdot \mathbf{w}) \mathbf{u}.$$

13.6   Exercises

(1) Show that if $\mathbf{a} \times \mathbf{u} = \mathbf{0}$ for all unit vectors $\mathbf{u}$, then $\mathbf{a} = \mathbf{0}$.

(2) If you only assume (13.31) holds for $\mathbf{u} = \mathbf{i}, \mathbf{j}, \mathbf{k}$, show that this implies (13.31) holds for all unit vectors $\mathbf{u}$.

(3) Let $m_1 = 5, m_2 = 1$, and $m_3 = 4$ where the masses are in kilograms and the distance is in meters. Suppose m_1 is located at $2\mathbf{i} - 3\mathbf{j} + \mathbf{k}$, m_2 is located at $\mathbf{i} - 3\mathbf{j} + 6\mathbf{k}$ and m_3 is located at $2\mathbf{i} + \mathbf{j} + 3\mathbf{k}$. Find the center of mass of these three masses.

(4) Let $m_1 = 2, m_2 = 3$, and $m_3 = 1$ where the masses are in kilograms and the distance is in meters. Suppose m_1 is located at $2\mathbf{i} - \mathbf{j} + \mathbf{k}$, m_2 is located at $\mathbf{i} - 2\mathbf{j} + \mathbf{k}$ and m_3 is located at $4\mathbf{i} + \mathbf{j} + 3\mathbf{k}$. Find the center of mass of these three masses.

(5) Find the angular velocity vector of a rigid body which rotates counter clockwise about the vector $i-2j+k$ at 40 revolutions per minute. Assume distance is measured in meters.

(6) Let $\{u_1, u_2, u_3\}$ be a right handed system with u_3 pointing in the direction of $i-2j+k$ and u_1 and u_2 being fixed with the body which is rotating at 40 revolutions per minute. Assuming all distances are in meters, find the constant speed of the point of the body located at $3u_1 + u_2 - u_3$ in meters per minute.

(7) Find the area of the triangle determined by the three points $(1, 2, 3)$, $(4, 2, 0)$ and $(-3, 2, 1)$.

(8) Find the area of the triangle determined by the three points $(1, 0, 3)$, $(4, 1, 0)$ and $(-3, 1, 1)$.

(9) Find the area of the triangle determined by the three points $(1, 2, 3)$, $(2, 3, 4)$ and $(0, 1, 2)$. Did something interesting happen here? What does it mean geometrically?

(10) Find the area of the parallelogram determined by the vectors $(1, 2, 3)$ and $(3, -2, 1)$.

(11) Find the area of the parallelogram determined by the vectors $(1, 0, 3)$ and $(4, -2, 1)$.

(12) Find the area of the parallelogram determined by the vectors $(1, -2, 2)$ and $(3, 1, 1)$.

(13) Find the volume of the parallelepiped determined by the vectors $i-7j-5k$, $i-2j-6k$, $3i+2j+3k$.

(14) Find the volume of the parallelepiped determined by the vectors $i+j-5k$, $i+5j-6k$, $3i+j+3k$.

(15) Find the volume of the parallelepiped determined by the vectors $i+6j+5k$, $i+5j-6k$, $3i+j+k$.

(16) Suppose a, b, and c are three vectors whose components are all integers. Can you conclude the volume of the parallelepiped determined from these three vectors will always be an integer?

(17) What does it mean geometrically if the box product of three vectors gives zero?

(18) ↑Find the equation of the plane through the three points
$$(1, 2, 3), (2, -3, 1), (1, 1, 7).$$

(19) It is desired to find an equation of a plane containing the two vectors a and b and the point 0. Using Problem 17, show an equation for this plane is
$$\begin{vmatrix} x & y & z \\ a_1 & a_2 & a_3 \\ b_1 & b_2 & b_3 \end{vmatrix} = 0$$

That is, the set of all (x, y, z) such that
$$x \begin{vmatrix} a_2 & a_3 \\ b_2 & b_3 \end{vmatrix} - y \begin{vmatrix} a_1 & a_3 \\ b_1 & b_3 \end{vmatrix} + z \begin{vmatrix} a_1 & a_2 \\ b_1 & b_2 \end{vmatrix} = 0$$

(20) Using the notion of the box product yielding either plus or minus the volume of the parallelepiped determined by the given three vectors, show that

$$(\mathbf{a} \times \mathbf{b}) \cdot \mathbf{c} = \mathbf{a} \cdot (\mathbf{b} \times \mathbf{c})$$

In other words, the dot and the cross can be switched as long as the order of the vectors remains the same. **Hint:** There are two ways to do this, by the coordinate description of the dot and cross product and by geometric reasoning.

(21) Is $\mathbf{a} \times (\mathbf{b} \times \mathbf{c}) = (\mathbf{a} \times \mathbf{b}) \times \mathbf{c}$? What is the meaning of $\mathbf{a} \times \mathbf{b} \times \mathbf{c}$? Explain. **Hint:** Try $(\mathbf{i} \times \mathbf{j}) \times \mathbf{j}$.

(22) Verify directly that the coordinate description of the cross product $\mathbf{a} \times \mathbf{b}$ has the property that it is perpendicular to both $\mathbf{a}$ and $\mathbf{b}$. Then show by direct computation that this coordinate description satisfies

$$|\mathbf{a} \times \mathbf{b}|^2 = |\mathbf{a}|^2 \cdot |\mathbf{b}|^2 - (\mathbf{a} \cdot \mathbf{b})^2 = |\mathbf{a}|^2 |\mathbf{b}|^2 \left(1 - \cos^2(\theta)\right)$$

where θ is the angle included between the two vectors. Explain why $|\mathbf{a} \times \mathbf{b}|$ has the correct magnitude. All that is missing is the material about the right hand rule. Verify directly from the coordinate description of the cross product that the right thing happens with regards to the vectors $\mathbf{i}, \mathbf{j}, \mathbf{k}$. Next verify that the distributive law holds for the coordinate description of the cross product. This gives another way to approach the cross product. First define it in terms of coordinates and then get the geometric properties from this.

(23) Discover a vector identity for $\mathbf{u} \times (\mathbf{v} \times \mathbf{w})$.

(24) Discover a vector identity for $(\mathbf{u} \times \mathbf{v}) \cdot (\mathbf{z} \times \mathbf{w})$.

(25) Discover a vector identity for $(\mathbf{u} \times \mathbf{v}) \times (\mathbf{z} \times \mathbf{w})$ in terms of box products.

(26) Simplify $(\mathbf{u} \times \mathbf{v}) \cdot (\mathbf{v} \times \mathbf{w}) \times (\mathbf{w} \times \mathbf{z})$.

(27) Simplify $|\mathbf{u} \times \mathbf{v}|^2 + (\mathbf{u} \cdot \mathbf{v})^2 - |\mathbf{u}|^2 |\mathbf{v}|^2$.

(28) Prove that $\varepsilon_{ijk}\varepsilon_{ijr} = 2\delta_{kr}$.

(29) If A is a 3×3 matrix such that $A = (\ \mathbf{u} \quad \mathbf{v} \quad \mathbf{w}\ )$ where these are the columns of the matrix A, show that $\det(A) = \varepsilon_{ijk}u_i v_j w_k$.

(30) If A is a 3×3 matrix, show $\varepsilon_{rps} \det(A) = \varepsilon_{ijk} A_{ri} A_{pj} A_{sk}$.

(31) When you have a rotating rigid body with angular velocity vector $\mathbf{\Omega}$ then the velocity, $\mathbf{u}'$ is given by $\mathbf{u}' = \mathbf{\Omega} \times \mathbf{u}$. It turns out that all the usual calculus rules such as the product rule hold. Also, $\mathbf{u}''$ is the acceleration. Show using the product rule that for $\mathbf{\Omega}$ a constant vector

$$\mathbf{u}'' = \mathbf{\Omega} \times (\mathbf{\Omega} \times \mathbf{u}).$$

It turns out this is the centripetal acceleration. Note how it involves cross products.

13.7   Planes

You have an idea of what a plane is already. It is the span of some vectors. However, it can also be considered geometrically in terms of a dot product. To find the equation of a plane, you need two things, a point contained in the plane and a vector normal to the plane. Let $\mathbf{p_0} = (x_0, y_0, z_0)$ denote the position vector of a point in the plane, let $\mathbf{p} = (x, y, z)$ be the position vector of an arbitrary point in the plane, and let $\mathbf{n}$ denote a vector normal to the plane. This means that

$$\mathbf{n} \cdot (\mathbf{p} - \mathbf{p_0}) = 0$$

whenever $\mathbf{p}$ is the position vector of a point in the plane. The following picture illustrates the geometry of this idea.

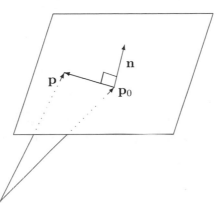

Expressed equivalently, the plane is just the set of all points $\mathbf{p}$ such that the vector $\mathbf{p} - \mathbf{p_0}$ is perpendicular to the given normal vector $\mathbf{n}$.

Example 13.20. Find the equation of the plane with normal vector $\mathbf{n} = (1, 2, 3)$ containing the point $(2, -1, 5)$.

From the above, the equation of this plane is just

$$(1, 2, 3) \cdot (x - 2, y + 1, z - 3) = x - 9 + 2y + 3z = 0$$

Example 13.21. $2x + 4y - 5z = 11$ is the equation of a plane. Find the normal vector and a point on this plane.

You can write this in the form $2\left(x - \frac{11}{2}\right) + 4(y - 0) + (-5)(z - 0) = 0$. Therefore, a normal vector to the plane is $2\mathbf{i} + 4\mathbf{j} - 5\mathbf{k}$ and a point in this plane is $\left(\frac{11}{2}, 0, 0\right)$. Of course there are many other points in the plane.

Definition 13.15. Suppose two planes intersect. The angle between the planes is defined to be the angle which is $\leq \pi/2$ between normal vectors to the respective planes.

Example 13.22. Find the angle between the two planes $x + 2y - z = 6$ and $3x + 2y - z = 7$.

The two normal vectors are $(1, 2, -1)$ and $(3, 2, -1)$. Therefore, the cosine of the angle desired is

$$\cos\theta = \frac{(1, 2, -1) \cdot (3, 2, -1)}{\sqrt{1^2 + 2^2 + (-1)^2}\sqrt{3^2 + 2^2 + (-1)^2}} = .872\,87$$

Now use a calculator or table to find what the angle is. $\cos\theta = .872\,87$. The solution is: $\{\theta = .509\,74\}$. This value is in radians.

Sometimes you need to find the equation of a plane which contains three points. Consider the following picture.

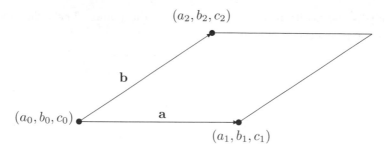

You have plenty of points but you need a normal. This can be obtained by taking $\mathbf{a}\times\mathbf{b}$ where $\mathbf{a} = (a_1 - a_0, b_1 - b_0, c_1 - c_0)$ and $\mathbf{b} = (a_2 - a_0, b_2 - b_0, c_2 - c_0)$.

Example 13.23. Find the equation of the plane which contains the three points
$$(1,2,1), (3,-1,2), \text{ and } (4,2,1).$$

You just need to get a normal vector to this plane. This can be done by taking the cross products of the two vectors
$$(3,-1,2) - (1,2,1) \text{ and } (4,2,1) - (1,2,1)$$
Thus a normal vector is $(2,-3,1) \times (3,0,0) = (0,3,9)$. Therefore, the equation of the plane is
$$0(x-1) + 3(y-2) + 9(z-1) = 0$$
or $3y + 9z = 15$ which is the same as $y + 3z = 5$. When you have what you think is the plane containing the three points, you ought to check it by seeing if it really does contain the three points.

Example 13.24. Find the equation of the plane which contains the three points
$$(1,2,1), (3,-1,2), \text{ and } (4,2,1).$$

You just need to get a normal vector to this plane. This can be done by taking the cross products of the two vectors
$$(3,-1,2) - (1,2,1) \text{ and } (4,2,1) - (1,2,1)$$
Thus a normal vector is $(2,-3,1) \times (3,0,0) = (0,3,9)$. Therefore, the equation of the plane is
$$0(x-1) + 3(y-2) + 9(z-1) = 0$$
or $3y + 9z = 15$ which is the same as $y + 3z = 5$.

Proposition 13.4. If $(a,b,c) \neq (0,0,0)$, then $ax + by + cz = d$ is the equation of a plane with normal vector $a\mathbf{i} + b\mathbf{j} + c\mathbf{k}$. Conversely, any plane can be written in this form.

Proof: One of a, b, c is nonzero. Suppose for example that $c \neq 0$. Then the equation can be written as

$$a\left(x - 0\right) + b\left(y - 0\right) + c\left(z - \frac{d}{c}\right) = 0$$

Therefore, $\left(0, 0, \frac{d}{c}\right)$ is a point on the plane and a normal vector is $a\mathbf{i} + b\mathbf{j} + c\mathbf{k}$. The converse follows from the above discussion involving the point and a normal vector. ∎

Example 13.25. Find the equation of the plane containing the points $(1, 2, 3)$ and the line $(0, 1, 1) + t\left(2, 1, 2\right) = (x, y, z)$.

There are several ways to do this. One is to find three points and use the above procedures. Let $t = 0$ and then let $t = 1$ to get two points on the line. This yields the three points $(1, 2, 3), (0, 1, 1)$, and $(2, 2, 3)$. Then a normal vector is obtained by fixing a point and taking the cross product of the differences of the other two points with that one. Thus in this case, fixing $(0, 1, 1)$, a normal vector is

$$(1, 1, 2) \times (2, 1, 2) = (0, 2, -1)$$

Therefore, an equation for the plane is

$$0\left(x - 0\right) + 2\left(y - 1\right) + (-1)\left(x - 3\right) = 0$$

Simplifying this yields

$$2y + 1 - x = 0$$

Example 13.26. Find the equation of the plane which contains the two lines, given by the following parametric expressions in which $t \in \mathbb{R}$.

$$(2t, 1 + t, 1 + 2t) = (x, y, z), \quad (2t + 2, 1, 3 + 2t) = (x, y, z)$$

Note first that you don't know there even is such a plane. However, if there is, you could find it by obtaining three points, two on one line and one on another and then using any of the above procedures for finding the plane. From the first line, two points are $(0, 1, 1)$ and $(2, 2, 3)$ while a third point can be obtained from second line, $(2, 1, 3)$. You need a normal vector and then use any of these points. To get a normal vector, form $(2, 0, 2) \times (2, 1, 2) = (\ 2, 0, 2)$. Therefore, the plane is $-2x + 0\left(y - 1\right) + 2\left(z - 1\right) = 0$. This reduces to $z - x = 1$. If there is a plane, this is it. Now you can simply verify that both of the lines are really in this plane. From the first, $(1 + 2t) - 2t = 1$ and the second, $(3 + 2t) - (2t + 2) = 1$ so both lines lie in the plane.

One way to understand how a plane looks is to connect the points where it intercepts the $x, y,$ and z axes. This allows you to visualize the plane somewhat and is a good way to sketch the plane. Not surprisingly these points are called intercepts.

Example 13.27. Sketch the plane which has intercepts $(2, 0, 0), (0, 3, 0)$, and $(0, 0, 4)$.

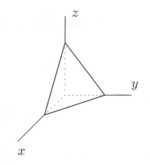

You see how connecting the intercepts gives a fairly good geometric description of the plane. These lines which connect the intercepts are also called the traces of the plane. Thus the line which joins $(0,3,0)$ to $(0,0,4)$ is the intersection of the plane with the yz plane. It is the trace on the yz plane.

Example 13.28. Identify the intercepts of the plane $3x - 4y + 5z = 11$.

The easy way to do this is to divide both sides by 11. Thus $\frac{x}{(11/3)} + \frac{y}{(-11/4)} + \frac{z}{(11/5)} = 1$. The intercepts are $(11/3, 0, 0), (0, -11/4, 0)$ and $(0, 0, 11/5)$. You can see this by letting both y and z equal to zero to find the point on the x axis which is intersected by the plane. The other axes are handled similarly.

13.8   Quadric Surfaces

In the above it was shown that the equation of an arbitrary plane is an equation of the form $ax + by + cz = d$. Such equations are called level surfaces. There are some standard level surfaces which involve certain variables being raised to a power of 2 which are sufficiently important that they are given names, usually involving the portentous semi-word "oid". These are graphed below using Maple, a computer algebra system.

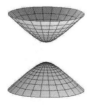

$$z^2/a^2 - x^2/b^2 - y^2/c^2 = 1$$

hyperboloid of two sheets

$$x^2/b^2 + y^2/c^2 - z^2/a^2 = 1$$

hyperboloid of one sheet

$$z = x^2/a^2 - y^2/b^2$$

hyperbolic paraboloid

$$z = x^2/a^2 + y^2/b^2$$

elliptic paraboloid

$$x^2/a^2 + y^2/b^2 + z^2/c^2 = 1$$

ellipsoid

$$z^2/a^2 = x^2/b^2 + y^2/c^2$$

elliptic cone

Why do the graphs of these level surfaces look the way they do? Consider first the hyperboloid of two sheets. The equation defining this surface can be written in the form

$$\frac{z^2}{a^2} - 1 = \frac{x^2}{b^2} + \frac{y^2}{c^2}.$$

Suppose you fix a value for z. What ordered pairs (x, y) will satisfy the equation? If $\frac{z^2}{a^2} < 1$, there is no such ordered pair because the above equation would require a negative number to equal a nonnegative one. This is why there is a gap and there are two sheets. If $\frac{z^2}{a^2} > 1$, then the above equation is the equation for an ellipse. That is why if you slice the graph by letting $z = z_0$ the result is an ellipse in the plane $z = z_0$.

Consider the hyperboloid of one sheet.

$$\frac{x^2}{b^2} + \frac{y^2}{c^2} = 1 + \frac{z^2}{a^2}.$$

This time, it does not matter what value z takes. The resulting equation for (x, y) is an ellipse.

Similar considerations apply to the elliptic paraboloid as long as $z > 0$ and the ellipsoid. The elliptic cone is like the hyperboloid of two sheets without the 1. Therefore, z can have any value. In case $z = 0$, $(x, y) = (0, 0)$. Viewed from the side, it appears straight, not curved like the hyperboloid of two sheets. This is because if (x, y, z) is a point on the surface, then if t is a scalar, it follows (tx, ty, tz) is also on this surface.

The most interesting of these graphs is the hyperbolic paraboloid[1], $z = \frac{x^2}{a^2} - \frac{y^2}{b^2}$. If $z > 0$, this is the equation of a hyperbola which opens to the right and left while if $z < 0$, it is a hyperbola which opens up and down. As z passes from positive to negative, the hyperbola changes type and this is what yields the shape shown in the picture.

Not surprisingly, you can find intercepts and traces of quadric surfaces just as with planes.

Example 13.29. Find the trace on the xy plane of the hyperbolic paraboloid, $z = x^2 - y^2$.

This occurs when $z = 0$ and so this reduces to $y^2 = x^2$. In other words, this trace is just the two straight lines, $y = x$ and $y = -x$.

Example 13.30. Find the intercepts of the ellipsoid, $x^2 + 2y^2 + 4z^2 = 9$.

To find the intercept on the x axis, let $y = z = 0$ and this yields $x = \pm 3$. Thus there are two intercepts, $(3, 0, 0)$ and $(-3, 0, 0)$. The other intercepts are left for you to find. You can see this is an aid in graphing the quadric surface. The surface is said to be bounded if there is some number C such that whenever (x, y, z) is a point on the surface $\sqrt{x^2 + y^2 + z^2} < C$. The surface is called unbounded if no such constant C exists. Ellipsoids are bounded but the other quadric surfaces are not bounded.

Example 13.31. Why is the hyperboloid of one sheet, $x^2 + 2y^2 - z^2 = 1$ unbounded?

Let z be very large. Does there correspond (x, y) such that (x, y, z) is a point on the hyperboloid of one sheet? Certainly. Simply pick any (x, y) on the ellipse $x^2 + 2y^2 = 1 + z^2$. Then $\sqrt{x^2 + y^2 + z^2}$ is large, at lest as large as z. Thus it is unbounded.

You can also find intersections between lines and surfaces.

Example 13.32. Find the points of intersection of the line

$$(x, y, z) = (1 + t, 1 + 2t, 1 + t)$$

with the surface $z = x^2 + y^2$.

---

[1] It is traditional to refer to this as a hyperbolic paraboloid. Not a parabolic hyperboloid.

First of all, there is no guarantee there is any intersection at all. But if it exists, you have only to solve the equation for t

$$1 + t = (1 + t)^2 + (1 + 2t)^2$$

This occurs at the two values of $t = -\frac{1}{2} + \frac{1}{10}\sqrt{5}, t = -\frac{1}{2} - \frac{1}{10}\sqrt{5}$. Therefore, the two points are

$$(1, 1, 1) + \left(-\frac{1}{2} + \frac{1}{10}\sqrt{5}\right)(1, 2, 1), \text{ and } (1, 1, 1) + \left(-\frac{1}{2} - \frac{1}{10}\sqrt{5}\right)(1, 2, 1)$$

That is

$$\left(\frac{1}{2} + \frac{1}{10}\sqrt{5}, \frac{1}{5}\sqrt{5}, \frac{1}{2} + \frac{1}{10}\sqrt{5}\right), \left(\frac{1}{2} - \frac{1}{10}\sqrt{5}, -\frac{1}{5}\sqrt{5}, \frac{1}{2} - \frac{1}{10}\sqrt{5}\right).$$

13.9 Excrcises

(1) Determine whether the lines $(1, 1, 2) + t(1, 0, 3)$ and $(4, 1, 3) + t(3, 0, 1)$ have a point of intersection. If they do not intersect, explain why they do not.

(2) Determine whether the lines $(1, 1, 2) + t(1, 0, 3)$ and $(4, 2, 3) + t(3, 0, 1)$ have a point of intersection.. If they do not intersect, explain why they do not.

(3) Find where the line $(1, 0, 1) + t(1, 2, 1)$ intersects the surface $x^2 + y^2 + z^2 = 9$ if possible. If there is no intersection, explain why.

(4) Find a parametric equation for the line through the points $(2, 3, 4, 5), (-2, 3, 0, 1)$.

(5) Find the equation of a line through $(1, 2, 3, 0)$ which has direction vector $(2, 1, 3, 1)$.

(6) Let $(x, y) = (2\cos(t), 2\sin(t))$ where $t \in [0, 2\pi]$. Describe the set of points encountered as t changes.

(7) Let $(x, y, z) = (2\cos(t), 2\sin(t), t)$ where $t \in \mathbb{R}$. Describe the set of points encountered as t changes.

(8) If there is a plane which contains the two lines,

$$(2t + 2, 1 + t, 3 + 2t) = (x, y, z)$$

and $(4 + t, 3 + 2t, 4 + t) = (x, y, z)$ find it. If there is no such plane tell why.

(9) If there is a plane which contains the two lines,

$$(2t + 4, 1 + t, 3 + 2t) = (x, y, z)$$

and $(4 + t, t + 1, 3 + t) = (x, y, z)$ find it. If there is no such plane tell why.

(10) Find the equation of the plane which contains the three points $(1, -2, 3), (2, 3, 4)$, and $(3, 1, 2)$.

(11) Find the equation of the plane which contains the three points $(1, 2, 3), (2, 0, 4)$, and $(3, 1, 2)$.

(12) Find the equation of the plane which contains the three points $(0, 2, 3), (2, 3, 4)$, and $(3, 5, 2)$.

(13) Find the equation of the plane which contains the three points $(1, 2, 3)$, $(0, 3, 4)$, and $(3, 6, 2)$.

(14) Find the equation of the plane having a normal vector $5\mathbf{i} + 2\mathbf{j} - 6\mathbf{k}$ which contains the point $(2, 1, 3)$.

(15) Find the equation of the plane having a normal vector $\mathbf{i} + 2\mathbf{j} - 4\mathbf{k}$ which contains the point $(2, 0, 1)$.

(16) Find the equation of the plane having a normal vector $2\mathbf{i} + \mathbf{j} - 6\mathbf{k}$ which contains the point $(1, 1, 2)$.

(17) Find the equation of the plane having a normal vector $\mathbf{i} + 2\mathbf{j} - 3\mathbf{k}$ which contains the point $(1, 0, 3)$.

(18) Determine the intercepts and sketch the plane $3x - 2y + z = 4$.

(19) Determine the intercepts and sketch the plane $x - 2y + z = 2$.

(20) Determine the intercepts and sketch the plane $x + y + z = 3$.

(21) Based on an analogy with the above pictures, sketch or otherwise describe the graph of $y = \frac{x^2}{a^2} - \frac{z^2}{b^2}$.

(22) Based on an analogy with the above pictures, sketch or otherwise describe the graph of $\frac{z^2}{b^2} + \frac{y^2}{c^2} = 1 + \frac{x^2}{a^2}$.

(23) The equation of a cone is $z^2 = x^2 + y^2$. Suppose this cone is intersected with the plane $z = ay + 1$, $a > 0$. Consider the projection of the intersection of the cone with this plane. This means $\left\{ (x, y) : (ay + 1)^2 = x^2 + y^2 \right\}$. Show this sometimes results in a parabola, sometimes a hyperbola, and sometimes an ellipse depending on a.

(24) Find the intercepts of the quadric surface $x^2 + 4y^2 - z^2 = 4$ and sketch the surface.

It does not intercept the z axis.

(25) Find the intercepts of the quadric surface $x^2 - \left(4y^2 + z^2\right) = 4$ and sketch the surface.

The intercepts are 2,-2 for x.

(26) Find the intersection of the line $(x, y, z) = (1 + t, t, 3t)$ with the surface $x^2/9 + y^2/4 + z^2/16 = 1$ if possible.

Chapter 14

Some Curvilinear Coordinate Systems

14.1 Polar Coordinates

So far points have been identified in terms of Cartesian coordinates but there are other ways of specifying points in two and three dimensional space. These other ways involve using a list of two or three numbers which have a totally different meaning than Cartesian coordinates to specify a point in two or three dimensional space. In general these lists of numbers which have a different meaning than Cartesian coordinates are called Curvilinear coordinates. Probably the simplest curvilinear coordinate system is that of **polar coordinates**. The idea is suggested in the following picture.

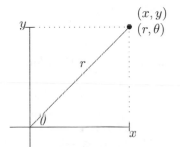

You see in this picture, the number r identifies the distance of the point from the origin, $(0,0)$ while θ is the angle shown between the positive x axis and the line from the origin to the point. This angle will always be given in radians and is in the interval $[0, 2\pi)$. Thus the given point, indicated by a small dot in the picture, can be described in terms of the Cartesian coordinates (x, y) or the polar coordinates (r, θ). How are the two coordinates systems related? From the picture,

$$x = r\cos(\theta), \; y = r\sin(\theta). \tag{14.1}$$

Example 14.1. The polar coordinates of a point in the plane are $\left(5, \frac{\pi}{6}\right)$. Find the Cartesian or rectangular coordinates of this point.

From (14.1), $x = 5\cos\left(\frac{\pi}{6}\right) = \frac{5}{2}\sqrt{3}$ and $y = 5\sin\left(\frac{\pi}{6}\right) = \frac{5}{2}$. Thus the Cartesian coordinates are $\left(\frac{5}{2}\sqrt{3}, \frac{5}{2}\right)$.

Example 14.2. Suppose the Cartesian coordinates of a point are $(3, 4)$. Find the polar coordinates.

Recall that r is the distance form $(0, 0)$ and so $r = 5 = \sqrt{3^2 + 4^2}$. It remains to identify the angle. Note the point is in the first quadrant. (Both the x and y values are positive.) Therefore, the angle is something between 0 and $\pi/2$ and also $3 = 5\cos(\theta)$, and $4 = 5\sin(\theta)$. Therefore, dividing yields $\tan(\theta) = 4/3$. At this point, use a calculator or a table of trigonometric functions to find that at least approximately, $\theta = .927\,295$ radians.

14.1.1   Graphs In Polar Coordinates

Just as in the case of rectangular coordinates, it is possible to use relations between the polar coordinates to specify points in the plane. The process of sketching their graphs is very similar to that used to sketch graphs of functions in rectangular coordinates. I will only consider the case where the relation between the polar coordinates is of the form, $r = f(\theta)$. To graph such a relation, you can make a table of the form

θ	r
θ_1	$f(\theta_1)$
θ_2	$f(\theta_2)$
$\vdots$	$\vdots$

and then graph the resulting points and connect them up with a curve. The following picture illustrates how to begin this process.

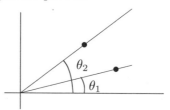

To obtain the point in the plane which goes with the pair $(\theta, f(\theta))$, you draw the ray through the origin which makes an angle of θ with the positive x axis. Then you move along this ray a distance of $f(\theta)$ to obtain the point. As in the case with rectangular coordinates, this process is tedious and is best done by a computer algebra system.

Example 14.3. Graph the polar equation $r = 1 + \cos\theta$.

To do this, I will use Maple. The command which produces the polar graph of this is: $>$ plot(1+cos(t),t=0..2*Pi,coords=polar); It tells Maple that r is given by

$1 + \cos(t)$ and that $t \in [0, 2\pi]$. The variable t is playing the role of θ. It is easier to type t than θ in Maple.

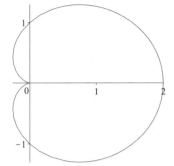

You can also see just from your knowledge of the trig. functions that the graph should look something like this. When $\theta = 0, r = 2$ and then as θ increases to $\pi/2$, you see that $\cos\theta$ decreases to 0. Thus the line from the origin to the point on the curve should get shorter as θ goes from 0 to $\pi/2$. Then from $\pi/2$ to π, $\cos\theta$ gets negative eventually equaling -1 at $\theta = \pi$. Thus $r = 0$ at this point. Viewing the graph, you see this is exactly what happens. The above function is called a **cardioid**.

Here is another example. This is the graph obtained from $r = 3 + \sin\left(\frac{7\theta}{6}\right)$.

Example 14.4. Graph $r = 3 + \sin\left(\frac{7\theta}{6}\right)$ for $\theta \in [0, 14\pi]$.

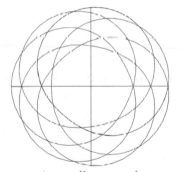

In polar coordinates people sometimes allow r to be negative. When this happens, it means that to obtain the point in the plane, you go in the opposite direction along the ray which starts at the origin and makes an angle of θ with the positive x axis. I do not believe the fussiness occasioned by this extra generality is justified by any sufficiently interesting application so no more will be said about this. It is mainly a fun way to obtain pretty pictures. Here is such an example.

Example 14.5. Graph $r = 1 + 2\cos\theta$ for $\theta \in [0, 2\pi]$.

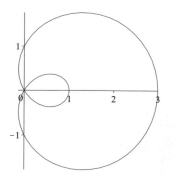

14.2   The Area In Polar Coordinates

How can you find the area of the region determined by $0 \le r \le f(\theta)$ for $\theta \in [a, b]$, assuming this is a well defined set of points in the plane? See Example 14.5 with $\theta \in [0, 2\pi]$ to see something which it would be better to avoid. I have in mind the situation where every ray through the origin having angle θ for $\theta \in [a, b]$ intersects the graph of $r = f(\theta)$ in exactly one point. To see how to find the area of such a region, consider the following picture.

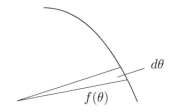

This is a representation of a small triangle obtained from two rays whose angles differ by only $d\theta$. What is the area of this triangle, dA? It would be

$$\frac{1}{2} \sin(d\theta) f(\theta)^2 \approx \frac{1}{2} f(\theta)^2 \, d\theta = dA$$

with the approximation getting better as the angle gets smaller. Thus the area should solve the initial value problem,

$$\frac{dA}{d\theta} = \frac{1}{2} f(\theta)^2, \ A(a) = 0.$$

Therefore, the total area would be given by the integral

$$\frac{1}{2} \int_a^b f(\theta)^2 \, d\theta. \tag{14.2}$$

Example 14.6. Find the area of the cardioid, $r = 1 + \cos\theta$ for $\theta \in [0, 2\pi]$.

From the graph of the cardioid presented earlier, you can see the region of interest satisfies the conditions above that every ray intersects the graph in only one point. Therefore, from (14.2) this area is

$$\frac{1}{2}\int_0^{2\pi}(1+\cos(\theta))^2\,d\theta = \frac{3}{2}\pi.$$

Example 14.7. Verify the area of a circle of radius a is πa^2.

The polar equation is just $r = a$ for $\theta \in [0, 2\pi]$. Therefore, the area should be

$$\frac{1}{2}\int_0^{2\pi}a^2\,d\theta = \pi a^2.$$

Example 14.8. Find the area of the region inside the cardioid, $r = 1 + \cos\theta$ and outside the circle, $r = 1$ for $\theta \in \left[-\frac{\pi}{2}, \frac{\pi}{2}\right]$.

As is usual in such cases, it is a good idea to graph the curves involved to get an idea what is wanted.

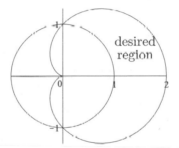

The area of this region would be the area of the part of the cardioid corresponding to $\theta \in \left[-\frac{\pi}{2}, \frac{\pi}{2}\right]$ minus the area of the part of the circle in the first quadrant. Thus the area is

$$\frac{1}{2}\int_{-\pi/2}^{\pi/2}(1+\cos(\theta))^2\,d\theta - \frac{1}{2}\int_{-\pi/2}^{\pi/2}1\,d\theta = \frac{1}{4}\pi + 2.$$

This example illustrates the following procedure for finding the area between the graphs of two curves given in polar coordinates.

Procedure 14.1. Suppose that for all $\theta \in [a, b]\,, 0 < g(\theta) < f(\theta)$. To find the area of the region defined in terms of polar coordinates by $g(\theta) < r < f(\theta)$, $\theta \in [a, b]$, you do the following.

$$\frac{1}{2}\int_a^b\left(f(\theta)^2 - g(\theta)^2\right)\,d\theta.$$

14.3 Exercises

(1) The following are the polar coordinates of points. Find the rectangular coordinates.

   (a) $\left(5, \frac{\pi}{6}\right)$
   (b) $\left(3, \frac{\pi}{3}\right)$
   (c) $\left(4, \frac{2\pi}{3}\right)$
   (d) $\left(2, \frac{3\pi}{4}\right)$
   (e) $\left(3, \frac{7\pi}{6}\right)$
   (f) $\left(8, \frac{11\pi}{6}\right)$

(2) The following are the rectangular coordinates of points. Find the polar coordinates of these points.

   (a) $\left(\frac{5}{2}\sqrt{2}, \frac{5}{2}\sqrt{2}\right)$
   (b) $\left(\frac{3}{2}, \frac{3}{2}\sqrt{3}\right)$
   (c) $\left(-\frac{5}{2}\sqrt{2}, \frac{5}{2}\sqrt{2}\right)$
   (d) $\left(-\frac{5}{2}, \frac{5}{2}\sqrt{3}\right)$
   (e) $\left(-\sqrt{3}, -1\right)$
   (f) $\left(\frac{3}{2}, -\frac{3}{2}\sqrt{3}\right)$

(3) In general it is a stupid idea to try to use algebra to invert and solve for a set of curvilinear coordinates such as polar or cylindrical coordinates in term of Cartesian coordinates. Not only is it often very difficult or even impossible to do it[1], but also it takes you in entirely the wrong direction because the whole point of introducing the new coordinates is to write everything in terms of these new coordinates and not in terms of Cartesian coordinates. However, sometimes this inversion can be done. Describe how to solve for r and θ in terms of x and y in polar coordinates.

(4) Suppose $r = \frac{a}{1+e\sin\theta}$ where $e \geq 0$. By changing to rectangular coordinates, show that this is either a parabola, an ellipse or a hyperbola. Determine the values of e which correspond to the various cases.

(5) In Example 14.4 suppose you graphed it for $\theta \in [0, k\pi]$ where k is a positive integer. What is the smallest value of k such that the graph will look exactly like the one presented in the example?

(6) Suppose you were to graph $r = 3 + \sin\left(\frac{m}{n}\theta\right)$ where m, n are integers. Can you give some description of what the graph will look like for $\theta \in [0, k\pi]$ for k a very large positive integer? How would things change if you did $r = 3 + \sin(\alpha\theta)$ where α is an irrational number?

(7) Graph $r = 1 + \sin\theta$ for $\theta \in [0, 2\pi]$.

(8) Graph $r = 2 + \sin\theta$ for $\theta \in [0, 2\pi]$.

(9) Graph $r = 1 + 2\sin\theta$ for $\theta \in [0, 2\pi]$.

---

[1]It is no problem for these simple cases of curvilinear coordinates. However, it is a major difficulty in general. Algebra is simply not adequate to solve systems of nonlinear equations.

(10) Graph $r = 2 + \sin(2\theta)$ for $\theta \in [0, 2\pi]$.

(11) Graph $r = 1 + \sin(2\theta)$ for $\theta \in [0, 2\pi]$.

(12) Graph $r = 1 + \sin(3\theta)$ for $\theta \in [0, 2\pi]$.

(13) Find the area of the bounded region determined by $r = 1 + \sin(3\theta)$ for $\theta \in [0, 2\pi]$.

(14) Find the area inside $r = 1 + \sin\theta$ and outside the circle $r = 1/2$.

(15) Find the area inside the circle $r = 1/2$ and outside the region defined by $r = 1 + \sin\theta$.

14.4 The Acceleration In Polar Coordinates

I assume that by now, the reader has encountered Newton's laws of motion, especially the second law which gives the relationship, force equals mass times acceleration. Sometimes you have information about forces which act not in the direction of the coordinate axes but in some other direction. When this is the case, it is often useful to express things in terms of different coordinates which are consistent with these directions. A good example of this is the force exerted by the sun on a planet. This force is always directed toward the sun and so the force vector changes as the planet moves. To discuss this, consider the following simple diagram in which two unit vectors $\mathbf{e}_r$ and $\mathbf{e}_\theta$ are shown.

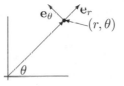

The vector $\mathbf{e}_r = (\cos\theta, \sin\theta)$ and the vector $\mathbf{e}_\theta = (-\sin\theta, \cos\theta)$. You should convince yourself that the picture above corresponds to this definition of the two vectors. Note that $\mathbf{e}_r$ is a unit vector pointing away from $\mathbf{0}$ and

$$\mathbf{e}_\theta = \frac{d\mathbf{e}_r}{d\theta}, \quad \mathbf{e}_r = -\frac{d\mathbf{e}_\theta}{d\theta}. \tag{14.3}$$

Now consider the position vector from $\mathbf{0}$ of a point in the plane, $\mathbf{r}(t)$. Then

$$\mathbf{r}(t) = r(t)\,\mathbf{e}_r\,(\theta(t))$$

where $r(t) = |\mathbf{r}(t)|$. Thus $r(t)$ is just the distance from the origin, $\mathbf{0}$ to the point. What is the velocity and acceleration? Using the chain rule,

$$\frac{d\mathbf{e}_r}{dt} = \frac{d\mathbf{e}_r}{d\theta}\theta'(t), \quad \frac{d\mathbf{e}_\theta}{dt} = \frac{d\mathbf{e}_\theta}{d\theta}\theta'(t)$$

and so from (14.3),

$$\frac{d\mathbf{e}_r}{dt} = \theta'(t)\,\mathbf{e}_\theta, \quad \frac{d\mathbf{e}_\theta}{dt} = -\theta'(t)\,\mathbf{e}_r \tag{14.4}$$

Using (14.4) as needed along with the product rule and the chain rule,

$$\mathbf{r}'\left(t\right) = r'\left(t\right)\mathbf{e}_r + r\left(t\right)\frac{d}{dt}\left(\mathbf{e}_r\left(\theta\left(t\right)\right)\right)$$
$$= r'\left(t\right)\mathbf{e}_r + r\left(t\right)\theta'\left(t\right)\mathbf{e}_\theta.$$

Next consider the acceleration.

$$\mathbf{r}''\left(t\right) = r''\left(t\right)\mathbf{e}_r + r'\left(t\right)\frac{d\mathbf{e}_r}{dt} + r'\left(t\right)\theta'\left(t\right)\mathbf{e}_\theta + r\left(t\right)\theta''\left(t\right)\mathbf{e}_\theta + r\left(t\right)\theta'\left(t\right)\frac{d}{dt}\left(\mathbf{e}_\theta\right)$$
$$= r''\left(t\right)\mathbf{e}_r + 2r'\left(t\right)\theta'\left(t\right)\mathbf{e}_\theta + r\left(t\right)\theta''\left(t\right)\mathbf{e}_\theta + r\left(t\right)\theta'\left(t\right)\left(-\mathbf{e}_r\right)\theta'\left(t\right)$$
$$= \left(r''\left(t\right) - r\left(t\right)\theta'\left(t\right)^2\right)\mathbf{e}_r + \left(2r'\left(t\right)\theta'\left(t\right) + r\left(t\right)\theta''\left(t\right)\right)\mathbf{e}_\theta. \tag{14.5}$$

This is a very profound formula. Consider the following examples.

Example 14.9. Suppose an object of mass m moves at a uniform speed s, around a circle of radius R. Find the force acting on the object.

By Newton's second law, the force acting on the object is $m\mathbf{r}''$. In this case, $r\left(t\right) = R$, a constant and since the speed is constant, $\theta'' = 0$. Therefore, the term in (14.5) corresponding to $\mathbf{e}_\theta$ equals zero and $m\mathbf{r}'' = -R\theta'\left(t\right)^2\mathbf{e}_r$. The speed of the object is s and so it moves s/R radians in unit time. Thus $\theta'\left(t\right) = s/R$ and so

$$m\mathbf{r}'' = -mR\left(\frac{s}{R}\right)^2\mathbf{e}_r = -m\frac{s^2}{R}\mathbf{e}_r.$$

This is the familiar formula for centripetal force from elementary physics, obtained as a very special case of (14.5).

Example 14.10. A platform rotates at a constant speed in the counter clockwise direction and an object of mass m moves from the center of the platform toward the edge at constant speed. What forces act on this object?

Let v denote the constant speed of the object moving toward the edge of the platform. Then

$$r'\left(t\right) = v,\ r''\left(t\right) = 0,\ \theta''\left(t\right) = 0,$$

while $\theta'\left(t\right) = \omega$, a positive constant. From (14.5)

$$m\mathbf{r}''\left(t\right) = -mr\left(t\right)\omega^2\mathbf{e}_r + m2v\omega\mathbf{e}_\theta.$$

Thus the object experiences centripetal force from the first term and also a funny force from the second term which is in the direction of rotation of the platform. You can observe this by experiment if you like. Go to a playground and have someone spin one of those merry go rounds while you ride it and move from the center toward the edge. The term $2r'\theta'$ is called the Coriolis force.

Suppose at each point of space $\mathbf{r}$ is associated a force $\mathbf{F}\left(\mathbf{r}\right)$ which a given object of mass m will experience if its position vector is $\mathbf{r}$. This is called a force field. a force field is a central force field if $\mathbf{F}\left(\mathbf{r}\right) = g\left(\mathbf{r}\right)\mathbf{e}_r$. Thus in a central force field the force an object experiences will always be directed toward or away from the origin,

0. The following simple lemma is very interesting because it says that in a central force field objects must move in a plane.

Lemma 14.1. *Suppose an object moves in three dimensions in such a way that the only force acting on the object is a central force. Then the motion of the object is in a plane.*

Proof: Let $\mathbf{r}(t)$ denote the position vector of the object. Then from the definition of a central force and Newton's second law,

$$m\mathbf{r}'' = g(\mathbf{r})\mathbf{r}.$$

Therefore, $m\mathbf{r}'' \times \mathbf{r} = m(\mathbf{r}' \times \mathbf{r})' = g(\mathbf{r})\mathbf{r} \times \mathbf{r} = \mathbf{0}$. Therefore, $(\mathbf{r}' \times \mathbf{r}) = \mathbf{n}$, a constant vector and so $\mathbf{r} \cdot \mathbf{n} = \mathbf{r} \cdot (\mathbf{r}' \times \mathbf{r}) = 0$ showing that $\mathbf{n}$ is a normal vector to a plane which contains $\mathbf{r}(t)$ for all t. ∎

14.5 Planetary Motion

Kepler's laws of planetary motion state, among other things, that planets move around the sun along an ellipse. These laws, discovered by Kepler, were shown by Newton to be consequences of his law of gravitation which states that the force acting on a mass m by a mass M is given by

$$\mathbf{F} = -GMm\left(\frac{1}{r^3}\right)\mathbf{r} = -GMm\left(\frac{1}{r^2}\right)\mathbf{e}_r$$

where r is the distance between centers of mass and $\mathbf{r}$ is the position vector from M to m. Here G is the gravitation constant. This is called an inverse square law. Gravity acts according to this law and so does electrostatic force. The constant G, is very small when usual units are used and it has been computed using a very delicate experiment. It is now accepted to be

$$6.67 \times 10^{-11} \text{ Newton meter}^2/\text{kilogram}^2.$$

The experiment involved a light source shining on a mirror attached to a quartz fiber from which was suspended a long rod with two equal masses at the ends which were attracted by two larger masses. The gravitation force between the suspended masses and the two large masses caused the fibre to twist ever so slightly and this twisting was measured by observing the deflection of the light reflected from the mirror on a scale placed some distance from the fibre. The constant was first measured successfully by Lord Cavendish in 1798. It has been measured repeatedly and one measurement was as recent as 2007. Experiments like these are major accomplishments.

In the following argument, M is the mass of the sun and m is the mass of the planet. (It could also be a comet or an asteroid.)

14.5.1 The Equal Area Rule, Kepler's Second Law

An object moves in three dimensions in such a way that the only force acting on the object is a central force. Then the object moves in a plane and the radius vector from the origin to the object sweeps out area at a constant rate. This is the equal area rule. In the context of planetary motion it is called Kepler's second law.

Lemma 14.1 says the object moves in a plane. From the assumption that the force field is a central force field, it follows from (14.5) that

$$2r'(t)\,\theta'(t) + r(t)\,\theta''(t) = 0$$

Multiply both sides of this equation by r. This yields

$$2rr'\theta' + r^2\theta'' = \left(r^2\theta'\right)' = 0. \tag{14.6}$$

Consequently,

$$r^2\theta' = c \tag{14.7}$$

for some constant C. Now consider the following picture.

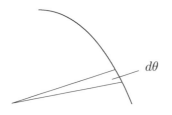

In this picture, $d\theta$ is the indicated angle and the two lines determining this angle are position vectors for the object at point t and point $t+dt$. The area of the sector, dA, is essentially $r^2 d\theta$ and so $dA = \frac{1}{2}r^2 d\theta$. Therefore,

$$\frac{dA}{dt} = \frac{1}{2}r^2\frac{d\theta}{dt} = \frac{c}{2}. \tag{14.8}$$

14.5.2 Inverse Square Law, Kepler's First Law

Consider the first of Kepler's laws, the one which states that planets move along ellipses. From Lemma 14.1, the motion is in a plane. Now from (14.5) and Newton's second law,

$$\left(r''(t) - r(t)\,\theta'(t)^2\right)\mathbf{e}_r + \left(2r'(t)\,\theta'(t) + r(t)\,\theta''(t)\right)\mathbf{e}_\theta$$

$$= -\frac{GMm}{m}\left(\frac{1}{r^2}\right)\mathbf{e}_r = -k\left(\frac{1}{r^2}\right)\mathbf{e}_r$$

Thus $k = GM$ and

$$r''(t) - r(t)\,\theta'(t)^2 = -k\left(\frac{1}{r^2}\right), \quad 2r'(t)\,\theta'(t) + r(t)\,\theta''(t) = 0. \tag{14.9}$$

As in (14.6), $\left(r^2\theta'\right)' = 0$ and so there exists a constant c, such that

$$r^2\theta' = c. \tag{14.10}$$

Now the other part of (14.9) and (14.10) implies

$$r''(t) - r(t)\,\theta'(t)^2 = r''(t) - r(t)\left(\frac{c^2}{r^4}\right) = -k\left(\frac{1}{r^2}\right). \tag{14.11}$$

It is only r as a function of θ which is of interest. Using the chain rule,

$$r' = \frac{dr}{d\theta}\frac{d\theta}{dt} = \frac{dr}{d\theta}\left(\frac{c}{r^2}\right) \tag{14.12}$$

and so also

$$
\begin{aligned}
r'' &= \frac{d^2r}{d\theta^2}\left(\frac{d\theta}{dt}\right)\left(\frac{c}{r^2}\right) + \frac{dr}{d\theta}(-2)(c)\left(r^{-3}\right)\frac{dr}{d\theta}\frac{d\theta}{dt} \\
&= \frac{d^2r}{d\theta^2}\left(\frac{c}{r^2}\right)^2 - 2\left(\frac{dr}{d\theta}\right)^2\left(\frac{c^2}{r^5}\right)
\end{aligned} \tag{14.13}
$$

Using (14.13) and (14.12) in (14.11) yields

$$\frac{d^2r}{d\theta^2}\left(\frac{c}{r^2}\right)^2 - 2\left(\frac{dr}{d\theta}\right)^2\left(\frac{c^2}{r^5}\right) - r(t)\left(\frac{c^2}{r^4}\right) = -k\left(\frac{1}{r^2}\right).$$

Now multiply both sides of this equation by r^4/c^2 to obtain

$$\frac{d^2r}{d\theta^2} - 2\left(\frac{dr}{d\theta}\right)^2\frac{1}{r} - r = \frac{-kr^2}{c^2}. \tag{14.14}$$

This is a nice differential equation for r as a function of θ but it is not clear what its solution is. It turns out to be convenient to define a new dependent variable, $\rho = r^{-1}$ so $r = \rho^{-1}$. Then

$$\frac{dr}{d\theta} = (-1)\rho^{-2}\frac{d\rho}{d\theta}, \quad \frac{d^2r}{d\theta^2} = 2\rho^{-3}\left(\frac{d\rho}{d\theta}\right)^2 + (-1)\rho^{-2}\frac{d^2\rho}{d\theta^2}.$$

Substituting this in to (14.14) yields

$$2\rho^{-3}\left(\frac{d\rho}{d\theta}\right)^2 + (-1)\rho^{-2}\frac{d^2\rho}{d\theta^2} - 2\left(\rho^{-2}\frac{d\rho}{d\theta}\right)^2\rho - \rho^{-1} = \frac{-k\rho^{-2}}{c^2}$$

which simplifies to

$$(-1)\rho^{-2}\frac{d^2\rho}{d\theta^2} - \rho^{-1} = \frac{-k\rho^{-2}}{c^2}$$

since those two terms which involve $\left(\frac{d\rho}{d\theta}\right)^2$ cancel. Now multiply both sides by $-\rho^2$ and this yields

$$\frac{d^2\rho}{d\theta^2} + \rho = \frac{k}{c^2}, \tag{14.15}$$

which is a much nicer differential equation. Let $R = \rho - \frac{k}{c^2}$. Then in terms of R, this differential equation is

$$\frac{d^2 R}{d\theta^2} + R = 0.$$

Multiply both sides by $\frac{dR}{d\theta}$.

$$\frac{1}{2} \frac{d}{d\theta} \left(\left(\frac{dR}{d\theta} \right)^2 + R^2 \right) = 0$$

and so

$$\left(\frac{dR}{d\theta} \right)^2 + R^2 = \delta^2 \tag{14.16}$$

for some $\delta > 0$. Therefore, there exists an angle $\psi = \psi(\theta)$ such that

$$R = \delta \sin(\psi), \quad \frac{dR}{d\theta} = \delta \cos(\psi)$$

because (14.16) says $\left(\frac{1}{\delta} \frac{dR}{d\theta}, \frac{1}{\delta} R \right)$ is a point on the unit circle. But differentiating, the first of the above equations,

$$\frac{dR}{d\theta} = \delta \cos(\psi) \frac{d\psi}{d\theta} = \delta \cos(\psi)$$

and so $\frac{d\psi}{d\theta} = 1$. Therefore, $\psi = \theta + \phi$. Choosing the coordinate system appropriately, you can assume $\phi = 0$. Therefore,

$$R = \rho - \frac{k}{c^2} = \frac{1}{r} - \frac{k}{c^2} = \delta \sin(\theta)$$

and so, solving for r,

$$r = \frac{1}{\left(\frac{k}{c^2} \right) + \delta \sin \theta} = \frac{c^2/k}{1 + (c^2/k)\delta \sin \theta} = \frac{p\varepsilon}{1 + \varepsilon \sin \theta}$$

where

$$\varepsilon = \left(c^2/k \right) \delta \text{ and } p = c^2/k\varepsilon. \tag{14.17}$$

Here all these constants are nonnegative.

Thus

$$r + \varepsilon r \sin \theta = \varepsilon p$$

and so $r = (\varepsilon p - \varepsilon y)$. Then squaring both sides,

$$x^2 + y^2 = (\varepsilon p - \varepsilon y)^2 = \varepsilon^2 p^2 - 2p\varepsilon^2 y + \varepsilon^2 y^2$$

And so

$$x^2 + \left(1 - \varepsilon^2 \right) y^2 = \varepsilon^2 p^2 - 2p\varepsilon^2 y. \tag{14.18}$$

In case $\varepsilon = 1$, this reduces to the equation of a parabola. If $\varepsilon < 1$, this reduces to the equation of an ellipse and if $\varepsilon > 1$, this is called a hyperbola. This proves that objects which are acted on only by a force of the form given in the above example move along hyperbolas, ellipses or circles. The case where $\varepsilon = 0$ corresponds to a circle. The constant ε is called the eccentricity. This is called Kepler's first law in the case of a planet.

14.5.3 *Kepler's Third Law*

Kepler's third law involves the time it takes for the planet to orbit the sun. From (14.18) you can complete the square and obtain

$$x^2 + \left(1 - \varepsilon^2\right)\left(y + \frac{p\varepsilon^2}{1 - \varepsilon^2}\right)^2 = \varepsilon^2 p^2 + \frac{p^2 \varepsilon^4}{(1 - \varepsilon^2)} = \frac{\varepsilon^2 p^2}{(1 - \varepsilon^2)},$$

and this yields

$$x^2 / \left(\frac{\varepsilon^2 p^2}{1 - \varepsilon^2}\right) + \left(y + \frac{p\varepsilon^2}{1 - \varepsilon^2}\right)^2 / \left(\frac{\varepsilon^2 p^2}{(1 - \varepsilon^2)^2}\right) = 1. \tag{14.19}$$

Now note this is the equation of an ellipse and that the diameter of this ellipse is

$$\frac{2\varepsilon p}{(1 - \varepsilon^2)} = 2u. \tag{14.20}$$

This follows because

$$\frac{\varepsilon^2 p^2}{(1 - \varepsilon^2)^2} \geq \frac{\varepsilon^2 p^2}{1 - \varepsilon^2}.$$

Now let T denote the time it takes for the planet to make one revolution about the sun. Using this formula, and (14.8) the following equation must hold.

$$\pi \underbrace{\frac{\varepsilon p}{\sqrt{1 - \varepsilon^2}} \frac{\varepsilon p}{(1 - \varepsilon^2)}}_{\text{area of ellipse}} = T \frac{c}{2}$$

Therefore,

$$T = \frac{2}{c} \frac{\pi c^2 p^2}{(1 - \varepsilon^2)^{3/2}}$$

and so

$$T^2 = \frac{4\pi^2 \varepsilon^4 p^4}{c^2 \left(1 - \varepsilon^2\right)^3}$$

Now using (14.17), recalling that $k = GM$, and (14.20),

$$T^2 = \frac{4\pi^2 \varepsilon^4 p^4}{k\varepsilon p \left(1 - \varepsilon^2\right)^3} = \frac{4\pi^2 \left(\varepsilon p\right)^3}{k \left(1 - \varepsilon^2\right)^3} = \frac{4\pi^2 a^3}{k} = \frac{4\pi^2 a^3}{GM}.$$

Written more memorably, this has shown

$$T^2 = \frac{4\pi^2}{GM} \left(\frac{\text{diameter of ellipse}}{2}\right)^3. \tag{14.21}$$

This relationship is known as Kepler's third law.

14.6   Exercises

(1) Suppose you know how the spherical coordinates of a moving point change
as a function of t. Can you figure out the velocity of the point? Specifically,
suppose $\phi(t) = t, \theta(t) = 1 + t$, and $\rho(t) = t$. Find the velocity of the object
in terms of Cartesian coordinates. **Hint:** You would need to find $x'(t), y'(t)$,
and $z'(t)$. Then in terms of Cartesian coordinates, the velocity would be
$x'(t)\mathbf{i} + y'(t)\mathbf{j} + z'(t)\mathbf{k}$.

(2) Find the length of the cardioid, $r = 1 + \cos\theta, \theta \in [0, 2\pi]$. **Hint:** A parametriza-
tion is $x(\theta) = (1 + \cos\theta)\cos\theta, y(\theta) = (1 + \cos\theta)\sin\theta$.

(3) In general, show that the length of the curve given in polar coordinates by
$r = f(\theta), \theta \in [a, b]$ equals $\int_a^b \sqrt{f'(\theta)^2 + f(\theta)^2}\,d\theta$.

(4) Using the above problem, find the lengths of graphs of the following polar
curves.

(a) $r = \theta, \ \theta \in [0, 3]$
(b) $r = 2\cos\theta, \ \theta \in [-\pi/2, \pi/2]$
(c) $r = 1 + \sin\theta, \ \theta \in [0, \pi/4]$
(d) $r = e^\theta, \ \theta \in [0, 2]$
(e) $r = \theta + 1, \ \theta \in [0, 1]$

(5) Suppose the curve given in polar coordinates by $r = f(\theta)$ for $\theta \in [a, b]$ is
rotated about the y axis. Find a formula for the resulting surface of revolution.
You should get

$$2\pi \int_a^b f(\theta)\cos(\theta)\sqrt{f'(\theta)^2 + f(\theta)^2}\,d\theta$$

(6) Using the result of the above problem, find the area of the surfaces obtained
by revolving the polar graphs about the y axis.

(a) $r = \theta\sec(\theta), \ \theta \in [0, 2]$
(b) $r = 2\cos\theta, \ \theta \in [-\pi/2, \pi/2]$
(c) $r = e^\theta, \ \theta \in [0, 2]$
(d) $r = (1 + \theta)\sec(\theta), \ \theta \in [0, 1]$

(7) Suppose an object moves in such a way that $r^2\theta'$ is a constant. Show that the
only force acting on the object is a central force.

(8) Explain why low pressure areas rotate counter clockwise in the Northern hemisphere and clockwise in the Southern hemisphere. **Hint:** Note that from the
point of view of an observer fixed in space above the North pole, the low
pressure area already has a counter clockwise rotation because of the rotation
of the earth and its spherical shape. Now consider (14.7). In the low pressure area stuff will move toward the center so r gets smaller. How are things
different in the Southern hemisphere?

(9) What are some physical assumptions which are made in the above derivation of Kepler's laws from Newton's laws of motion?

(10) The orbit of the earth is pretty nearly circular and the distance from the sun to the earth is about 149×10^6 kilometers. Using (14.21) and the above value of the universal gravitation constant, determine the mass of the sun. The earth goes around it in 365 days. (Actually it is 365.256 days.)

(11) It is desired to place a satellite above the equator of the earth which will rotate about the center of mass of the earth every 24 hours. Is it necessary that the orbit be circular? What if you want the satellite to stay above the same point on the earth at all times? If the orbit is to be circular and the satellite is to stay above the same point, at what distance from the center of mass of the earth should the satellite be? You may use that the mass of the earth is 5.98×10^{24} kilograms. Such a satellite is called geosynchronous.

14.7 Spherical And Cylindrical Coordinates

Now consider two three dimensional generalizations of polar coordinates. The following picture serves as motivation for the definition of these two other coordinate systems.

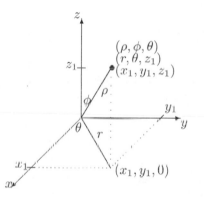

In this picture, ρ is the distance between the origin, the point whose Cartesian coordinates are $(0,0,0)$ and the point indicated by a dot and labelled as (x_1, y_1, z_1), (r, θ, z_1), and (ρ, ϕ, θ). The angle between the positive z axis and the line between the origin and the point indicated by a dot is denoted by ϕ, and θ, is the angle between the positive x axis and the line joining the origin to the point $(x_1, y_1, 0)$ as shown, while r is the length of this line. Thus r and θ determine a point in the plane determined by letting $z = 0$ and r and θ are the usual polar coordinates. Thus $r \geq 0$ and $\theta \in [0, 2\pi)$. Letting z_1 denote the usual z coordinate of a point in three dimensions, like the one shown as a dot, (r, θ, z_1) are the cylindrical coordinates of the dotted point. The spherical coordinates are determined by (ρ, ϕ, θ). When ρ is specified, this indicates that the point of interest is on some sphere of radius ρ

which is centered at the origin. Then when ϕ is given, the location of the point is narrowed down to a circle and finally, θ determines which point is on this circle. Let $\phi \in [0, \pi], \theta \in [0, 2\pi)$, and $\rho \in [0, \infty)$. The picture shows how to relate these new coordinate systems to Cartesian coordinates. For Cylindrical coordinates,

$$x = r \cos(\theta),$$
$$y = r \sin(\theta),$$
$$z = z$$

and for spherical coordinates,

$$x = \rho \sin(\phi) \cos(\theta),$$
$$y = \rho \sin(\phi) \sin(\theta),$$
$$z = \rho \cos(\phi).$$

Spherical coordinates should be especially interesting to you because you live on the surface of a sphere. This has been known for several hundred years. You may also know that the standard way to determine position on the earth is to give the longitude and latitude. The latitude corresponds to ϕ and the longitude corresponds to θ.[2]

Example 14.11. Express the surface $z = \frac{1}{\sqrt{3}}\sqrt{x^2 + y^2}$ in spherical coordinates.

This is

$$\rho \cos(\phi) = \frac{1}{\sqrt{3}}\sqrt{(\rho \sin(\phi) \cos(\theta))^2 + (\rho \sin(\phi) \sin(\theta))^2} = \frac{1}{3}\sqrt{3}\rho \sin \phi.$$

Therefore, this reduces to

$$\tan \phi = \sqrt{3}$$

and so this is just $\phi = \pi/3$.

Example 14.12. Express the surface $y = x$ in terms of spherical coordinates.

This says $\rho \sin(\phi) \sin(\theta) = \rho \sin(\phi) \cos(\theta)$. Thus $\sin \theta = \cos \theta$. You could also write $\tan \theta = 1$.

Example 14.13. Express the surface $x^2 + y^2 = 4$ in cylindrical coordinates.

This says $r^2 \cos^2 \theta + r^2 \sin^2 \theta = 4$. Thus $r = 2$.

---

[2] Actually latitude is determined on maps and in navigation by measuring the angle from the equator rather than the pole but it is essentially the same idea that we have presented here.

14.8   Exercises

(1) The following are the cylindrical coordinates of points, (r, θ, z). Find the rectangular and spherical coordinates.

  (a) $\left(5, \frac{5\pi}{6}, -3\right)$
  (b) $\left(3, \frac{\pi}{3}, 4\right)$
  (c) $\left(4, \frac{2\pi}{3}, 1\right)$
  (d) $\left(2, \frac{3\pi}{4}, -2\right)$
  (e) $\left(3, \frac{3\pi}{2}, -1\right)$
  (f) $\left(8, \frac{11\pi}{6}, -11\right)$

(2) The following are the rectangular coordinates of points, (x, y, z). Find the cylindrical and spherical coordinates of these points.

  (a) $\left(\frac{5}{2}\sqrt{2}, \frac{5}{2}\sqrt{2}, -3\right)$
  (b) $\left(\frac{3}{2}, \frac{3}{2}\sqrt{3}, 2\right)$
  (c) $\left(-\frac{5}{2}\sqrt{2}, \frac{5}{2}\sqrt{2}, 11\right)$
  (d) $\left(-\frac{5}{2}, \frac{5}{2}\sqrt{3}, 23\right)$
  (e) $\left(-\sqrt{3}, -1, -5\right)$
  (f) $\left(\frac{3}{2}, -\frac{3}{2}\sqrt{3}, -7\right)$

(3) The following are spherical coordinates of points in the form (ρ, ϕ, θ). Find the rectangular and cylindrical coordinates. The rectangular coordinates (x, y, z) and cylindrical coordinates (r, θ, z) are as listed.

  (a) $\left(4, \frac{\pi}{4}, \frac{5\pi}{6}\right)$
  (b) $\left(2, \frac{\pi}{3}, \frac{2\pi}{3}\right)$
  (c) $\left(3, \frac{5\pi}{6}, \frac{3\pi}{2}\right)$
  (d) $\left(4, \frac{\pi}{2}, \frac{7\pi}{4}\right)$
  (e) $\left(4, \frac{2\pi}{3}, \frac{\pi}{6}\right)$
  (f) $\left(4, \frac{3\pi}{4}, \frac{5\pi}{3}\right)$

(4) The following are rectangular coordinates of points, (x, y, z). Find the spherical and cylindrical coordinates.

  (a) $\left(\sqrt{2}, \sqrt{6}, 2\sqrt{2}\right)$
  (b) $\left(-\frac{1}{2}\sqrt{3}, \frac{3}{2}, 1\right)$
  (c) $\left(-\frac{3}{4}\sqrt{2}, \frac{3}{4}\sqrt{2}, -\frac{3}{2}\sqrt{3}\right)$
  (d) $\left(-\sqrt{3}, 1, 2\sqrt{3}\right)$
  (e) $\left(-\frac{1}{4}\sqrt{2}, \frac{1}{4}\sqrt{6}, -\frac{1}{2}\sqrt{2}\right)$

(5) A point has Cartesian coordinates $(1, 2, 3)$. Find its spherical and cylindrical coordinates using a calculator or other electronic gadget.

(6) Describe the following surface in rectangular coordinates. $\phi = \pi/4$ where ϕ is the polar angle in spherical coordinates.

(7) Describe the following surface. $\theta = \pi/4$ where θ is the angle measured from the positive x axis in spherical coordinates.

(8) Describe the following surface in rectangular coordinates. $r = 5$ where r is one of the cylindrical coordinates.

(9) Describe the following surface in rectangular coordinates. $\rho = 4$ where ρ is the distance to the origin.

(10) Give the cone, $z = \sqrt{x^2 + y^2}$ in cylindrical coordinates and in spherical coordinates.

(11) Write the following in spherical coordinates.

  (a) $z = x^2 + y^2$.
  (b) $x^2 - y^2 = 1$.
  (c) $z^2 + x^2 + y^2 = 6$.
  (d) $z = \sqrt{x^2 + y^2}$.
  (e) $y = x$.
  (f) $z = x$.

(12) Write the following in cylindrical coordinates.

  (a) $z = x^2 + y^2$.
  (b) $x^2 - y^2 = 1$.
  (c) $z^2 + x^2 + y^2 = 6$.
  (d) $z = \sqrt{x^2 + y^2}$.
  (e) $y = x$.
  (f) $z = x$.

Appendix A

Basic Plane Geometry

A.1 Similar Triangles And Parallel Lines

Definition A.1. Two triangles are similar if they have the same angles. For example, in the following picture, the two triangles are similar because the angles are the same.

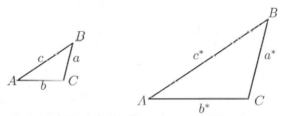

The fundamental axiom for similar triangles is the following.

Axiom A.1. *If two triangles are similar then the ratios of corresponding parts are the same.*

For example in the above picture, this says that

$$\frac{a}{b} = \frac{a^*}{b^*}$$

Definition A.2. Two lines in the plane are said to be parallel if no matter how far they are extended, they never intersect.

Definition A.3. If two lines l_1 and l_2 are parallel and if they are intersected by a line, l_3, the alternate interior angles are shown in the following picture labeled as α.

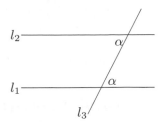

As suggested by the above picture, the following axiom will be used.

Axiom A.2. *If l_1 and l_2 are parallel lines intersected by l_3, then alternate interior angles are equal.*

Definition A.4. An angle is a right angle if when either side is extended, the new angle formed by the extension equals the original angle.

Axiom A.3. *Suppose l_1 and l_2 both intersect a third line, l_3 in a right angle. Then l_1 and l_2 are parallel.*

Definition A.5. A right triangle is one in which one of the angles is a right angle.

Axiom A.4. *Given a straight line and a point, there exists a straight line which contains the point and intersects the given line in two right angles. This line is called perpendicular to the given line.*

Theorem A.1. *Let α, β, and γ be the angles of a right triangle with γ the right angle. Then if the angles, α and β are placed next to each other, the resulting angle is a right angle.*

Proof: Consider the following picture.

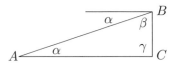

In the picture the top horizontal line is obtained from Axiom A.4. It is a line perpendicular to the line determined by the line segment joining B and C which passes through the point B. Thus from Axiom A.3 this line is parallel to the line joining A and B and by Axiom A.2 the angle between the line joining A and B and this new line is α as shown in the picture. Therefore, the angle formed by placing α and β together is a right angle as claimed. ∎

Definition A.6. When an angle α is placed next to an angle β as shown above, then the resulting angle is denoted by $\alpha + \beta$. A right angle is said to have 90° or to be a 90° angle.

With this definition, Theorem A.1 says the sum of the two non 90° angles in a right triangle is 90°.

In a right triangle the long side is called the hypotenuse. The similar triangles axiom can be used to prove the Pythagorean theorem.

Theorem A.2. *(Pythagoras) In a right triangle the square of the length of the hypotenuse equals the sum of the squares of the lengths of the other two sides.*

Proof: Consider the following picture in which the large triangle is a right triangle and D is the point where the line through C perpendicular to the line from A to B intersects the line from A to B. Then c is defined to be the length of the line from A to B, a is the length of the line from B to C, and b is the length of the line from A to C. Denote by $\overline{DB}$ the length of the line from D to B.

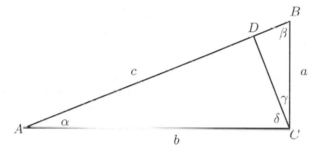

Then from Theorem A.1, $\delta + \gamma = 90°$ and $\beta + \gamma = 90°$. Therefore, $\delta = \beta$. Also from this same theorem, $\alpha + \delta = 90°$ and so $\alpha = \gamma$. Therefore, the three triangles shown in the picture are all similar. By Axiom A.1,

$$\frac{c}{a} = \frac{a}{\overline{DB}}, \text{ and } \frac{c}{b} = \frac{b}{c - \overline{DB}}.$$

Therefore, $c\overline{DB} = a^2$ and

$$c\left(c - \overline{DB}\right) = b^2$$

so

$$c^2 = c\overline{DB} + b^2$$
$$= a^2 + b^2.$$

This proves the Pythagorean theorem.[1] ∎

This theorem implies there should exist some such number which deserves to be called $\sqrt{a^2 + b^2}$.

---

[1] This theorem is due to Pythagoras who lived about 572-497 B.C. This was during the Babylonian captivity of the Jews. Thus Pythagoras was probably a contemporary of the prophet Daniel, sometime before Ezra and Nehemiah. Alexander the great would not come along for more than 100 years. There was, however, an even earlier Greek mathematician named Thales, 624-547 B.C. who also did fundamental work in geometry. Greek geometry was organized and published by Euclid about 300 B.C. The above proof is due to him.

A.2 Distance Formula And Trigonometric Functions

As just explained, points in the plane may be identified by giving a pair of numbers. Suppose there are two points in the plane and it is desired to find the distance between them. There are actually many ways used to measure this distance but the best way, and the only way used in this book is determined by the Pythagorean theorem. Consider the following picture.

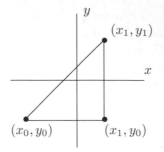

In this picture, the distance between the points denoted by (x_0, y_0) and (x_1, y_1) should be the square root of the sum of the squares of the lengths of the two sides. The length of the side on the bottom is $|x_0 - x_1|$ while the length of the side on the right is $|y_0 - y_1|$. Therefore, by the Pythagorean theorem the distance between the two indicated points is $\sqrt{(x_0 - x_1)^2 + (y_0 - y_1)^2}$. Note you could write either

$$\sqrt{(x_1 - x_0)^2 + (y_1 - y_0)^2} \text{ or } \sqrt{(x_0 - x_1)^2 + (y_1 - y_0)^2}$$

and it would make no difference in the resulting number. The distance between the two points is written as $|(x_0, y_0) - (x_1, y_1)|$ or sometimes when P_0 is the point determined by (x_0, y_0) and P_1 is the point determined by (x_1, y_1), as $d(P_0, P_1)$ or $|P_0 P|$.

The trigonometric functions cos and sin are defined next. Consider the following picture in which the small circle has radius 1, the large circle has radius R, and the right side of each of the two triangles is perpendicular to the bottom side which lies on the x axis.

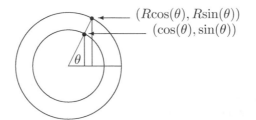

By Theorem A.1 on Page 402 the two triangles have the same angles and so they are similar. Now define by $(\cos\theta, \sin\theta)$ the coordinates of the top vertex of the smaller triangle. Therefore, it follows the coordinates of the top vertex of the larger triangle are as shown. This shows the following definition is well defined.

Definition A.7. For θ an angle, define $\cos\theta$ and $\sin\theta$ as follows. Place the vertex of the angle (The vertex is the point.) at the point whose coordinates are $(0,0)$ in such a way that one side of the angle lies on the positive x axis and the other side extends upward. Extend this other side until it intersects a circle of radius R. Then the point of intersection, is given as $(R\cos\theta, R\sin\theta)$. In particular, this specifies $\cos\theta$ and $\sin\theta$ by simply letting $R = 1$.

Proposition A.1. For any angle θ, $\cos^2\theta + \sin^2\theta = 1$.

Proof: This follows immediately from the above definition and the distance formula. Since $(\cos\theta, \sin\theta)$ is a point on the circle which has radius 1, the distance of this point to $(0,0)$ equals 1. Thus the above identity holds. ■

The other trigonometric functions are defined as follows.

$$\tan\theta \equiv \frac{\sin\theta}{\cos\theta}, \cot\theta \equiv \frac{\cos\theta}{\sin\theta}, \sec\theta \equiv \frac{1}{\cos\theta}, \csc\theta \equiv \frac{1}{\sin\theta}. \tag{A.1}$$

It is also important to understand these functions in terms of a right triangle. Consider the following picture of a right triangle.

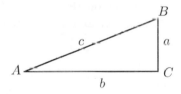

You should verify $\sin A \equiv a/c$, $\cos A \equiv b/c$, $\tan A \equiv a/b$, $\sec A \equiv c/b$, and $\csc A \equiv c/a$.

Having defined the cos and sin there is a very important generalization of the Pythagorean theorem known as the law of cosines. Consider the following picture of a triangle in which a, b and c are the lengths of the sides and A, B, and C denote the angles indicated.

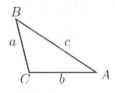

The law of cosines is the following.

Theorem A.3. *Let ABC be a triangle as shown above. Then*

$$c^2 = a^2 + b^2 - 2ab\cos C$$

Proof: Situate the triangle so the vertex of the angle C, is on the point whose coordinates are $(0,0)$ and so the side opposite the vertex, B is on the positive x

axis as shown in the above picture. Then from the definition of the $\cos C$, the coordinates of the vertex, B are $(a \cos C, a \sin C)$ while it is clear the coordinates of A are $(b, 0)$. Therefore, from the distance formula, and Proposition A.1,

$$c^2 = (a \cos C - b)^2 + a^2 \sin^2 C$$
$$= a^2 \cos^2 C - 2ab \cos C + b^2 + a^2 \sin^2 C$$
$$= a^2 + b^2 - 2ab \cos C$$

as claimed. ∎

Corollary A.1. *Let ABC be any triangle as shown above. Then the length of any side is no longer than the sum of the lengths of the other two sides.*

Proof: This follows immediately from the law of cosines. From Proposition A.1, $|\cos \theta| \leq 1$ and so $c^2 = a^2 + b^2 - 2ab \cos C \leq a^2 + b^2 + 2ab = (a+b)^2$. ∎

Corollary A.2. *Suppose T and T' are two triangles such that one angle is the same in the two triangles and in each triangle, the sides forming that angle are equal. Then the corresponding sides are proportional.*

Proof: Let $T = ABC$ with the two equal sides being AC and AB. Let T' be labeled in the same way but with primes on the letters. Thus the angle at A is equal to the angle at A'. The following picture is descriptive of the situation.

Denote by $a, a',\ b, b', c$ and c' the sides indicated in the picture. Then by the law of cosines,

$$a^2 = b^2 + c^2 - 2bc \cos A$$
$$= 2b^2 - 2b^2 \cos A$$

and so $a/b = \sqrt{2(1 - \cos A)}$. Similar reasoning shows $a'/b' = \sqrt{2(1 - \cos A)}$ and so

$$a/b = a'/b'.$$

Similarly, $a/c = a'/c'$. By assumption $c/b = 1 = c'/b'$. ∎

Such triangles in which two sides are equal are called isosceles.

Appendix B

The Fundamental Theorem Of Algebra

The fundamental theorem of algebra states that every nonconstant polynomial having coefficients in $\mathbb{C}$ has a zero in $\mathbb{C}$. If $\mathbb{C}$ is replaced by $\mathbb{R}$, this is not true because of the example, $x^2 + 1 = 0$. This theorem is a very remarkable result and notwithstanding its title, all the easiest proofs of it depend on either analysis or topology. It was studied by many people in the eighteenth century. Gauss gave a proof in 1797 which had a loose end. The first correct proof was given by Argand in 1806. A good discussion of the history of this theorem as well as many different ways to prove it are found in the Wikipedia article on the web. Just google fundamental theorem of algebra. The proof given here follows Rudin [23]. See also Hardy [15] for a similar proof, more discussion and references. The easiest proof is found in the theory of complex analysis and takes a few lines.

Recall Corollary 1.2 which says every nonzero complex number has k k^{th} roots.

Now here is a definition of what it means for a sequence of complex numbers to converge. You will note it is the **same** definition given for a sequence of real numbers, only here the absolute value refers to the absolute value of a complex number.

Definition B.1. Let $\{z_n\}$ be a sequence of complex numbers. Then

$$\lim_{n \to \infty} z_n = z$$

means: For every $\varepsilon > 0$ there exists n_ε such that if $n \geq n_\varepsilon$, then

$$|z_n - z| < \varepsilon.$$

Now here is a useful observation.

Proposition B.1. Suppose $\{a_n + ib_n\}$ is a sequence of complex numbers. Then it converges to $a + ib$ if and only if $\lim_{n \to \infty} a_n = a$ and $\lim_{n \to \infty} b_n = b$.

Proof: I will leave the details to you. Recall

$$|(a_n + ib_n) - (a + ib)| \equiv \sqrt{(a - a_n)^2 + (b - b_n)^2}.$$

This implies

$$\max\left(|a_n - a|, |b_n - b|\right) \leq |(a_n + ib_n) - (a + ib)| \leq |a_n - a| + |b_n - b|$$

Now the conclusion of the proposition follows right away. ■

Next is an important existence theorem.

Theorem B.1. *Denote by $[a, b] + i\,[c, d]$ the set of complex numbers $x + iy$ where $x \in [a, b]$ and $y \in [c, d]$. Suppose $\{z_n\}$ is a sequence of complex numbers in this set. Then there exists a complex number $z \in [a, b] + i\,[c, d]$ and a subsequence $\{z_{n_k}\}$ such that*

$$\lim_{k \to \infty} z_{n_k} = z.$$

Proof: Let $z_n = x_n + iy_n$ and consider first the sequence $\{x_n\} \subseteq [a, b]$. Consider the two intervals $\left[a, \frac{a+b}{2}\right], \left[\frac{a+b}{2}, b\right]$, each having length $(b - a)/2$. Then in one of these intervals, perhaps both, there are x_n for infinitely many values of n. Pick one of the two intervals for which this is so. Then divide it in half. One of the two halves has x_n for infinitely many values of n. Pick the half for which this is so. Divide it in half and pick the half which has the property that it contains x_n for infinitely many n. Continue this way. Thus there is a sequence of intervals $\{I_n\}$ where the length of the nth interval is no larger than $(b - a) \, 2^{-n+1}$ and each of these intervals contains x_n for infinitely many values of n. Now pick $x_{k_1} \in I_1$. If $x_{k_1}, \cdots, x_{k_n}$ have been chosen, let $x_{k_{n+1}}$ be in I_{n+1} such that $k_{n+1} > k_n$. This can be done because each interval contains x_n for infinitely many n. Then $\{x_{n_k}\}_{k=1}^{\infty}$ is a Cauchy sequence because for $k, l > n$,

$$|x_{n_k} - x_{n_l}| < (b - a) \, 2^{-n+1}.$$

By Theorem 2.16 and Theorem 2.23, this sequence converges to some $x \in [a, b]$. Now consider the sequence $\{y_{n_k}\}_{k=1}^{\infty} \subseteq [c, d]$. By the same reasoning there is a subsequence $\left\{y_{n_{k_l}}\right\}_{l=1}^{\infty}$ which converges to $y \in [c, d]$. It follows from Theorem 2.13 that $\left\{x_{n_{k_l}}\right\}_{l=1}^{\infty}$ also converges to x. By Proposition B.1

$$\lim_{l \to \infty} x_{n_{k_l}} + iy_{n_{k_l}} = x + iy \in [a, b] + i\,[c, d]. \qquad\blacksquare$$

Lemma B.1. *If $p(z) = \sum_{k=0}^{n} a_k z^k$ is a polynomial and if $\lim_{n \to \infty} z_n = z$, then*

$$\lim_{n \to \infty} |p(z_n)| = |p(z)|.$$

Proof: $z_n = z_n - z + z$ and so by the binomial theorem,

$$p(z_n) = a_0 + \sum_{k=1}^{n} a_k \sum_{j=0}^{k} \binom{k}{j} z^{k-j} (z_n - z)^j$$

$$= a_0 + \sum_{k=1}^{n} a_k \left(z^k + \sum_{j=1}^{k} \binom{k}{j} z^{k-j} (z_n - z)^j \right)$$

$$= p(z) + \sum_{k=1}^{n} a_k \sum_{j=1}^{k} \binom{k}{j} z^{k-j} (z_n - z)^j \equiv p(z) + e(z_n - z)$$

where $\lim_{n \to \infty} e(z_n - z) = 0$. It follows from the triangle inequality

$$\left| |p(z_n)| - |p(z)| \right| \le |p(z_n) - p(z)| \le |e(z_n - z)|$$

and so $\lim_{n \to \infty} |p(z_n)| = |p(z)|$ as claimed. ∎

Lemma B.2. *Let $p(z)$ be a polynomial with complex coefficients as above. Then $|p(z)|$ achieves its minimum value on any set of the form $[-R, R] + i[-R, R]$.*

Proof: $|p(z)|$ is bounded below by 0 and so

$$\lambda \equiv \inf \{|p(z)| : z \in [-R, R] + i[-R, R]\}$$

exists. By Proposition 10.1 there exists a sequence $\{z_n\}$ of points of $[-R, R] + i[-R, R]$ which satisfy

$$\lambda = \lim_{n \to \infty} |p(z_n)|$$

This is called a minimizing sequence. Then by Theorem B.1 there is a subsequence $\{z_{n_k}\}$ which converges to a point $z \in [-R, R] + i[-R, R]$. Then by Lemma B.1,

$$\lambda = \lim_{k \to \infty} |p(z_{n_k})| = |p(z)|.$$ ∎

Theorem B.2. *(Fundamental theorem of Algebra) Let $p(z)$ be a nonconstant polynomial. Then there exists $z \in \mathbb{C}$ such that $p(z) = 0$.*

Proof: Let

$$p(z) = \sum_{k=0}^{n} a_k z^k$$

where $a_n \ne 0$, $n > 0$. Then

$$|p(z)| \ge |a_n| |z|^n - \sum_{k=0}^{n-1} |a_k| |z|^k$$

and so

$$\lim_{|z| \to \infty} |p(z)| = \infty. \tag{B.1}$$

Now let

$$\lambda \equiv \inf \{|p(z)| : z \in \mathbb{C}\}.$$

By (B.1), there exists an $R > 0$ such that if $z \notin [-R, R] + i[-R, R]$, it follows that $|p(z)| > \lambda + 1$. Therefore,

$$\lambda = \inf \{|p(z)| : z \in \mathbb{C}\} = \inf \{|p(z)| : z \in [-R, R] + i[-R, R]\}.$$

By Lemma B.2, there exists $w \in [-R, R] + i[-R, R]$ such that

$$\lambda = |p(w)|$$

I want to argue $\lambda = 0$. Suppose it is greater than 0. Then consider

$$q(z) \equiv \frac{p(z+w)}{p(w)}.$$

It follows $q(z)$ is of the form

$$q(z) = 1 + c_k z^k + \cdots + c_n z^n$$

where $c_k \neq 0$, because $q(0) = 1$. It is also true that

$$|q(z)| = \frac{|p(z+w)|}{|p(w)|} \geq 1$$

by the assumption that $|p(w)|$ is the smallest value of $|p(z)|$. Now let $\theta \in \mathbb{C}$ be a complex number with $|\theta| = 1$ and

$$\theta c_k w^k = -|w|^k |c_k|.$$

(If

$$w \neq 0, \theta = \frac{-|w^k| |c_k|}{w^k c_k}$$

and if $w = 0$, $\theta = 1$ will work.) Next let $\eta^k = \theta$ and let t be a small positive number.

$$q(t\eta w) \equiv 1 - t^k |w|^k |c_k| + \cdots + c_n t^n (\eta w)^n$$

which is of the form

$$1 - t^k |w|^k |c_k| + t^k (g(t, w))$$

where $\lim_{t \to 0} g(t, w) = 0$. Letting t be small enough,

$$|g(t, w)| < |w|^k |c_k| / 2$$

and so for such t,

$$|q(t\eta w)| < 1 - t^k |w|^k |c_k| + t^k |w|^k |c_k| / 2 < 1,$$

a contradiction to $|q(z)| \geq 1$. ∎

Appendix C

Newton's Laws Of Motion

Vectors were discussed earlier. When the position vector of an object depends on time, we write it in the form $\mathbf{r}(t)$. Thus at time t, the position of the object is the vector $\mathbf{r}(t)$. One can consider the Cartesian coordinates of this object in the usual way. Thus these are of the form

$$\mathbf{r}(t) = (x(t), y(t), z(t)).$$

The derivative and second derivatives are defined in terms of differentiating the component functions. Thus

$$\mathbf{r}'(t) \equiv (x'(t), y'(t), z'(t)),$$
$$\mathbf{r}''(t) \equiv (x''(t), y''(t), z''(t))$$

Definition C.1. Let $\mathbf{r}(t)$ denote the position of an object. Then the acceleration of the object is defined to be $\mathbf{r}''(t)$.

Newton's[2] first law is: "Every body persists in its state of rest or of uniform motion in a straight line unless it is compelled to change that state by forces impressed on it."

Newton's second law is:

$$\mathbf{F} = m\mathbf{a} \tag{C.1}$$

where $\mathbf{a}$ is the acceleration and m is the mass of the object.

Newton's third law states: "To every action there is always opposed an equal reaction; or, the mutual actions of two bodies upon each other are always equal, and directed to contrary parts."

---

[2]Isaac Newton 1642-1727 is often credited with inventing calculus although this is not correct since most of the ideas were in existence earlier. However, he made major contributions to the subject partly in order to study physics and astronomy. He formulated the laws of gravity, made major contributions to optics, and stated the fundamental laws of mechanics listed here. He invented a version of the binomial theorem when he was only 23 years old and built a reflecting telescope. He showed that Kepler's laws for the motion of the planets came from calculus and his laws of gravitation. In 1686 he published an important book, Principia, in which many of his ideas are found. Newton was also very interested in theology and had strong views on the nature of God which were based on his study of the Bible and early Christian writings. He finished his life as Master of the Mint.

Of these laws, only the second two are independent of each other, the first law being implied by the second. The third law says roughly that if you apply a force to something, the thing applies the same force back.

The second law is the one of most interest. Note that the statement of this law depends on the concept of the derivative because the acceleration is defined as a derivative. Newton used calculus and these laws to solve profound problems involving the motion of the planets and other problems in mechanics. The next example involves the concept that if you know the force along with the initial velocity and initial position, then you can determine the position.

Example C.1. Let $r(t)$ denote the position of an object of mass 2 kilograms at time t and suppose the force acting on the object is given by $\mathbf{F}(t) = \left(t, 1 - t^2, 2e^{-t}\right)$. Suppose $\mathbf{r}(0) = (1, 0, 1)$ meters, and $\mathbf{r}'(0) = (0, 1, 1)$ meters/sec. Find $\mathbf{r}(t)$.

By Newton's second law, $2\mathbf{r}''(t) = \mathbf{F}(t) = \left(t, 1 - t^2, 2e^{-t}\right)$ and so

$$\mathbf{r}''(t) = \left(t/2, \left(1 - t^2\right)/2, e^{-t}\right).$$

Therefore the velocity is given by

$$\mathbf{r}'(t) = \left(\frac{t^2}{4}, \frac{t - t^3/3}{2}, -e^{-t}\right) + \mathbf{c}$$

where $\mathbf{c}$ is a constant vector which must be determined from the initial condition given for the velocity. Thus letting $\mathbf{c} = (c_1, c_2, c_3)$,

$$(0, 1, 1) = (0, 0, -1) + (c_1, c_2, c_3)$$

which requires $c_1 = 0, c_2 = 1$, and $c_3 = 2$. Therefore, the velocity is found.

$$\mathbf{r}'(t) = \left(\frac{t^2}{4}, \frac{t - t^3/3}{2} + 1, -e^{-t} + 2\right).$$

Now from this, the displacement must equal

$$\mathbf{r}(t) = \left(\frac{t^3}{12}, \frac{t^2/2 - t^4/12}{2} + t, e^{-t} + 2t\right) + (C_1, C_2, C_3)$$

where the constant vector (C_1, C_2, C_3) must be determined from the initial condition for the displacement. Thus

$$\mathbf{r}(0) = (1, 0, 1) = (0, 0, 1) + (C_1, C_2, C_3)$$

which means $C_1 = 1, C_2 = 0$, and $C_3 = 0$. Therefore, the displacement has also been found.

$$\mathbf{r}(t) = \left(\frac{t^3}{12} + 1, \frac{t^2/2 - t^4/12}{2} + t, e^{-t} + 2t\right) \text{ meters.}$$

Actually, in applications of this sort of thing acceleration does not usually come to you as a nice given function written in terms of simple functions you understand. Rather, it comes as measurements taken by instruments and the position is continuously being updated based on this information. Another situation which often occurs is the case when the forces on the object depend not just on time but also on the position or velocity of the object.

Example C.2. An artillery piece is fired at ground level on a level plain. The angle of elevation is $\pi/6$ radians and the speed of the shell is 400 meters per second. How far does the shell fly before hitting the ground?

Neglect air resistance in this problem. Also let the direction of flight be along the positive x axis. Thus the initial velocity is the vector $400 \cos(\pi/6)\mathbf{i} + 400 \sin(\pi/6)\mathbf{j}$ while the only force experienced by the shell after leaving the artillery piece is the force of gravity, $-mg\mathbf{j}$ where m is the mass of the shell. The acceleration of gravity equals 9.8 meters per sec^2 and so the following needs to be solved.

$$m\mathbf{r}''(t) = -mg\mathbf{j}, \ \mathbf{r}(0) = (0,0), \mathbf{r}'(0) = 400 \cos(\pi/6)\mathbf{i} + 400 \sin(\pi/6)\mathbf{j}.$$

Denoting $\mathbf{r}(t)$ as $(x(t), y(t))$,

$$x''(t) = 0, \ y''(t) = -g.$$

Therefore, $y'(t) = -gt + C$ and from the information on the initial velocity,

$$C = 400 \sin(\pi/6) = 200.$$

Thus

$$y(t) = -4.9t^2 + 200t + D.$$

$D - 0$ because the artillery piece is fired at ground level which requires both x and y to equal zero at this time. Similarly, $x'(t) = 400 \cos(\pi/6)$ so $x(t) = 400 \cos(\pi/6)t = 200\sqrt{3}t$. The shell hits the ground when $y = 0$ and this occurs when $-4.9t^2 + 200t = 0$. Thus $t = 40.816\,326\,530\,6$ seconds and so at this time,

$$x = 200\sqrt{3}\,(40.816\,326\,530\,6) - 14139.190\,265\,0 \text{ meters}.$$

The next example is more complicated because it also takes in to account air resistance. We do not live in a vacuum.

Example C.3. A lump of "blue ice" escapes the lavatory of a jet flying at 600 miles per hour at an altitude of 30,000 feet. This blue ice weighs 64 pounds near the earth and experiences a force of air resistance equal to $(-.1)\mathbf{r}'(t)$ pounds. Find the position and velocity of the blue ice as a function of time measured in seconds. Also find the velocity when the lump hits the ground. Such lumps have been known to surprise people on the ground.

The first thing needed is to obtain information which involves consistent units. The blue ice weighs 32 pounds near the earth. Thus 32 pounds is the force exerted by gravity on the lump and so its mass must be given by Newton's second law as follows.

$$64 = m \times 32.$$

Thus $m = 2$ slugs. The slug is the unit of mass in the system involving feet and pounds. The jet is flying at 600 miles per hour. I want to change this to feet per second. Thus it flies at

$$\frac{600 \times 5280}{60 \times 60} = 880 \text{ feet per second.}$$

The explanation for this is that there are 5280 feet in a mile and so it goes 600×5280 feet in one hour. There are 60×60 seconds in an hour. The position of the lump of blue ice will be computed from a point on the ground directly beneath the airplane at the instant the blue ice escapes and regard the airplane as moving in the direction of the positive x axis. Thus the initial displacement is

$$\mathbf{r}(0) = (0, 30000) \text{ feet}$$

and the initial velocity is

$$\mathbf{r}'(0) = (880, 0) \text{ feet/sec.}$$

The force of gravity is

$$(0, -64) \text{ pounds}$$

and the force due to air resistance is

$$(-.1) \mathbf{r}'(t) \text{ pounds.}$$

Newtons second law yields the following initial value problem for $\mathbf{r}(t) = (r_1(t), r_2(t))$.

$$2(r_1''(t), r_2''(t)) = (-.1)(r_1'(t), r_2'(t)) + (0, -64), \ (r_1(0), r_2(0)) = (0, 30000),$$
$$(r_1'(0), r_2'(0)) = (880, 0)$$

Therefore,

$$
\begin{aligned}
&2r_1''(t) + (.1) r_1'(t) = 0 \\
&2r_2''(t) + (.1) r_2'(t) = -64 \\
&r_1(0) = 0, \ r_2(0) = 30000 \\
&r_1'(0) = 880, \ r_2'(0) = 0
\end{aligned}
\tag{C.2}
$$

Using Theorem 9.2, on these equations, yields

$$r_1(t) = -\frac{1}{(.1)} 2 \left(\exp\left(\frac{-(.1)}{2} t \right) \right) (880) + \left(\frac{2}{(.1)} (880) \right)$$
$$= -17600.0 \exp(-.05t) + 17600.0$$

and

$$r_2(t) = (-64/(.1)) t - \frac{1}{(.1)} 2 \left(\exp\left(-\frac{(.1)}{2} t \right) \right) \left(\frac{64}{(.1)} \right) + \left(30000 + \frac{2}{(.1)} \left(\frac{64}{(.1)} \right) \right)$$
$$= -640.0t - 12800.0 \exp(-.05t) + 42800.0$$

This gives the coordinates of the position. What of the velocity? Differentiating these or using the steps in the above derivation,

$$r_1' (t) = 880.0 \exp(-.05t),$$
$$r_2' (t) = -640.0 + 640.0 \exp(-.05t). \tag{C.3}$$

To determine the velocity when the blue ice hits the ground, it is necessary to find the value of t when this event takes place and then to use (C.3) to determine the velocity. It hits ground when $r_2 (t) = 0$. Thus it suffices to solve the equation,

$$0 = -640.0t - 12800.0 \exp(-.05t) + 42800.0.$$

This is a fairly hard equation to solve using the methods of algebra. In fact, I do not have a good way to find this value of t using algebra. However if plugging in various values of t using a calculator or by graphing and zooming, you eventually find that when $t = 66.14$,

$$-640.0(66.14) - 12800.0 \exp(-.05(66.14)) + 42800.0 = 1.588 \text{ feet.}$$

This is close enough to hitting the ground and so plugging in this value for t yields the approximate velocity,

$$(880.0 \exp(-.05(66.14)), -640.0 + 640.0 \exp(-.05(66.14))) = (32.23, -616.56).$$

Notice how, because of air resistance the component of velocity in the horizontal direction is only about 32 feet per second even though this component started out at 880 feet per second while the component in the vertical direction is -616 feet per second even though this component started off at 0 feet per second. You see that air resistance can be very important so it is not enough to pretend, as is often done in beginning physics courses that everything takes place in a vacuum. Actually, this problem used several physical simplifications. It was assumed the force acting on the lump of blue ice by gravity was constant. This is not really true because it actually depends on the distance between the center of mass of the earth and the center of mass of the lump. It was also assumed the air resistance is proportional to the velocity. This is an over simplification when high speeds are involved. However, increasingly correct models can be studied in a systematic way as above.

C.1   Impulse And Momentum

Work and energy involve a force acting on an object for some distance. Impulse involves a force which acts on an object for an interval of time.

Definition C.2. Let $\mathbf{F}$ be a force which acts on an object during the time interval $[a, b]$. The **impulse** of this force is

$$\int_a^b \mathbf{F}(t) \, dt.$$

This is defined as

$$\left(\int_a^b F_1(t)\,dt, \int_a^b F_2(t)\,dt, \int_a^b F_3(t)\,dt \right).$$

The **linear momentum** of an object of mass m and velocity $\mathbf{v}$ is defined as

$$\text{Linear momentum } = m\mathbf{v}.$$

The notion of impulse and momentum are related in the following theorem.

Theorem C.1. *Let $\mathbf{F}$ be a force acting on an object of mass m. Then the impulse equals the change in momentum. More precisely,*

$$\int_a^b \mathbf{F}(t)\,dt = m\mathbf{v}(b) - m\mathbf{v}(a).$$

Proof: This is really just the fundamental theorem of calculus and Newton's second law applied to the components of $\mathbf{F}$.

$$\int_a^b \mathbf{F}(t)\,dt = \int_a^b m\frac{d\mathbf{v}}{dt}\,dt = m\mathbf{v}(b) - m\mathbf{v}(a) \qquad (C.4)$$

Now suppose two point masses, A and B collide. Newton's third law says the force exerted by mass A on mass B is equal in magnitude but opposite in direction to the force exerted by mass B on mass A. Letting the collision take place in the time interval $[a, b]$ and denoting the two masses by m_A and m_B and their velocities by $\mathbf{v}_A$ and $\mathbf{v}_B$ it follows that

$$m_A \mathbf{v}_A(b) - m_A \mathbf{v}_A(a) = \int_a^b (\text{Force of } B \text{ on } A)\,dt$$

and

$$
\begin{aligned}
m_B \mathbf{v}_B(b) - m_B \mathbf{v}_B(a) &= \int_a^b (\text{Force of } A \text{ on } B)\,dt \\
&= -\int_a^b (\text{Force of } B \text{ on } A)\,dt \\
&= -\left(m_A \mathbf{v}_A(b) - m_A \mathbf{v}_A(a) \right)
\end{aligned}
$$

and this shows

$$m_B \mathbf{v}_B(b) + m_A \mathbf{v}_A(b) = m_B \mathbf{v}_B(a) + m_A \mathbf{v}_A(a).$$

In other words, in a collision between two masses the total linear momentum before the collision equals the total linear momentum after the collision. This is known as the conservation of linear momentum. ∎

C.2 Kinetic Energy

Before considering this, here is a simple lemma.

Lemma C.3. *Let* $\mathbf{v}(t)$ *be a vector valued function. Then*

$$\frac{d}{dt}|\mathbf{v}(t)|^2 = 2(\mathbf{v}(t), \mathbf{v}'(t))$$

Proof: Let $\mathbf{v}(t) = (v_1(t), v_2(t), v_3(t))$. Then it follows that

$$|\mathbf{v}(t)|^2 = v_1^2(t) + v_2^2(t) + v_3^2(t)$$

and so

$$
\begin{aligned}
\frac{d}{dt}|\mathbf{v}(t)|^2 &= 2v_1(t)v_1'(t) + 2v_2(t)v_2'(t) + 2v_2(t)v_3'(t) \\
&= 2(\mathbf{v}(t), \mathbf{v}'(t)).
\end{aligned}
$$ ∎

Newton's second law is also the basis for the notion of **kinetic energy**. When a force is exerted on an object which causes the object to move, it follows that the force is doing work which manifests itself in a change of velocity of the object. How is the total work done on the object by the force related to the final velocity of the object? By Newton's second law, and letting $\mathbf{v}$ be the velocity,

$$\mathbf{F}(t) = m\mathbf{v}'(t).$$

Now in a small increment of time, $(t, t + dt)$, the work done on the object would be approximately equal to

$$dW = \mathbf{F}(t) \cdot \mathbf{v}(t)\, dt. \tag{C.5}$$

If no work has been done at time $t = 0$, then (C.5) implies

$$\frac{dW}{dt} = \mathbf{F} \cdot \mathbf{v}, \; W(0) = 0.$$

Hence,

$$\frac{dW}{dt} = m\mathbf{v}'(t) \cdot \mathbf{v}(t) = \frac{m}{2}\frac{d}{dt}|\mathbf{v}(t)|^2.$$

Therefore, the total work done up to time t would be $W(t) = \frac{m}{2}|\mathbf{v}(t)|^2 - \frac{m}{2}|\mathbf{v}_0|^2$ where $|\mathbf{v}_0|$ denotes the initial speed of the object. This difference represents the change in the kinetic energy.

C.3   Exercises

(1) Show that the solution to $\mathbf{v}' + r\mathbf{v} = \mathbf{c}$ with the initial condition, $\mathbf{v}(0) = \mathbf{v}_0$ is
$\mathbf{v}(t) = \left(\mathbf{v}_0 - \frac{\mathbf{c}}{r}\right)e^{-rt} + (\mathbf{c}/r)$. If $\mathbf{v}$ is velocity and $r = k/m$ where k is a constant for air resistance and m is the mass, and $\mathbf{c} = \mathbf{f}/m$, argue from Newton's second law that this is the equation for finding the velocity, $\mathbf{v}$ of an object acted on by air resistance proportional to the velocity and a constant force $\mathbf{f}$, possibly from gravity. Does there exist a terminal velocity? What is it? **Hint:** To find the solution to the equation, multiply both sides by e^{rt}. Verify that then $\frac{d}{dt}\left(e^{rt}\mathbf{v}\right) = \mathbf{c}e^{rt}$. Then integrating both sides, $e^{rt}\mathbf{v}(t) = \frac{1}{r}\mathbf{c}e^{rt} + \mathbf{C}$. Now you need to find $\mathbf{C}$ from using the initial condition which states $\mathbf{v}(0) = \mathbf{v}_0$.

(2) Suppose that the air resistance is proportional to the velocity but it is desired to find the constant of proportionality. Describe how you could find this constant.

(3) Suppose an object having mass equal to 5 kilograms experiences a time dependent force $\mathbf{F}(t) = e^{-t}\mathbf{i} + \cos(t)\mathbf{j} + t^2\mathbf{k}$ meters per sec^2. Suppose also that the object is at the point $(0, 1, 1)$ meters at time $t = 0$ and that its initial velocity at this time is $\mathbf{v} = \mathbf{i} + \mathbf{j} - \mathbf{k}$ meters per sec. Find the position of the object as a function of t.

(4) Suppose the force $\mathbf{F}$ acting on an object is always perpendicular to the velocity of the object. Thus $\mathbf{F} \cdot \mathbf{v} = 0$. Show that the Kinetic energy of the object is constant. Such forces are sometimes called forces of constraint because they do not contribute to the speed of the object, only its direction.

(5) A cannon is fired at an angle, θ from ground level on a vast plain. The speed of the ball as it leaves the mouth of the cannon is known to be s meters per second. Neglecting air resistance, find a formula for how far the cannon ball goes before hitting the ground. Show that the maximum range for the cannon ball is achieved when $\theta = \pi/4$.

(6) Suppose in the context of Problem 5 that the cannon ball has mass 10 kilograms and it experiences a force of air resistance which is $.01\mathbf{v}$ Newtons where $\mathbf{v}$ is the velocity in meters per second. The acceleration of gravity is 9.8 meters per sec^2. Also suppose that the initial speed is 100 meters per second. Find a formula for the displacement, $\mathbf{r}(t)$ of the cannon ball. If the angle of elevation equals $\pi/4$, use a calculator or other means to estimate the time before the cannon ball hits the ground.

(7) Show that Newton's first law can be obtained from the second law.

(8) Show that if $\mathbf{v}'(t) = \mathbf{0}$, for all $t \in (a, b)$, then there exists a constant vector $\mathbf{z}$ independent of t such that $\mathbf{v}(t) = \mathbf{z}$ for all t.

(9) Suppose an object moves in three dimensional space in such a way that the only force acting on the object is directed toward a single fixed point in three dimensional space. Verify that the motion of the object takes place in a plane. **Hint:** Let $\mathbf{r}(t)$ denote the position vector of the object from the fixed point. Then the force acting on the object must be of the form $g(\mathbf{r}(t))\mathbf{r}(t)$ and by

Newton's second law, this equals $m\mathbf{r}''(t)$. Therefore,

$$m\mathbf{r}'' \times \mathbf{r} = g(\mathbf{r})\mathbf{r} \times \mathbf{r} = \mathbf{0}.$$

Now argue that $\mathbf{r}'' \times \mathbf{r} = (\mathbf{r}' \times \mathbf{r})'$, showing that $(\mathbf{r}' \times \mathbf{r})$ must equal a constant vector $\mathbf{z}$. Therefore, what can be said about $\mathbf{z}$ and $\mathbf{r}$?

(10) Suppose the only forces acting on an object are the force of gravity, $-mg\mathbf{k}$ and a force $\mathbf{F}$ which is perpendicular to the motion of the object. Thus $\mathbf{F} \cdot \mathbf{v} = \mathbf{0}$. Show that the total energy of the object,

$$E \equiv \frac{1}{2}m|\mathbf{v}|^2 + mgz$$

is constant. Here $\mathbf{v}$ is the velocity and the first term is the kinetic energy while the second is the potential energy. **Hint:** Use Newton's second law to show that the time derivative of the above expression equals zero.

(11) Using Problem 10, suppose an object slides down a frictionless inclined plane from a height of 100 feet. When it reaches the bottom, how fast will it be going? Assume it starts from rest.

(12) The ballistic pendulum is an interesting device which is used to determine the speed of a bullet. It is a large massive block of wood hanging from a long string. A rifle is fired into the block of wood which then moves. The speed of the bullet can be determined from measuring how high the block of wood rises. Explain how this can be done and why. **Hint:** Let v be the speed of the bullet which has mass m and let the block of wood have mass M. By conservation of momentum $mv = (m + M)V$ where V is the speed of the block of wood immediately after the collision. Thus the energy is $\frac{1}{2}(m + M)V^2$ and this block of wood rises to a height of h. Now use Problem 10.

(13) In the experiment of Problem 12, show that the kinetic energy before the collision is greater than the kinetic energy after the collision. Thus linear momentum is conserved but energy is not. Such a collision is called inelastic.

(14) There is a popular toy consisting of identical steel balls suspended from strings of equal length as illustrated in the following picture.

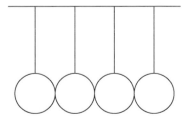

The ball at the right is lifted and allowed to swing. When it collides with the other balls, the ball on the left is observed to swing away from the others

with the same speed the ball on the right had when it collided. Why does
this happen? Why do not two or more of the stationary balls start to move,
perhaps at a slower speed? This is an example of an elastic collision because
energy is conserved. Of course this could change if you fixed things so the
balls would stick to each other.

Bibliography

Apostol, T. M., *Calculus,* second edition, Wiley, 1967.

Apostol T. *Calculus Volume II,* second edition, Wiley, 1969.

Apostol, T. M., *Mathematical Analysis,* Addison Wesley Publishing Co., 1974.

Baker, Roger, *Linear Algebra,* Rinton Press, 2001.

Bartle R.G., *A Modern Theory of Integration,* Grad. Studies in Math., Amer. Math. Society, Providence, RI, 2000.

Chahal J.S., *Historical Perspective of Mathematics 2000 B.C. - 2000 A.D. Kendrick Press, Inc. (2007)*

Davis H. and Snidor A., *Vector Analysis,* Wm. C. Brown, 1995.

D'Angelo, J. and West D. *Mathematical Thinking Problem Solving and Proofs,* Prentice Hall, 1997.

Edwards C.H. *Advanced Calculus of Several Variables,* Dover, 1994.

Euclid, *The Thirteen Books of the Elements,* Dover, 1956.

Fitzpatrick P. M., *Advanced Calculus a Course in Mathematical Analysis,* PWS Publishing Company, 1996.

Fleming W., *Functions of Several Variables,* Springer Verlag, 1976.

Greenberg, M., *Advanced Engineering Mathematics,* second edition, Prentice Hall, 1998.

Gurtin M. *An Introduction to Continuum Mechanics,* Academic Press, 1981.

Hardy G., *A Course Of Pure Mathematics,* tenth edition, Cambridge University Press, 1992.

Horn R. and Johnson C. *Matrix Analysis,* Cambridge University Press, 1985.

Karlin S. and Taylor H. *A First Course in Stochastic Processes,* Academic Press, 1975.

Kuttler K. L., *Basic Analysis,* Rinton.

Kuttler K.L., *Modern Analysis,* CRC Press, 1998.

Lang S. *Real and Functional Analysis,* third edition, Springer Verlag, 1993.

Nobel B. and Daniel J. *Applied Linear Algebra,* Prentice Hall, 1977.

Rose, David, A., The College Math Journal, vol. 22, No.2 March 1991.

Rudin, W., *Principles of Mathematical Analysis,* third edition, McGraw Hill, 1976

Rudin W., *Real and Complex Analysis,* third edition, McGraw-Hill, 1987.

Salas S. and Hille E., *Calculus One and Several Variables,* Wiley, 1990.

Sears and Zemansky, *University Physics,* third edition, Addison Wesley, 1963.

Tierney John, *Calculus and Analytic Geometry,* fourth edition, Allyn and Bacon, Boston, 1969.

Yosida K., *Functional Analysis,* Springer Verlag, 1978.

Appendix D

Answers To Selected Exercises

Exercises 1.6

(3) Find $\cos\theta$ and $\sin\theta$ for

$\theta \in \left\{ \frac{2\pi}{3}, \frac{3\pi}{4}, \frac{5\pi}{6}, \pi, \frac{7\pi}{6}, \frac{5\pi}{4}, \frac{4\pi}{3}, \frac{3\pi}{2}, \frac{5\pi}{3}, \frac{7\pi}{4}, \frac{11\pi}{6}, 2\pi \right\}$.

$\cos\frac{2\pi}{3} = -\frac{1}{2}$,

$\cos\frac{3\pi}{4} = -\frac{1}{2}\sqrt{2}$,

$\cos\frac{5\pi}{6} = -\frac{1}{2}\sqrt{3}$,

$\cos\pi = -1$

$\cos\frac{7\pi}{6} = -\frac{1}{2}\sqrt{3}$,

$\cos\frac{5\pi}{4} = -\frac{1}{2}\sqrt{2}$,

$\cos\frac{4\pi}{3} = -\frac{1}{2}$,

$\cos\frac{3\pi}{2} = 0$, $\cos\frac{5\pi}{3} = \frac{1}{2}$

$\cos\frac{7\pi}{4} = \frac{1}{2}\sqrt{2}$,

$\cos\frac{11\pi}{6}$, $\cos 2\pi = 1$

$\sin\frac{2\pi}{3} = \frac{1}{2}\sqrt{3}$, $\sin\frac{3\pi}{4} = \frac{1}{2}\sqrt{2}$,

$\sin\frac{5\pi}{6} = \frac{1}{2}$, $\sin\pi = 0$

$\sin\frac{7\pi}{6} = -\frac{1}{2}$, $\sin\frac{5\pi}{4} = -\frac{1}{2}\sqrt{2}$,

$\sin\frac{4\pi}{3} = -\frac{1}{2}\sqrt{3}$, $\sin\frac{3\pi}{2} = -1$,

$\sin\frac{5\pi}{3} = -\frac{1}{2}\sqrt{3}$

$\sin\frac{7\pi}{4} = -\frac{1}{2}\sqrt{2}$, $\sin\frac{11\pi}{6} = -\frac{1}{2}$,

$\sin 2\pi = 0$

(4) $\cos^2\theta + \sin^2\theta = 1$

$\cos^2\theta - \sin^2\theta = \cos 2\theta$

Add these and divide by 2. This gives $\cos^2\theta = \frac{1+\cos 2\theta}{2}$

Subtract them and divide by 2. This gives $\sin^2\theta = \frac{1-\cos 2\theta}{2}$

(6) $1 + \cot^2\theta = \frac{\sin^2\theta}{\sin^2\theta} + \frac{\cos^2\theta}{\sin^2\theta} = \frac{1}{\sin^2\theta} = \csc^2\theta$

$1 + \tan^2\theta = \frac{\cos^2\theta}{\cos^2\theta} + \frac{\sin^2\theta}{\cos^2\theta} = \frac{1}{\cos^2\theta} = \sec^2\theta$

(8) Prove that $\sin x \sin y = \frac{1}{2}\left(\cos\left(x-y\right) - \cos\left(x+y\right)\right)$.

The right equals

$\frac{1}{2}\left(\cos x \cos y + \sin x \sin y - \left[\cos x \cos y - \sin x \sin y\right]\right) = \sin x \sin y$

(9) Prove that $\cos x \cos y = \frac{1}{2}\left(\cos\left(x+y\right) + \cos\left(x-y\right)\right)$.

The right equals

$\frac{1}{2}\left(\cos x \cos y - \sin x \sin y + \cos x \cos y + \sin x \sin y\right) = \cos x \cos y$

(10) $\left(\sin x - \sin y\right) = 2 \sin\left(\frac{x-y}{2}\right)\sin\left(\frac{x+y}{2}\right)$

(11) Suppose $\sin x = a$ where $0 < a < 1$. Find all possible values for

(a) $\tan x,\ \frac{a}{\pm\sqrt{1-a^2}}$

(b) $\cot x,\ \frac{\pm\sqrt{1-a^2}}{a}$

(c) $\sec x,\ \frac{1}{\pm\sqrt{1-a^2}}$

(d) $\csc x,\ \frac{1}{a}$

(e) $\cos x,\ \pm\sqrt{1-a^2}$

(12) Solve the equations and give all solutions.

(a) $\sin\left(3x\right) = \frac{1}{2}$,

$x = \frac{\pi}{18} + \frac{2k\pi}{3}, k$ an integer, $\frac{5\pi}{18} + \frac{2k\pi}{3}, k$ an integer

(b) $\cos\left(5x\right) = \frac{\sqrt{3}}{2}$,

$x = \frac{\pi}{30} + \frac{2k\pi}{5}, k$ an integer, $\frac{-\pi}{30} + \frac{2k\pi}{5}, k$ an integer

(c) $\tan\left(x\right) = \sqrt{3}$,

$\frac{\pi}{3} + 2k\pi, k$ an integer, $\frac{4\pi}{3} + 2k\pi, k$ an integer

(d) $\sec\left(x\right) = 2$,

$\frac{\pi}{4} + 2k\pi, k$ an integer, $-\frac{\pi}{4} + 2k\pi, k$ an integer

(e) $\sin\left(x+7\right) = \frac{\sqrt{2}}{2}$,

$-7 + \frac{\pi}{4} + 2k\pi, k$ an integer, $-7 + \frac{3\pi}{4} + 2k\pi, k$ an integer

(f) $\cos^2\left(x\right) = \frac{1}{2}$,

Any of $\frac{\pi}{4}, -\frac{\pi}{4}, \frac{3\pi}{4}, \frac{5\pi}{4}$ or a multiple of 2π added to any of these.

(g) $\sin^4\left(x\right) = 4$ There are no solutions to this. $|\sin x| \le 1$

(13) Sketch a graph of $y = \sin x$.

(14) Sketch a graph of $y = \cos x$.

(15) Sketch a graph of $y = \sin 2x$.

(16) Sketch a graph of $y = \tan x$.

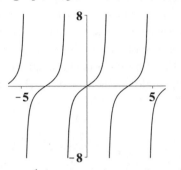

(17) $\sin x \cos y = \frac{1}{2} \left(\sin(x+y) + \sin(x-y) \right)$

(18) Graph $y = \cos^2 x$.

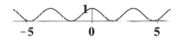

(20) Graph $y = \sin x + \sqrt{3} \cos x$.

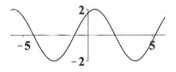

(21) $x = 2k\pi, \frac{\pi}{3} + 2k\pi$, k an integer.

(23) $c = 7$

(24) $\frac{3}{2}\sqrt{3}\sqrt{2}$

(26) $\frac{\tan\theta + \tan\beta}{1 - \tan\theta \tan\beta}$

(27) $\frac{2\tan\theta}{1-\tan^2\theta}$

(28) $\tan\frac{\theta}{2} = -\frac{1}{\tan\theta} \pm \frac{1}{\tan\theta}\sqrt{1+\tan^2\theta}$

(29) You could simply use the straight line which has the same slope as the given line.

(32) This can be done by considering cases. One case is the sides of the inscribed angle are on either side of a diameter. Then you just apply the above result to the to angles which subtend the two arcs. The other case is when the inscribed angle does not have the sides on either side of a diameter. In this case, you consider the given angle as the difference of two angles which have a side on a diameter. Each of these angles in the difference is the length of a subtended arc divided by $2a$ and so the given angle is the length of the subtended arc divided by $2a$.

Exercises 1.8

(1) The directrix is $y = \frac{47}{8} - \frac{1}{8} = \frac{23}{4}$

     The focus is $\left(-\frac{3}{4}, 6\right)$

(2) Sketch a graph of the ellipse whose equation is $\frac{(x-1)^2}{4} + \frac{(y-2)^2}{9} = 1$.

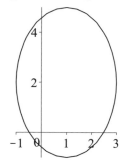

(3) Sketch a graph of the ellipse whose equation is $\frac{(x-1)^2}{9} + \frac{(y-2)^2}{4} = 1$.

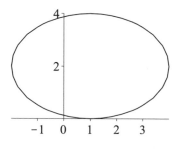

(4) Sketch a graph of the hyperbola, $\frac{x^2}{4} - \frac{y^2}{9} = 1$.

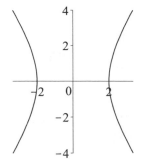

(5) Sketch a graph of the hyperbola, $\frac{y^2}{4} - \frac{x^2}{9} = 1$.

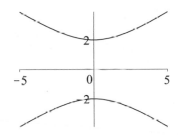

(6) Focus points are

$$\left(1 + \sqrt{13}, 2\right), \left(1 - \sqrt{13}, 2\right)$$

(7) $\sqrt{b^2 + \frac{b^2 x^2}{a^2}} - \frac{bx}{a} = \dfrac{b^2}{\sqrt{b^2 + \frac{b^2 x^2}{a^2}} + \frac{bx}{a}}$ which converges to 0 as $|x| \to \infty$.

(8) 6. This is the length of the long axis.

(9) Hyperbola. The focus points are
$(2 + \sqrt{13}, -1), (2 - \sqrt{13}, -1)$

(10) $\frac{(x-1)^2}{17} + \frac{(y-2)^2}{\frac{17}{4}} = 1$. The focus points are $\left(1 + \sqrt{17 - \frac{17}{4}}, 2\right)$,

$\left(1 - \sqrt{17 - \frac{17}{4}}, 2\right)$

(11) The focus is $\left(-15 + \frac{1}{8}, 2\right) = \left(-\frac{119}{8}, 2\right)$ and directrix is $x = -15 - \frac{1}{8} = -\frac{121}{8}$

(12) The focus is $\left(\frac{22}{15} - \frac{5}{12}, \frac{1}{3}\right)$ and the directrix is $x = \frac{22}{15} + \frac{5}{12}$

(13) $-60x^2 + 56x + 8xy + 196 + 56y - 60y^2 = 0$

(14) $\frac{1}{2}x^2 + \frac{1}{2}y^2 - xy + x + y - \frac{1}{2} = 0$

(15) $x = 2 + 2\cos t, \ y = 2\sqrt{2}\sin(t) - 1$

Exercises 1.10

(1) $(5 + i9)^{-1} = \frac{5}{106} - \frac{9}{106}i$

(2) Let $z = 2 + i7$ and let $w = 3 - i8$. Find $zw, z + w, z^2$, and w/z.
$\frac{w}{z} = \frac{3 - i8}{2 + i7} = -\frac{50}{53} - \frac{37}{53}i$
$wz = (3 - i8)(2 + i7) = 62 + 5i$

$z + w = 3 - i8 + 2 + i7 = 5 - i$

$(2 + i7)^2 = -45 + 28i$

(3) These solutions are the fourth roots of -16

$$2\left(\cos\left(\frac{\pi + 2k\pi}{4}\right) + i\sin\left(\frac{\pi + 2k\pi}{4}\right)\right)$$

for $k = 0, 1, 2, 3$.

(4) The cube roots of 8 are $-1 + i\sqrt{3}, -1 - i\sqrt{3}, 2$

The four fourth roots of 16 are $2, 2i, -2, -2i$

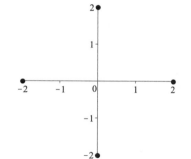

(5) If $z = 0$, let $w = 1$. If $z \neq 0$, let $w = \frac{\bar{z}}{|z|}$.

(6) $[r(\cos t + i\sin t)]^{-n} = \frac{1}{[r(\cos t + i\sin t)]^n}$

$= \frac{1}{r^n(\cos(nt) + i\sin(nt))} = r^{-n}(\cos(nt) - i\sin(nt))$

(7) $\cos(5x) = \cos^5 x - 10\cos^3 x\sin^2 x + 5\cos x\sin^4 x$

$\sin(5x) = 5\cos^4 x\sin x - 10\cos^2 x\sin^3 x + \sin^5 x$

(8) $r_1(\cos\theta + i\sin\theta)r_2(\cos\phi + i\sin\phi) =$

$r_1r_2[(\cos\theta)\cos\phi - (\sin\theta)\sin\phi] +$

$ir_1r_2((\sin\theta)\cos\phi + (\cos\theta)\sin\phi)$

$= r_1r_2(\cos(\theta + \phi) + i\sin(\theta + \phi))$

(9) $x^3 + 8 = (x + 2)\left(x - \left(-1 + i\frac{\sqrt{3}}{2}\right)\right) \cdot$

$\left(x - \left(-1 - i\frac{\sqrt{3}}{2}\right)\right)$

(10) $x^3 + 27 = (3 + x)(x^2 - 3x + 9)$

(11) Completely factor $x^4 + 16$ as a product of linear factors.

$x^4 + 16 =$

$$\left(x - 2\left(\tfrac{\sqrt{2}}{2} + i\tfrac{\sqrt{2}}{2}\right)\right)\left(x - 2\left(\tfrac{\sqrt{2}}{2} - i\tfrac{\sqrt{2}}{2}\right)\right)$$

$$\left(x - 2\left(-\tfrac{\sqrt{2}}{2} + i\tfrac{\sqrt{2}}{2}\right)\right)\left(x - 2\left(-\tfrac{\sqrt{2}}{2} - i\tfrac{\sqrt{2}}{2}\right)\right)$$

(12) $x^4 + 16 = \left(x^2 - 2x\sqrt{2} + 4\right)\left(x^2 + 2x\sqrt{2} + 4\right)$

(18) Yes. When you have something like $\sin^n \theta$ you can write it in the form $\left(\tfrac{1}{2}\left(\cos\theta + i\sin\theta\right) - \left(\cos\theta - i\sin\theta\right)\right)^n$ and then use the binomial theorem followed by De Moivre's theorem to obtain an expression like the above. Similar considerations apply to $\cos^n \theta$.

Exercises 1.12

(1) $x = 6/5, y = 2/5, z = -1/10$.

(2) $x = \tfrac{1}{2} - t\tfrac{1}{2}, \; y = \tfrac{3}{4} + \tfrac{1}{4}t$ where $t \in \mathbb{R}$.

(3) $y = -1 + 4t, \; x - 2 - 4t$.

(4) In this case, the system is inconsistent.

(5) $x = t, w = -3t + 1, y = -2t + \tfrac{4}{3}, z = \tfrac{1}{3}$.

(6) $x_4 = 1 - 6t, x_3 = -1 + 7t, x_2 = 4t, x_1 = 3 - 9t, t \in \mathbb{R}$.

(7) $x_4 = -\tfrac{1}{2} + t\tfrac{3}{2}, x_2 = \tfrac{3}{2} - t\tfrac{1}{2}, x_1 = \tfrac{5}{2} + \tfrac{1}{2}t - 2s$, where $s, t \in \mathbb{R}$.

(8) $y = t, x = -2t + 1, z = 1$

(9) $x = t, y = 2t - 4, z = -\tfrac{1}{4}t + \tfrac{1}{2}$

(10) $z = -1, y = 2, x = -1$

(11) $y = 2, x = 1, z = -5$

(12) $x = -1, y = -5, z = 4$

(13) $y = 2t - 2, z - \tfrac{1}{2}t - \tfrac{1}{2}, x = t$

(14) There is no solution to this system.

(15) because legitimate row operations were not used.

(16) $G = 60, S = 50, B = 200, I = 90$.

(17) $x + y + z = 3, x + y + z = 5$

(18) $x + y = 3, x + y = 3, x - y = 1$.

(19) You would choose h such that $12 - 3h = 0$ so that $h = 4$. This will work nicely.

(20) You could choose $4 - 2h \neq 0$ but even if $h = 2$, then this system is consistent.

(21) Most values of h work. In fact you only need $h \neq 3$ and you will be able to solve the system. However, even if $h = 3$, there will be a solution. Namely $x = 4 - t, y = t$.

(22) If k is anything and $4 - 2h \neq 0$ you will be able to solve this uniquely. If $h = 2$ and $k = 7$ there will be no solutions. If $h = 2$ and $k = 4$ there will be infinitely many solutions.

Exercises 2.2

(1) $2 > \tfrac{1}{t}$ and $t \neq 0$.

(2) Give the domains of the following functions.

(a) $f(x) = \tfrac{x+3}{3x-2}, \; x \neq 2/3$

(b) $f(x) = \sqrt{x^2 - 4}, |x| \geq 2$

(c) $f(x) = \sqrt{4 - x^2}, |x| \leq 2$

(d) $f(x) = \sqrt{\frac{x-4}{3x+5}}, (-\infty, -5/3) \cup [4, \infty)$

(e) $f(x) = \sqrt{\frac{x^2-4}{x+1}}, [-2, -1) \cup [2, \infty)$

(3) $f^{-1}(t) = \sqrt[3]{t} - 1$.

(4) $a_4 = -1 + 2 \times 3 + 3 = 8$

(5) $f^{-1}(x) = -\frac{x}{x-1}$.

(6) Yes it does because the function is one to one.

(7) This function is not one to one. It has no inverse. Consider 1.5 it comes from two different values of t, one for $t > 1$ and one for $t < 1$.

(8) Yes. This is easy to see. If $f(g(x)) = f(g(x_1))$ then since f is one to one, $g(x) = g(x_1)$ and now since g is one to one, $x = x_1$.

(9) If $f : \mathbb{R} \to \mathbb{R}$ and $g : \mathbb{R} \to \mathbb{R}$ are two one to one functions, which of the following are necessarily one to one on their domains? Explain why or why not by giving a proof or an example.

(a) $f + g$ Maybe not one to one. If $f(x) = x$ and $g(x) = -x$ their sum is not one to one.

(b) fg Maybe not one to one. You could have $f(x) = g(x) = x$.

(c) f^3 This is one to one.

(d) f/g Maybe not one to one. Let $f(x) = x$ and $g(x) = x$.

(10) Draw the graph of the function $f(x) = x^3 + 1$.

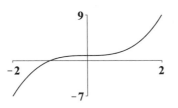

(11) Draw the graph of the function $f(x) = x^2 + 2x + 2$.

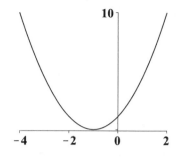

(12) Draw the graph of the function $f(x) = \frac{x}{1+x}$.

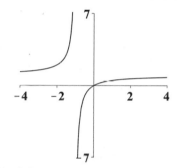

(13) Find the following:

(a) $\cot(\arcsin(x)) = \frac{\sqrt{(1-x^2)}}{x}$

(b) $\sec(\arcsin(x)) = \frac{1}{\sqrt{(1-x^2)}}$

(c) $\csc(\arcsin(x)) = \frac{1}{x}$

(d) $\cos(\arcsin(x)) = \sqrt{(1-x^2)}$

(14) Using Problem 13 and the formulas for the trig functions of a sum of angles, find the following.

(a) $\cot(\arcsin(2x)) = \frac{1}{2}\frac{\sqrt{(1-4x^2)}}{x}$

(b) $\sec(\arcsin(x+y)) = \frac{1}{\sqrt{(1-x^2-2xy-y^2)}}$

(c) $\csc(\arcsin(x^2)) = \frac{1}{x^2}$

(d) $\cos(2\arcsin(x)) = 2\cos^2(\arcsin(x)) - 1 = 1 - 2x^2$

(e) $\tan(\arcsin(x) + \arcsin(y)) = \frac{x\sqrt{(1-y^2)}+y\sqrt{(1-x^2)}}{\sqrt{(1-x^2)}\sqrt{(1-y^2)}-xy}$

(15) Find the following.

(a) $\tan(\arccos(x)) = \frac{\sqrt{(1-x^2)}}{x}$

(b) $\cot(\arccos(x)) = \frac{x}{\sqrt{(1-x^2)}}$

(c) $\sin(\arccos(x)) = \sqrt{(1-x^2)}$

(d) $\csc(\arccos(x)) = \frac{1}{\sqrt{(1-x^2)}}$

(e) $\sec(\arccos(x)) = \frac{1}{x}$

(16) Find the following.

(a) $\cot(\arccos(2x)) = 2\frac{x}{\sqrt{(1-4x^2)}}$

(b) $\sec(\arccos(x+y)) = \frac{1}{x+y}$

(c) $\csc(\arccos(x^2)) = \frac{1}{\sqrt{(1-x^4)}}$

(d) $\cos(\arcsin(x) + \arccos(y)) =$
$y\sqrt{(1-x^2)} - x\sqrt{(1-y^2)}$

(e) $\tan(\arcsin(x) + \arccos(y)) =$
$\frac{xy + \sqrt{(1-x^2)}\sqrt{(1-y^2)}}{y\sqrt{(1-x^2)} - x\sqrt{(1-y^2)}}$

(17) Find

(a) $\cos(\arctan(x)) = \frac{1}{\sqrt{(x^2+1)}}$

(b) $\cot(\arctan(x)) = \frac{1}{x}$

(c) $\sin(\arctan(x)) = \frac{x}{\sqrt{(x^2+1)}}$

(d) $\csc(\arctan(x)) = \frac{\sqrt{(x^2+1)}}{x}$

(e) $\sec(\arctan(x)) = \sqrt{(x^2+1)}$

(18) Find the following.

(a) $\cot(\arctan(2x)) = \frac{1}{2x}$

(b) $\sec(\arctan(x+y)) = \sqrt{(1+x^2+2xy+y^2)}$

(c) $\csc(\arccos(x^2)) = \frac{1}{\sqrt{(1-x^4)}}$

(d) $\cos(2\arctan(x) + \arcsin(y))$
$= \left[\frac{2}{x^2+1} - 1\right]\sqrt{(1-y^2)} - \left(2\frac{x}{x^2+1}\right)y$

(19) $b_k = a_{2k} = 2^{-k}$.

(20) Where is the function increasing?

(a) $[0,1], [2,3]$

(b) $[-2,-1], [3,4]$

(c) $[\frac{1}{2},1], [2,3], [4,4.5]$

Exercises 2.6

(1) $\delta \leq \varepsilon/2$

(3) Let $f(x) = x^2 + 1$. Show f is continuous at $x = 4$.
$|f(x) - f(4)| = |x^2 + 1 - 17| = |x^2 - 16| = |x+4||x-4| \leq 5|x-4|$ provided $|x-4| < 1$. Then for a given $\varepsilon > 0$ let $\delta = \min(1, \varepsilon/5)$. Then if $|x-4| < \delta$, everything holds and $|f(x) - f(4)| < 5\frac{\varepsilon}{5} = \varepsilon$.

(4) Let $f(x) = 2x^2 + 1$. Show f is continuous at $x = 1$.
$\left|2x^2 + 1 - 3\right| = 2\left|x + 1\right|\left|x - 1\right| \leq 10\left|x - 1\right|$ provided $\left|x - 1\right| < 2$. This is because if this condition holds, then $\left|x + 1\right| \leq \left|x - 1\right| + 2 \leq 4$. If $\varepsilon > 0$ is given, let $\delta \leq \min\left(2, \frac{\varepsilon}{10}\right)$.

(6) $\left|\left|2x + 3\right| - \left|2y + 3\right|\right| \leq 2\left|x - y\right|$ and so given $\varepsilon > 0$ you can simply let $\delta = \varepsilon/2$.

(7) The denominator never vanishes so this follows from Theorem 2.1.

(10) This follows right away from Theorem 2.1. To see $f(x) = x$ is continuous, for a given $\varepsilon > 0$ you can take $\delta = \varepsilon$.

(11) This is continuous at 0 and every other point $n\pi$ for n an integer at which $\sin x = 0$.

(12) Would $\delta = 1/4$ work for a given $\varepsilon > 0$? This shows that the idea that a continuous function is one for which you can draw the graph without taking the pencil off the paper is a lot of nonsense.
The function is continuous because given $\varepsilon > 0$ you have only to let $\delta = 1/2$. Then if $\left|y - x\right| < \delta$ for x an integer, it follows $y = x$ and so $\left|f(x) - f(y)\right| = 0 < \varepsilon$.

(13) They are continuous at points where they make sense. In the case of tan, it is not continuous at all the points where $\cos = 0$ but is continuous everywhere else. Thus tan is continuous at all points other than $(2k + 1)\pi/2$ for k an integer, same is true for sec. cot is continuous other than at the points $n\pi$ for n an integer. The same is true for csc.

(14) $f(x) = 1$ on the irrationals and -1 on the rationals. Then $\left|f\right|$ is continuous everywhere although f is not continuous anywhere.

(15) Let $f(x) = 1$ on the rationals and 0 on the irrationals and let g be 0 on the rationals and 1 on the irrationals. Then the product of these functions equals 0 so the product is continuous but neither function is continuous.

(16) f is 1 on the irrationals and -1 on the rationals while g is 1 on the rationals and -1 on the irrationals. Then $f + g = 0$ so it is continuous but neither function is.

(17) Let f be 1 on the rationals and -1 on the irrationals. Let $g = f$. Then $f/g = 1$ which is continuous.

(18) It is continuous on its domain which is everywhere that the tangent makes sense, all x not equal to $(2k + 1)\frac{\pi}{2}$ for k an integer.

(19) It is continuous whenever $\sin x \neq (2k + 1)\frac{\pi}{2}$ for k an integer. However, $(2k + 1)\frac{\pi}{2}$ is never in the interval $[-1, 1]$ which is where sin has its values. Therefore, the given function must be continuous everywhere.

Exercises 2.8

(1) $y = 1/x$ works.

(2) $y = x$ works.

(3) $f(x) = x$ if $x \neq 0, 1$. $f(0) = f(1) = 1/2$.

(4) $y = -1 + x$ seems to work.

(5) Let $f(x) = x^5 + ax^4 + bx^3 + cx^2 + dx + e$ where $a, b, c, d,$ and e are numbers. Show there exists x such that $f(x) = 0$.

$f(a) < 0$ for a large and negative because

$f(x) = x^5 \left(1 + \frac{1}{x}a + \frac{1}{x^2}b + \frac{1}{x^3}c + \frac{1}{x^4}d + \frac{1}{x^5}e\right)$ and the sum of the terms which are divided by a power of x converges to 0 as $x \to -\infty$. Therefore, if x is large and negative $f(x) < 0$. Similarly, if x is large and positive, $f(x) > 0$ and so by intermediate value theorem, there exists x such that $f(x) = 0$.

(6) You could consider $f(x) = 1$ on $[0, 1)$ and $f(x) = 5 - x$ on $[1, 2]$. This is clearly one to one but is increasing on $[0,1)$ and decreasing on $[1,2]$.

(7) $f(0) < 0$ and $f(8) > 0$. Therefore, there exists a point x such that $f(x) = 0$.

(11) Consider $f(\theta) = T(\theta) - T(\theta + \pi)$. This is a continuous function of θ. Suppose $f(\theta) = T(\theta) - T(\theta + \pi) > 0$. Then

$$f(\theta + \pi) = T(\theta + \pi) - T(\theta + 2\pi) = T(\theta + \pi) - T(\theta) < 0$$

Hence there exists $\alpha \in (\theta, \theta + \pi)$ such that $f(\alpha) = 0$. Hence

$$T(\alpha) = T(\alpha + \pi)$$

(12) Let $g(t)$ be the gallons in the tank and let $m(t)$ be the miles driven as a function of t. Then consider $f(t) = g(t) - m(t)$. $f(0) > 0$ and $f(T) < 0$ where T is the time when the car runs out of gas. Therefore, there exists $t \in (0, T)$ such that $f(t) = 0$.

Exercises 2.10

(1) Find the following limits if possible

    (a) $\lim_{x \to 0+} \frac{|x|}{x} = 1$

    (b) $\lim_{x \to 0+} \frac{x}{|x|} = 1$

    (c) $\lim_{x \to 0-} \frac{|x|}{x} = -1$

    (d) $\lim_{x \to 4} \frac{x^2 - 16}{x + 4} = 0$

    (e) $\lim_{x \to 3} \frac{x^2 - 9}{x + 3} = 0$

    (f) $\lim_{x \to -2} \frac{x^2 - 4}{x - 2} = 0$

    (g) $\lim_{x \to \infty} \frac{x}{1 + x^2} = 0$

    (h) $\lim_{x \to \infty} -2\frac{x}{1 + x^2} = 0$

(2) $\lim_{h \to 0} \frac{\frac{1}{(x+h)^3} - \frac{1}{x^3}}{h} = -\frac{3}{x^4}$

(3) $\lim_{x \to 4} \frac{\sqrt[4]{x} - \sqrt{2}}{\sqrt{x} - 2} = \frac{1}{2}\sqrt[4]{4} - \frac{1}{4}\sqrt{2}$

(4) $\lim_{x \to \infty} \frac{\sqrt[5]{3x} + \sqrt[4]{x} + 7\sqrt{x}}{\sqrt{3x + 1}} = \frac{7}{3}\sqrt{3}$

(5) $\lim_{x \to \infty} \frac{(x - 3)^{20}(2x + 1)^{30}}{(2x^2 + 7)^{25}} = 32$

(6) $\lim_{x \to 2} \frac{x^2 - 4}{x^3 + 3x^2 - 9x - 2} = \frac{4}{15}$

(7) -7

(10) $\lim_{h \to 0} \frac{(x+h)^2 - x^2}{h} = 2x$

(11) This follows because every point of the given sets is a limit point and so the two definitions are the same in this case.

(12) $\lim_{h\to 0} \frac{(x+h)^3 - x^3}{h} = 3x^2$

(13) $\lim_{h\to 0} \frac{\frac{1}{x+h} - \frac{1}{x}}{h} = -\frac{1}{x^2}$

(14) $\lim_{x\to -3} \frac{x^3 + 27}{x+3} = 27$

(15) $\lim_{h\to 0} \frac{\sqrt{(3+h)^2} - 3}{h} = 1$

(16) $x > 0$

(17) $\frac{1}{3\left(\sqrt[3]{x}\right)^2}$

(20) $\lim_{x\to 0} \left(\lim_{y\to 0} f(x, y)\right) = \lim_{x\to 0} (1) = 1$
$\lim_{y\to 0} \left(\lim_{x\to 0} f(x, y)\right) = \lim_{y\to 0} (-1) = -1$
It shows you cannot necessarily interchange limits.

Exercises 2.12

(1) Find $\lim_{n\to\infty} \frac{n}{3n+4} = \frac{1}{3}$

(2) Find $\lim_{n\to\infty} \frac{n^3 - 100n}{2n^3 + 79n^2} = \frac{1}{2}$

(3) Find $\lim_{n\to\infty} \frac{3n^4 + 7n + 1000}{n^4 + 1} = 3$

(4) For $a_k, b_k \neq 0$, find $\lim_{n\to\infty} \frac{a_k n^k + a_{k-1} n^{k-1} + \cdots + a_1 n + a_0}{b_k n^k + b_{k-1} n^{k-1} + \cdots + b_1 n + b_0} = \frac{a_k}{b_k}$

(5) For $b_k \neq 0$, find $\lim_{n\to\infty} \frac{a_k 1 n^k 1 + \cdots + a_1 n + a_0}{b_k n^k + b_{k-1} n^{k-1} + \cdots + b_1 n + b_0} - 0$

(6) Find $\lim_{n\to\infty} \frac{2^n + 7(5^n)}{4^n + 2(5^n)} = \frac{7}{2}$

(7) Find $\lim_{n\to\infty} n \tan \frac{1}{n} = 1$

(8) Find $\lim_{n\to\infty} n \sin \frac{2}{n} = 2$

(9) Find $\lim_{n\to\infty} \sqrt{\left(n \sin \frac{9}{n}\right)} = 3$

(10) Find $\lim_{n\to\infty} \left(\sqrt{(n^2 + 6n)} \; n\right) - 3$

(11) Find $\lim_{n\to\infty} \sum_{k=1}^{n} \frac{1}{10^k} = \frac{1}{9}$

(12) From the definition of the absolute value for complex numbers,
$\max(|x_n - x|, |y_n - y|) \leq |x_n + iy_n - (x + iy)|$
$\leq |x_n - x| + |y_n - y|$ The assertion follows from this.

(14) $x = \frac{34}{99}$

(15) Yes. It is continuous. Let $\varepsilon > 0$ be given and let $\delta = 1$. Then if $y \in D(f)$ and $|y - 9| < \delta$, it follows $y = 9$ and so $f(y) = f(9)$ so $|f(y) - f(9)| = 0 < \varepsilon$.

(16) If this is not so, then $x > c$ and so letting $c < b < x$, it follows that for all n large enough, $x_n > b$ which does not happen.

(17) It doesn't have to be unique. For example,

$$\frac{1}{2} = .5 = .499\overline{9}$$

(18) This cannot possibly exist because the sequence is not even bounded.

(19) The assertion in the hint is obvious because if this is not so, then $a + \varepsilon$ would be a lower bound larger than the greatest lower bound. Then there exists

$a_n \in [a, a + \varepsilon)$ and since the sequence is decreasing, whenever $m > n$, it follows $a_m \in [a, a + \varepsilon)$ and this says $a = \lim_{m \to \infty} a_m$.

(20) Try $\{(-1)^n\}$.

Exercises 2.14

(1) Let $\varepsilon > 0$. Then let $\delta = \varepsilon/K$. This works in the definition of uniform continuity.

(2) $|y_n - z| \leq |x_n - y_n| + |x_n - z|$ and both of these terms converge to 0.

(3) $\left| \frac{1}{x} - \frac{1}{y} \right| = \left| \frac{y-x}{xy} \right| \leq |y - x|$. If $\varepsilon > 0$ is given, let $\delta = \varepsilon$ and this works for the definition of uniform continuity.

(4) $||f(x)| - |f(y)|| \leq |f(x) - f(y)|$. If f is uniformly continuous then so is $|f|$. In fact, the same δ which wors for f will work for $|f|$.

$|f|$ can be uniformly continuous without f even being continuous. Consider $f(x) = 1$ on the rationals and -1 on the irrationals.

(5) Let M be as in the hint. Then if $L < M$, there must exist x_n such that $f(x_n) > L$ since otherwise, M is not really the sup. Let L_n be an increasing sequence which converges to M and let x_n be as described. Then $f(x_n) \to M$. Since K is sequentially compact, there exists a subsequence $x_{n_k} \to x \in K$. Then by continuity of f,

$$M = \lim_{k \to \infty} f(x_{n_k}) = f(x)$$

so the function achieves its maximum. A similar argument shows f achieves its minimum.

Exercises 2.16

(2) No such function

(3) No such function

(5) $A_n = P\left(\frac{(1+r)^{n+1} - 1}{r} \right)$

(6) $A_n = 2^n - 1$

(8) $h(0) \leq 0, h(1) \geq 0$. If either of these gives equality, you are done. If they both are strict inequalities, then the result follows from the intermediate value theorem.

(14) $\sum_{k=2}^{n} \frac{1}{k(k-1)} = \sum_{k=2}^{n} \left(\frac{1}{k-1} - \frac{1}{k} \right)$

$= \frac{1}{2} - \frac{1}{n-1} \leq \frac{1}{2}$. Hence the A_n must converge because it is an increasing sequence which is bounded above.

(15) $\sum_{k=1}^{n} \frac{1}{k^2} \leq 1 + \sum_{k=2}^{n} \frac{1}{k(k-1)} \leq 1 + \frac{3}{2}$ and since these are increasing and bounded above, the sequence must converge.

(17) $\frac{9}{2} - \frac{3}{2}\sqrt{5}$

(21) Yes. $||a_n| - |a|| \leq |a_n - a|$ which is given to converge to 0.

(22) Show the following converge to 0.

(a) $\frac{n^5}{1.01^n}$

$0 \leq \frac{n^5}{1.01^n} \leq \dfrac{n^5}{\sum_{k=0}^{6} \binom{n}{k} (.01)^k}$ and this in the denominator is a polynomial

in n of degree 6. Therefore, the sequence converges to 0.

(b) $\frac{10^n}{n!}$

This is done as follows. For $n > 20$, $n! \geq n10^n$. Hence for such n,

$$\frac{10^n}{n!} \leq \frac{10^n}{n10^n} = \frac{1}{n}$$

(24) Prove $\lim_{n \to \infty} \sqrt[n]{n} = 1$. **Hint:** Let $e_n \equiv \sqrt[n]{n} - 1$ so that $(1 + e_n)^n = n$. Now observe that $e_n > 0$ and use the binomial theorem or Example 1.2 to conclude $1 + ne_n + \frac{n(n-1)}{2} e_n^2 \leq n$. This nice approach to establishing this limit using only elementary algebra is in Rudin [23].

Use the hint.

$$1 + ne_n + \frac{n(n-1)}{2} e_n^2 \leq (1 + e_n)^n = n$$

and so

$$e_n^2 \leq \frac{n}{\left(\frac{n(n-1)}{2} \right)}$$

The right side converges to 0 and so by the squeezing theorem, $e_n \to 0$.

Exercises 3.3

(1) Find intervals on which the function is increasing.

   (a) $(-\infty, 1), (2, \infty)$
   (b) $(-\infty, -1), (3, \infty)$
   (c) $(-\infty, 1), (2, 3), (4, \infty)$

(2) Find derivatives of the following functions.

   (a) $6x + 12x^2 - 7$
   (b) $-\frac{3x^2}{(x^3+1)^2}$
   (c) $(4x^3 + 1)(x^2 + 7x) + (x^4 + x + 5)(5x^4 + 7)$
   (d) 1
   (e) $8(x^2 + 2)^3 x$
   (f) $-(x^3 + 2x + 1)^{-2}(3x^2 + 2)$

(3) $-630x^4 + 240x^2 + 12$

(4) $-210x^4 + 60x^2 + 6$

(5) Find $f'(x)$ for the given functions

   (a) $\frac{(9x^2-1)(3x^2+1)-6x(3x^3-x-1)}{(3x^2+1)^2}$
   (b) $\frac{(9x^2+3)(x^2+1)-2x(3x^3+3x-1)}{(x^2+1)^2}$

(c) $\dfrac{8x\left(x^2+5\right)^3\left(x^2+1\right)^2-4x\left(x^2+1\right)\left(x^2+5\right)^4}{\left(x^2+1\right)^4}$

(d) $6\left(4x+7x^3\right)^5\left(4+21x^2\right)$

(e) $3\left(-2x^3+x+3\right)^2\left(-6x+1\right)\left(-2x^2+3x-1\right)$
$\left(-2x^3+x+3\right)^3\left(-4x+3\right)$

$+$

(6) $\dfrac{4x}{\sqrt{4x^2+1}}$

(7) $\dfrac{2x}{\left(\sqrt[3]{3x^2+1}\right)^2}$

(8) $-15\left(5-3x\right)^4$

(9) -42

(11) Not necessarily. Consider for example a function with a point like $y=|x|$ at $x=0$. There is no derivative at the point where $x=0$.

(12) $f\left(x\right)=1$ at rationals and -1 at irrationals.

(15) $h'\left(0\right)=1$

(16) $\left(f_1 f_2\cdots f_n\right)'=\sum_{k=1}^{n}f_1\cdots f'_k\cdots f_n$

(17) $k, f\left(x\right)=kx$

(18) $f\left(x\right)=7x+x^2/2$

Exercises 3.5

(1) No.

(2) $x=16$ and $y=16$.

(3) $4,4$.

(4) $2\sqrt{6}, \frac{4}{3}\sqrt{6}$

(5) $\$12.275$

(6) About 6 times a year with $500/6=83$ in each order.

(7) You look at the points of (a,b) at which the derivative equals 0 and then you also look at the end points. Pick the best answer.

(8) The maximum occurs at $\frac{5}{2}\sqrt{2}$ and the minimum is at -5.

(9) The maximum happens when it all goes to the square and the minimum happens when $x=\dfrac{9L}{9+4\sqrt{3}}$ and $y=4L\dfrac{\sqrt{3}}{9+4\sqrt{3}}$.

(10) $r=\sqrt[3]{30}$. Then $h=3\sqrt[3]{30}$.

(11) The maximum happens at 10 and the minimum happens at 4.

(13) $x=10$. Thus one side is 10 and the other 20.

(14) 106×318.

(15) $x=4\sqrt{2}, y=4\sqrt{2}$

(16) $3\sqrt{2}\times2\sqrt{2}$.

(17) $\pi\frac{2048}{9}\sqrt{3}$

(19) tan odd, cot odd, sec even csc odd.

(20) $x=0, x=\frac{2}{5}$.

(21) $x=1$

(22) The point where $x=5$ is a local minimum and the point where $x=-5$ is a

local maximum.

(23) 125×350.

(24) $\$ 95\,200$ is maximum profit at rent of $\$1040$.

(25) $(1, \sqrt{23})$.

(26) $x = \frac{200}{3}$. The dimmest point is then $\frac{200}{3}$ from the dim light.

(27) 512π

(28) $\frac{120}{11}$ miles along the shore from the closest city to the river.

(29) $4 + \frac{8}{5} = \frac{28}{5}$ feet from the corner.

Exercises 3.7

(1) $\$140$.

(3) Find $c \in (a, b)$

   (a) $1/4$

   (b) $\pm 1/\sqrt{3}$

   (c) $\sqrt{2} + 1$

(4) $\sqrt{ab}$

(6) The function $x \to |x| + x$ is increasing and the function $x \to x^7$ is strictly increasing and so the given function is strictly increasing. Hence there is at most one solution to the equation $f(x) = 0$. There does exist a solution because $f(-10) < 0$ and $f(10) > 0$. By intermediate value theorem, there exists a solution to $f(x) = 0$.

(7) No. This does not follow because this function is not differentiable on the open interval. In fact there is no point where the derivative equals 0.

(8) The derivative is $2ax + b$ and if this is set equal to a number, there is exactly one solution.

(9) $f'(x) > 0$ and so the function is increasing.

(11) In each case, the functions achieve positive and negative values and the derivative is positive.

(14) No. It can't be because it does not have the intermediate value property.

Exercises 3.9

(1) Here is a graph of a function. Identify intervals on which the function is concave up.

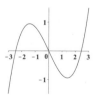

$(0, 3)$

(3) Sketch the graph of the function, $f(x) = x^3 - 3x + 1$ showing the intervals on which the function is concave up and down and identifying the intervals on which the function is increasing.

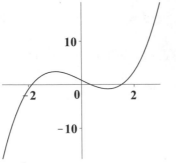

$f'(x) = 3x^2 - 3$ and so critical points are $-1, 1$ These are points where the slope of the tangent line is 0. $f''(x) = 6x$ so $x = 0$ is a possible inflection point. In fact it is an inflection point because the second derivative changes sign there.

(4) Find intervals on which the function, $f(x) = \sqrt{1 - x^2}$ is increasing and intervals on which it is concave up and concave down. Sketch a graph of the function

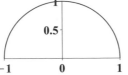

It is clear from the graph what this function does.

(5) Sketch the graphs of $y = x^4, y = x^3$, and $y = -x^4$. What do these graphs tell you about the case when the second derivative equals zero?

When the second derivative equals 0 the test fails. $y = x^4$ has a local minimum at 0 and $y = -x^4$ has a local maximum there while $y = x^3$ has neither a local minimum nor a local maximum.

(6) Sketch the graph of $f(x) = 1/(1 + x^2)$ showing the intervals on which the function is increasing or decreasing and the intervals on which the graph is concave up and concave down.

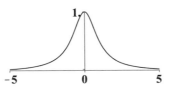

The derivative is $-2\frac{x}{(x^2+1)^2}$ so this has a single critical point at $x = 0$. The second derivative is $2\frac{3x^2-1}{(x^2+1)^3}$ and so there are two possible inflection points at

$\pm(1/\sqrt{3})$. These are indeed inflection points because the second derivative changes sign at these points.

(7) Sketch the graph of $f(x) = x/(1+x^2)$ showing the intervals on which the function is increasing or decreasing and the intervals on which the graph is concave up and concave down.

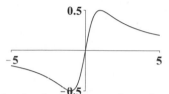

$\frac{1-x^2}{(x^2+1)^2}$ is the derivative and so there are two critical points where the slope of the tangent linear are 0 and these occur where $x = -1, 1$. The function is increasing between these and decreasing on the the other two intervals as shown in the picture. The second derivative is $2x\frac{x^2-3}{(x^2+1)^3}$. There are three possible inflection points, $0, -\sqrt{3}, \sqrt{3}$. Since the second derivatives change sign at these points, they are all real inflection points.

(9) Yes

(10) $x = \pm\frac{1}{\sqrt{3}}$

(11) Identify the intervals on which the following functions are concave up.

(a) $\sqrt{|x|}$ This is never concave up.

(b) $|x|x$, $(-\infty, \infty)$.

(c) $\frac{x}{1+x}$, $x < -1$.

(d) $x^3 - 2x$, $x > 0$.

(e) $\frac{1+x^2}{1-x^2}$, $|x| < 1$.

(f) $-x^3 + 3x$, $x < 0$.

(g) $x^4 + 2x^3 - x^2 + 3x$, $\left(-\frac{1}{2} - \frac{1}{6}\sqrt{15}, -\frac{1}{2} + \frac{1}{6}\sqrt{15}\right)$

(12) $(1, \infty)$.

(13) The exact graph:

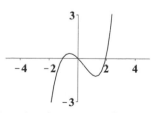

(15) $y''(0) = -1$ and so the function is concave down.

(16) Sketch the following rational functions.

(a) $\frac{x^2+1}{x-1}$, $x < -1$

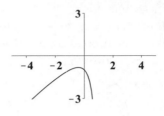

(b) $\frac{x+2}{x^2-1}$

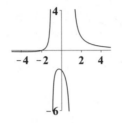

(c) $\frac{x^2}{1-x^2}$

Exercises 4.2

(3) $m \sin^{m-1}(x) \cos(x)$

(4) $30 \sin^5(5x)$

(5) $28 \tan^6(4x)$

(6) $\frac{6 \sec^2(2x) \tan^3(3x) - 9 \tan^2(3x) \sec^2(3x)}{\tan^6(3x)}$

(7) Find derivatives.

    (a) $-3 \frac{\cos x}{1 - 2\cos^2 x + \cos^4 x}$

    (b) $2(\tan x) \sec^2(x)$

    (c) $4 \sin x \cos^3 x - 2 \sin x \cos x$

    (d) $-4 \sin x \cos x$

    (e) $-\frac{4 \cos^2 x - 5}{\cos^6 x}$

    (f) $2x \tan^2(x) + 2x^2 \tan(x) \sec^2(x)$

(8) $\left(k\pi, \frac{2k+1}{2}\pi\right)$, k an integer.

(9) $\left(\frac{(2k+1)\pi}{3}, \frac{(2k+2)\pi}{3}\right)$, k an integer.

(10) $\left(\left(\sqrt[3]{3}\right)^2 + \left(\sqrt[3]{5}\right)^2\right)^{3/2}$

(11) $\left(2^{2/3} + 5^{2/3}\right)^{3/2}$

(12) $\emptyset$

(13) $\emptyset$

(14) $x = \sqrt{7}$.

(15) $x' = -\frac{625}{8}$

Exercises 4.4

(2) Simplify

    (a) $2 + \log_4(x)$

    (b) $3 + 3\log_3(x)$

    (c) 1

(3) Simplify

    (a) 0

    (b) 1

    (c) $\log_a(x)$

    (d) $\frac{5}{3}$

(4) Find the derivatives of the following functions. You may want to do this by looking at the definition of the derivative in some cases.

    (a) $\frac{2}{x \ln 5}$

    (b) $\frac{5}{(5x+1) \ln 3}$

    (c) $\frac{1}{2x(\ln(6))}$

    (d) $\frac{x}{(x^2+1) \ln 3}$

(5) Find the derivatives of the following functions. You may want to do this by looking at the definition of the derivative in some cases.

    (a) $(2)\left(3^{x^2} x \ln 3\right)$

    (b) $6 \times 8^x \ln 2$

    (c) $2 \ln(5) \, 5^{2x+7}$

    (d) $3 \ln 7 \left(7^{x^3}\right) x^2$

    (e) $3^{\sqrt{(x^2+1)}} \frac{1}{\sqrt{(x^2+1)}} x \ln 3$

(7) $x = \frac{8}{5}$

(8) $x = 24 + 16\sqrt{2},\ 24 - 16\sqrt{2}$.

(9) $x = -9 \frac{\ln 5}{2 \ln 5 - \ln 7}$

(10) $-3 + \log_4(x)$.

(11) $x = \frac{\ln 4 + \ln 9}{\ln 4 - \ln 9}$

(13) The second derivative is $\frac{1}{\ln(b)}\left(\frac{-1}{x^2}\right) < 0$

(14) $x = -e^{-1}$

(20) Find derivatives.

    (a) $3\cos(3x)\,5^{6x} + 6\sin(3x)\ln(5)\,5^{6x}$

    (b) $2\sec^2(2x)\,3^{3x+1} + 3\tan(2x)\ln(3)\,3^{3x+1}$

    (c) $2\cos(2x)\cos(3x)\tan(4x)\,2^{3x} -$
        $3\sin(2x)\sin(3x)\tan(4x)\,2^{3x} +$
        $\sin(2x)\cos(3x)\,4\sec^2(4x)\,2^{3x} +$
        $\sin(2x)\cos(3x)\tan(4x)\,3\ln(2)\,3^{3x}$

(21) $b = -\alpha$ and $a = \alpha^2 - \beta^2$

Exercises 5.2

(1) In each of the following, find $\frac{dy}{dx}$.

    (a) $(\cos x)\,e^{\sin x}$

    (b) $\frac{1}{2\sqrt{(7+x^2+\sin x)}}\,(2x + \cos x)$

    (c) $2\frac{x}{x^2+1}$

    (d) $2\left(\cos\left(\ln\left(x^2+1\right)\right)\right)\frac{x}{x^2+1}$

    (e) $\frac{\cos x}{\sin x+3}$

    (f) $\frac{1}{\cos^2 x\,\tan x}$

    (g) $2\left(\sin\left(\ln\left(\tan x\right)\right)\right)\frac{\cos(\ln(\tan x))}{\cos x\,\sin x}$

    (h) $\cos\left((x+\tan x)^6\right)\left[6\,(x+\tan x)^5\right](1+\sec^2(x))$

    (i) $2\left(\cos\left(x^2+7\right)\right)\frac{x}{\sin(x^2+7)}$

    (j) $-2\left(\sin\left(x^2\right)\right)\frac{x}{\cos^2(\cos(x^2))}$

    (k) $\frac{\cos x}{(\sin x+6)\ln 2}$

    (l) $2\left(\cos\left(\frac{\ln(x^2+1)}{\ln 3}\right)\right)\frac{x}{(x^2+1)\ln 3}$

    (m) $\frac{1}{2}\frac{3x^2\sin x+12x^2-(\cos x)x^3-7\cos x}{\sqrt{(x^3+7)}\left(\sqrt{(\sin x+4)}\right)^3}$

    (n) $3^{\tan(\sin(x))}\left(\cos x\right)\frac{\ln 3}{\cos^2(\sin x)}$

    (o) $6\left(\frac{x^2+2x}{\tan(x^2+1)}\right)^5 \cdot$
        $\left(\frac{(2x+2)\tan(x^2+1)-\sec^2(x^2+1)2x(x^2+2x)}{\tan^2(x^2+1)}\right)$

(2) In each of the following, assume the relation defines y as a function of x for values of x and y of interest and use the process of implicit differentiation to find $y'(x)$.

    (a) $y' = \frac{-y^2+3x^2}{2xy+\cos y}$

    (b) $y' = \frac{\cos y^2-4x^3}{y(-3y+2x\sin y^2)}$

    (c) $y' = \frac{-y\sin x+2\left(\tan y\sin x^2\right)x}{-\cos x+\cos x^2+\cos x^2\tan^2 y}$

(d) $y' = \dfrac{3x^2y - 12x\left(x^2+y^2\right)^5}{12(x^2+y^2)^5y - x^3}$

(e) $y' = y\dfrac{y^6-1}{3xy^6 - 2x^2y + 4y^5 - x + (\sin y)y^{10} + 2(\sin y)y^5x + (\sin y)x^2}$

(f) $y' = \dfrac{-(\sin y)x + 3\sqrt{(x^2+y^4)}}{2(\sin y)y^3 + (\cos y)x^2 + (\cos y)y^4}$

(g) $y' = -y^2\dfrac{y^2\cos x + 2xy - 2\times 2^{x^2}x\ln 2}{3y^3\sin x + 2x^2y^2 - 2^{x^2}y - 1}$

(h) $y' = -y\dfrac{1 + y^3(\cos xy)x\ln 3}{x(2y^2\sin xy\ln 3 + y^3(\cos xy)x\ln 3 + 1 - 2y^2\ln 3)}$

(i) $y' = -\left(\cos\left(\tan\left(xy^2\right)\right)\right)\cdot$

$\dfrac{y}{2(\cos(\tan(xy^2)))x + 3y - 3y\sin^2 xy^2}$

(j) $y' = 3\dfrac{\sin^2(\tan y) - 1}{(\sin(\sec(\tan(y))))\sin(\tan y))(1 + \tan^2 y)}$

(4) If $f(g(x)) = f(g(x'))$ then since f is one to one, $g(x) = g(x')$. Then since g is one to one, $x = x'$. Thus $f\circ g$ is one to one.

(5) Using Problem 4 show that the following functions are one to one and find the derivative of the inverse function at the indicated point.

(a) $\left(f^{-1}\right)'\left(e^2\right) = 1/f'(1) = \frac{1}{3e^2}$

(b) $\left(f^{-1}\right)'(0) = 1/f'(x) = 1/4$

(c) $\left(f^{-1}\right)'(f(0)) - 1/f'(0) = 1/2$

(d) $\left(f^{-1}\right)'(f(0)) = 1/f'(0) = -1/2.$

Exercises 5.4

(1) $\left(f^{-1}\right)'(2) = \frac{1}{3}.$

(2) $\left(f^{-1}\right)'(3) - \frac{1}{f'(0)} = \frac{1}{7}.$

(4) The problem is that you might be dividing by 0.

(5) No

(6) $\left(f^{-1}\right)(y) = (y-1)^{1/3}$ and so

$\left(f^{-1}\right)'(y) = \dfrac{1}{3\left(\sqrt[3]{(y-1)}\right)^2}$

(7) Find the derivatives of the following functions.

(a) $2x\dfrac{\cos x^2\ln\left(1+x^2\right) + \left(\cos x^2\ln\left(1+x^2\right)\right)x^2 + \sin x^2}{1+x^2}$

(b) $2\frac{x}{1+x^2}$

(c) $6\left(x^3+1\right)^5 3x^2\sin\left(x^2+7\right)$

$+\left(x^3+1\right)^6\cos\left(x^2+7\right)2x$

(d) $2\left(9\left(\sin\left(x^2+7\right)\right)x +\right.$

$\left.\left(\cos\left(x^2+7\right)\right)x^3 + \cos\left(x^2+7\right)\right)\cdot$

$\dfrac{x}{(\sin(x^2+7))(x^3+1)}$

(e) $2\left(\sin\left(\sin\left(1+x^2\right)\right)\cos\left(1+x^2\right)\right)\cdot$

$\dfrac{x}{\cos^2(\sec(\sin(1+x^2)))\cos^2(\sin(1+x^2))}$

(f) $4\left(1 - \cos^2\left(x^2+5\right)\right)^{\sqrt 7}\sqrt 7\left(\tan\left(x^2+5\right)\right)x$

(8) Use logarithmic differentiation to differentiate the following functions.

(a) $\sec^2(x)\left(2+\sin\left(x^2+6\right)\right)^{(\tan(x)-1)}$.
$$\left(2\ln\left(2+\sin\left(x^2+6\right)\right)\right.$$
$$+\ln\left(2+\sin\left(x^2+6\right)\right)\sin\left(x^2+6\right)$$
$$\left.+2\left(\sin x\cos\left(x^2+6\right)\right)x\cos x\right)$$

(b) $x^x(\ln x+1)$

(c) $2\left(x^x\right)^x x\ln x+\left(x^x\right)^x x$

(d) $\sec\left(x^2+4\right)\left(\sec^2\left(\left(x^4+4\right)\right)\right)^{\cos x}$.
$$\left(-\sin(x)\ln\frac{\sec^2\left(x^2-2x+2\right)}{\left(x^2+2x+2\right)}\cos\left(x^4+4\right)\right.$$
$$\left.+8\left(\cos(x)\sin\left(x^4+4\right)\right)x^3\right)$$

(e) $\left(\sin^2(x)\right)^{\tan(x)}\left(\ln\left(\sin^2(x)\right)\sec^2(x)+2\right)$

(9) 2500π

(10) $2\pi-\frac{2}{3}\sqrt{6}\pi$

(11) $20\sqrt{2}$

(12) $\theta=\pi/3$

(13) Width: $\frac{L}{6-\sqrt{3}}$,Height of rectangle: $\frac{1}{2}L\left(\frac{3-\sqrt{3}}{6-\sqrt{3}}\right)$

(14) $\frac{11}{75}$ radians per second.

Exercises 5.7

(2) Simplify the following.

(a) $\sin(\arctan(1))=\frac{1}{2}\sqrt{2}$

(b) $\cos\left(\arctan\sqrt{3}\right)=\frac{1}{2}$

(c) $\tan\left(\arcsin\left(\sqrt{3}/2\right)\right)=\sqrt{3}$

(d) $\sec\left(\arcsin\left(\sqrt{2}/2\right)\right)=\sqrt{2}$

(e) $\tan(\arcsin(1/2))=\frac{1}{3}\sqrt{3}$

(f) $\cos\left(\arctan\left(\sqrt{3}\right)-\arcsin\left(\sqrt{2}/2\right)\right)$
$$=\frac{1}{4}\sqrt{6}\left(1+\frac{1}{3}\sqrt{3}\right)$$
Also $\frac{1}{4}\sqrt{2}+\frac{1}{4}\sqrt{3}\sqrt{2}$

(g) $\sin(\arctan(x))=\frac{x}{\sqrt{(1+x^2)}}$

(h) $\cos(\arcsin(x))=\sqrt{(1-x^2)}$

(i) $\cos(2\arcsin(x))=1-2x^2$

(j) $\sec(2\arctan(x))=\frac{1+x^2}{1-x^2}$.

(k) $x\frac{-2\sqrt{(1-x^2)}-x+2x^3}{-1+2x^2+2x^3\sqrt{(1-x^2)}}$

(l) $\frac{1}{\sqrt{1-x^2}(2x^2-1)-2x^2\sqrt{1-x^2}}$

(3) Find the derivatives and give the domains of the following functions.

(a) $\dfrac{2x+3}{\sqrt{(1-x^4-6x^3-9x^2)}}$

(b) $\dfrac{3}{26+9x^2+30x}$

(c) $\dfrac{1}{(2x+1)\sqrt{x}\sqrt{(x+1)}}$

(d) $\dfrac{1}{\cosh x}$

(e) $3\dfrac{\cosh(\tan(3x))}{\cos^2 3x}$

(f) $3\left(\sinh\left(\csc\left(3x\right)\right)\right)\left(-\cot\left(3x\right)\csc\left(3x\right)\right)$

(g) $\dfrac{3\tan(3x)\sec(3x)}{\cosh^2(\sec(3x))}$

(4) $\dfrac{1}{a^2+b^2x^2}$

(5) 0

(6) Use the process of implicit differentiation to find y' in the following examples.

(a) $-y\left(2x + 2x^3y^4 + y\sqrt{(1 - x^4y^2)}\right) \cdot$

$\left(x^2 + x^4y^4 + 2y\sqrt{(1 - x^4y^2)}x\right.$

$- (\cosh y)\sqrt{(1 - x^4y^2)}$

$\left.- (\cosh y)\sqrt{(1 - x^4y^2)}x^2y^4\right)^{-1}$

(b) $-\dfrac{yx^4+y+2x}{(x^4+1)x}$

(c) $y' = \dfrac{1}{2}\dfrac{-y^4(\cosh x)\sqrt{(1-x^4\ \ 2x^2y^2-y^4)}+2x}{y\left(2y^2(\sinh x)\sqrt{(1-x^4-2x^2y^2-y^4)}-1\right)}$

(d) $-\dfrac{1}{2}\dfrac{-\tanh(x^2+y^2)-2x^2+2\left(\tanh^2\left(x^2+y^2\right)\right)x^2}{xy(-1+\tanh^2(x^2+y^2))}$

(7) Find the derivatives of the following functions.

(a) $2\dfrac{x}{\sqrt{(50+x^4+14x^2)}}$

(b) $\dfrac{1}{1-x^2}$

(c) $2\left(\cos\left(\operatorname{arcsinh}\left(x^2 + 2\right)\right)\right)\dfrac{x}{\sqrt{(5\mid x^4+4x^2)}}$

(d) $\dfrac{\operatorname{goo}(\tanh x)}{\cosh^2 x}$

(e) $2x\sinh\left(\sin\left(\cos x\right)\right) - x^2\cosh\left(\sin\left(\cos x\right)\right)\cos\left(\cos x\right)\sin x$

(f) $3\left(\cosh^{\left(\sqrt{6}-1\right)} x^3\right)\sqrt{6}\left(\sinh x^3\right)x^2$

(g) $\left(1 + x^4\right)^{\sin x-1}\left(\cos x\ln\left(1 + x^4\right) + \right.$
$\left.\left(\cos x\ln\left(1 + x^4\right)\right)x^4 + 4\left(\sin x\right)x^3\right)$

(8) $\dfrac{\pi}{2}$

(10) $\dfrac{1}{2}\left(\ln\left(1 + x\right) - \ln\left(1 - x\right)\right)$

(11) $\ln\left(x + \sqrt{(1 + x^2)}\right).$

(14) $y = \ln\left(x + \sqrt{(x^2 - 1)}\right).$

(16) $-65\sqrt{2},\ 650$

Exercises 6.2

(1) Find the limits.

(a) $\lim_{x \to 0} \frac{3x - 4 \sin 3x}{\tan 3x} = -3$

(b) $\lim_{x \to \frac{\pi}{2}-} (\tan x)^{x - (\pi/2)} = 1$

(c) $\lim_{x \to 1} \frac{\arctan(4x - 4)}{\arcsin(4x - 4)} = 1$

(d) $\lim_{x \to 0} \frac{\arctan 3x - 3x}{x^3} = -9$

(e) $\lim_{x \to 0+} \frac{9^{\sec x - 1} - 1}{3^{\sec x - 1} - 1} = 2$

(f) $\lim_{x \to 0} \frac{3x + \sin 4x}{\tan 2x} = \frac{7}{2}$

(g) $\lim_{x \to \pi/2} \frac{\ln(\sin x)}{x - (\pi/2)} = 0$

(h) $\lim_{x \to 0} \frac{\cosh 2x - 1}{x^2} = 2$

(i) $\lim_{x \to 0} \frac{-\arctan x + x}{x^3} = \frac{1}{3}$

(j) $\lim_{x \to 0} \frac{x^8 \sin \frac{1}{x}}{\sin 3x} = 0$

(k) $\lim_{x \to \infty} (1 + 5^x)^{\frac{2}{x}} = 25$

(l) $\lim_{x \to 0} \frac{-2x + 3 \sin x}{x} = 1$

(m) $\lim_{x \to 1} \frac{\ln(\cos(x - 1))}{(x - 1)^2} = -\frac{1}{2}$

(n) $\lim_{x \to 0+} \sin^{\frac{1}{x}} x = 0$

(o) $\lim_{x \to 0} (\csc 5x - \cot 5x) = 0$

(p) $\lim_{x \to 0+} \frac{3^{\sin x} - 1}{2^{\sin x} - 1} = \frac{\ln 3}{\ln 2}$

(q) $\lim_{x \to 0+} (4x)^{x^2} = 1$

(r) $\lim_{x \to \infty} \frac{x^{10}}{(1.01)^x} = 0$

(s) $\lim_{x \to 0} (\cos 4x)^{(1/x^2)} = e^{-8}$

(2) Find the following limits.

(a) $\lim_{x \to 0+} \frac{1 - \sqrt{\cos 2x}}{\sin^4(4\sqrt{x})} = \frac{1}{256}$

(b) $\lim_{x \to 0} \frac{2^{x^2} - 2^{5x}}{\sin\left(\frac{x^2}{5}\right) - \sin(3x)} = \frac{5}{3} \ln 2$

(c) $\lim_{n \to \infty} n \left(\sqrt[n]{7} - 1 \right) = \ln 7$

(d) $\lim_{x \to \infty} \left(\frac{3x + 2}{5x - 9} \right)^{x^2} = 0$

(e) $\lim_{x \to \infty} \left(\frac{3x + 2}{5x - 9} \right)^{1/x} = 1$

(f) $\lim_{n \to \infty} \left(\cos \frac{2x}{\sqrt{n}} \right)^n = e^{-2x^2}$

(g) $\lim_{n \to \infty} \left(\cos \frac{2x}{\sqrt{5n}} \right)^n = e^{-\frac{2}{5}x^2}$

(h) $\lim_{x \to 3} \frac{x^x - 27}{x - 3} = 27 \ln 3 + 27$

(i) $\lim_{n \to \infty} \cos \left(\pi \frac{\sqrt{4n^2 + 13n}}{n} \right) = 1$

(j) $\lim_{x \to \infty} \left(\sqrt[3]{x^3 + 7x^2} - \sqrt{x^2 - 11x} \right) = \frac{47}{6}$

(k) $\lim_{x \to \infty} \left(\sqrt[5]{x^5 + 7x^4} - \sqrt[3]{x^3 - 11x^2} \right) = \frac{76}{15}$

(1) $\lim_{x\to\infty} \left(\frac{5x^2+7}{2x^2-11}\right)^{\frac{x}{1-x}} = \frac{2}{5}$

(m) $\lim_{x\to\infty} \left(\frac{5x^2+7}{2x^2-11}\right)^{\frac{x\ln x}{1-x}} = 0$

(n) $\lim_{x\to 0+} \frac{\ln\left(e^{2x^2}+7\sqrt{x}\right)}{\sinh\left(\sqrt{x}\right)} = 7$

(o) $\lim_{x\to 0+} \frac{\sqrt[7]{x}-\sqrt[5]{x}}{\sqrt[9]{x}-\sqrt[11]{x}} = 0$

(3) Find the following limits.

 (a) $\lim_{x\to 0+} (1+3x)^{\cot 2x} = e^{\frac{3}{2}}$

 (b) $\lim_{x\to 0} \frac{\sin x - x}{x^2} = 0$

 (c) $\lim_{x\to 0} \frac{\sin x - x}{x^3} = -\frac{1}{6}$

 (d) $\lim_{x\to 0} \frac{\tan(\sin x)-\sin(\tan x)}{x^7} = \frac{1}{30}$

 (e) $\lim_{x\to 0} \frac{\tan(\sin 2x)-\sin(\tan 2x)}{x^7} = \frac{64}{15}$

 (f) $\lim_{x\to 0} \frac{\sin\left(x^2\right)-\sin^2(x)}{x^4} = \frac{1}{3}$

 (g) $\lim_{x\to 0} \frac{e^{-\left(1/x^2\right)}}{x} = 0$

 (h) $\lim_{x\to 0} \left(\frac{1}{x} - \cot(x)\right) = 0$

 (i) $\lim_{x\to 0} \frac{\cos(\sin x)-1}{x^2} = -\frac{1}{2}$

 (j) $\lim_{x\to\infty} \left(x^2 \left(4x^4+7\right)^{1/2} - 2x^4\right) = \frac{7}{4}$

 (k) $\lim_{x\to 0} \frac{\cos(x)-\cos(4x)}{\tan(x^2)} = \frac{15}{2}$

 (l) $\lim_{x\to 0} \frac{\arctan(3x)}{x} = 3$

 (m) $\lim_{x\to\infty} \left[\left(x^9+5x^6\right)^{1/3} - x^3\right] = \frac{5}{3}$

(4) $P = 1409.376\,18$.

(5) $5.308\,579\,63 \times 10^{-6}$ There is essentially no difference.

(6) Let $a > 1$. Find

$$\lim_{x\to\infty} \left(\frac{1}{x}\frac{a^x-1}{a-1}\right)^{1/x}.$$

This is a case of a problem which appeared on the 1956 Putnam exam. **Hint:** Consider the ln of the function and split it up. It isn't too bad if you do this. Take the ln of this.

$$\frac{1}{x}\ln\left(\frac{1}{x}\right) + \frac{1}{x}\left[\ln\left(a^x-1\right) - \ln\left(a-1\right)\right]$$

The first term converges to 0. The term $\frac{\ln(a-1)}{x}$ also converges to 0. This leaves only

$$\lim_{x\to\infty} \frac{\ln\left(a^x-1\right)}{x} = \lim_{x\to\infty} a^x \frac{\ln a}{a^x-1} = \ln a.$$

Thus the original limit equals a.

Exercises 6.4

(1) 6

(2) -1800

(3) $\frac{1}{16}$.

(4) $-\frac{4}{3}$

(5) -44

(6) $\frac{1}{5}\sqrt{2}\left(\sqrt[3]{15}\right)^2$

(7) -52.0

(8) 10

(9) $h = 27$.

(10) $y' = \frac{1}{6\pi}$.

(12) 60π

(13) $-\frac{5}{6}$

(14) $\frac{2}{5}\sqrt{21}$

(15) $5\sqrt{3}$

(16) $-\frac{1}{40}$

(17) $V' = \frac{-P'V + kT'}{P}$

(18) $y' = \frac{25}{9\pi}$.

(19) $\theta' = -\frac{16}{15}$

(20) $y' = -\frac{2}{3}$

(21) $-\left(\frac{1}{2}r\sqrt{2} + \frac{1}{2}\sqrt{4l^2 - 2r^2}\right)r\sqrt{2}\frac{\omega}{\sqrt{4l^2 - 2r^2}}$

Exercises 6.6

(1) Find the maximum and minimum values for the following functions defined on the given intervals.

   (a) The maximum value is 13 at 4. The minimum value is $-9.088\,662\,11$ at $1 + \frac{1}{3}\sqrt{6}$

   (b) Maximum is at 2 and equals $1.386\,294\,36$

   Minimum is at $1/2$ and equals $.559\,615\,788$

   (c) The maximum is 1030 at 10 and the minimum is -4 at -1.

   (d) The minimum is $-1/8$ at -1 and the maximum is $5/8$ at 1.

   (e) The maximum occurs at $-\frac{\sqrt{3}}{3}$ and is $.375\,466\,62$ while the minimum occurs at $\frac{\sqrt{3}}{3}$ and is $-.375\,466\,621$.

   (f) The maximum occurs at 0 and is 0. The minimum occurs at $1, -1$ and is $-.557\,407\,725$.

   (g) The maximum occurs at 2 and -2 and equals 9. The minimum occurs at 1 and -1 and equals 0.

   (h) The maximum is at 2 and equals $2.302\,585\,09$. The minimum is at 1 and -1 and equals 0.

   (i) The maximum is at 2 and equals 4. The minimum is at -2 and equals -12.

    (j) The maximum is at -2 and equals 16. The minimum is at 3/2 and equals 3.75.

    (k) The minimum is at -4 and equals -28. The maximum is at 3/2 and equals 2.25.

    (l) The minimum is 2 and it occurs at 1.

(2) $3\sqrt[3]{\pi}\left(\sqrt[3]{60}\right)^2$

(3) $4, 4$

(4) $\frac{\sqrt{2}}{2} = y$ and $x = \sqrt{2}$.

(5) $2r^2$

(6) The triangle with smallest area is the one which corresponds to $\left(\sqrt{2}, \frac{\sqrt{2}}{2}\right)$. The area of this triangle is then 2.

(7) The maximum point is $\left(e, \frac{1}{e}\right)$.

(8) Maximum occurs at 4.84.

(9) $\left(\frac{2}{\sqrt{3}}, \frac{2}{3}\sqrt{6}\right)$

(10) 45000 square feet.

(11) $\left(10 - \frac{10}{3}\sqrt{3}\right)\left(\frac{20}{3}\sqrt{3}\right)\left(20 + \frac{20}{3}\sqrt{3}\right) = 1539.6$

(12) $\theta = \pi/6$

(13) $r = \sqrt{3}\sqrt[6]{150}$, $h = \frac{90}{\left(\sqrt{3}\sqrt[6]{150}\right)^2} = \frac{1}{5}\left(\sqrt[3]{150}\right)^2$.

(14) The maximum area occurs when $x = 10$ so that all the wire goes to the circle. The minimum area occurs when $x = 10\frac{\pi}{\pi+4} = 4.399$. Thus part of it goes to the square and part to the circle. A little more than half goes to the square.

(15) He walks $\frac{5}{4}$ miles down the road before heading through the woods.

(16) He walks $10 - \frac{1}{15}\sqrt{15}$ miles down the road.

(17) $\frac{1}{6}\sqrt{6}$ down the shore towards the refinery.

(18) $\sqrt{\left(\left(\sqrt[3]{150}\right)^2 + 25\right)} + \sqrt{\left(6\sqrt[3]{150} + 36\right)} = 15.5348806$

(20) $4\frac{\pi+2}{4+3\pi}, 2\frac{4\pi+4+\pi^2}{4+3\pi}$

(22) $b = \frac{2}{\sqrt{3}}a$, This is an isocelese right triangle.

Exercises 6.8

(1) Here is the picture of two approximations. The case where $f'' < 0$ is similar.

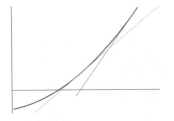

(2) Here is a picture which shows how the method diverges.

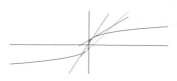

(3) .739 085 133
(4) 4. 493 409 53
(5) 1. 732 050 81
(6) This converges to $c^{1/p}$.

Exercises 6.9

(1) Not a multiple of π.
(2) $-3(x+4)^2 + 3(y+1)^2 = 1$ and so the curve is a hyperbola and its vertices are at
$$\left(-4, -1 + \tfrac{1}{3}\sqrt{3}\right), \left(-4, -1 - \tfrac{1}{3}\sqrt{3}\right).$$

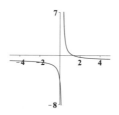

(3)
(4) $\frac{3}{5}$
(5) 3
(6) $(2k+1)\frac{\pi}{2}$.
(7) $(x-2)^2 - (y-1)^2 = 1$ and so the curve is a hyperbola and its vertices are at $(3,1), (3,1)$.
(9) 1
(10) $-2(x-2)^2 + y - 3 = 0$ and so the curve is a parabola and its vertex is at the point $(2,3)$.
(11) The function is negative for large negative x and positive for large positive x and so it follows it must equal zero somewhere by the intermediate value theorem.
(12) $f^{-1}(t) = -\frac{1}{3}\sqrt[3]{(9t - 27)}$
(13) $\left(-\frac{1}{2} + \frac{1}{2}i\sqrt{3}\right), \left(-\frac{1}{2} - \frac{1}{2}i\sqrt{3}\right), 1$
(14) $-\frac{117}{17} - \frac{26}{17}i$
(15) These are $\frac{-3x}{\sqrt{5}}, \frac{\sqrt{5+9x^2}}{\sqrt{5}}, \frac{\sqrt{5}}{\sqrt{5+9x^2}}, \frac{\sqrt{5+9x^2}}{-3x}, \frac{\sqrt{5}}{-3x}$

(16) $2(x-3)^2 + 4(y+2)^2 = 1$ and so the curve is an ellipse and its two axes have lengths $\frac{1}{2}\sqrt{2}$ and $\frac{1}{2}$. The center is $(3,-2)$.

(17) $\frac{5x^4+8}{\sqrt{(1-x^{10}-16x^6-64x^2)}}$

(18) The second derivative of this is $-25\sin 5x$. We need to find where this is negative. This happens when $5x \in ((2k+1)\pi, (2k+2)\pi)$ where k is an integer. Then the function is concave on the intervals $\left(\frac{(2k+1)}{5}\pi, \frac{(2k+2)}{5}\pi\right)$.

(19) Let f, g be continuous on $[a,b]$ and differentiable on (a,b). Then there exists $x \in (a,b)$ such that $f'(x)(g(b)-g(a)) = g'(x)(f(b)-f(a))$.

(20) $y' = -y^2 \frac{6x+5}{2+9yx^2+15yx}$

(21) The numbers are x and $10-x$. I want to maximize $x(10-x)$. Taking the derivative and setting equal to 0 yields the numbers are both 5.

(22) $y' = -2y \frac{8y\cos 4x + 2y(\cos 4x)x^2 + x}{8y\sin 4x + 2y(\sin 4x)x^2 + 4\ln(4+x^2) + (\ln(4+x^2))x^2}$

(23) $-\frac{36(\cos x^2)x^2 + 6(\cos x^2)x + 1 - 18\sin x^2}{36x^2+12x+1}$

(24) $x' = -\frac{3}{5}$

(25) The length is $2x + y$ where $y = 287/x$ and so the function to minimize is $2x + 287/x$. Then take the derivative and set equal to 0. $x = \frac{1}{2}\sqrt{2}\sqrt{287}$, And so the least possible length of fencing is $575\sqrt{2}\sqrt{287}$.

(26) -3

(27) $-5\left(\begin{array}{c}\cos\left(\cos\left(\ln(x+1)\left(x^4-x^3+x^2-x+1\right)\right)\right)\cdot\\ \sin\left(\ln(x+1)\left(x^4-x^3+x^2-x+1\right)\right)\end{array}\right)\frac{x^4}{1+x^5}$

(28) $6^{x(x+6)}(2x+6)\ln 6$

(29) $x > \frac{1}{2}$ increasing. Concave up on $(-\infty, -10) \cup (-3, \infty)$

(30) $\frac{10}{x^2+1}x$

(31) You note that $f(0) = 3$ and so the desired derivative is $\frac{1}{f'(0)} = \frac{1}{8}$.

(32) $\frac{8}{7}$

(33) $10\frac{x}{\sqrt{(50+25x^4+70x^2)}}$

(34) 3

(35) $\frac{x}{\sqrt{1-x^2}}$

(36) $\sqrt{34}$

(37) $e^{-\frac{256}{3}}$

(38) $(x^2+1)^{\sin 3x - 1}\left(3(\cos 3x \ln(x^2+1))x^2 + 3\cos 3x \ln(x^2+1) + 2(\sin 3x)x\right)$

(39) $\lim_{x\to 0}(\cos(3x))^{\frac{6}{x^2}} = e^{-27}$

(40) $x = \frac{-4\ln 8 + 5\ln 4}{5\ln 8 - \ln 4}$

(41) $\lim_{x\to 0}(\cos(9x^2))^{\frac{4}{x^4}} = e^{-162}$

(42) $-4x$

(43) $8(\cos(2x^4+3))\frac{x^3}{\sqrt{(2-\cos^2(2x^4+3))}}$

(44) $\lim_{x\to 0+}(1+4x)^{\cot(4x)} = e$

(46) $2\frac{\cos(\operatorname{arcsinh}2x)}{\sqrt{(1+4x^2)}}$

(47) $\frac{9}{2} \frac{\cos 9x}{(\sin 9x + 8) \ln 2}$

(48) $\lim_{x \to 0} \frac{6x + 9x^2 + 2 \tan 6x}{\sin 2x + \tan 2x^2} = 9$

(49) $\frac{2}{\sqrt{(-12 - 4x^2 - 14x)}}$

(50) $\frac{6}{\cosh 6x}$

Exercises 7.3

(1) Find the indicated antiderivatives.

   (a) $\frac{1}{6} \left(\sqrt{(2x - 3)} \right)^3 + \frac{3}{2} \sqrt{(2x - 3)} + C$

   (b) $\frac{(3x^2 + 6)^6}{36} + C$

   (c) $-\frac{1}{2} \cos x^2 + C$

   (d) $\frac{1}{8} \sin^4 2x + C$

   (e) $\frac{1}{2} \operatorname{arcsinh}(2x)$

(2) Solve the initial value problems.

   (a) $y = -\frac{2}{3} + \frac{1}{3} \sqrt{(2x - 3)}x + \sqrt{(2x - 3)}$

   (b) $y = 3 - 5 \times 6^4 + \frac{5(3x^2 + 6)^6}{36}$

   (c) $y = \frac{3}{2} - \cos^2 x^3 + \frac{1}{2} \cos 2$

   (d) $y = 1 + \frac{1}{3} \left(\ln \left(\sqrt{3}x + \sqrt{(3x^2 + 1)} \right) \right) \sqrt{3} - \frac{1}{3} \left(\ln \left(2 + \sqrt{3} \right) \right) \sqrt{3}$

   (e) $y = \ln |\sec x + \tan x| + 3$

   (f) $y = \frac{1}{2} \ln \left(\csc x^2 - \cot x^2 \right) + 1 - \frac{1}{2} \ln \left(\csc 1 - \cot 1 \right)$

(3) $y = \frac{3}{4} \ln \left(7 + t^4 \right) + \left(2 - \frac{3}{4} \ln (8) \right)$

(4) $y = -\frac{1}{4} \cos 2t^2 + \left(1 + \frac{1}{4} \cos 2 \right)$

(5) $y = \ln |\sec t + \tan t| + \left(-2 - \ln (\sec 1 + \tan 1) \right)$

(6) Find the indicated antiderivatives.

   (a) $\frac{1}{3} \ln |\sec 3x + \tan 3x| + C$

   (b) $\frac{1}{6} \sec^2 3x + C$

   (c) $\frac{1}{15} \sqrt{15} \arctan \left(\frac{1}{3} x \sqrt{15} \right) + C$

   (d) $\frac{1}{2} \arcsin \left(\frac{2}{5} \sqrt{5} x \right) + C$

   (e) $-\frac{3}{5} \sqrt{5} \arctan \frac{\sqrt{5}}{\sqrt{(-5 + 4x^2)}} + C$

(7) Find the indicated antiderivatives.

   (a) $\frac{1}{2} \sinh \left(x^2 + 1 \right) + C$

   (b) $\frac{1}{4 \ln 5} 5^{x^4} + C$

   (c) $-\frac{1}{\ln 7} 7^{\cos x} + C$

   (d) $-\frac{1}{2} \cos x^2 + C$

   (e) $\frac{1}{14} x^4 \left(\sqrt{(2x^2 + 1)} \right)^3 - \frac{1}{35} x^2 \left(\sqrt{(2x^2 + 1)} \right)^3$

   $+ \frac{1}{105} \left(\sqrt{(2x^2 + 1)} \right)^3 + C$

(8) $\frac{1}{2}x - \frac{1}{4}\sin 2x + C$

(9) Find the indicated antiderivatives.

    (a) $\int \frac{\ln x}{x}\,dx = \frac{1}{2}\ln^2 x + C$

    (b) $\int \frac{x^3}{3+x^4}\,dx = \frac{1}{4}\ln\left(3 + x^4\right) + C$

    (c) $\arctan(1 + x) + C$

    (d) $\arcsin \frac{1}{2}x + C$

    (e) $\frac{1}{3}\operatorname{arcsec}\left|\frac{x}{3}\right| + C$

    (f) $\frac{1}{4}\ln^2 x^2 + C$

    (g) $\frac{1}{18}x^2\sqrt{(6x^2 + 5)} - \frac{5}{54}\sqrt{(6x^2 + 5)} + C$

    (h) $\frac{1}{84}\left(\sqrt[3]{(6x + 4)}\right)^7 - \frac{1}{12}\left(\sqrt[3]{(6x + 4)}\right)^4 + C$

(10) Find the indicated antiderivatives.

    (a) $\frac{1}{10}\left(\sqrt{(2x + 4)}\right)^5 - \frac{2}{3}\left(\sqrt{(2x + 4)}\right)^3 + C$

    (b) $-\frac{4}{27}\left(\sqrt{(3x + 2)}\right)^3 + \frac{2}{45}\left(\sqrt{(3x + 2)}\right)^5 + C$

    (c) $\frac{1}{5}\arcsin \frac{5}{6}x + C$

    (d) $\frac{1}{2}\arcsin \frac{2}{3}x + C$

    (e) $\frac{1}{2}\operatorname{arcsinh}2x + C$

    (f) $\frac{2}{9}\sqrt{(3x - 1)} + \frac{2}{27}\left(\sqrt{(3x - 1)}\right)^3 + C$

    (g) $-\frac{2}{25}\sqrt{(5x + 1)} + \frac{2}{75}\left(\sqrt{(5x + 1)}\right)^3 + C$

    (h) $\frac{3}{4}\operatorname{arcsec}\left|\frac{3}{2}x\right| + C$

    (i) $\frac{1}{2}\operatorname{arcsinh}\frac{2}{3}x + C$

(11) $2\sqrt{x} - 3\sqrt[3]{x} + 6\sqrt[6]{x} - 6\ln\left(1 + \sqrt[6]{x}\right) + C$

(12) $f(x) = kx.$

(13) $f(x) = e^{kx}.$

(14) $f(x) = k\ln x$

(15) $f(x) = 7x + \frac{3x^2}{2}$

Exercises 7.5

(1) Find the following antiderivatives.

    (a) $-\frac{1}{3}x^3 e^{-3x} - \frac{1}{3}x^2 e^{-3x} - \frac{2}{9}xe^{-3x} - \frac{2}{27}e^{-3x} + C$

    (b) $x^4\sin x + 4x^3\cos x - 12x^2\sin x + 24\sin x - 24x\cos x + C$

    (c) $x^5 e^x - 5x^4 e^x + 20x^3 e^x - 60x^2 e^x + 120xe^x - 120e^x + C$

    (d) $-\frac{1}{2}x^6\cos 2x + \frac{3}{2}x^5\sin 2x +$
        $\frac{15}{4}x^4\cos 2x - \frac{15}{2}x^3\sin 2x -$
        $\frac{45}{4}x^2\cos 2x + \frac{45}{8}\cos 2x +$
        $\frac{45}{4}x\sin 2x + C$

    (e) $\frac{1}{2}\cos x^2 + \frac{1}{2}x^2\sin x^2 + C$

(2) Find the following antiderivatives.

(a) $-\frac{1}{3}xe^{-3x} - \frac{1}{9}e^{-3x} + C$

(b) $\int \frac{1}{x(\ln(|x|))^2}\, dx = -\frac{1}{\ln|x|} + C$

(c) $-\frac{4}{3}\left(\sqrt{(2-x)}\right)^3 + \frac{2}{5}\left(\sqrt{(2-x)}\right)^5 + C$

(d) $\left(\ln^2|x| - 2\ln|x| + 2\right)x + C$

(e) $\frac{1}{2}\cos x^2 + \frac{1}{2}x^2\sin x^2 + C$

(4) $\frac{e^x}{1+x} + C$

(6) Find the following antiderivatives.

(a) $\frac{1}{4}x^4\arctan x - \frac{1}{12}x^3 + \frac{1}{4}x - \frac{1}{4}\arctan x + C$

(b) $\frac{1}{4}x^4\ln x - \frac{1}{16}x^4 + C$

(c) $-x^2\cos x + 2\cos x + 2x\sin x + C$

(d) $x^2\sin x - 2\sin x + 2x\cos x + C$

(e) $\frac{1}{2}x^2\arcsin x + \frac{1}{4}x\sqrt{(1-x^2)} - \frac{1}{4}\arcsin x + C$

(f) $-\frac{1}{10}\cos 5x - \frac{1}{2}\cos x + C$

(g) $\frac{1}{2}x^2 e^{x^2} - \frac{1}{2}e^{x^2} + C$

(h) $\frac{1}{2}\cos x^2 + \frac{1}{2}x^2\sin x^2 + C$

(7) Find the antiderivatives

(a) $-x^2\cos x + 2\cos x + 2x\sin x + C$

(b) $-x^3\cos x + 3x^2\sin x - 6\sin x + 6x\cos x + C$

(c) $\frac{x^3}{\ln 7}7^x - \frac{6}{\ln^4 7}7^x - \frac{3}{\ln^2 7}x^2 7^x + \frac{6}{\ln^3 7}x 7^x + C$

(d) $\frac{1}{3}x^3\ln x - \frac{1}{9}x^3 + C$

(e) $x^2 e^x + 2x e^x + 2e^x + C$

(f) $\frac{x^3}{\ln 2}2^x - \frac{6}{\ln^4 2}2^x - \frac{3}{\ln^2 2}x^2 2^x + \frac{6}{\ln^3 2}x 2^x + C$

(g) $\frac{1}{6}\sec^3 2x + C$

(h) $\frac{x^2}{\ln 7}7^x - \frac{2}{\ln^2 7}x 7^x + \frac{2}{\ln^3 7}7^x + C$

(8) Solve the initial value problem $y'(x) = f(x)$, $\lim_{x\to 0+} y(x) = 1$ where $f(x)$ is each of the integrands in Problem 7.

(a) $y = -x^2\cos x + 2\cos x + 2x\sin x - 1$

(b) $y = -x^3\cos x + 3x^2\sin x - 6\sin x + 6x\cos x + 1$

(c) $y = \frac{x^3}{\ln 7}7^x - \frac{6}{\ln^4 7}7^x - \frac{3}{\ln^2 7}x^2 7^x + \frac{6}{\ln^3 7}x 7^x + \frac{6+\ln^4 7}{\ln^4 7}$

(d) $y = \frac{1}{3}x^3\ln x - \frac{1}{9}x^3 + 1$

(e) $y = x^2 e^x + 2x e^x + 2e^x - 1$

(f) $y = \frac{x^3}{\ln 2}2^x - \frac{6}{\ln^4 2}2^x - \frac{3}{\ln^2 2}x^2 2^x + \frac{6}{\ln^3 2}x 2^x + \frac{6+\ln^4 2}{\ln^4 2}$

(g) $y = \frac{1}{6}\sec^3 2x + \frac{5}{6}$

(h) $y = \frac{x^2}{\ln 7}7^x - \frac{2}{\ln^2 7}x 7^x + \frac{2}{\ln^3 7}7^x + \frac{-2+\ln^3 7}{\ln^3 7}$

(9) Solve the initial value problem $y'(x) = f(x)$, $\lim_{x\to 0+} y(x) = 2$ where $f(x)$ is each of the integrands in Problem 6.

(a) $y = \frac{1}{4}x^4 \arctan x - \frac{1}{12}x^3 + \frac{1}{4}x - \frac{1}{4}\arctan x + 2$

(b) $y = \frac{1}{4}x^4 \ln x - \frac{1}{16}x^4 + 2$

(c) $y = -x^2 \cos x + 2\cos x + 2x \sin x$

(d) $y = x^2 \sin x - 2\sin x + 2x \cos x + 2$

(e) $y = \frac{1}{2}x^2 \arcsin x + \frac{1}{4}x\sqrt{(1-x^2)} - \frac{1}{4}\arcsin x + 2$

(f) $y = -\frac{1}{10}\cos 5x - \frac{1}{2}\cos x + \frac{13}{5}$

(g) $y = \frac{1}{2}x^2 e^{x^2} - \frac{1}{2}e^{x^2} + \frac{5}{2}$

(h) $y = \frac{1}{2}\cos x^2 + \frac{1}{2}x^2 \sin x^2 + \frac{3}{2}$

(11) $y = \sin t - t\cos t + (2 - \sin 1 + \cos 1)$

(12) $y = 2 + \frac{1}{2}\tan(x)\sec(x) + \frac{1}{2}\ln|\sec x + \tan x|$

(13) Find the antiderivatives.

(a) $\int x \cos(x^2)\, dx = \frac{1}{2}\sin(x^2) + C$

(b) $\int \sin(\sqrt{x})\, dx = 2\sin\sqrt{x} - 2\sqrt{x}\cos\sqrt{x} + C$

(c) $\int \ln(|\sin(x)|)\cos(x)\, dx = (\ln|\sin x| - 1)\sin x + C$

(d) $\int \cos^4(x)\, dx = \frac{3}{8}x + \frac{1}{4}\sin 2x + \frac{1}{32}\sin 4x + C$

(e) $\int \arcsin(x)\, dx = x \arcsin x + \sqrt{(1-x^2)} + C$

(f) $\int \sec^3(x)\tan(x)\, dx = \frac{1}{3}\sec^3 x + C$

(g) $\int \tan^2(x)\sec(x)\, dx = \frac{1}{2}\sec(x)\tan^3(x) + \frac{1}{2}\sin x - \frac{1}{2}\ln|\sec x + \tan x| + C$

Exercises 7.7

(1) Find the antiderivatives.

(a) $\int \frac{x}{\sqrt{4-x^2}}\, dx = -\sqrt{(4-x^2)} + C$

(b) $\int \frac{3}{\sqrt{36-25x^2}}\, dx = \frac{3}{5}\arcsin\frac{5}{6}x + C$

(c) $\int \frac{3}{\sqrt{16-25x^2}}\, dx = \frac{3}{5}\arcsin\frac{5}{4}x + C$

(d) $\int \frac{1}{\sqrt{4-9x^2}}\, dx = \frac{1}{3}\arcsin\frac{3}{2}x + C$

(e) $\int \frac{1}{\sqrt{36-x^2}}\, dx = \arcsin\frac{1}{6}x + C$

(f) $\frac{1}{4}x\left(\sqrt{(9-16x^2)}\right)^3 + \frac{27}{8}x\sqrt{(9-16x^2)} + \frac{243}{32}\arcsin\frac{4}{3}x + C$

(g) $\frac{1}{6}x\left(\sqrt{(16-x^2)}\right)^5 + \frac{10}{3}x\left(\sqrt{(16-x^2)}\right)^3 + 80x\sqrt{(16-x^2)} + 1280\arcsin\frac{1}{4}x + C$

(h) $\frac{1}{2}x\sqrt{(25-36x^2)} + \frac{25}{12}\arcsin\frac{6}{5}x + C$

(i) $\frac{1}{4}x\left(\sqrt{(4-9x^2)}\right)^3 + \frac{3}{2}x\sqrt{(4-9x^2)} + 2\arcsin\frac{3}{2}x + C$

(j) $\frac{1}{2}x\sqrt{(1-9x^2)} + \frac{1}{6}\arcsin 3x + C$

(2) Find the antiderivatives.

(a) $\frac{1}{2}x\sqrt{(-25+36x^2)} - \frac{25}{12}\ln\left(6x + \sqrt{(-25+36x^2)}\right) + C$

(b) $\frac{1}{2}x\sqrt{(x^2-4)} - 2\ln\left(x+\sqrt{(x^2-4)}\right) + C$

(c) $\frac{1}{4}x\left(\sqrt{(16x^2-9)}\right)^3 - \frac{27}{8}x\sqrt{(16x^2-9)}$

     $+\frac{243}{32}\ln\left(4x+\sqrt{(16x^2-9)}\right) + C$

(d) $\frac{1}{2}x\sqrt{(25x^2-16)} - \frac{8}{5}\ln\left(5x+\sqrt{(25x^2-16)}\right) + C$

(3) Find the antiderivatives.

(a) $\frac{1}{5}\arctan\left(\frac{1}{5}x-\frac{1}{5}\right) + C$

(b) $\frac{9}{2}\ln\left(\frac{\sqrt{(x^2+9)}}{3}+\frac{x}{3}\right) + \frac{1}{2}x\sqrt{(x^2+9)} + C$

(c) $\int\sqrt{4x^2+25}\,dx = \frac{1}{2}x\sqrt{(4x^2+25)} + \frac{25}{4}\operatorname{arcsinh}\frac{2}{5}x + C$

(d) $\int x\sqrt{4x^4+9}\,dx = \frac{1}{4}x^2\sqrt{(4x^4+9)} + \frac{9}{8}\operatorname{arcsinh}\frac{2}{3}x^2 + C$

(e) $\int x^3\sqrt{4x^4+9}\,dx = \frac{1}{24}\left(\sqrt{(4x^4+9)}\right)^3 + C$

(f) $\frac{(x-3)}{32(241+25x^2-150x)} + \frac{1}{640}\arctan\left(\frac{5(x-3)}{4}\right) + C$

(g) $\frac{1}{30}\arctan\left(\frac{5}{6}x-\frac{5}{2}\right) + C$

(h) $\frac{1}{4}x\left(\sqrt{(25x^2+9)}\right)^3 + \frac{27}{8}x\sqrt{(25x^2+9)}$

     $+\frac{243}{40}\operatorname{arcsinh}\frac{5}{3}x + C$

(i) $\frac{1}{20}\arctan\frac{4}{5}x + C$

(4) Find the antiderivatives. **Hint: Complete the square.**

(a) $-\frac{1}{4}\ln\left(2x+4+\sqrt{(4x^2+16x+15)}\right) + C$

(b) $\frac{1}{4}(2x+6)\sqrt{(x^2+6x)}$

     $-\frac{9}{2}\ln\left(x+3+\sqrt{(x^2+6x)}\right) + C$

(c) $\arcsin\left(\frac{3}{2}x+3\right) + C$

(d) $3\arcsin\left(\frac{1}{2}x+\frac{3}{2}\right) + C$

(e) $\frac{1}{4}\arcsin\left(\frac{4}{5}+\frac{4}{5}x\right) + C$

(f) $\frac{1}{16}(8x+16)\sqrt{(4x^2+16x+7)}$

     $-\frac{9}{4}\ln\left(2x+4+\sqrt{(4x^2+16x+7)}\right) + C$

Exercises 7.10

(1) $b^2 - 4ac < 0$.

(2) Find the partial fractions expansion of the following rational functions.

(a) $\frac{2x+7}{(x+1)^2(x+2)} = \frac{5}{(x+1)^2} - \frac{3}{x+1} + \frac{3}{x+2}$

(b) $\frac{5x+1}{(x^2+1)(2x+3)} = -\frac{2}{2x+3} + \frac{x+1}{x^2+1}$

(c) $\frac{5x+1}{(x^2+1)^2(2x+3)} = -\frac{8}{13(2x+3)} + \frac{2}{13}\frac{-3+2x}{x^2+1} + \frac{x+1}{(x^2+1)^2}$

(d) $\frac{5x^4+10x^2+3+4x^3+6x}{(x+1)(x^2+1)^2} = \frac{2}{x+1} + \frac{1+3x}{x^2+1} + 2\frac{x}{(x^2+1)^2}$

(3) Find the antiderivatives

(a) $\int \frac{x^5+4x^4+5x^3+2x^2+2x+7}{(x+1)^2(x+2)}\,dx = \frac{1}{3}x^3 - \frac{5}{x+1} - 3\ln(x+1) + 3\ln(x+2) + C$

(b) $\int \frac{5x+1}{(x^2+1)(2x+3)}\,dx = -\ln(2x+3) + \frac{1}{2}\ln(x^2+1) + \arctan x + C$

(c) $\int \frac{5x+1}{(x^2+1)^2(2x+3)} = -\frac{4}{13}\ln(2x+3) + \frac{2}{13}\ln(x^2+1) + \frac{1}{26}\arctan x + \frac{1}{4}\frac{2x-2}{x^2+1} + C$

(4) Each of $\cot\theta, \tan\theta, \sec\theta$, and $\csc\theta$ is a rational function of $\cos\theta$ and $\sin\theta$. Use the technique of substituting $u = \tan\left(\frac{\theta}{2}\right)$ to find antiderivatives for each of these.

Recall $du = \frac{1}{2}(1+u^2)\,d\theta$ with this substitution and also

$$\cos\theta = \frac{1-u^2}{1+u^2}, \quad \sin\theta = \pm\frac{2u}{1+u^2}$$

Then

$\int \cot\theta\,d\theta = \ln|\sin\theta| + C$

$\int \tan\theta\,d\theta = \ln|\sec\theta| + C$

$\int \sec\theta = \ln(\sec\theta + \tan\theta) + C$

$\int \csc\theta = \ln(\csc\theta - \cot\theta) + C$

(5) $\frac{1-\sin\theta}{\cos\theta} + \theta + C$

(6) $-\frac{2}{3}\sqrt{3}\arctan\left(\frac{1}{3}\sqrt{3}\tan\left(\frac{\theta}{2}\right)\right) + \theta + C$

(7) $\ln|\sec x + \tan x| + C$

(8) $\ln|\cot x - \csc x| + C$

(9) Find the antiderivatives.

(a) $\frac{5}{6}\ln(6x+1) + 2\ln(x-1) + C$

(b) $\frac{5}{2}x^2 + \ln(5x+3) - \ln(x-2) + \frac{1}{2}\ln(2x-1) + C$

(c) $-\frac{2}{3}\ln(3x+1) + \ln(x-1) + \ln(2x-1) + C$

(d) $\frac{1}{10}\ln(5x+3) + \frac{1}{2}\ln(x+1) + C$

(10) Find the antiderivatives

(a) $-\frac{5}{x+1} + 3\ln(x+1) - \frac{6}{5}\sqrt{15}\arctan\left(\frac{1}{30}(10x+10)\sqrt{15}\right) + C$

(b) $\ln(3x+2) - \frac{1}{3}\sqrt{6}\arctan\frac{1}{12}(6x-12)\sqrt{6} + C$

(c) $\ln(3x+1) + \frac{2}{3}\sqrt{6}\arctan\frac{1}{12}(6x-6)\sqrt{6} + C$

(11) Solve the initial value problem, $y' = f(x), y(0) = 1$ for $f(x)$ equal to each of the integrands in Poblem 10.

(a) $y = -\frac{5}{x+1} + 3\ln(x+1)$
$\quad -\frac{6}{5}\sqrt{15}\arctan\frac{1}{30}(10x+10)\sqrt{15}$
$\quad +6 + \frac{6}{5}\sqrt{15}\arctan\frac{1}{3}\sqrt{15}$

(b) $y = \ln(3x+2) - \frac{1}{3}\sqrt{6}\arctan\frac{1}{12}(6x-12)\sqrt{6}$
$\quad +1 - \ln 2 - \frac{1}{3}\sqrt{6}\arctan\sqrt{6}$

(c) $y = \ln(3x+1) +$
$\quad \frac{2}{3}\sqrt{6}\arctan\frac{1}{12}(6x-6)\sqrt{6}$
$\quad +1 + \frac{2}{3}\sqrt{6}\arctan\frac{1}{2}\sqrt{6}$

(12) *Find the antiderivatives.

(a) $\frac{1}{12}\frac{6x+12}{3x^2+12x+13} + \frac{1}{6}\sqrt{3}\arctan\frac{1}{6}(6x+12)\sqrt{3} + C$

(b) $\frac{1}{40}\frac{10x+10}{5x^2+10x+7} + \frac{1}{40}\sqrt{10}\arctan\frac{1}{20}(10x+10)\sqrt{10} + C$

(c) $\frac{1}{60}\frac{-20+10x}{5x^2-20x+23} + \frac{1}{90}\sqrt{15}\arctan\frac{1}{30}(-20+10x)\sqrt{15} + C$

(13) Solve the initial value problem, $y' = f(x), y(0) = 1$ for $f(x)$ equal to each of the integrands in Problem 12.

(a) $y = \frac{12}{13} - \frac{1}{6}\sqrt{3}\arctan 2\sqrt{3} +$

$\frac{1}{12}\frac{6x+12}{3x^2+12x+13} + \frac{1}{6}\sqrt{3}\arctan\frac{1}{6}(6x+12)\sqrt{3}$

(b) $y = \frac{1}{40}\frac{10x+10}{5x^2+10x+7} +$

$\frac{1}{40}\sqrt{10}\arctan\frac{1}{20}(10x+10)\sqrt{10}$

$-\frac{1}{28} - \frac{1}{40}\sqrt{10}\arctan\frac{1}{2}\sqrt{10}$

(c) $y = \frac{1}{60}\frac{-20+10x}{5x^2-20x+23} +$

$\frac{1}{90}\sqrt{15}\arctan\frac{1}{30}(-20+10x)\sqrt{15}$

$+\frac{1}{69} + \frac{1}{90}\sqrt{15}\arctan\frac{2}{3}\sqrt{15}$

Exercises 8.3

(1) $\frac{125}{6}$

(2) $\frac{125}{3}$

(3) $\frac{243}{2}$

(4) $\frac{5}{2}$

(5) 4

(6) $-3 + \frac{5}{8}\pi^2$

(7) $\frac{9}{2}$

(8) $\frac{10}{3}\sqrt{5} - \frac{2}{3}$

(10) $k \approx .206$

(11) $\ln(2)$

(12) $\frac{1}{2}$

(13) $\sqrt{2} - 1$

(14) $e^{2\pi} - 1$

(15) $4\sqrt{2}$

(16) π

(17) 1

(18) $\frac{1}{12}\ln 5 + \frac{1}{12}\ln 7 + \frac{1}{12}\ln 11$

(19) $\frac{1}{2}\sin\pi^2 - \frac{1}{2}\pi^2\cos\pi^2 + 30\pi$

(20) $\frac{7}{2}\pi + 2$

(21) $\frac{8}{5} + \ln 5 + \ln 13 - 2\arctan 5 + \frac{1}{2}\pi$

(22) $2 + \frac{1}{2}\ln 7 + \frac{1}{2}\ln 5 + \arctan 2 - \frac{1}{2}\ln 3.$

(23) 2.35087127

(24) $.322188173$

(25) $.594388602$

(26) $4\pi - \frac{96}{25} - 8\arcsin\frac{3}{5}$

(28) $\frac{1}{2}\int_a^b (r(\theta))^2 \, d\theta$

(29) $\frac{1}{2}\int_0^{2\pi} (1 - \cos\theta)^2 \, d\theta = \frac{3}{2}\pi$

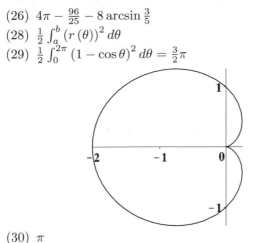

(30) π

Exercises 8.5

(1) $\frac{4}{3}\pi b a^2, 4\pi^2 a^2 b$

(2) $\frac{4}{3}\left(R^2 - r^2\right)\sqrt{(R^2 - r^2)}\pi$

(5) $\frac{1}{2}\pi^2$

(6) $\frac{4}{3}r^3\sqrt{3}$

(7) $\frac{1}{10}\pi$

(8) $\frac{1}{48}r^3\sqrt{3}$

(9) 64

(10) $\frac{16}{3}$

(11) $\frac{64}{3}$

(12) $\frac{32}{3}$

(13) 8

(14) 12π

(15) $\frac{2}{15}\pi$

(16) $9\pi\ln 3 - 4\pi$

(17) $\frac{35}{6}\pi$

(18) $5\pi\arctan 2 - 2\pi$

(19) $9\pi\arctan 2 - 2\pi - \pi\ln 5$

(20) $2\pi^2$

(21) $2\pi^2 + 4\pi$

(22) $20\pi - 2\pi^2$

(23) $-\frac{1}{2}\pi^2 + 8\pi$

(24) $4\pi^3 - 4\pi^2$

(25) $\frac{14}{9}\pi^2 + \frac{4}{3}\pi + \frac{2}{9}\pi^3$

(26) $\frac{23}{30}\pi$

(27) $4\pi + 2\pi^2$

Exercises 8.7

(1) $\ln\left(\sqrt{2}+1\right)$

(2) $\int_0^1 2\pi x \sec\left(x\right) dx$

(3) $2\sqrt{3}-\frac{4}{3}$

(4) $\frac{10}{27}\sqrt{10}\pi - \frac{1}{27}\pi$

(5) $\frac{1}{3}\pi\left(\frac{3}{2}\sqrt{10}+\frac{1}{2}\ln\left(\sqrt{10}+3\right)\right)$

(6) $2\pi\sqrt{5}+\pi\ln\left(2+\sqrt{5}\right)-\pi\sqrt{2}-\pi\ln\left(\sqrt{2}+1\right)$

(7) $\int_1^2 2\pi\ln\left(x\right)\sqrt{1+\frac{1}{x^2}}dx$

(8) $\sinh\left(1\right)$

(9) $6+\frac{1}{16}\ln 2$

(10) $\frac{451}{24}\pi$

(11) $3+\frac{1}{8}\ln 2$

(12) $\frac{115}{12}\pi$

(13) $\frac{1}{4}\pi\left(-1+e^4+4e^2\right)e^{-2}$

(14) $\pi\left(-5+3e^2+2e\right)e^{-1}$

(15) $\frac{1}{4}\frac{6a^2+\ln 2}{a}$

(16) $\frac{9}{16}\pi\sqrt{5}+\frac{1}{32}\pi\ln\left(-2+\sqrt{5}\right)$

(17) $\frac{9}{16}\pi\sqrt{5}-\frac{1}{64}\pi\ln\left(9+4\sqrt{5}\right)$

(18) $\frac{44}{15}\pi\sqrt{2}$

(19) $-\pi\sqrt{2}-\pi\ln\left(\sqrt{2}+1\right)$

(20) $\pi b\left(b^2-a^2\right)^{-1/2}\left(2\sqrt{\left(b^2-a^2+a^6\right)}\sqrt{\left(b^2-a^2\right)}\right.$

   $\left.+a^6\ln\left(\sqrt{\left(b-a\right)}\sqrt{\left(b+a\right)}+\sqrt{\left(b^2-a^2+a^6\right)}\right)\right)$

(21) $2\sqrt{3}-\frac{4}{3}\sqrt{2}$

(22) $\frac{16}{5}\pi\sqrt{2}-\frac{4}{5}\pi$

(23) $\ln\left(1+\sin 1\right)-\ln\left(\cos 1\right)$

(24) $2\pi^3+2\pi^2$

(25) Area: $\int_1^R 2\pi\frac{1}{x}\sqrt{1+\frac{1}{x^4}}dx = \int_1^R 2\pi\sqrt{\frac{x^4+1}{x^6}}dx \geq \int_1^R 2\pi\frac{1}{x}dx = 2\pi\ln R$

   Volume: $\int_1^R \pi\left(\frac{1}{x^2}\right)dx = \pi\frac{R-1}{R}$

   Note there is a finite limit as $R\to\infty$ for the volume but the limit does not exist for the area.

Exercises 8.9

(1) $440\,000\,000$ foot pounds.

(2) $400\,000\pi$.

(3) $\int_0^8 \left(50\right)40\sqrt{40y-y^2}\left(50-y\right)dy$

(4) $62.5\left(10\,000\pi\right)$

(5) 9 foot pounds.

(6) 1000

(7) 20 000

(8) $2.60416667 \times 10^{10}$ pounds.

(9) The bucket is not empty at the top. The work done is 4000 foot pounds.

(10) 2500 and the bucket weighs 20 pounds at the top.

(11) $\frac{55\,625}{4}\pi$

(12) 785398.163 pounds

(13) 785398.163 pounds

(14) 1444.89

(15) 3141592.65

(16) $62.5\left(\frac{1}{3}y^2\sqrt{(100-y^2)}+100y\arcsin\frac{1}{10}y\right.$
$\left.+\frac{200}{3}\sqrt{(100-y^2)}-\frac{2000}{3}\right)$

(17) 8.26 feet approximately.

(18) 5.256×10^8 foot pounds

Exercises 9.3

(1) $\frac{1}{\ln(x)}\left(\frac{1}{x}-\frac{(\ln(x))^{\ln(x)}}{x}\right)$

(3) $y(x) = \tan x$

(5) $T = \frac{\ln(1/2)}{-k^2} = \frac{\ln(2)}{k^2}$

(6) 67.5 years

(8) $T = 11.552\,453$

(9) $v(t) = -\frac{1}{k}\left(\frac{g}{m}-kv_0\right)e^{-kt}+\frac{g}{mk}$

(10) $A(t) = -\frac{1}{r}-\frac{1}{r^2+1}\cos t-\frac{r}{r^2+1}\sin t+\frac{r^3+2r+r^2+1}{r(r^2+1)}e^{rt}$ In case $r < 0$, the last term disappears as $t \to \infty$ and the behavior of the function is just like the first three terms on the right in the above.

(11) $t = 6.301\,338\,01$ years.

(12) $T = 69.314\,718\,1$ years.

(13) $k^2 = 1.732\,867\,95 \times 10^{-2}$

(14) 12 days

(15) 1648.721\,27

(16) $A = 969$

(17) $A = \frac{r}{k^2}+Ce^{-k^2t}$

(18) 9952.812\,88

(19) 61%.

(20) $M = A_2\frac{A_3A_2+A_1A_2-2A_1A_3}{A_2^2-A_1A_3}$

(21) $y = (2x-1)x$

(22) Solve the following initial value problems involving homogeneous differential equations.

(a) $y = x\tan\left(\ln(x)+\frac{\pi}{4}\right)$

(b) $y = x\arcsin(x\sin(1))$

(c) $\frac{1}{3}\left(\frac{y}{x}\right)^3+\frac{1}{2}\left(\frac{y}{x}\right)^2 = \ln x+\frac{5}{6}$

(23) There is no differentiable solution to this functional equation.

Exercises 9.5

(5) $9.983 \times 10^{-2} \leq \sin .1 \leq 9.9834 \times 10^{-2}$

$.995 \leq \cos(.1) \leq .9951$

(7) $y(t) = -.01 \cos(5t)$.

(8) $e^{-\sqrt{19}t} \left(.01 \cos\left(\frac{9}{2}t\right) + (.00968) \sin\left(\frac{9}{2}t\right) \right)$

(10) $y(t) = \left(\frac{5}{58}\sqrt{29} + \frac{1}{2} \right) \exp\left(\frac{1}{2}\left(-5 + \sqrt{29}\right)t \right)$
$+ \frac{1}{58}\left(-5 + \sqrt{29}\right)\sqrt{29} \exp\left(-\frac{1}{2}\left(5 + \sqrt{29}\right)t\right)$

(11) $y(t) = e^{-t}\sin t$

(15) Not necessarily.

Review Exercises 9.6

(1) $-\frac{1}{5}x^2 e^{-5x} - \frac{2}{25}xe^{-5x} - \frac{2}{125}e^{-5x} + C, \ (d = 2)$

(2) $\frac{1}{8}\cos x^8 + \frac{1}{8}x^8 \sin x^8 + C$

(3) $\frac{1}{2}x\sqrt{(4 - 9x^2)} + \frac{2}{3}\arcsin\frac{3}{2}x + C$

(4) $\frac{1}{3}\sqrt{3}\arcsin\left(\frac{1}{2}x\sqrt{3}\right) + C$

(5) $\arctan(1 + x)$

(6) $-\frac{8}{25}\sqrt{(4 + 5x)} + \frac{2}{75}\left(\sqrt{(4 + 5x)}\right)^3 + C$

(7) $-\frac{1}{2}x^2 \arctan x + \frac{1}{2}x - \frac{1}{2}\arctan x + C$

(8) $e^{\pi} - 1$

(9) $-\pi + \frac{1}{2}\pi^2$

(10) $-\frac{1}{3}\ln|\sec 3x - \tan 3x| + C$

(11) $C_1 e^{(5 + 2\sqrt{7})t} + C_2 e^{e(5 - 2\sqrt{7})t}$

(12) $\frac{1}{2}x\sqrt{(x^2 - 1)} - \frac{1}{2}\ln\left(x + \sqrt{(x - 1)}\sqrt{(1 + x)}\right) + C$

(13) $\frac{1}{6}\sec^2 3x + C$

(14) $\frac{1}{16}\frac{x}{4x^2 + 8} + \frac{1}{128}\sqrt{2}\arctan\frac{1}{2}x\sqrt{2} + C$

(15) $\frac{1}{10}\sqrt{10}\arctan\left(\frac{1}{5}x\sqrt{10}\right) + C$

(16) $\frac{3}{2}x\sqrt{(1 + x^2)} + \frac{3}{2}\ln 3 + \frac{3}{2}\ln\left(x + \sqrt{(1 + x^2)}\right) + C$

(17) $\int_{-3}^{3} \frac{\sqrt{3}}{4}\left(2\sqrt{(9 - x^2)}\right)^2 dx = 36\sqrt{3}$

(18) $62.5\left(\frac{65}{12}(7)^4 \pi\right)$

(19) $\frac{1}{7}\sqrt{7}\arctan\frac{1}{7}\sqrt{7}x + \frac{1}{7}\ln(7x + 4) + C$

(20) $\frac{1}{512}\pi\left(8208\sqrt{257} - \ln\left(16 + \sqrt{257}\right)\right)$

(21) $\frac{1}{25}\left(\sec(5x)\tan(5x) + \ln|\sec(5x) + \tan(5x)|\right) + C$

(22) $\frac{9}{2}$

(23) $(\ln|x| - 1)x + C$

(24) $\ln\left(\sqrt{2} + 1\right)$

(25) $\frac{1}{3}\left(\begin{array}{c} \frac{2}{9}\left(\sqrt{(x^3 + 3)}\right)^9 - \frac{18}{7}\left(\sqrt{(x^3 + 3)}\right)^7 + \\ \frac{54}{5}\left(\sqrt{(x^3 + 3)}\right)^5 - 18\left(\sqrt{(x^3 + 3)}\right)^3 \end{array} \right) + C$

(26) $\frac{1}{18}\left(2-3x^2\right)^{12}+C$

(27) $C_1 e^t \cos\left(\sqrt{7}t\right)+C_2 e^t \sin\left(\sqrt{7}t\right)$

(28) $\ln\left(x+2\right)+6\ln\left(x+5\right)+C$

(29) $3-\frac{1}{4}\cos^2\frac{1}{2}x^8+\frac{1}{4}\cos^2\frac{1}{2}$

(30) $\frac{1}{2}x^2+x+\ln\left(x+2\right)+\frac{1}{7}\ln\left(7x+2\right)+C$

(31) $x\arcsin 3x+\frac{1}{3}\sqrt{(1-9x^2)}+C$

(32) $-\frac{1}{26}\pi\sqrt{2}\sqrt{13}\left(\begin{array}{l}-130-\sqrt{2}\sqrt{13}\ln\left(5+\sqrt{2}\sqrt{13}\right)\\+2\sqrt{13}+\left(\ln\left(\sqrt{2}+1\right)\right)\sqrt{2}\sqrt{13}\end{array}\right)$

(33) $32\pi\sqrt{5}$

(34) $C_1 e^{-4t}+C_2 te^{-4t}$

(35) $x\arctan x-\frac{1}{2}\ln\left(1+x^2\right)$

(36) $\frac{1}{2}\sqrt{2}\ln\left(x\sqrt{2}+\sqrt{(2+2x^2)}\right)+C$

(37) $\frac{1}{3}\ln y-\frac{1}{3}\ln\left(3+2y\right)=t-\frac{1}{3}\ln 5.$

(38) $\pi\left(3\pi-1-5\ln 2\right)$

(39) $6\sqrt{2}$

(40) $\frac{1}{4}\ln 2+\frac{3}{2}$

Exercises 10.2

(3) $U\left(f,P\right)=\frac{1799}{216}$

$L\left(f,P\right)=\frac{907}{216}$

(6) $.63452381=\frac{4}{5}\frac{1}{4}+\frac{2}{3}\frac{1}{4}+\frac{4}{7}\frac{1}{4}$ | $\frac{1}{2}\frac{1}{4}\le\ln 2$

$\le 1\frac{1}{4}+\frac{4}{5}\frac{1}{4}+\frac{2}{3}\frac{1}{4}+\frac{4}{7}\frac{1}{4}=1.75952381$

(11) Using the above problem, compute the following definite integrals.

(a) $\int_0^1 \cos\left(x\right)dx=\sin 1$

(b) $\int_0^1 \sec^2\left(x\right)dx=\tan 1$

(c) $\int_0^2 \sqrt{2x^2+1}\,dx=3+\frac{1}{4}\sqrt{2}\ln\left(2\sqrt{2}+3\right)$

(d) $\int_{-2}^2 \sqrt{4-x^2}\,dx=2\pi$

(e) $\int_0^{\pi/2} x\sin\left(2x\right)dx=\frac{1}{4}\pi$

(f) $\int_1^2 \frac{x^2+1+x}{(x+1)(x^2+1)}dx=\frac{1}{2}\ln 3+\frac{1}{4}\ln 5+\frac{1}{2}\arctan 2-\frac{3}{4}\ln 2-\frac{1}{8}\pi$

(g) $\int_0^1 \frac{3x+2}{(x+1)(2x+1)}dx=\ln 2+\frac{1}{2}\ln 3$

(12) $\int_0^\pi 2\pi\left(x+1\right)\sin\left(x\right)dx=4\pi+2\pi^2$

(13) Recognize the limits.

(a) $\lim_{n\to\infty}\sum_{k=1}^n\left(\frac{k}{n}\right)^2\frac{1}{n}=\frac{1}{3}$

(b) $\lim_{n\to\infty}\sum_{k=1}^n\cos\left(\frac{k\pi}{2n}\right)\frac{1}{n}=\frac{2}{\pi}$

(c) $\lim_{n\to\infty}\sum_{k=1}^n\frac{1}{n+k}=\ln 2$

Exercises 10.7

(1) $3x^2\left(\frac{\left(x^3\right)^5+7}{\left(x^3\right)^7+87\left(x^3\right)^6+1}\right)-2x\left(\frac{\left(x^2\right)^5+7}{\left(x^2\right)^7+87\left(x^2\right)^6+1}\right)$

(2) Here is the picture.

(3) $.691\,219\,891$

(4) $.786\,700\,13$

(7) $5000 \times 2000 + \int_0^{921} 5\sqrt{1 + (.01\sinh(.01x))^2}\,.$
$(-\cosh.01x + 5000)\,dx$

(8) $F'(x) = \frac{1}{a^2 + a^2x^2}a + \frac{1}{a^2 + \left(\frac{a}{x}\right)^2}\left(\frac{-a}{x^2}\right) = 0$ and so F must be a constant.

(9) $y(x) = 5 + \int_{10}^x \frac{t^7 + 1}{t^6 + 97t^5 + 7}\,dt$

(13) Yes it is.

(15) Find the limits.

    (a) $\frac{2}{\pi}$

    (b) $\ln 2$

    (c) $\frac{4}{3}$

    (d) $\frac{1}{4}\pi$

Exercises 10.9

(1) $\ln(2) = \int_1^2 \frac{1}{t}\,dt$ and $\frac{1}{2} \cdot 1 \le \int_1^2 \frac{1}{t}\,dt \le \frac{1}{1} \cdot 1 = 1$

(2) $.695\,634\,921$

(5) $\log\left(x^{1/7}\right) = \frac{1}{7}\log(x)$. You look up $\log(x)$ in your table and then divide by 7. The result is the logarithm of the number you want so you go back to the table and approximate this number.

Exercises 10.11

(1) Find the integrals.

    (a) $\int_0^4 xe^{-3x}\,dx = -\frac{13}{9}e^{-12} + \frac{1}{9}$

    (b) $\int_2^3 \frac{1}{x(\ln(|x|))^2}\,dx = -\frac{\ln 2 - \ln 3}{\ln 3 \ln 2}$

    (c) $\int_0^1 x\sqrt{2 - x}\,dx = -\frac{14}{15} + \frac{16}{15}\sqrt{2}$

    (d) $3\ln^2 3 - 6\ln 3 + 2 - 2\ln^2 2 + 4\ln 2$

    (e) $\frac{1}{2}\cos\pi^2 + \frac{1}{2}\pi^2\sin\pi^2 - \frac{1}{2}$

(2) Find $\int_1^2 x\ln(x^2)\,dx = 4\ln 2 - \frac{3}{2}$

(3) $-\frac{1}{2}e\cos 1 + \frac{1}{2}e\sin 1 + \frac{1}{2}$

(4) $\frac{\ln 2 - 3\ln 2 \tan^2\frac{1}{2} + 4\tan\frac{1}{2}}{(1 + \ln^2 2)\left(1 + \tan^2\frac{1}{2}\right)}$

(5) $-4\sin 2 + 6\cos 2 + 6$

(6) Find the integrals.

    (a) $9 - \frac{8}{3}\sqrt{7}$

    (b) $\frac{1}{6}\left[\frac{54^6}{6} - \frac{18^6}{6}\right]$

(c) $-\frac{1}{2}\cos\pi^2 + \frac{1}{2}$

(d) $\frac{1}{8}$

(e) $\frac{1}{2}\sinh^{-1}(14)$ Also $-\frac{1}{2}\ln\left(-14 + \sqrt{197}\right)$

(7) Find the integrals.

(a) $\int_0^{\pi/9}\sec(3x)\,dx = \frac{1}{3}\ln\left(2 + \sqrt{3}\right)$

(b) $\int_0^{\pi/9}\sec^2(3x)\tan(3x)\,dx = \frac{1}{2}$

(c) $\int_0^5 \frac{1}{3+5x^2}\,dx = \frac{1}{15}\left(\arctan\frac{5}{3}\sqrt{15}\right)\sqrt{15}$

(d) $\int_0^1 \frac{1}{\sqrt{5-4x^2}}\,dx = \frac{1}{2}\arcsin\frac{2}{5}\sqrt{5}$

(e) $-\frac{3}{5}\left(\arctan\frac{1}{139}\sqrt{139}\sqrt{5}\right)\sqrt{5}$
$\quad +\frac{3}{5}\left(\arctan\frac{1}{11}\sqrt{11}\sqrt{5}\right)\sqrt{5}$

(8) Find the integrals.

(a) $\frac{1}{4}\left(e^{20} - 1 - e^{11} + e^9\right)e^{-10}$

(b) $\frac{1}{4}\left(\frac{5^{16}}{\ln(5)} - \frac{1}{\ln(5)}\right)$

(c) 0

(d) $-\frac{1}{2}\cos\left(\pi^2\right) + \frac{1}{2}$

(e) $\frac{981}{35} - \frac{11}{70}\sqrt{3}$

(9) $\frac{1}{8}\pi - \frac{1}{4}$

(10) $4\sqrt{2}$

(11) Find the following integrals.

(a) $-\frac{1}{3}\cos\left(\pi^3\right) + \frac{1}{3}\cos 1$

(b) $\frac{1}{2}\ln 37 - \frac{1}{2}\ln 2$

(c) $\frac{1}{8}\pi$

(d) $\frac{400}{21}\sqrt{5} - \frac{44}{105}\sqrt{2}$

Exercises 10.13

(4) $\frac{1}{p-1}$ if $p > 1$. The integral diverges if $p \leq 1$.

(5) $\frac{\ln^{(-p+1)}(2)}{p-1}$ if $p > 1$. The integral diverges if $p \leq 1$.

(6) It exists.

(7) Integral converges.

(8) The given integral exists.

(9) This integral exists by comparison to $\int_1^\infty \frac{1}{x^{5/2}}\,dx$

(10) This obviously exists because the function is continuous on $[0, 1]$ if it is defined as 0 at $t = 0$.

(11) This fails to exist.

(12) Does not exist.

(13) Does not exist.

(14) 2

(15) $\int_0^1 \frac{1}{\sqrt[3]{1-x}}\,dx = \frac{3}{2}$

(16) $\int_0^{\pi/2} \frac{\cos x}{\sqrt{1-\sin x}} dx = 2$

(17) $\frac{1}{12}\pi$

(18) 1

(19) $1, \frac{1}{2}\pi, \int_0^\infty 2\pi\sqrt{1+e^{-2x}}e^{-x}dx$ converges.

(22) $\Gamma\left(\frac{5}{2}\right) = \frac{3}{2}\Gamma\left(\frac{3}{2}\right) = \frac{3}{2}\frac{1}{2}\Gamma\left(\frac{1}{2}\right) = \frac{3}{4}\sqrt{\pi}.$

(24) $\int_0^1 t^\alpha \ln t \, dt = -\frac{1}{2\alpha+1+\alpha^2}.$

(25) $\frac{1}{\alpha}$

(28) Volume $= \int_1^\infty \pi \frac{1}{x^2} dx = \pi$

Exercises 11.2

(2) $\cos(x) = 1 - \frac{1}{2}x^2 + \frac{1}{24}x^4 - \frac{1}{720}x^6 + \frac{f^{(7)}(y)x^7}{7!},$

.877 582

(3) $-3 + 11(x-1) + 12(x-1)^2 + 6(x-1)^3 + (x-1)^4$

The remainder equals 0 after four terms so equality must hold.

(4) $-5 - 8(x+1) + 13(x+1)^2 - 10(x+1)^3 + 3(x+1)^4$

The remainder equals 0 after four terms so equality must hold.

(5) $\sum_{k=0}^\infty \frac{x^{2k}}{(2k)!} = \cosh x$

(6) $\sum_{k=0}^\infty \frac{x^{2k+1}}{(2k+1)!} = \sinh x$

(7) $\sum_{k=0}^\infty (-1)^k \frac{x^{k+1}}{k+1} = \ln(1+x)$

(8) $\sum_{k=0}^\infty \frac{-x^{k+1}}{k+1} = \ln(1-x)$

(9) $\ln(5) \approx \sum_{k=0}^8 \frac{2\left(\frac{2}{3}\right)^{2k+1}}{2k+1} = 1.609\,358\,19$

(11) $\sin(x) = \lim_{n\to\infty} \sum_{k=1}^n (-1)^{n-1} \frac{x^{2n-1}}{(2n-1)!}$

(14) The series converges on $[-1,1]$.

(16) Series converges on $(-1,1]$

(17) Series converges on $[-1,1)$

(18) $\sum_{k=0}^n \frac{x^k}{k!}$

(19) $\sum_{k=0}^n (-1)^n \frac{x^{2k+1}}{(2k+1)!}$

Exercises 11.4

(1) Determine whether the following series converge and give reasons for your answers.

(a) $\sum_{n=1}^\infty \frac{1}{\sqrt{n^2+n+1}}$ Diverges by comparison with $\sum 1/n$.

(b) $\sum_{n=1}^\infty \left(\sqrt{n+1} - \sqrt{n}\right)$ Diverges by comparison with $\sum 1/\sqrt{n}$

(c) $\sum_{n=1}^\infty \frac{(2n)!}{(n!)^2}$ $\lim_{n\to\infty} \frac{(2n)!}{(n!)^2} = \infty$ so this series diverges.

(d) $\sum_{n=1}^\infty \frac{1}{2n+2}$ Diverges by comparison with $\sum 1/n$

(e) $\sum_{n=1}^\infty \left(\frac{n}{n+1}\right)^n$ Diverges because $\lim_{n\to\infty} \left(\frac{n}{n+1}\right)^n = e^{-1} \neq 0.$

(2) Determine whether the following series converge and give reasons for your answers.

(a) $\sum_{n=1}^{\infty} \frac{\ln(k^5)}{k}$ This diverges.

(b) $\sum_{n=1}^{\infty} \frac{\ln(k^5)}{k^{1.01}}$ This converges.

(c) $\sum_{n=1}^{\infty} \sin\left(\frac{1}{n}\right)$ This diverges by comparision with $\sum 1/n$.

(d) $\sum_{n=1}^{\infty} \tan\left(\frac{1}{n^2}\right)$ This converges by comparison with $\sum 1/n^2$

(e) $\sum_{n=1}^{\infty} \cos\left(\frac{1}{n^2}\right)$ This diverges because $\lim_{n\to\infty} \cos\left(\frac{1}{n^2}\right) = 1 \neq 0$

(f) $\sum_{n=1}^{\infty} \sin\left(\frac{\sqrt{n}}{n^2+1}\right)$ This converges by comparison with $\sum 1/n^{3/2}$.

(3) Determine whether the following series converge and give reasons for your answers.

(a) $\sum_{n=1}^{\infty} \frac{2^n+n}{n2^n}$ Diverges.

(b) $\sum_{n=1}^{\infty} \frac{2^n+n}{n^2 2^n}$ Converges by comparison to $\sum 1/n^2$

(c) $\sum_{n=1}^{\infty} \frac{n}{2n+1}$ Diverges because the nth term fails to converge to 0.

(d) $\sum_{n=1}^{\infty} \frac{\ln n}{n^2}$ Converges

(7) Converges if $p > 1$, diverges if $p \leq 1$.

(8) For $p > 1$ this converges and if $p \leq 1$ it diverges.

(9) Converges.

(10) No

Exercises 11.7

(1) Determine whether the following series converge absolutely, conditionally, or not at all and give reasons for your answers.

(a) $\sum_{n=1}^{\infty} (-1)^n \frac{1}{\sqrt{n^2+n+1}}$ It fails to converge absolutely by a comparison with $\sum \frac{1}{n}$ but it converges by alternating series test.

(b) $\sum_{n=1}^{\infty} (-1)^n \left(\sqrt{n+1} - \sqrt{n}\right)$ Converges conditionally by alternating series test and comparison with $\sum 1/\sqrt{n}$

(c) $\sum_{n=1}^{\infty} (-1)^n \frac{(n!)^2}{(2n)!}$ Converges absolutely by the ratio test.

(d) $\sum_{n=1}^{\infty} (-1)^n \frac{(2n)!}{(n!)^2}$ Diverges.

(e) $\sum_{n=1}^{\infty} \frac{(-1)^n}{2n+2}$ Converges conditionally.

(f) $\sum_{n=1}^{\infty} (-1)^n \left(\frac{n}{n+1}\right)^n$ Diverges because $\lim_{n\to\infty} \left(\frac{n}{n+1}\right)^n = e^{-1} \neq 0$.

(g) $\sum_{n=1}^{\infty} (-1)^n \left(\frac{n}{n+1}\right)^{n^2}$ Converges absolutely by the root test.

(2) Determine convergence.

(a) $\sum_{n=1}^{\infty} (-1)^n \frac{\ln(k^5)}{k}$ Converges conditionally.

(b) $\sum_{n=1}^{\infty} (-1)^n \frac{\ln(k^5)}{k^{1.01}}$ Converges absolutely

(c) $\sum_{n=1}^{\infty} (-1)^n \frac{10^n}{(1.01)^n}$ Diverges

(d) $\sum_{n=1}^{\infty} (-1)^n \sin\left(\frac{1}{n}\right)$ Converges conditionally by alternating series test and a comparison test with $\sum 1/n$.

(e) $\sum_{n=1}^{\infty} (-1)^n \tan\left(\frac{1}{n^2}\right)$ Converges absolutely by comparison with $\sum 1/n^2$.

(f) $\sum_{n=1}^{\infty} (-1)^n \cos\left(\frac{1}{n^2}\right)$ Diverges because the nth term fails to converge to 0.

(g) $\sum_{n=1}^{\infty} (-1)^n \sin\left(\frac{\sqrt{n}}{n^2+1}\right)$ Converges absolutely by comparison with $\sum \frac{1}{n^{3/2}}$

(3) Determine convergence.

(a) $\sum_{n=1}^{\infty} (-1)^n \frac{2^n+n}{n2^n}$ converges by alternating series test.

(b) $\sum_{n=1}^{\infty} (-1)^n \frac{2^n+n}{n^2 2^n}$ Converges absolutely by comparison test.

(c) $\sum_{n=1}^{\infty} (-1)^n \frac{n}{2n+1}$ Diverges since n^{th} term fails to converge to 0.

(d) $\sum_{n=1}^{\infty} (-1)^n \frac{10^n}{n!}$ Converges absolutely by ratio test.

(e) $\sum_{n=1}^{\infty} (-1)^n \frac{n^{100}}{1.01^n}$ Converges absolutely.

(f) $\sum_{n=1}^{\infty} (-1)^n \frac{\ln n}{n^2}$ Converges absolutely.

(g) $\sum_{n=1}^{\infty} (-1)^n \frac{3^n}{n^3}$ Diverges.

(h) $\sum_{n=1}^{\infty} (-1)^n \frac{n^3}{3^n}$ Converges absolutely by root test.

(i) $\sum_{n=1}^{\infty} (-1)^n \frac{n^3}{n!}$ Converges absolutely by ratio test.

(j) $\sum_{n=1}^{\infty} (-1)^n \frac{n!}{n^{100}}$ Diverges.

(4) Find the exact values of the following infinite series if they converge.

(a) $\sum_{k=3}^{\infty} \frac{1}{k(k-2)} = \frac{3}{4}$

(b) $\sum_{k=1}^{\infty} \frac{1}{k(k+1)} = 1$

(c) $\sum_{k=3}^{\infty} \frac{1}{(k+1)(k-2)} = \frac{11}{18}$

(d) $\sum_{k=1}^{\infty} \left(\frac{1}{\sqrt{k}} - \frac{1}{\sqrt{k+1}}\right) = 1$

(e) $\sum_{n=1}^{\infty} \ln\left(\frac{(n+1)^2}{n(n+2)}\right)$ diverges.

(5) Yes.

(6) The ratio test gives either absolute convergence or spectacular divergence or fails entirely.

(7) It says nothing about divergence.

(8) This is not possible.

(9) No. Consider $\sum (-1)^n \frac{1}{n^{1/3}}$ and $\sum (-1)^n \frac{1}{n^{1/3}}$

(10) Yes for the first case. No in the second case.

(13) How large should n be?

(a) $n > 999$

(b) $n > 31$.

(c) $n > 999$.

(d) $n > 10^6 - 1$

(e) $n > 10^6 - 1$.

Exercises 11.9

(1) Radius of convergence?

(a) 2

(b) 1/3

(c) 0

(d) ∞

(e) ∞

(2) Find $\sum_{k=1}^{\infty} k2^{-k} = 2$

(3) Find $\sum_{k=1}^{\infty} k^2 3^{-k} = \frac{3}{2}$

(4) $\ln 2$

(5) $-\ln \frac{2}{3}$

(6) It diverges for all x.

(7) $\sum_{k=0}^{\infty} (-1)^k x^{2k}, |x| < 1$

(8) $\sum_{k=0}^{\infty} (-1)^k \frac{x^k}{(2k+1)!}$. This series converges for all values of x by the ratio test. However, the function only makes sense for $x \geq 0$.

(9) $\sum_{k=0}^{\infty} \frac{x^k}{k!}$

(10) $\sum_{k=0}^{\infty} \frac{(-t)^k}{k!}$

(11) $x - \frac{1}{12}x^4 + \frac{1}{504}x^7 + \cdots$

(19) $\sum_{k=0}^{\infty} (-1)^k \frac{(x^2)^{2k+1}}{(2k+1)!}$

(20) $x^2 - \frac{1}{3}x^4 + \frac{2}{45}x^6 - \frac{1}{315}x^8 + \frac{2}{14\,175}x^{10}$

(21) $\sum_{k=0}^{\infty} \left(\frac{-\frac{1}{2}}{k!} \right) (-1)^k x^{2k}$

(22) $\int_0^1 \left(1 + x^2 + \frac{x^4}{2!} \right) dx = \frac{43}{30} = 1.433\,333\,33$

(23) $.313\,311\,688$

(24) $\frac{1}{30}$.

Note what happens if you plug in small values of x. Say $x = .000001$

$$\frac{\tan(\sin(.000001)) - \sin(\tan(.000001))}{(.000001)^7} = 0$$

Exercises 12.7

(2) Compute the following

(a) $(17, 16, 3, 32)$

(b) $(-7, 4, 2)$

(c) $(-1, 0, -11, 7)$

(d) $(-7, 8, 13, -8)$

(e) $(2, 6, -1, 16)$

(3) $\frac{x-2}{-4} = y - 2 = \frac{z-4}{-3}$

(4) $\frac{x-1}{-3} = \frac{y-2}{-1} = \frac{z-4}{-3}$

(5) Symmetric equations for a line are given. Find parametric equations of the line.

(a) $x = 3t - 1, y = \frac{2t-3}{2}, z = t - 7$

(b) $x = \frac{3t+1}{2}, y = \frac{6t-3}{2}, z = t + 7$

(c) $x = 3t - 1, y = \frac{t-3}{2}, z = \frac{t+1}{2}$

(d) $x = -\frac{3}{2}t + \frac{1}{2}, y = \frac{3}{2} - t, z = t - 1$

(e) $x = 3t + 1, y = \frac{5t+3}{2}, z = t - 2$

(f) $x = 3t - 1, y = -5t + 3, z = t - 1$

(6) Parametric equations for a line are given. Find symmetric equations for the line if possible. If it is not possible to do it explain why.

(a) $\frac{x-1}{2} = \frac{3-y}{1} = \frac{z-5}{3}$

(b) $x - 1 = 3 - y = \frac{z-5}{-3}$

(c) $\frac{x-1}{2} = y - 3 = \frac{z-5}{3}$

(d) This line does not have symmetric equations because y does not depend on t.

(e) $1 - x = \frac{y-3}{2} = \frac{z-5}{-3}$

(f) $x = 3 - y = z - 1$

(7) The first point given is a point contained in the line. The second point given is a direction vector for the line. Find parametric equations for the line determined by this information.

(a) $(1, 2, 1) + t(2, 0, 3)$

(b) $(1, 0, 1) + t(1, 1, 3)$

(c) $(1, 2, 0) + t(1, 1, 0)$

(d) $(1, 0, -6) + t(-2, -1, 3)$

(e) $(-1, -2, -1) + t(2, 1, -1)$

(f) $(0, 0, 0) + t(2, -3, 1)$

(8) Determine a direction vector for this line.

(a) $(2, -1, 3)$

(b) $(1, 3, -1)$

(c) $(1, 4, -3)$

(d) $(2, -3, 3)$

(e) $(2, 2, 1)$

(f) $(1, 3, 1)$

(9) Identify the direction vector.

(a) $(0, 1, 0) + t(2, 0, 2)$

(b) $(0, 1, 1) + t(2, 4, -1)$

(c) $(1, 1, 0) + t(-1, 0, 2)$

(d) $(0, 1, 3) + t(0, 2, -3)$

(e) $(0, 1, 0) + t(0, 5, 2)$

(f) $(0, 1, 2) + t(2, -1, 0)$

(10) Here is the picture.

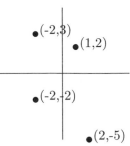

(11) No. They are of different sizes so you can't add them.

(14) $(x - 1)^2 + (y - 2)^2 + (z - 3)^2 = 16$

(15) $x^2 - 2xx_0 + x_0^2 + y_0^2 - l^2 = 2(y_0 - l)y$ The other case is similar.

(16) $(x - x_0)^2 + (y - y_0)^2 + (z - z_0)^2 = r^2$.

(17) It is a square centered at $(0,0)$ with sides parallel to the axes.

(18) It is a diamond shaped thing centered at $(0,0)$. The sides are lines from $(1,0)$ to $(0,1)$ to $(-1,0)$ to $(0,-1)$ to $(1,0)$.

(20) $0 = 2y(y_1 - y_2) + 2x(x_1 - x_2) + 2z(z_1 - z_2) + x_2^2 + y_2^2 + z_2^2 - x_1^2 - y_1^2 - z_1^2$

(21) $\frac{x^2}{\frac{1}{4}r^2} + \frac{y^2}{\frac{1}{4}r^2 - a^2} = 1$

(22) $\frac{x^2}{\frac{1}{4}r^2} - \frac{y^2}{r^2a^2 - \frac{1}{4}r^4} = 1$

(23) $0 = -2xx_2 + x_2^2 - 2yy_2 + y_2^2 + 2xx_1 - x_1^2 + 2yy_1 - y_1^2$

Exercises 12.0

(1) $(400, 40)$

(2) $(800, 80)$

(3) $\left(\frac{3}{10}\sqrt{11}, -\frac{1}{10}\right)$.

(4) $\left(150\sqrt{2} + 50, 150\sqrt{2}\right)$

(5) 9.6 degrees, 295.8

(6) The plane will not make it.

(7) $\left(155, 150 + 75\sqrt{3}\right)$

(8) Two hours.

(9) 1/3 miles down stream.

(10) $(\cos(.84), \sin(.84))$ In the second part, he could not do it.

(11) $(-3, 2, -5)$.

(12) $(-2, 2, -1)$

(13) $(3, 0, 0) = 3\mathbf{i} + 0\mathbf{j} + 0\mathbf{k}$

(14) $2\mathbf{i} + 2\mathbf{j} + 2\mathbf{k}$

(15) $\left(\frac{5}{2}\sqrt{3} + 5\sqrt{2}, \frac{5}{2} - 5\sqrt{2}\right)$

(16) $T = 250$ pounds.

Exercises 13.3

(1) 17

(4) $\cos\left(\theta\right) = \frac{(3,-1,-1)\cdot(1,4,2)}{\sqrt{11}\sqrt{1+16+4}}$

$= -\frac{1}{77}\sqrt{11}\sqrt{21} = -.197\,385\,508$

$\theta = 1.769\,486\,57$ radians.

(5) $\cos\theta = \frac{(1,-2,1)\cdot(1,2,-7)}{\sqrt{1+4+1}\sqrt{1+4+49}} = -.555\,555\,556$

and so $\theta = .981\,765\,356$.

(6) $\left(-\frac{5}{14}, -\frac{5}{7}, -\frac{15}{14}\right)$

(7) $\left(-\frac{1}{2}, 0, -\frac{3}{2}\right)$

(8) $\left(-\frac{1}{14}, -\frac{1}{7}, -\frac{3}{14}, 0\right)$

(9) No, it makes no sense.

(11) $40\cos\left(40°\right)100 = 3064.177\,77$

(12) $40\cos\left(30°\right)200 = 6928.203\,23$

(13) $20\cos\left(45°\right)300 = 4242.640\,69$

(14) $2000\sqrt{3}$

(15) 30

(16) $50\sqrt{2}$

(17) $-10\sqrt{2}$

(23) $\int_0^1 f\left(x\right)g\left(x\right)dx \le \left(\int_0^1 f\left(x\right)^2 dx\right)^{1/2}\left(\int_0^1 g\left(x\right)^2 dx\right)^{1/2}$

Exercises 13.6

(3) $\frac{1}{10}\left(20, -14, 23\right)$

(4) $\frac{1}{6}\left(11, -7, 8\right)$

(5) $\frac{(1,-2,1)}{\sqrt{6}}80\pi$

(6) $80\pi\left|\mathbf{u}_1 - \mathbf{u}_2\right|$.

(7) 9.

(8) $\frac{1}{2}\sqrt{293}$.

(9) 0 The three points are on a line.

(10) $8\sqrt{3}$

(11) $\sqrt{161}$.

(12) $\sqrt{90}$.

(13) 113

(14) 70

(15) 173

(16) Yes. It involves multiplying and adding integers.

(17) It means one of the vectors is a linear combination of the other two.

(18) $-22x + 33 - 4y - z = 0$

(21) $(\mathbf{i} \times \mathbf{j}) \times \mathbf{j} = \mathbf{k} \times \mathbf{j} = -\mathbf{i}$. However, $\mathbf{i} \times (\mathbf{j} \times \mathbf{j}) = \mathbf{0}$. The cross product is not associative and so $\mathbf{a} \times \mathbf{b} \times \mathbf{c}$ equals meaningless gobledeegook.

(23) $\mathbf{u} \times (\mathbf{v} \times \mathbf{w}) = (\mathbf{u} \cdot \mathbf{w})\mathbf{v} - (\mathbf{u} \cdot \mathbf{v})\mathbf{w}$

(24) $(\mathbf{u} \cdot \mathbf{z})(\mathbf{v} \cdot \mathbf{w}) - (\mathbf{u} \cdot \mathbf{w})(\mathbf{v} \cdot \mathbf{z})$

(25) $[\mathbf{u}, \mathbf{v}, \mathbf{w}]\mathbf{z} - [\mathbf{u}, \mathbf{v}, \mathbf{z}]\mathbf{w}$

(26) $(\mathbf{u} \times \mathbf{v}) \cdot (\mathbf{v} \times \mathbf{w}) \times (\mathbf{w} \times \mathbf{z}) = [\mathbf{v}, \mathbf{w}, \mathbf{z}][\mathbf{u}, \mathbf{v}, \mathbf{w}]$

(27) 0

Exercises 13.9

(1) $(1, 1, 2)$

(2) No intersection.

(3) $\left(\frac{2}{3} + \frac{1}{6}\sqrt{46}, -\frac{2}{3} + \frac{1}{3}\sqrt{46}, \frac{2}{3} + \frac{1}{6}\sqrt{46}\right)$
    $\left(\frac{2}{3} - \frac{1}{6}\sqrt{46}, -\frac{2}{3} - \frac{1}{3}\sqrt{46}, \frac{2}{3} - \frac{1}{6}\sqrt{46}\right)$

(4) $(2, 3, 4, 5) + t\,(-4, 0, -4, -4) = (x, y, z)$.

(5) $(1, 2, 3, 0) + t\,(2, 1, 3, 1) = (x_1, x_2, x_3, x_4)$

(6) This is a circle of radius 2 centered at $(0, 0)$.

(7) This is a helix of radius 2.

(8) There is no such plane.

(9) $-x + 1 + z = 0$

(10) $-8x + 35 + 3y - 7z = 0$

(11) $x - 6 + y + z = 0$

(12) $-4x + 5y - 19 + 3z = 0$

(13) $5x - 21 - y + 6z = 0$

(14) $5x + 6 + 2y - 6z = 0$

(15) $x + 2 + 2y - 4z = 0$.

(16) $2x + 9 + y - 6z = 0$.

(17) $x + 8 + 2y - 3z = 0$.

(18) Intercepts: $4/3, -2, 4$

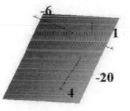

(19) Intercepts: $2, -1, 2$.

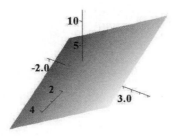

(20) Intercepts: $3, 3, 3$

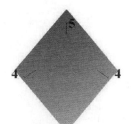

(21) Hyperbolic paraboloid oriented in the y direction rather than the z direction

(22) Hyperboloid of one sheet oriented in the x direction.

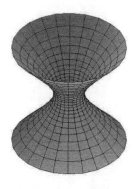

(24)

(25) Here is the picture

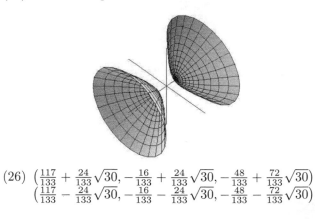

(26) $\left(\frac{117}{133} + \frac{24}{133} \sqrt{30}, -\frac{16}{133} + \frac{24}{133} \sqrt{30}, -\frac{48}{133} + \frac{72}{133} \sqrt{30} \right)$
$\left(\frac{117}{133} - \frac{24}{133} \sqrt{30}, -\frac{16}{133} - \frac{24}{133} \sqrt{30}, -\frac{48}{133} - \frac{72}{133} \sqrt{30} \right)$

Exercises 14.3

(1) The following are the polar coordinates of points. Find the rectangular coordinates.

(a) $x = \frac{5}{2}\sqrt{3}, y = \frac{5}{2}$

(b) $x = \frac{3}{2}, y = \frac{3}{2}\sqrt{3}$

(c) $x = -2, y = 2\sqrt{3}$

(d) $x = -\sqrt{2}, y = \sqrt{2}$

(e) $x = -\frac{3}{2}\sqrt{3}, y = -\frac{3}{2}$

(f) $x = 4\sqrt{3}, y = -4$

(2) The following are the rectangular coordinates of points. Find the polar coordinates of these points.

(a) $r = 5, \theta = \frac{\pi}{4}$

(b) $r = 3, \theta = \pi/3$

(c) $r = 5, \theta = 3\pi/4$

(d) $r = 5, \theta = \frac{2}{3}\pi$

(e) $r = 2, \theta = \frac{4}{3}\pi$

(f) $r = 3, \theta = 5\pi/3$

(4) Suppose $r = \frac{a}{1+e\sin\theta}$ where $e \geq 0$. By changing to rectangular coordinates, show this is either a parabola, an ellipse or a hyperbola. Determine the values of e which correspond to the various cases.

$r + er\sin\theta = a$ and so $\sqrt{x^2 + y^2} = a - ey$. Now square both sides to get

$$x^2 + y^2 = a^2 - 2aey + e^2y^2$$

Hence

$$x^2 + \left(1 - e^2\right)y^2 + ey - a^2$$

If $e = 1$, it is obviously a parabola. If $e = 0$ it is obviously a circle. For e between 0 and 1 the curve is an ellipse. If $e > 1$, it is a hyperbola.

(5) $k = 12$

(6) No repeats if α is irrational. Eventual repeats if α is rational.

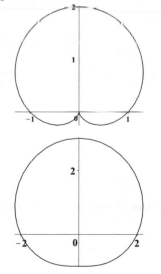

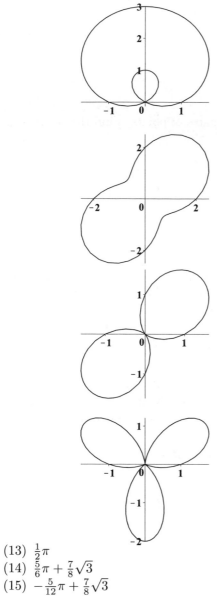

(13) $\frac{1}{2}\pi$

(14) $\frac{5}{6}\pi + \frac{7}{8}\sqrt{3}$

(15) $-\frac{5}{12}\pi + \frac{7}{8}\sqrt{3}$

Exercises 14.6

(2) 8

(4) Using the above problem, find the lengths of graphs of the following polar curves.

(a) $\frac{3}{2}\sqrt{10} - \frac{1}{2}\ln\left(-3 + \sqrt{10}\right)$

(b) 2π

(c) $\sqrt{\left(4+2\sqrt{2}\right)\sqrt{2}-2\sqrt{\left(4+2\sqrt{2}\right)}+2\sqrt{2}}$

(d) $\sqrt{2}e^2 - \sqrt{2}$

(e) $\sqrt{5} - \frac{1}{2}\ln\left(-2+\sqrt{5}\right) - \frac{1}{2}\sqrt{2} - \frac{1}{2}\ln\left(\sqrt{2}+1\right)$

(6) Find the areas of surface of revolution about y axis.

(a) $2\pi\left(\frac{5}{3}\sqrt{5} - \frac{1}{3}\right)$

(b) $4\pi^2$

(c) $2\pi\left(\frac{2}{5}\sqrt{2}e^4\cos 2 + \frac{1}{5}\sqrt{2}e^4\sin 2 - \frac{2}{5}\sqrt{2}\right)$

(d) $2\pi\left(\frac{7}{6}\sqrt{2} - \frac{1}{2}\ln\left(\sqrt{2}-1\right) - \frac{1}{3}\right)$

(9) One was the planet experiences a central force which is an inverse square law.

(10) $1.968\,699\,79 \times 10^{30}$ kilograms.

(11) $= 4.225\,047\,43 \times 10^7$ meters

Exercises 14.8

(1) Find the rectangular and spherical coordinates.

(a) $\left(-\frac{5}{2}\sqrt{3}, \frac{5}{2}, -3\right), \left(\sqrt{34}, \frac{5}{6}\pi, \arccos\left(\frac{-3}{\sqrt{34}}\right)\right)$

(b) $\left(\frac{3}{2}, \frac{3}{2}\sqrt{3}, 4\right), \left(5, \frac{1}{3}\pi, \arccos\frac{4}{5}\right)$

(c) $\left(-2, 2\sqrt{3}, 1\right), \left(\sqrt{17}, \frac{2}{3}\pi, \arccos\left(\frac{1}{17}\sqrt{17}\right)\right)$

(d) $\left(-\sqrt{2}, \sqrt{2}, -2\right), \left(\sqrt{8}, \frac{3\pi}{4}, \arccos\left(\frac{-2}{\sqrt{8}}\right)\right)$

(e) $\left(0, -3, -1\right), \left(\sqrt{10}, \frac{3\pi}{2}, \arccos\left(\frac{-1}{\sqrt{10}}\right)\right)$

(f) $\left(4\sqrt{3}, -4, -11\right), \left(\sqrt{185}, \frac{11\pi}{6}, \arccos\left(\frac{-11}{\sqrt{185}}\right)\right)$

(2) Find the cylindrical and spherical coordinates of these points.

(a) $\left(5, \frac{1}{4}\pi, -3\right), \left(\sqrt{34}, \frac{\pi}{4}, \arccos\left(\frac{-3}{\sqrt{34}}\right)\right)$

(b) $\left(3, \frac{1}{3}\pi, 2\right), \left(\sqrt{13}, \frac{\pi}{3}, \arccos\left(\frac{2}{\sqrt{13}}\right)\right)$

(c) $\left(5, \frac{3}{4}\pi, 11\right), \left(\sqrt{130}, \frac{3\pi}{4}, \arccos\left(\frac{11}{\sqrt{130}}\right)\right)$

(d) $\left(5, \frac{2}{3}\pi, 23\right), \left(\sqrt{538}, \frac{2\pi}{3}, \arccos\left(\frac{23}{\sqrt{538}}\right)\right)$

(e) $\left(2, \frac{7\pi}{6}, 2\right), \left(\sqrt{29}, \frac{7\pi}{6}, \arccos\left(\frac{-5}{\sqrt{29}}\right)\right)$

(f) $\left(3, \frac{5}{3}\pi, 2\right), \left(\sqrt{53}, \frac{5\pi}{3}, \arccos\left(-\frac{7}{53}\sqrt{53}\right)\right)$

(3) Find the rectangular and cylindrical coordinates.

(a) $\left(-\sqrt{2}\sqrt{3}, \sqrt{2}, 2\sqrt{2}\right), \left(2\sqrt{2}, \frac{5}{6}\pi, 2\sqrt{2}\right)$

(b) $\left(-\frac{1}{2}\sqrt{3}, \frac{3}{2}, 1\right), \left(\sqrt{3}, \frac{2}{3}\pi, 1\right)$

(c) $\left(0, -\frac{3}{2}, -\frac{3}{2}\sqrt{3}\right), \left(\frac{3}{2}, \frac{3}{2}\pi, -\frac{3}{2}\sqrt{3}\right)$

(d) $\left(2\sqrt{2}, -2\sqrt{2}, 0\right), \left(4, \frac{7}{4}\pi, 0\right)$

(e) $\left(3, \sqrt{3}, -2\right), \left(2\sqrt{3}, \frac{1}{6}\pi, -\frac{3}{2}\right)$

(f) $\left(\sqrt{2}, -\sqrt{2}\sqrt{3}, -2\sqrt{2}\right), \left(2\sqrt{2}, \frac{5}{3}\pi, -\frac{3}{2}\sqrt{2}\right)$

(4) Find the spherical and cylindrical coordinates.

(a) $\left(4, \frac{1}{4}\pi, \frac{\pi}{3}\right), \left(\sqrt{10}, \frac{\pi}{3}, 2\sqrt{2}\right)$

(b) $\left(2, \frac{1}{3}\pi, \frac{5\pi}{3}\right), \left(\sqrt{3}, \frac{1}{3}\pi, 1\right)$

(c) $\left(3, \frac{4\pi}{6}, \frac{7\pi}{4}\right), \left(\frac{3}{2}\sqrt{3}, \frac{3\pi}{4}, -\frac{3}{2}\sqrt{3}\right)$

(d) $\left(4, \frac{1}{6}\pi, \frac{5\pi}{6}\right), \left(2, \frac{5\pi}{6}, 2\sqrt{3}\right)$

(e) $\left(1, \frac{3\pi}{4}, \frac{4\pi}{6}\right), \left(\frac{1}{2}\sqrt{2}, \frac{4\pi}{6}, -\frac{1}{2}\sqrt{2}\right)$

(5) $\left(\sqrt{14}, .640\,522\,3, 1.\,107\,148\,72\right), \left(\sqrt{5}, 1.\,107\,14, 3\right)$

(6) $z = \sqrt{x^2 + y^2}$

(8) $x^2 + y^2 = 25.$

(9) $x^2 + y^2 + z^2 = 16.$

(10) $z = r,\ \phi = \pi/4.$

(11) Write the following in spherical coordinates.

(a) $\cos \phi = \rho \sin^2 (\phi)$

(b) $\left(\rho \sin (\phi) \cos (\theta)\right)^2 - \left(\rho \sin (\phi) \sin (\theta)\right)^2 = 1$

(c) $\rho = \sqrt{6}$

(d) $\phi = \pi/4$

(e) $\theta = \pi/4, \theta = \frac{5}{4}\pi$

(f) $\rho \cos \phi = \rho \sin (\phi) \cos (\theta)$

(12) Write the following in cylindrical coordinates.

(a) $z = r^2$

(b) $r^2 \left(\cos^2 (\theta) - \sin^2 (\theta)\right) = 1$

(c) $z^2 + r^2 = 6$

(d) $z = r$

(e) $\tan (\theta) = 1$

(f) $z = r \cos \theta$

Index

Calculus: Theory and Applications Volume I